PHOTONICS

PHOTONICS

Scientific Foundations, Technology and Applications

Nanophotonic Structures and Materials

Volume II

Edited by

DAVID L. ANDREWS
School of Chemical Sciences
University of East Anglia
Norwich, UK

Copyright © 2015 by John Wiley & Sons, Inc. All rights reserved.

Published by John Wiley & Sons, Inc., Hoboken, New Jersey.
Published simultaneously in Canada.

No part of this publication may be reproduced, stored in a retrieval system, or transmitted in any form or by any means, electronic, mechanical, photocopying, recording, scanning, or otherwise, except as permitted under Section 107 or 108 of the 1976 United States Copyright Act, without either the prior written permission of the Publisher, or authorization through payment of the appropriate per-copy fee to the Copyright Clearance Center, Inc., 222 Rosewood Drive, Danvers, MA 01923, (978) 750-8400, fax (978) 750-4470, or on the web at www.copyright.com. Requests to the Publisher for permission should be addressed to the Permissions Department, John Wiley & Sons, Inc., 111 River Street, Hoboken, NJ 07030, (201) 748-6011, fax (201) 748-6008, or online at http://www.wiley.com/go/permission.

Limit of Liability/Disclaimer of Warranty: While the publisher and author have used their best efforts in preparing this book, they make no representations or warranties with respect to the accuracy or completeness of the contents of this book and specifically disclaim any implied warranties of merchantability or fitness for a particular purpose. No warranty may be created or extended by sales representatives or written sales materials. The advice and strategies contained herein may not be suitable for your situation. You should consult with a professional where appropriate. Neither the publisher nor author shall be liable for any loss of profit or any other commercial damages, including but not limited to special, incidental, consequential, or other damages.

For general information on our other products and services or for technical support, please contact our Customer Care Department within the United States at (800) 762-2974, outside the United States at (317) 572-3993 or fax (317) 572-4002.

Wiley also publishes its books in a variety of electronic formats. Some content that appears in print may not be available in electronic formats. For more information about Wiley products, visit our web site at www.wiley.com.

Library of Congress Cataloging-in-Publication Data:

Nanophotonic structures and materials / edited by David L. Andrews.
    pages cm. – (Photonics ; volume II)
  Includes bibliographical references and index.
  ISBN 978-1-118-22551-6 (cloth)
 1. Nanophotonics.   2. Nanostructured materials–Optical properties.   I. Andrews, David L., 1952-
  TA1530.N346 2015
  621.36′5–dc23
                                                                                 2014041370

Printed in the United States of America

10 9 8 7 6 5 4 3 2 1

CONTENTS

List of Contributors   ix

Preface   xi

1 Silicon Photonics   1
Wim Bogaerts

   1.1 Introduction, 1
   1.2 Applications, 1
   1.3 Optical Functions, 3
   1.4 Silicon Photonics Technology, 10
   1.5 Conclusion, 15
   References, 15

2 Cavity Photonics   21
J. Mørk, P. T. Kristensen, P. Kaer, M. Heuck, Y. Yu, and N. Gregersen

   2.1 Introduction, 21
   2.2 Cavity Fundamentals, 22
   2.3 Cavity-Based Switches, 26
   2.4 Emitters in Cavities, 32
   2.5 Nanocavity Lasers and LEDs, 42

2.6 Summary, 46
Acknowledgments, 47
References, 47

3   Metamaterials: State-of-the Art and Future Directions   53

Natalia M. Litchinitser and Vladimir M. Shalaev

3.1 Introduction, 53
3.2 Negative-Index Materials, 54
3.3 Magnetic Metamaterials, 59
3.4 Graded-Index Transition Metamaterials, 62
3.5 Transformation Optics, 70
3.6 Metasurfaces, 75
References, 78

4   Quantum Nanoplasmonics   85

Mark I. Stockman

4.1 Introduction, 85
4.2 Spaser and Nanoplasmonics with Gain, 86
4.3 Adiabatic Hot-Electron Nanoscopy, 118
Acknowledgments, 125
References, 125

5   Dielectric Photonic Crystals   133

Robert H. Lipson

5.1 Introduction, 133
5.2 Fundamentals, 134
5.3 Fabrication Methods and Materials, 145
5.4 Applications, 154
5.5 Conclusions, 159
References, 159

6   Quantum Dots   169

Stanley Tsao and Manijeh Razeghi

6.1 Introduction, 169
6.2 Quantum Dots for Infrared Detection, 175
6.3 Quantum Dot Growth, 179
6.4 Device Fabrication and Measurement Procedures, 184
6.5 Gallium Arsenide–Based Quantum Dot Detectors, 186
6.6 Indium Phosphide-Based Quantum Dot Detectors, 198
6.7 Colloidal Quantum Dots, 215
6.8 Conclusion, 216
References, 217

7 Magnetic Control of Spin in Molecular Photonics   221
Eitan Ehrenfreund and Z. Valy Vardeny

7.1 Introduction, 221
7.2 A Survey of the Magneto-Electroluminescence in OLEDs, 222
7.3 Organic MEL at Small Magnetic Fields; Compass Effect, 232
7.4 Magnetic Field Effect on Excited State Spectroscopies in Organic Semiconductor Films, 236
7.5 Basic Quantum Mechanical Models Based on Spin-Mixing Manipulation by Magnetic Fields, 246
7.6 Summary, 254
Acknowledgments, 255
References, 255

8 Thin-Film Molecular Nanophotonics   261
Tetsuzo Yoshimura

8.1 Introduction, 261
8.2 Molecular Assembling for Nanoscale Tailored Structures, 262
8.3 Molecular Layer Deposition, 264
8.4 Organic Multiple Quantum Dots (MQDs), 267
8.5 Self-Organized Lightwave Network, 283
8.6 Proposed Applications, 292
8.7 Summary, 305
References, 305

9 Light-Harvesting Materials for Organic Electronics   311
Damien Joly, Juan Luis Delgado, Carmen Atienza, and Nazario Martín

9.1 Introduction, 311
9.2 Photoinduced Electron Transfer (PET) in Artificial Photosynthetic Systems, 313
9.3 Fullerenes for Organic Photovoltaics, 323
9.4 Molecular Wires, 330
9.5 Conclusions, 335
Acknowledgments, 335
References, 336

10 Recent Advances in Metal Oxide-Based Photoelectrochemical Hydrogen Production   343
Bob C. Fitzmorris and Jin Z. Zhang

10.1 Introduction, 343
10.2 Materials for PEC Hydrogen Production, 346
10.3 Conclusion, 362
References, 363

11 Optical Control of Cold Atoms and Artificial Electromagnetism   **371**
Gediminas Juzeliūnas and Patrik Öhberg

11.1 Introduction, 371
11.2 Atomic Bose–Einstein Condensates, 372
11.3 Optical Forces on Atoms, 376
References, 393

Index   **401**

LIST OF CONTRIBUTORS

Carmen Atienza Departamento de Química Orgánica, Facultad de Química, Universidad Complutense, Madrid, Spain

Wim Bogaerts Ghent University, imec, Department of Information Technology, Gent, Belgium

Juan Luis Delgado IMDEA-Nanoscience, Campus de Cantoblanco, Madrid, Spain

Eitan Ehrenfreund Department of Physics and Solid State Institute, Technion-Israel Institute of Technology, Haifa, Israel

Bob C. Fitzmorris Department of Chemistry and Biochemistry, University of California, Santa Cruz, CA, USA

Damien Joly IMDEA-Nanoscience, Campus de Cantoblanco, Madrid, Spain

Gediminas Juzeliūnas Institute of Theoretical Physics and Astronomy, Vilnius University, Vilnius, Lithuania

Robert H. Lipson Department of Chemistry, University of Victoria, Victoria, BC, Canada

Natalia M. Litchinitser University at Buffalo, The State University of New York, Buffalo, NY, USA

Nazario Martín IMDEA-Nanoscience, Campus de Cantoblanco, Madrid, Spain Departamento de Química Orgánica, Facultad de Química, Universidad Complutense, Madrid, Spain

Jesper Mørk DTU Fotonik, Technical University of Denmark, Kongens Lyngby, Denmark

Patrik Öhberg SUPA, Institute of Photonics and Quantum Sciences, Heriot-Watt University, Edinburgh, UK

Manijeh Razeghi Center for Quantum Devices, Department of Electrical Engineering and Computer Science, Northwestern University, Evanston, IL, USA

Vladimir M. Shalaev Purdue University, West Lafayette, IN, USA

Mark I. Stockman Ludwig Maximilian University, Munich, Germany
Max Plank Institute for Quantum Optics, Garching at Munich, Germany
Georgia State University, Atlanta, GA, USA

Stanley Tsao Center for Quantum Devices, Department of Electrical Engineering and Computer Science, Northwestern University, Evanston, IL, USA

Z. Valy Vardeny Department of Physics and Astronomy, University of Utah, Salt Lake City, UT, USA

Tetsuzo Yoshimura School of Computer Science, Tokyo University of Technology, Tokyo, Japan

Jin Z. Zhang Department of Chemistry and Biochemistry, University of California, Santa Cruz, CA, USA

PREFACE

Since its inception, the term "photonics" has been applied to increasingly wide realms of application, with connotations that distinguish it from the broader-brush terms "optics" or "the science of light." The briefest glance at the topics covered in these volumes shows that such applications now extend well beyond an obvious usage of the term to signify phenomena or mechanistic descriptions involving photons. Those who first coined the word partly intended it to convey an aspiration that new areas of science and technology, based on microscale optical elements, would one day develop into a comprehensive range of commercial applications as familiar and distinctive as electronics. The fulfilment of that hope is amply showcased in the four present volumes, whose purpose is to capture the range and extent of photonics science and technology.

It is interesting to reflect that in the early 1960s, the very first lasers were usually bench-top devices whose only function was to emit light. In the period of growth that followed, most technical effort was initially devoted to increasing laser stability and output levels, often with scant regard for possibilities that might be presented by truly photon-based processes at lower intensities. The first nonlinear optical processes were observed within a couple of years of the first laser development, while quantum optics at first grew slowly in the background, then began to flourish more spectacularly several years later. A case can be made that the term "photonics" itself first came into real prominence in 1982, when the trade publication that had previously been entitled *Optical Spectra* changed its name to *Photonics Spectra*. At that time the term still had an exotic and somewhat contrived ring to it, but it acquired a new respectability and wider acceptance with the publication of Bahaa Saleh and Malvin Teich's definitive treatise, *Fundamentals of Photonics*, in 1991. With the passage of time, the increasing pace of development has been characterized by the striking progress in miniaturization and integration of optical components, paving the way for fulfilment of the early promise. As the laser industry has evolved, parallel growth in the optical fiber industry has helped spur the continued push toward the long-sought goal of total integration in optical devices.

Throughout the commissioning, compiling, and editing that have led to the publication of these new volumes, it has been my delight and privilege to work with many of the world's top scientists. The quality of the product attests to their commitment and willingness to devote precious time to writing chapters that glow with authoritative expertise. I also owe personal thanks to the ever-professional and dependable staff of Wiley, without whose support this project would never have come to fruition. It seems fitting that the culmination of all this work is a sequence of books published at the very dawning of the UNESCO International Year of Light. Photonics is shaping the world in which we live, more day by day, and is now ready to take its place alongside electronics, reshaping modern society as never before.

David L. Andrews

Norwich, U.K., July 2014

1

SILICON PHOTONICS

WIM BOGAERTS

Ghent University, imec, Department of Information Technology, Gent, Belgium

1.1 INTRODUCTION

Since the beginning of the century, silicon photonics has grown from a niche research field to a field with strong industrial interest and several near-future applications [1, 2]. This rapid growth can be attributed to several unique characteristics of silicon photonics. First of all, the use of silicon makes it possible to make photonic integrated circuits (PICs) with much smaller building blocks than in other material systems [3]. This enables smaller chips, but also more complex photonic circuits. Also, silicon is the base material for electronic circuits, and huge investments in manufacturing technology can be put to work to make photonic circuits. This offers a route to high volume, low cost photonic circuits that could be applied in many applications in sensing [4, 5] and optical communication [6, 7].

In this chapter, we will discuss current state of the art in silicon photonics. We will look a bit closer in the applications, and from that we derive the functions needed on the chip. Finally, we discuss the technology implementations.

1.2 APPLICATIONS

1.2.1 Interconnects

Integrated photonics has been mainly used for applications in optical communication, especially in telecom backbone and metro networks. The advent of silicon photonics

---

Photonics: Scientific Foundations, Technology and Applications, Volume II, First Edition.
Edited by David L. Andrews.
© 2015 John Wiley & Sons, Inc. Published 2015 by John Wiley & Sons, Inc.

2 SILICON PHOTONICS

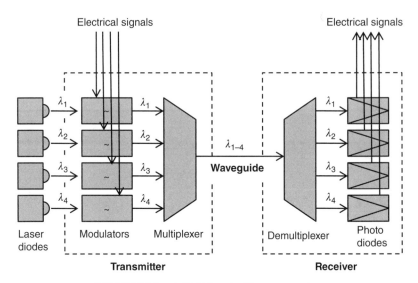

FIGURE 1.1 A WDM optical interconnection.

and its potential for low cost and low power transceiver chips has opened up new, shorter-range interconnect applications in high-performance computing and datacenters [8,9]. Silicon photonic chips might turn out to be a game-changer in interconnects on an even smaller scale: it is the first technology that can offer an attractive solution to solve the off-chip bandwidth bottleneck [10].

Typical optical links involve light sources, signal modulators, a waveguide medium, and a photodetector. These individual functions are described in Section 1.3. In the case of wavelength-division multiplexing (WDM), signals are encoded onto different carrier wavelengths, which are multiplexed into the same waveguide. This technique, illustrated in Figure 1.1, is widely used to increase the bandwidth of optical links. As we will see in Section 1.3.2, silicon photonics can implement WDM filters with a very small footprint.

1.2.2 Sensors and Spectroscopy

Another application field where silicon photonics can enable unique capabilities is that of sensing. As we will see later in this chapter, silicon waveguides can be extremely sensitive to different effects, such as temperature, cladding index [11], strain [12] and deposition of layers [13]. Especially the latter is important, as proper surface chemistry enables selective response to specific effects or molecules, enabling biosensors [5] or specific gas sensors [14]. In addition to high sensitivity, the technology offers integration of many sensor functions on a single chip, potentially with the inclusion of the read-out circuitry. Some examples of photonic sensors that could be integrated on a silicon chip are shown in Figure 1.2: A ring resonator could be

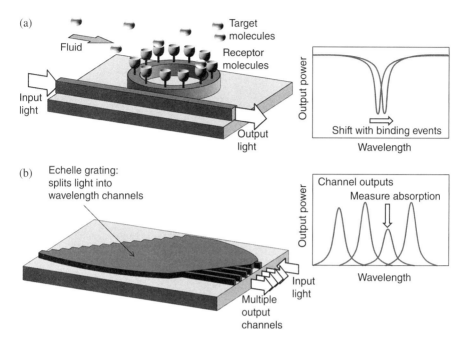

FIGURE 1.2 Two examples of silicon-photonics-based sensor systems. (a) A ring-resonator-based biosensor [4] and (b) an on-chip spectrometer [15].

used to capture selective molecular binding events and thus measure concentrations of specific (bio)molecules in a medium. Or, wavelength filters of multiplexers could be used to make a spectrometer that could be used for a variety of spectroscopic measurement systems [15].

1.3 OPTICAL FUNCTIONS

A PIC can accommodate many different optical functions. The most common functions have to do with transport of light, wavelength filtering and coupling to off-chip elements and fibers. These are called passive functions, as light is typically not altered in the process. Active functions involve electro-optic elements such as light sources, signal modulators, and photodetectors. We will discuss these functions and the state of the art in terms of performance that has been demonstrated in silicon photonics. The actual technology is discussed in Section 1.4

1.3.1 Waveguides and Routing

The key optical function on a PIC is guiding light between parts on a chip. An optical waveguide consists of a high-index core surrounded by a lower-index cladding. The

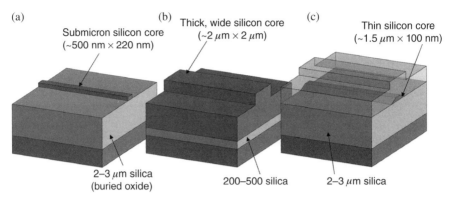

FIGURE 1.3 Silicon waveguides: (a) small-core photonic wires [17], (b) large-core rib waveguides [20], and (c) oxidized waveguide [23].

higher the index contrast, the more compact you can make the waveguide core. As it is, silicon has a very high refractive index in the regime where it is transparent (wavelength > 1.2 μm). This way, it is possible to make high contrast waveguides with core dimensions down to 200–500 nm, using a cladding oxide ($n = 1.45$) or air ($n = 1.0$) [3, 16]. Such waveguides, often called photonic wires, can have bend radii of only a few micrometers with low loss [17].

Apart from photonic wires, it is also possible to use silicon for large-core waveguides. Such waveguides are defined in silicon of several micrometers thick [18–20], and to obtain single-mode condition, they are only partially etched. Such waveguide is shown, together with a photonic wire waveguide, in Figure 1.3. Because the index contrast between the unetched core and the etched cladding is relatively low, such waveguides still require a large bend radius.

The key performance metic for optical waveguides is the propagation loss. Typical photonic wires have a loss of 1–2 dB/cm [16, 17, 21]. Large-core waveguides have a lower loss, on the order of 0.1 dB/cm [20]. To obtain lower loss in the small-code waveguide system, one can also use a shallow-etched rib waveguide geometry, which can reduce the losses with a factor of 3–4, but again with a penalty of larger bend radius [22].

Because waveguide losses are largely caused by scattering at etched sidewalls, alternative definition techniques can reduce the losses. For instance, waveguides can be defined by oxidation, which provides a smooth sidewall surface [23].

The high contrast and submicron dimensions of silicon photonic wire waveguides give them a rather strong dispersion. While the effective index of a 450 nm × 220 nm wire is around 2.4 (at 1550 nm wavelength), its group index is around 4.3. The tight confinement also makes these waveguides very susceptible to small variations, both in geometry and material parameters. A very small deviation of the width or height will have a significant effect on the effective index, to the extent that for some functions, nanometer-scale precision is required. Large-core waveguides, on the other hand, are much less sensitive to geometrical variations.

Small-core photonic wires are very birefringent. The TE and TM polarization have a very different effective index. While it is in theory possible to make a non-birefringent waveguide [24], the tolerances on the waveguide dimensions is extremely stringent. Therefore, photonic wire waveguide circuits are typically optimized to function at a single polarization only.

1.3.2 Wavelength Filtering

Wavelength filtering is one of the more common functions in photonic circuit. It is important for multiplexing and demultiplexing WDM channels, but wavelength filters are also often used for sensing applications. Wavelength filters can be implemented in various ways, but their operation is always based on interferometry. Different types of wavelength filters are shown in Figure 1.4.

Mach–Zehnder interferometers (MZIs) are by far the simplest filter in concept, and as any other waveguide component, they can benefit from small bends to obtain a reduced footprint. Also, the large group index of silicon wires will help, as the optical delay in an MZI will scale with the inverse of the group index.

Ring resonators are a very widely used structure in silicon photonics, because the rings can again be very compact [27]. In a ring, the light self-interferes to build up a resonance, so a much sharper filter characteristic can be obtained, compared to an MZI. The sharpness of the resonance is largely determined by the ring's losses, so the quality of the waveguide is important. Rings and Mach–Zehnders can actually be cascaded into more complex spectral filter structures to obtain sharper filter characteristics or flat transmission bands [26].

Instead of a single delay path, it is possible to use multiple delays. An arrayed waveguide grating (AWG) distributes light over many waveguides with a different delay, and by recombining the light in spherical phase front, the refocusing can be directed as a function of wavelength: this way, different wavelengths are routed to different output waveguides [28]. The same device can of course be used to combine wavelength channels into a single waveguide. Similarly, an echelle grating introduces multiple delays by etching a set of reflective facets in an unpatterned region of the silicon [25].

All these different filter types can be implemented in both small-core and large-core silicon waveguides, but of course the performance will differ. In large-core waveguides, the devices will have a much larger footprint, because both the waveguides and the bends are much larger. On the other hand, large-core waveguides will make it easier to have wavelength filters fabricated to match the design specifications. This is because these waveguides are much less sensitive to small geometric variations, and the effective index of the waveguides can be better controlled.

Photonic-wire-based wavelength filters can suffer easily from variability in the manufacturing process, where geometric changes of the order of 1 nm can already induce a wavelength shift of the same magnitude [29]. This makes it difficult to make absolute wavelength filters without some tuning or active feedback mechanisms, such as integrated heaters [30–32].

6 SILICON PHOTONICS

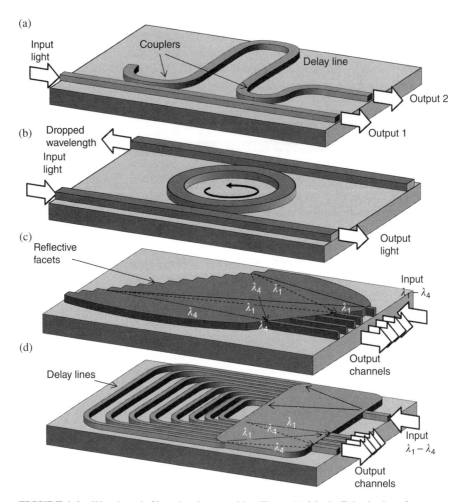

FIGURE 1.4 Wavelength filters implemented in silicon: (a) Mach–Zehnder interferometer, (b) ring resonator, (c) Echelle grating or planar concave grating [25], and (d) arrayed waveguide grating [26].

1.3.3 Coupling to Fiber

An essential function for many applications is efficient coupling to and from optical fibers. As single-mode fibers are still the standard transport medium for light, this means the optical mode in the silicon waveguide should be converted to a fiber mode. Several mechanisms are possible and the most common are illustrated in Figure 1.5.

For large-core silicon waveguides, the mismatch in mode profile is relatively small, and often a simple taper structure suffices to have an efficient butt-coupling between waveguides and fiber [33]. Instead of full-size single-mode fibers, the use of ultrahigh NA fibers, with a mode field diameter of about 3 μm, can relax the requirements for

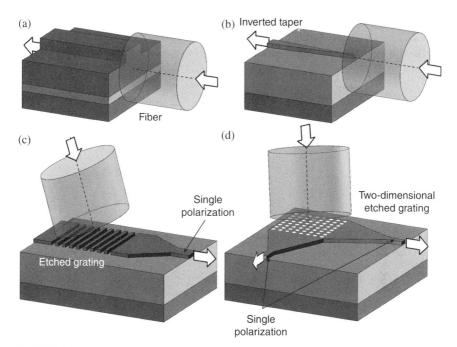

FIGURE 1.5 Coupling schemes for silicon photonic chips to and from optical fiber. (a) Edge coupling for large-core waveguides, (b) an inverted taper for edge coupling of photonic wires, (c) 1D grating coupler [34], and (d) a 2D grating coupler with polarization splitting [35].

the taper. The fibers can be mounted on the same substrate by etching V-grooves, which control the horizontal and vertical alignment.

For small-core silicon photonic wires, the mode mismatch is very large. Direct butt coupling would result in a coupling efficiency of only 0.1%. Expanding the mode of the wire waveguide in the lateral direction can be accomplished by defining a taper. However, the challenge is expanding the waveguide mode in the vertical direction. This can be accomplished by tapering the waveguide to a very narrow width, effectively pushing the mode towards cutoff and expelling it from the waveguide core. Such an inverted taper structure in the silicon can then be accompanied by a larger-core waveguide in a lower-index material (e.g., silicon oxynitride [24] or polymer [36] with a cross section that is matched to an ultra-high numerical aperture (UHNA) fiber). Such coupling structures have been demonstrated to achieve 95% coupling efficiency.

An important drawback of edge coupling mechanisms is that this can only be done after the wafer processing is complete and the devices have been diced, and if needed, facet-polished. This means that wafer-scale testing of devices is not possible.

The alternative to edge coupling is vertical coupling. For this, the most used component is a grating coupler [34]. A grating coupler is a diffraction grating that diffracts light from a fiber into an on-chip waveguide, or vice versa. Depending on the fabrication process, such gratings can have efficiencies upward of 50% [37, 38].

However, because of the birefringence of the silicon waveguides, such a grating coupler couples only to one fiber polarization. To couple both polarizations, a 2D grating can be used, which performs the same function for both fiber polarizations at the same time [35].

By diffracting the light immediately in a thin silicon waveguide, the grating coupler already performs a mode compression in the vertical direction. In the lateral direction, a taper can be used, or the grating coupler can be designed to have a focusing function [37, 39]. This can even be done in two dimensions [37, 40]. This way, a very compact footprint coupling structure is possible that allows for wafer-scale optical probing of a photonic chip.

The main drawback of grating couplers is that they have a limited wavelength bandwidth, as the operating principle is based on a diffractive grating. Typical 3 dB bandwidth is of the order of 60–80 nm.

1.3.4 Electro-Optic and Opto-Electronic Conversion

For active photonic circuits, efficient conversion of electrical signals to optical signals is needed. For electro-optic modulation, silicon is actually a poor material: it has no intrinsic electro-optic effects. Instead, indirect effects are needed to obtain a modulation of the optical properties. Such mechanisms include mechanical structures based on MEMS [41], thermal tuning using heaters [30, 42, 43], and electro-optic cladding materials such as liquid crystals [44]. These mechanisms are typically limited in operation speed and are used for tuning and switching rather than for signal modulation.

For high speed modulation, the most-adopted mechanism is carrier manipulation. The refractive index of silicon depends on the concentration of free electrons and holes [45]. The density of carriers can be electrically influenced by incorporating a junction or capacitor in the core of the waveguide [46]. The larger the spatial overlap of the carrier accumulation or depletion region with the optical mode of the waveguide, the stronger the modulation effect. The junction or capacitor can bet either horizontally or vertically oriented and are electrically connected to the metal interconnect layers, as illustrated in Figure 1.6.

The modulation of the carrier density will modulate the absorption in the waveguide, but the stronger effect is the modulation of the real part of the effective index [45]. Carrier-based silicon modulators are therefore more effective as a phase modulator rather than an amplitude modulator. To obtain amplitude modulation from a phase modulator, the phase shifter section should be embedded inside an interferometer or a resonator [46].

Carrier-based silicon modulators have now been demonstrated with quite good efficiency [46] and modulation speeds over 40 Gbps. Modulators can be implemented both in ring resonator configuration [47,48] and Mach–Zehnder configuration [49–51].

For the conversion of optical signals back into electrical signals, the common solution is the use of photodetectors. As silicon is transparent at the wavelengths used for telecommunication, another material should be introduced. This can be III-V semiconductor, which can be bonded onto the silicon [52], but the commonly accepted

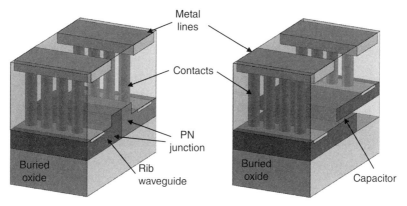

FIGURE 1.6 Different modulators in silicon. (a) P–N junction geometry and (b) capacitor geometry.

material for photodetectors in silicon photonic circuits is germanium, as this can be epitaxially grown on silicon [53]. Germanium photodetectors have been demonstrated with good high speed performance (over 40 Gbps [54]), but typically suffer from higher dark currents that their III-V counterparts. Also, avalanche photodiodes have also been made in this technology, with very good bandwidth-gain products [55]. while Germanium detectors are more compact when integrated with thin-silicon waveguides, it is also possible to integrate them with thick silicon rib waveguides [56].

1.3.5 Lasers

The most challenging function to integrate in a silicon photonic circuit is obviously the light source. Silicon has an indirect bandgap, so light emission is not very efficient compared to nonradiative recombination mechanisms. Therefore, alternative ways to integrate a light source are needed.

The most obvious integration method for a light source is to just use an external laser (or multiple lasers) and connect them to the silicon chip either through flip-chip-like integration [37] or using a fiber connection. The advantage of this approach is that the laser simply acts as an external optical power supply, and that it can be electrically and thermally decoupled from the silicon chip.

In large-core silicon waveguides, it is possible to integrate lasers into the same waveguide substrate using flip-chip assembly techniques. By defining proper alignment structures, the laser mode can be aligned with the silicon waveguide core [57]. The mode size of the silicon waveguide can be adapted to match that of the laser, to obtain a good coupling efficiency.

To integrate laser directly into small-core silicon waveguide circuits, a suitable laser material needs to be integrated. This can be accomplished by incorporating suitable material (mainly III–V semiconductor stacks of quantum wells) on the chip. Direct epitaxy is not trivial, so the most promising approach at the moment is bonding [58]. Using various variations of this technique, lasers with outputs of several milliwatts have been demonstrated [59], including DBR lasers [60], ring lasers [61], and

10   SILICON PHOTONICS

widely tunable lasers. Also, microlasers with output power of the order of 100 μW have been demonstrated using disk geometries [62] or photonic crystals.

While bonding has already shown promising results and is the likely technology to provide short-term availability of on-chip lasers in silicon waveguides, the longer-term objective is to have epitaxially grown lasers monolithically integrated into the chip. III–V epitaxy on silicon is not yet sufficiently advanced to realize this objective, but lasers have been demonstrated in epitaxially grown germanium on silicon. This did require extensive bandgap engineering to favor recombination in the direct bandgap [63], but it offers a future route for integrated lasers on silicon.

1.4 SILICON PHOTONICS TECHNOLOGY

In this section, we go deeper into the technology that is used to make silicon photonic components and circuits, and the impact that technology has on the functionality of the devices discussed in the previous section.

1.4.1 Passive Circuits

Passive silicon waveguide circuits, which include waveguides, wavelength filters, and coupling structures, are usually fabricated with a simple lithographic process, both in large-core platforms as small-core photonic wire-based platforms. A schematic process flow is depicted in Figure 1.7 [17]. The base substrate material is silicon-on-insulator (SOI), which consists of a silicon substrate, and oxide buffer layer, and a single-crystal top layer in which the waveguide circuit will be defined. The thickness of the layer stack depends on the requirements of small-core or large-core silicon waveguides. Small-core waveguides have a top layer of 200–500 nm, and require a buried oxide of 2 μm or more to decouple the waveguide from the substrate. Large-core waveguides have a thicker waveguide layer, and because the confinement is larger in the thicker silicon, they can have a thinner buffer oxide layer, on the order of 400–100 nm.

The pattern of the waveguide circuit is defined in a photosensitive resist using a lithography step (optical or e-beam lithography). After development, the pattern is transferred into the silicon layer. This is typically done using reactive ion etching or a

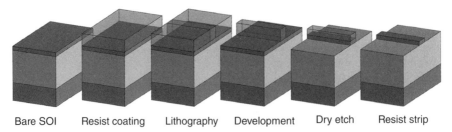

FIGURE 1.7   Fabrication process for passive waveguides. From Reference 17.

similar plasma-based etch. The etch can be either completely through the silicon (for strip waveguides) or partially in the silicon (for rib waveguides). After patterning, the resist layer and hard masks can be stripped. Multiple patterns with different etch depths can be applied subsequently, for instance, to implement shallow-etched grating couplers in deep-etched strip waveguides [26].

The details of such a fabrication process have a significant effect on the performance of the devices. As already mentioned, small geometry variations will induce a change in the effective index of the waveguides. While the linewidth and the thickness of the silicon waveguide have the largest impact, secondary factors such as sidewall angle after the etch can impact bend losses or unwanted polarization conversion [64]. The patterning itself also introduces imperfections. The lithography process should have sufficiently high resolution. While for large-core waveguides contact lithography or i-line lithography is sufficient, small-core waveguides require deep UV lithography at 248 nm [3, 65] or 193 nm [17]. E-beam lithography is also used, but it has limited scalability with respect to circuit size and industrial deployment.

Even with high-end lithography, the pattern definition can induce unwanted effects. For example, (optical) proximity effects will affect feature dimensions if neighboring features are present. In some cases, corrections can be applied, but it is difficult to guarantee uniform dimensions. For example, a waveguide might get a different linewidth in a directional coupler because of the presence of a second waveguide [3]. Also, the etch process might be influenced by the local and global density of patterns.

Controlling uniformity in silicon photonic waveguide circuits is especially challenging in small-core waveguides. Uniformity needs to be managed within a single chip, between chips on a wafer and between wafers and lots. High-end lithography tools already enable nanometer-level control of the linewidth [29]. However, a more difficult problem is the control of the thickness of the silicon layer. While SOI wafers can be manufactured with precision of a few percent, this still means variations on the order of 10 nm. To correct this, it is possible to use local processing of the wafer to make the top layer more uniform [66].

SOI is not the only material to make silicon waveguides. It is also possible to manufacture high quality waveguides in deposited layers, such as amorphous silicon. The advantage of such materials is that they can be used to add waveguides to substrates that have already active structures on them, such as electronics. While amorphous silicon has a somewhat higher optical absorption than crystalline silicon, good waveguides have been demonstrated [67, 68]. Polycrystalline waveguides have a higher propagation loss due to scattering and absorption at grain boundaries [69].

Deposited silicon (amorphous or poly) can also be used in combination with the crystalline waveguide circuits to provide special functions, or add a degree of freedom in defining local geometry where higher absorption losses have a lower impact [38].

1.4.2 Modulators

To implement carrier-based modulators, a junction or capacitor needs to be incorporated in the waveguide core. The most common way to define a junction is by ion implantation, where P-doping and N-doping are subsequently implanted after a patterning step [46]. Of course, the alignment of the P and N patterns is crucial to obtain good electrical properties as well as a good overlap with the optical mode [70]. Multiple junctions can be used to enhance this overlap [51], or the junctions can be arranged in a finger pattern [70]. However, larger junction area might improve the modulation efficiency (larger phase shift for the same bias voltage) but it will also increase the electrical capacitance of the modulator, and thus limit the modulation bandwidth. Higher implantation doses will also improve the modulation efficiency, but at the expense of a higher optical loss.

Similar to a junction, a capacitor can be embedded into the waveguide. This is not straightforward to implement in a horizontal geometry, but the capacitor can be created with a vertical silicon-oxide-silicon stack [71, 72]. With a thin oxide layer, a capacitor can accumulate more charge than a reverse-biased p–n junction, and it can have a higher modulation efficiency. Again, this can come at the expense of RC-limited modulation bandwidth.

Both in the junction as the capacitor it is required to activate the implanted dopants. This can be done by applying a short high temperature anneal (1000C). Depending on the doping profile, this anneal will also induce some diffusion, so a good trade-off between anneal conditions and performance is needed [73].

Electrically contacting the modulators requires the use of a compatible metallization process. While different materials can be used, the preferred solution is to use Complementary Metal Oxide Semiconductor (CMOS)-compatible contacting. This usually involves a high dose implantation step, followed by a local silicidation. CMOS-compatible contacts to silicide are typically implemented in a tungsten (W) damascene process, involving etching, deposition and a chemical-mechanical polishing step. Metal wiring is done using aluminium or copper.

1.4.3 Active Tuning

While carrier modulators can reach operational speeds up to 40 Gbps, they are not the most effective method for slow tuning or switching. The effect is not that strong, and the implants and carriers induce excess optical loss.

To implement active tuning, the most straightforward technique is to incorporate heaters close to the waveguide structures. Heaters consist of an electrically contacted resistor, and they can be implemented in the metal layers of the modulator metallization, but also in dedicated resistive layers of tungsten [42] or titanium [74]. It is also possible to use the silicon or silicide layers as a resistor, positioned next to the optical waveguide instead of above it [75]. While often having different electrical properties, most heater mechanisms have similar efficiency, expressed in induced phase shift per dissipated power, on the order of 30 mW/π.

The main influence on tuning power is the thermal environment, rather than the electrical or optical properties of the device. The tuning power can be reduced dramatically by thermally insulating the optical element together with its heater from the surroundings. This can be done by etching deep trenches around the device [74].

1.4.4 Photodetectors

Photodetectors often have the most extensive processing, as they require the introduction of Germanium onto the chip. Germanium can be grown epitaxially on silicon, but it takes care to avoid the emergence of dislocations in positions where photocarriers are generated or collected, as recombination or trapping of carriers could reduce the efficiency of the detector.

Dislocations can be avoided or trapped by growing the germanium in a confined area. This way, dislocations are diverted to the sidewalls of the confined area where they cause no harm to the device. The epitaxy imposes a thermal budget on the rest of the processing, such as the anneal step for the modulator implants. This needs to be executed in advance.

Alternatively, amorphous germanium can be deposited and crystallized at a later stage using solid or liquid phase epitaxy [63]. This is an alternative way of avoiding dislocations. It does require a high temperature step to perform the regrowth.

1.4.5 Lasers

For lasers, we already discussed that bonding is the most likely solution for the short-term implementation of lasers on silicon [58]. The technique consists of bonding III–V epitaxial stacks onto the processed silicon photonic wafers, after which the bonded material is thinned down to a film of 100–2000 nm thick, depending on the applications. Multiple dies, or even entire III–V wafers, can be bonded on silicon photonics wafer. After bonding and thinning, the thin films of III–V material can be processed with wafer-scale processes similar to those used for the silicon processes. A schematic process is shown in Figure 1.8.

The bonding itself can be done in different ways. Molecular bonding uses Van Der Waals forces between two surfaces to attach the III–V material to the silicon circuit. For this, it is required that both surfaces are close to atomically flat. This technique has been used to demonstrate Fabry–Perot lasers [76], DFB-lasers [59], ring lasers [61] as well as small microdisk lasers [77]. An advantage of molecular bonding is that there is a close contact between the silicon layer and the III–V material, which provides a decent thermal contact to sink the dissipated power in the laser. Still, careful thermal design is needed.

An alternative to molecular bonding is adhesive bonding, which uses a glue layer to attach the III–V material to silicon [78]. Because of the intermediate spin-coated glue layer, this technique is more tolerant toward surface flatness. Bonding layers of 30 nm up to 1000 nm can be used, by tuning the viscosity of the adhesive, which is typically Dynamic Vapour Sorption-Benzocyclobutene (DVS-BCB). Similar to molecular bonding, different types of lasers have been demonstrated [62].

While epitaxial III–V materials are not yet sufficiently mature, germanium can be grown onto silicon substrates. Germanium has an indirect bandgap, but also a direct conduction band valley with a slightly higher energy than the indirect valley. By applying strain in the germanium layer during the epitaxial growth, the difference in energy can be partially compensated [63]. Subsequently, strong N-type doping

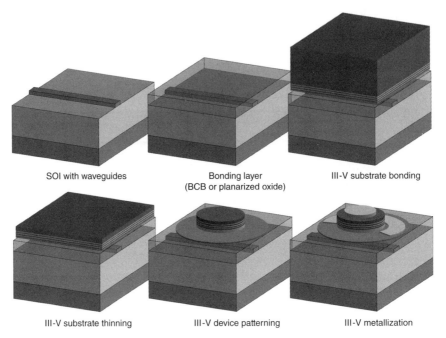

FIGURE 1.8 Fabrication of bonded lasers. From Reference 58.

can saturate the indirect conduction band valley, favoring recombination of injected electron–hole pairs via the direct conduction band valley [63]. A germanium laser based on this principle has been demonstrated [79].

Processing an efficient laser on silicon will still require significant effort. While it would definitely increase the flexibility of silicon photonic circuits, the currently most relevant technique is still the use of an outside discrete source, flipchipped on or fiber-pigtailed to the silicon chip.

1.4.6 Photonic–Electronic Integration

While silicon photonics can provide most optical functions on a chip, many of these functions require electronic control (e.g., modulators, photodetectors, and tuning elements). As circuits will become more complex, more electronics is needed. Depending on the applications, the electronics is needed to control the optical functions of the silicon chip, but in many cases, especially interconnect, the photonics is introduced to augment the performance of the electronics (e.g., high speed interconnects). The requirements of the photonic–electronic integration might be very different depending on this relation.

Different mechanisms exist for photonic–electronic cointegration. When no dense integration is needed, the simplest solution can be wirebonding, or simple flipchipping. When dense interconnects are needed, and especially operating at high speed, flipchipping with microbumps is an option.

Even denser integration is possible using through-silicon vias (TSV). 3D stacking using TSVs can enable very dense vertical interconnects with very little electrical parasitics. However, such dense integration can introduce problems with thermal crosstalk. TSVs can be implemented in different metals, but typically the standard CMOS materials such as W and Cu are used [80].

As SOI is not only a suitable material for photonics, but also for electronics, it is possible to integrate photonics and electronics side by side in the silicon layer [37]. This has the advantage that there can be a very tight integration between electronics and photonics, but on the other hand, it means that there is competition for floor space between both technologies, which might in the end result in a much larger chip. Also, process requirements can be very different for both technologies, so it is not always possible to reconcile the two technologies.

Instead of using crystalline silicon, it is also possible to use deposited silicon, which can be amorphous or polycrystalline. Layers of this material can be deposited on top of electronics layers, or anywhere in the stack of such wafers. This way, it is possible to implement photonic circuits together with the metal interconnect layers, without having a competition of floor space. The main issue is that amorphous silicon has a poorer performance than crystalline silicon, and that it is difficult to implement electro-optic modulators in it.

To use a crystalline silicon substrate for both electronics and photonics without having a competition for floor space, it is possible to use both sides of the wafer, one for photonics and one for electronics, and use TSVs to connect both layers. This approach introduces difficulties in processing, as the patterned backside needs to be protected during the processing of the front side. Also, TSVs need to pass through the entire wafer, which requires large and deep holes.

1.5 CONCLUSION

In this chapter, we tried to sketch an overview of the field of silicon photonics. We structured the technology from the applications downward, looking at the optical functions that are needed and the technologies that can implement them. In this approach, we mainly stuck to what is considered to be the mainstream silicon photonics, which is being developed by silicon fabs in various places in the world [37, 65, 80].

Still, the field of silicon photonics is much richer than this, with a lot of research exploring the introduction of new materials, design of waveguide structures, and novel applications.

Given the potential of silicon photonics, we can expect to see it deployed in actual products and industry in the coming years.

REFERENCES

[1] R. Soref, "The past, present, and future of silicon photonics," *J. Sel. Top. Quantum Electron.* **12**(6), 1678–1687 (2006).

[2] G. T. Reed and A. P. Knights, *Silicon Photonics: The State of the Art* (Wiley Online Library, 2008).

[3] W. Bogaerts, R. Baets, P. Dumon, V. Wiaux, S. Beckx, D. Taillaert, B. Luyssaert, J. Van Campenhout, P. Bienstman, and D. Van Thourhout, "Nanophotonic waveguides

in silicon-on-insulator fabricated with CMOS technology," *J. Lightwave Technol.* **23**(1), 401–412 (2005).

[4] K. De Vos, J. Girones, T. Claes, Y. De Koninck, S. Popelka, E. Schacht, R. Baets, and P. Bienstman, "Multiplexed antibody detection with an array of silicon-on-insulator microring resonators," *IEEE Photonics J.* **1**(4), 225–235 (2009).

[5] M. Iqbal, M. Gleeson, B. Spaugh, F. Tybor, W. Gunn, M. Hochberg, T. Baehr-Jones, R. Bailey, and L. Gunn, "Label-free biosensor arrays based on silicon ring resonators and high-speed optical scanning instrumentation," *J. Sel. Top. Quantum Electron.* **16**(3), 654–661 (2010).

[6] A. Mekis, S. Abdalla, P. De Dobbelaere, D. Foltz, S. Gloeckner, S. Hovey, S. Jackson, Y. Liang, M. Mack, G. Masini, et al., "SPIE. Scaling CMOS photonics transceivers beyond 100 Gb/s," in *Proc SPIE*, Conference on Optoelectronic Integrated Circuits XIV, San Francisco, CA, January 25–26, 2012, p. 8265.

[7] T. Pinguet, B. Analui, E. Balmater, D. Guckenberger, M. Harrison, R. Koumans, D. Kucharski, Y. Liang, G. Masini, A. Mekis, et al., "Monolithically integrated high-speed CMOS photonic transceivers," Group IV Photonics 2008, pp. 362–364.

[8] J. Shah, "High performance Silicon Photonics technology for ubiquitous communications: intrachip to data warehouses," Optical Interconnects Conference 2012, (2012), p. 112.

[9] Y. A. Vlasov, "Silicon CMOS-integrated nano-photonics for computer and data communications beyond 100G," *Commun. Mag. IEEE* **50**(2), s67–s72 (2012).

[10] S. Beamer, C. Sun, Y. Kwon, A. Joshi, C. Batten, V. Stojanović, and K. Asanović, "Re-architecting DRAM memory systems with monolithically integrated silicon photonics," in *ACM SIGARCH Computer Architecture News*, (ACM, 2010), Vol. 38(3), pp. 129–140.

[11] K. De Vos, I. Bartolozzi, E. Schacht, P. Bienstman, and R. Baets, "Silicon-on-insulator microring resonator for sensitive and label-free biosensing," *Opt. Express* **15**(12), 7610–7615 (2007).

[12] D. Taillaert, W. Van Paepegem, J. Vlekken, and R. Baets, "A thin foil optical strain gage based on silicon-on-insulator microresonators," in *Proc. SPIE.*, Vol. 6619 of Third European Workshop on Optical Fibre Sensors (EWOFS), 2007, p. 661914.

[13] M. S. Luchansky, A. L. Washburn, T. A. Martin, M. Iqbal, L. C. Gunn, and R. C. Bailey, "Characterization of the evanescent field profile and bound mass sensitivity of a label-free silicon photonic microring resonator biosensing platform," *Biosens. Bioelectron.* **26**(4), 1283–1291 (2010).

[14] N. A. Yebo, P. Lommens, Z. Hens, and R. Baets, "An integrated optic ethanol vapor sensor based on a silicon-on-insulator microring resonator coated with a porous ZnO film," *Opt. Express* **18**(11), 11859–11866 (2010).

[15] E. Ryckeboer, A. Gassenq, N. Hattasan, K. Kuyken, L. Cerutti, J.-B. Rodriguez, E. Tournie, G. Roelkens, W. Bogaerts, and R. Baets, "Integrated spectrometer and integrated detectors on silicon-on-insulator for short-wave infrared applications," in *CLEO: Science and Innovations* (Optical Society of America, 2012), p. CTu1A.3.

[16] Y. A. Vlasov, and S. J. McNab, "Losses in single-mode silicon-on-insulator strip waveguides and bends," *Opt. Express* **12**(8), 1622–1631 (2004).

[17] S. K. Selvaraja, P. Jaenen, W. Bogaerts, D. Van Thourhout, P. Dumon, and R. Baets, "Fabrication of photonic wire and crystal circuits in silicon-on-insulator using 193 nm optical lithography," *J. Lightwave Technol.* **27**(18), 4076–4083 (2009).

[18] A. G. Rickman and G. T. Reed, "Silicon-on-insulator optical rib waveguides: loss, mode characteristics, bends and y-junctions," IEE *Proc. : Optoelectron.* **141**(6), 391–393 (1994).

[19] S. Bidnyk, D. Feng, A. Balakrishnan, M. Pearson, G. M., H. Liang, W. Qian, C.-C. Kung, J. Fong, J. Yin, and M. Asghari, "Silicon-on-insulator-based planar circuit for passive optical network applications," *IEEE Photonics Technol. Lett.* **18**(22), 2392–2394 (2006).

[20] U. Fischer, T. Zinke, J. R. Kropp, F. Arndt, and K. Petermann, "0.1 dB/cm waveguide losses in single-mode SOI rib waveguides," *IEEE Photonics Technol. Lett.* **8**(5), 647–648 (1996).

[21] M. Gnan, S. Thoms, D. S. Macintyre, R. M. De La Rue, and M. Sorel, "Fabrication of low-loss photonic wires in silicon-on-insulator using hydrogen silsesquioxane electron-beam resist," *Electron. Lett.* **44**(2), 115–116 (2008).

[22] W. Bogaerts and S. K. Selvaraja, "Compact single-mode silicon hybrid rib/strip waveguide with adiabatic bends," *IEEE Photon J.* **3**(3), 422–432 (2011).

[23] M. A. Webster, R. M. Pafchek, G. Sukumaran, and T. L. Koch, "Low-loss quasi-planar ridge waveguides formed on thin silicon-on-insulator," *Appl. Phys. Lett.* **87**, 231108 (2005).

[24] T. Shoji, T. Tsuchizawa, T. Watanabe, K. Yamada, and H. Morita, "Low loss mode size converter from 0.3 μm square Si waveguides to singlemode fibres," *Electron. Lett.* **38**(25), 1669–1700 (2002).

[25] J. Brouckaert, W. Bogaerts, S. Selvaraja, P. Dumon, R. Baets, and D. Van Thourhout, "Planar concave grating demultiplexer with high reflective bragg reflector facets," *IEEE Photonics Technol. Lett.* **20**(4), 309–311 (2008).

[26] W. Bogaerts, S. Selvaraja, P. Dumon, J. Brouckaert, K. De Vos, D. Van Thourhout, and R. Baets, "Silicon-on-insulator spectral filters fabricated with CMOS technology," *J. Sel. Top. Quantum Electron.* **16**(1), 33–44 (2010).

[27] W. Bogaerts, P. De Heyn, T. Van Vaerenbergh, K. De Vos, S. Kumar Selvaraja, T. Claes, P. Dumon, P. Bienstman, D. Van Thourhout, and R. Baets, "Silicon microring resonators," *Laser Photon Rev.* **6**(1), 47–73 (2012).

[28] C. Dragone, "An NxN optical multiplexer using a planar arrangement of two star couplers," *IEEE Photonics Technol. Lett.* **3**(9), 812–814 (1991).

[29] S. K. Selvaraja, W. Bogaerts, P. Dumon, D. Van Thourhout, and R. Baets, "Subnanometer linewidth uniformity in silicon nanophotonic waveguide devices using CMOS fabrication technology," *J. Sel. Top. Quantum Electron.* **16**(1), 316–324 (2010).

[30] F. Gan, T. Barwicz, M. A. Popovic, M. S. Dahlem, C. W. Holzwarth, P. T. Rakich, H. I. Smith, E. P. Ippen, and F. X. Kartner, "Maximizing the thermo-optic tuning range of silicon photonic structures," in *Photonics in Switching 2007*, pp. 67–68.

[31] D. Dai, L. Yang, and S. He, "Ultrasmall thermally tunable microring resonator with a submicrometer heater on Si nanowires," *J. Lightwave Technol.* **26**(5–8), 704–709 (2008).

[32] I. Kiyat, A. Aydinli, and N. Dagli, "Low-power thermooptical tuning of SOI resonator switch," *IEEE Photonics Technol. Lett.* **18**(1–4), 364–366 (2006).

[33] C. Kopp, S. Bernabe, B. Bakir, J. Fedeli, R. Orobtchouk, F. Schrank, H. Porte, L. Zimmermann, and T. Tekin, "Silicon photonic circuits: on-CMOS integration, fiber optical coupling, and packaging," *IEEE J. Sel. Top. Quantum Electron.* **17**(3), 498–509 (2011).

[34] D. Taillaert, W. Bogaerts, P. Bienstman, T. Krauss, P. Van Daele, I. Moerman, S. Verstuyft, K. De Mesel, and R. Baets, "An out-of-plane grating coupler for efficient butt-coupling between compact planar waveguides and single-mode fibers," *J. Quantum Electron.* **38**(7), 949–955 (2002).

[35] D. Taillaert, H. Chong, P. Borel, L. Frandsen, R. De La Rue, and R. Baets, "A compact two-dimensional grating coupler used as a polarization splitter," *IEEE Photonics Technol. Lett.* **15**(9), 1249–1251 (2003).

[36] S. J. McNab, N. Moll, and Y. A. Vlasov, "Ultra-low loss photonic integrated circuit with membrane-type photonic crystal waveguides," *Opt. Express* **11**(22), 2927–2939 (2003).

[37] A. Mekis, S. Gloeckner, G. Masini, A. Narasimha, T. Pinguet, S. Sahni, and P. De Dobbe-laere, "A grating-coupler-enabled CMOS photonics platform," *IEEE J. Sel. Top. Quantum Electron.* **17**(3), 597–608 (2011).

[38] D. Vermeulen, S. Selvaraja, P. Verheyen, G. Lepage, W. Bogaerts, P. Absil, D. Van Thourhout, and G. Roelkens, "High-efficiency fiber-to-chip grating couplers realized using an advanced CMOS-compatible silicon-on-insulator platform," *Opt. Express* **18**(17), 18278–18283 (2010).

[39] F. Van Laere, T. Claes, J. Schrauwen, S. Scheerlinck, W. Bogaerts, D. Taillaert, L. O'Faolain, D. Van Thourhout, and R. Baets, "Compact focusing grating couplers for silicon-on-insulator integrated circuits," *Photonics Technol. Lett.* **19**(23), 1919–1921 (2007).

[40] F. van Laere, W. Bogaerts, P. Dumon, G. Roelkens, D. Van Thourhout, and R. Baets, "Focusing polarization diversity grating couplers in silicon-on-insulator," *J. Lightwave Technol.* **27**(5), 612–618 (2009).

[41] K. Van Acoleyen, J. Roels, P. Mechet, T. Claes, D. Van Thourhout, and R. Baets, "Ultra-compact phase modulator based on a cascade of NEMS-operated slot waveguides fabricated in silicon-on-insulator," *IEEE Photonics J.* **4**(3), 779–788 (2012).

[42] A. Masood, M. Pantouvaki, D. Goossens, G. Lepage, P. Verheyen, D. Van Thourhout, P. Absil, and W. Bogaerts, "CMOS-compatible Tungsten heaters for silicon photonic waveguides," Group IV Photonics 2012, pp. 234–236.

[43] T. Barwicz, M. A. Popovic, F. Gan, M. S. Dahlem, C. W. Holzwarth, P. T. Rakich, E. P. Ippen, F. X. Kartner, and H. I. Smith, "Reconfigurable silicon photonic circuits for telecommunication applications," *Proc. SPIE* **6872**, 68720Z–68720Z–12 (2008).

[44] W. De Cort, J. Beeckman, T. Claes, K. Neyts, and R. Baets, "Wide tuning of silicon-on-insulator ring resonators with a liquid crystal cladding," *Opt. Lett.* **36**(19), 3876–3878 (2011).

[45] R. Soref and B. Bennett, "Electrooptical effects in silicon," *J. Quantum Electron.* **23**(1), 123–129 (1987).

[46] G. T. Reed, G. Mashanovich, F. Y. Gardes, and D. J. Thomson, "Silicon optical modulators [Review]," *Nature Photonics* **4**(8), 518–526 (2010).

[47] M. R. Watts, D. C. Trotter, R. W. Young, and A. L. Lentine, "Ultralow power silicon microdisk modulators and switches," Group IV Photonics 2008, pp. 4–6.

[48] Q. Xu, S. Manipatruni, B. Schmidt, J. Shakya, and M. Lipson, "12.5 Gbit/s carrier-injection-based silicon micro-ring silicon modulators," *Opt. Express* **15**(2), 430–436 (2007).

[49] W. M. J. Green, M. J. Rooks, L. Sekaric, and Y. A. Vlasov, "Ultra-compact, low RF power, 10 Gb/s silicon Mach-Zehnder modulator," *Opt. Express* **15**(25), 17106–17113 (2007).

[50] L. Liao, A. Liu, J. Basak, H. Nguyen, M. Paniccia, D. Rubin, Y. Chetrit, R. Cohen, and N. Izhaky, "40 Gbit/s silicon optical modulator for highspeed applications," *Electron. Lett.* **43**(22) (2007).

[51] D. Marris-Morini, L. Vivien, J. M. Fédéli, E. Cassan, P. Lyan, and S. Laval, "Low loss

and high speed silicon optical modulator based on a lateral carrier depletion structure," *Opt. Express* **16**(1), 334–339 (2008).

[52] J. Brouckaert, G. Roelkens, D. Van Thourhout, and R. Baets, "Compact InAlAs/InGaAs metal-semiconductor-metal photodetectors integrated on silicon-on-insulator waveguides," *IEEE Photon Technol. Lett.* **19**(19), 1484–1486 (2007).

[53] L. Colace, M. Balbi, G. Masini, G. Assanto, H. C. Luan, and L. C. Kimerling, "Ge on Si p-i-n photodiodes operating at 10 Gbit/s ," *Appl. Phys. Lett.* **88**(10) (2006).

[54] L. Vivien, J. Osmond, J.-M. Fédéli, D. Marris-Morini, P. Crozat, J.-F. Damlencourt, E. Cassan, Y. Lecunff, and S. Laval, "42 GHz p.i.n Germanium photodetector integrated in a silicon-on-insulator waveguide," *Opt. Express* **17**(8), 6252–6257 (2009).

[55] Y. Kang, H.-D. Liu, M. Morse, M. Paniccia, M. Zadka, S. Litski, G. Sarid, A. Pauchard, Y.-H. Kuo, H.-W. Chen, et al., "Monolithic germanium/silicon avalanche photodiodes with 340 GHz gain-bandwidth product," *Nat. Photonics* **3**(1), 59–63 (2009).

[56] D. Feng, S. Liao, P. Dong, H. Feng, N.-N. Liang, D. Zheng, C.-C. Kung, J. Fong, R. Shafiha, J. Cunningham, et al., "High-speed Ge photodetector monolithically integrated with large cross-section silicon-on-insulator waveguide," *Appl. Phys. Lett.* **95**(26), 261105–261105-3 (2009).

[57] J. Bruns, T. Mitze, M. Schnarrenberger, L. Zimmermann, K. Voigt, M. Krieg, J. Kreil, K. Janiak, T. Hartwich, and K. Petermann, "SOI-based optical board technology," *AEU Int. J. Electron. Commun.* **61**(3), 158–162 (2007).

[58] G. Roelkens, L. Liu, D. Liang, R. Jones, A. Fang, B. Koch, and J. Bowers, "III-V/silicon photonics for on-chip and inter-chip optical interconnects," *Laser Photon Rev.* **4**(6), 751–779 (2010).

[59] B. Koch, A. Fang, E. Lively, R. Jones, O. Cohen, D. Blumenthal, and J. Bowers, "Mode locked and distributed feedback silicon evanescent lasers," *Laser Photon Rev.* **3**(4), 355–369 (2009).

[60] A. Fang, B. Koch, R. Jones, E. Lively, D. Liang, Y.-H. Kuo, and J. Bowers, "A distributed bragg reflector silicon evanescent laser," *IEEE PhotonicsTechnol. Lett.* **20**(20), 1667–1669 (2008).

[61] A. Fang, B. Koch, K.-G. Gan, H. Park, R. Jones, O. Cohen, M. Paniccia, D. Blumenthal, and J. Bowers, "A racetrack mode-locked silicon evanescent laser," *Opt. Express* **16**(2), 1393–1398 (2008).

[62] J. Van Campenhout, P. Rojo Romeo, P. Regreny, C. Seassal, D. Van Thourhout, S. Verstruyft, L. Di Ciocco, J.-M. Fedeli, C. Lagahe, and R. Baets, "Electrically pumped InP-based microdisk lasers integrated with a nanophotonic silicon-on-insulator waveguide circuit," *Opt. Express* **15**(11), 6744–6749 (2007).

[63] J. Liu, X. Sun, D. Pan, X. Wang, L. Kimerling, T. Koch, and J. Michel, "Tensile-strained, n-type Ge as a gain medium for monolithic laser integration on Si," *Opt. Express* **15**(18), 11272–11277 (2007).

[64] S. K. Selvaraja, W. Bogaerts, and D. Van Thourhout, "Loss reduction in silicon nanophotonic waveguide micro-bends through etch profile improvement," *Opt. Comm.* **284**(8), 2141–2144 (2011).

[65] K.-W. Ang, T.-Y. Liow, Q. Fang, M. Yu, F. Ren, S. Zhu, J. Zhang, J. Ng, J. Song, Y. Xiong, et al., "Silicon photonics technologies for monolithic electronic-photonic integrated circuit (EPIC) applications: current progress and future outlook," in *Electron Devices Meeting (IEDM)*, 2009 IEEE International, pp. 1–4.

[66] S. Selvaraja, L. Fernandez, M. Vanslembrouck, J.-L. Everaert, P. Dumon, J. Van Campenhout, W. Bogaerts, and P. Absil, "Si photonic device uniformity improvement using

wafer-scale location specific processing," Photonics Conference (IPC), 2012 IEEE, pp. 725–726.

[67] D. Sparacin, R. Sun, A. Agarwal, M. Beals, J. Michel, L. Kimerling, T. Conway, A. Pomerene, D. Carothers, M. Grove, et al., "Low-loss amorphous silicon channel waveguides for integrated photonics," Group IV Photonics 2006, pp. 255–257.

[68] S. Selvaraja, E. Sleeckx, M. Schaekers, W. Bogaerts, D. Van Thourhout, P. Dumon, and R. Baets, "Low-loss amorphous silicon-on-insulator technology for photonic integrated circuitry," *Opt. Comm.* **282**(9), 1767–1770 (2009).

[69] S. Zhu, Q. Fang, M. B. Yu, G. Q. Lo, and D. L. Kwong, "Propagation losses in undoped and n-doped polycrystalline silicon wire waveguides," *Opt. Express* **17**(23), 20891–20899 (2009).

[70] Y Hui, M. Pantouvaki, J. Van Campenhout, D. Korn, K. Komorowska, P. Dumon, Y. Li, P. Verheyen, P. Absil, L. Alloatti, et al., "Performance tradeoff between lateral and interdigitated doping patterns for high speed carrier-depletion based silicon modulators," *Opt. Express* **20**(12), 12926–12938 (2012).

[71] J. C. Rosenberg, W. M. J. Green, S. Assefa, D. M. Gill, T. Barwicz, M. Yang, S. M. Shank, and Y A. Vlasov, "A 25 Gbps silicon microring modulator based on an interleaved junction," *Opt. Express* **20**(24), 26411–26423 (2012).

[72] J. Van Campenhout, M. Pantouvaki, P. Verheyen, S. Selvaraja, G. Lepage, H. Yu, W. Lee, J. Wouters, D. Goossens, M. Moelants, et al., "Low-voltage, low-loss, multi-Gb/s silicon micro-ring modulator based on a MOS capacitor," in *Optical Fiber Communication Conference* (Optical Society of America, 2012), p. OM2E.4.

[73] H. Yu, W. Bogaerts, and A. De Keersgieter, "Optimization of ion implantation condition for depletion-type silicon optical modulators," *J. Quentum. Electron.* **46**(12), 1763–1768 (2010).

[74] P. Dong, W. Qian, H. Liang, R. Shafiiha, D. Feng, G. Li, J. Cunningham, A. Krishnamoorthy, and M. Asghari, "Thermally tunable silicon racetrack resonators with ultralow tuning power," *Opt. Express* **18**(19), 20298–20304 (2010).

[75] J. Van Campenhout, W. M. J. Green, S. Assefa, and Y. A. Vlasov, "Integrated NiSi waveguide heaters for CMOS-compatible silicon thermooptic devices," *Opt. Lett.* **35**(7), 1013–1015 (2010).

[76] M. Lamponi, S. Keyvaninia, F. Pommereau, R. Brenot, G. de Valicourt, F. Lelarge, G. Roelkens, D. Van Thourhout, S. Messaoudene, J.-M. Fedeli, et al., "Heterogeneously integrated InP/SOI laser using double tapered single-mode waveguides through adhesive die to wafer bonding," Group IV Photonics 2010, pp. 22–24.

[77] T. Spuesens, F. Mandorlo, P. Rojo-Romeo, P. R6greny, N. Olivier, J.-M. Fédil, and D. Van Thourhout, "Compact integration of optical sources and detectors on SOI for optical interconnects fabricated in a 200 mm CMOS pilot line," *J. Lightwave Technol.* **30**(11), 1764–1770 (2012).

[78] G. Roelkens, J. Brouckaert, D. Van Thourhout, R. Baets, R. Notzel, and M. Smit, "Adhesive bonding of InP/InGaAsP dies to processed silicon-on-insulator wafers using DVS-bis-benzocyclobutene," *J. Electrochem. Soc.* **153**(12), G1015–G1019 (2006).

[79] R. Camacho-Aguilera, Y. Cai, N. Patel, J. Bessette, M. Romagnoli, L. Kimerling, and J. Michel, "An electrically pumped germanium laser," *Opt. Express* **20**(10), 11316–11320 (2012).

[80] J. Fedeli, E. Augendre, J. Hartmann, L. Vivien, P. Grosse, V. Mazzocchi, W. Bogaerts, D. Van Thourhout, and F. Schrank, "Photonics and electronics integration in the HELIOS project," Group IV Photonics 2010, pp. 356–358.

2

CAVITY PHOTONICS

J. Mørk, P. T. Kristensen, P. Kaer, M. Heuck, Y. Yu, and N. Gregersen

DTU Fotonik, Technical University of Denmark, Kogens Lyngby, Denmark

2.1 INTRODUCTION

The optical cavity is a key device in many branches of physics and has numerous commercial applications. To name but a few, it provides the feedback required for an optical emitter to undergo a transition to lasing, it provides the filtering required for high-resolution optical spectroscopy, and it enables the strong coupling of light and matter states allowing the experimental exploration of some of the most fundamental and mind-challenging predictions of quantum mechanics. In this chapter we provide an overview of some of the exciting research topics within cavity photonics, focusing on micro- and nanoscale cavity structures [1], that is, cavities with dimensions on the order of a few wavelengths to subwavelength, which are enabled by recent progress in nanofabrication techniques. While the development of low-dimensional semiconductor structures such as quantum wells and quantum dots have enabled control of the electronic properties of matter, nanocavities provide control of the optical modes and the dispersion at a very fundamental level. The simultaneous control of electronic and optical degrees of freedom thus allows manipulating the light–matter coupling—with numerous possibilities for exploring fundamental physics as well as engineering advanced devices.

We do not pretend to give a complete overview of the entire field of cavity photonics, which is currently developing at a rapid pace. Important subjects not covered in

---

Photonics: Scientific Foundations, Technology and Applications, Volume II, First Edition.
Edited by David L. Andrews.
© 2015 John Wiley & Sons, Inc. Published 2015 by John Wiley & Sons, Inc.

this chapter include microring structures, which constitute an important class of cavities with a range of different filtering applications and interesting nonlinear optics properties, as well as the field of cavity optomechanics, in which light in optical cavities is coupled to mechanical degrees of freedom to investigate novel physical phenomena [2]. Rather, we focus on the properties of nanocavities that are relevant to their application in all-optical switching, lasers and light emitting diodes (LEDs), as well as cavity quantum electrodynamics (QED).

We start in Section 2.2 by introducing different structures that have been used to realize optical cavities with sizes on the order of a wavelength, in particular focusing on micropillar and photonic crystal cavities. Assuming that the reader is familiar with the basic properties of Fabry–Perot resonators, we then discuss how these cavities often, but not always, can be represented by effective Fabry–Perot cavities in which case one can benefit from the well-established understanding of the properties and possibilities of such cavities.

Section 2.3 is devoted to cavity-based switches. If the cavity embeds a nonlinear material in which the refractive index depends on the intensity of the optical field inside the cavity, the transmission through the cavity can be controlled optically. This provides the means for implementing an optical switch, and the system also allows for bistability and displays a rich dynamical behavior. While optical switching and bistability is a classical subject within photonics [3], the emergence of ultrasmall cavities with dimensions on the order of hundreds of nanometers opens new possibilities such as switching at energies approaching the attojoule range [4].

In Section 2.4 we discuss the properties of emitters embedded in cavities. What is peculiar about nanocavity structures is that the spontaneous and stimulated emission into the cavity mode may be enhanced relative to the emission into free space. This so-called Purcell effect is introduced in Section 2.4.1 and its consequences are discussed. Section 2.4.2 is devoted to the QED of a single emitter, such as a semiconductor quantum dot, embedded in a nanocavity. We discuss the weak coupling regime, where the spontaneous emission is Purcell enhanced, and the strong coupling regime, where the coupling of the emitter and the cavity may lead to so-called vacuum Rabi oscillations as indicated by a splitting of the cavity resonance. The Jaynes–Cummings model [5] for analyzing these effects is presented in Section 2.4.3 and the regimes of different qualitative behavior are discussed, that is, Purcell enhancement versus strong coupling effects. In Section 2.4.3 we discuss the properties of LEDs and lasers exploiting the Purcell effect to enhance the light emission. This analysis pertains to the semi-classical regime where numerous emitters (material transitions) are in resonance with the cavity. In this case the spontaneous emission of the material can be amplified and, if the cavity feedback is sufficiently strong, lasing may take place. Finally, a brief summary is presented in Section 2.6.

2.2 CAVITY FUNDAMENTALS

Figure 2.1 illustrates two important kinds of cavities that can realize confinement of the optical field to volumes on the order of or smaller than the cube of the wavelength in the medium and at the same time store the field for many optical cycles before it leaks out. The *micropillar* cavity [6] in Figure 2.1a confines light in the transverse

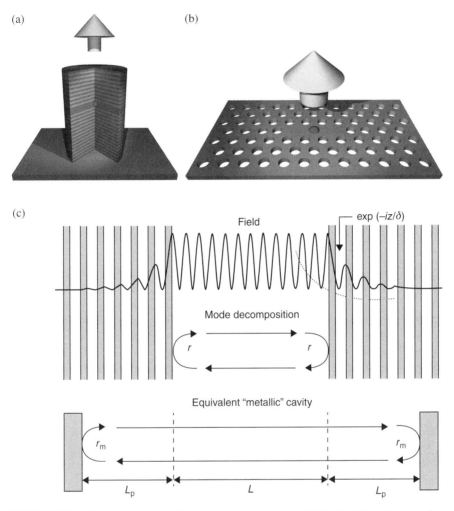

FIGURE 2.1 Cavity structures and representation in terms of effective Fabry–Perot cavity: (a) micropillar cavity; (b) photonic crystal defect cavity; (c) illustration of field distribution in cavity with mirrors composed of period Bragg structures and corresponding equivalent Fabry–Perot cavity. Reproduced with permission from Reference 14.

direction by the well-known effect of total internal reflection from a high index to a low index medium (the effect also employed in optical fibers), while the mirrors are implemented using periodic index modulations known as Bragg stacks, relying on the build-up of a strong reflected field at certain wavelengths by the constructive addition of scattering from multiple interfaces. To obtain a high mirror reflectivity, multiple quarter wavelength layers of alternating refractive index are employed, and the light is guided out, primarily, through the top by using a larger number of layer pairs in the bottom mirror. The *photonic crystal defect* cavity [7] in Figure 2.1b is a membrane structure embedded in air or a low index material. In extended cavities, confinement transverse to the plane is due to total internal reflection, as in slab waveguides. The in-plane confinement is achieved using the so-called photonic crystal structure in which the holes form a periodic lattice that leads to a band of frequencies, the bandgap, where propagation of electromagnetic waves in the material is impossible. Instead, the field localizes in the nonperiodic defect region, which may be in the form of a single point defect, where one of the holes is smaller than the others, or an extended waveguide-like line defect, as illustrated in Figure 2.1b. The ability of such cavities to filter and store the field depends strongly on the detailed geometry of the defect region. For example, the displacement of holes within a fraction of a wavelength has been shown to be a resource for strongly improving the filtering and storage properties of the cavity [8,9].

The quality (Q) factor is an important characteristic of a cavity. It is defined as the ratio of the resonance frequency ω_c (or wavelength λ_c) to the width (FWHM) $\Delta\omega_c$ of the resonance

$$Q = \frac{\lambda_c}{\Delta\lambda_c} = \frac{\omega_c}{\Delta\omega_c}, \qquad (2.1)$$

so that $1/Q$ is the fractional bandwidth of the resonance. The optical energy in the cavity decays exponentially with rate $\kappa = \Delta\omega_c$ and the quality factor may also be defined as the number of optical periods (cycles) that elapses before the energy has decayed by a factor $e^{-2\pi}$. The corresponding half-time for the decay thus corresponds to $\ln(2)/(2\pi)Q \simeq 0.11Q$ cycles of the electromagnetic field. At a wavelength of 1 μm the optical period is 3.3 fs, and a quality factor of 10,000 therefore means that the electromagnetic energy decays to half its initial value within 3.7 ps. For micropillar structures based on GaAs/AlAs layers, quality factors of more than 150,000 have been obtained for diameters on the order of 4 μm [10], while photonic crystal nanocavities have achieved quality factors exceeding 2.5 million [11], corresponding to a cavity lifetime of more than 2 ns.

The optical field in a cavity may be conveniently described in terms of the cavity *modes*. Like the bound states of the hydrogen atom, the cavity modes represent the electromagnetic field distributions that may naturally be supported by the cavity. The analogy, however, is deceptive: in contrast to the hydrogen atom, the modes of an optical cavity are inherently leaky and decay in time at a rate set by the Q factor, as discussed above. Mathematically, the cavity modes may be unambiguously defined as eigenfunctions to Maxwell's equations with outgoing wave boundary conditions (the Silver–Müller radiation condition). These boundary conditions render the differential equation problem non-Hermitian and leads to complex eigenfrequencies $\tilde{\omega}_c = \omega_c - i\gamma_c$ with a nonzero imaginary part as expected from decaying modes. From the discussion above, we find that $Q = \omega_c/2\gamma_c$. Because of the boundary condition, the spatial part of the field diverges at large distances, which means that the modes cannot be normalized by the traditional inner product, which is commonly adopted for Hermitian eigenvalue problems. Instead, the cavity modes should be normalized by an alternative inner product [12], which rigorously accounts for the long-distance behavior. Despite the added complexity due to the non-Hermiticity of the eigenvalue problem, many properties of the cavity modes may be understood *qualitatively* from a simplified picture in which the cavity mode is nonzero only inside the cavity region. As an example, the so-called effective mode volume is a physically appealing quantity, which is easily introduced for modes that vanish outside the cavity. Unfortunately, the cavity modes are nonzero everywhere and in general there is no way of introducing an unambiguous and physically meaningful cut-off. Nevertheless, an effective mode volume can indeed be introduced for cavity modes in a rigorous and unambiguous way [13], as we discuss in connection with the so-called Purcell effect in Section 2.4.1.

A number of important properties of the modes of micropillar cavities and photonic crystal defect cavities can be well accounted for by describing the structure as an effective one-dimensional Fabry–Perot cavity with discrete mirrors, as illustrated in Figure 2.1c. This even holds for cavities with sizes on the order of the wavelength [14], if one introduces an effective length L_{FP}, which accounts for the finite penetration depth into the distributed mirrors terminating the Fabry–Perot cavity [14]

$$L_{FP} = L + 2L_P, \quad L_P = \frac{-\lambda_c^2}{4\pi n_g} \frac{\partial \phi}{\partial \lambda_c}. \tag{2.2}$$

Here, n_g is the group index and the cavity resonance wavelength λ_c is determined by the condition of constructive interference upon a round trip in the cavity. In other words, the total phase Φ upon a cavity round trip should equal an integer p times 2π

$$\Phi(\lambda_c) = k_0 n_{eff} 2L + 2\phi(\lambda_c) = 2p\pi, \quad p \in \mathbb{Z}, \tag{2.3}$$

where n_{eff} is the effective index of the homogeneous material distributed over the length L. The phase $\phi(\lambda_c)$ is defined as the phase of the modal reflection coefficient of the guided mode, $r(\lambda_c) = |r| \exp\{i\phi(\lambda_c)\}$ and needs to be evaluated for a specific structure in order to calculate the corresponding effective penetration length L_P. Assuming that the cavity mode experiences losses only due to partial transmission through the mirrors, that is, below unity values of the power reflection coefficient $R = |r|^2$, the corresponding quality factor Q of the mode can be approximated as [14]

$$Q \approx \frac{k_0}{1-R} n_g (L + 2L_P). \tag{2.4}$$

Intuitively, the equivalence of a micropillar or a photonic crystal line defect cavity with a Fabry–Perot cavity would be expected to work well when the physical cavity length is much larger than the wavelength and a transverse mode can be identified. However, a comprehensive study [14] shows that the equivalence works quite well even for very small cavities down to the size of a one-hole defect cavity, and that even in this case the light confinement can be understood as an interference effect formed by the back-and-forth bouncing of a transverse mode in a waveguide [14]. In order to exploit the simple result in Eq. (2.4) one needs, however, to calculate the effective reflection coefficient R and the penetration length L_P, and this requires advanced numerical techniques [14, 15].

Equation (2.4) is the basis for optimizing the cavity quality factor. A larger mirror reflection coefficient obviously increases the quality factor as does a longer (effective) cavity length by virtue of the longer propagation time between the cavity mirrors, which are the only sources of loss in this simple model. Reference 14 discusses the importance of mode matching at the waveguide mirror interface, leading to a characterization of different strategies toward optimizing the quality factor using either abrupt, tapered, or matched interfaces, the latter also being referred to as a heterostructure interface [16]. The use of Fourier analysis in a plane parallel to the membrane has been important in identifying designs that lead to high quality factors [17]. The key idea is to gently confine the field in order not to induce plane wave components that cannot be confined to the slab by total internal reflection. The expression (2.4) shows that the Q factor also depends on the group index of the material between the mirrors. This dependence arises because the propagation time and thereby the storage time depends inversely on the group velocity. The strong dispersion of photonic crystal waveguides allows for a drastic change in the group index, realizing a regime of ultraslow light propagation [18]. Photonic crystal-based cavities thus offer a unique possibility for realizing ultrahigh quality factors despite the small structure size. For photonic crystal line defect cavities, a comprehensive description of the mode properties has been given in Reference 19, which considers the transition from a one-hole defect cavity to a cavity with 21 holes and analyzes the emergence and properties of the various modes. The paper also includes a discussion of how to distinguish Fabry–Perot-like modes from modes sharing characteristics with distributed feedback (DFB) modes.

2.3 CAVITY-BASED SWITCHES

The resonant character of a cavity may be used to control the transmission of an injected monochromatic signal. A change of the cavity resonance wavelength by a fraction $1/Q$ reduces the power transmission to 20% of its original value and efficient switches requiring small external perturbation can therefore be realized with high Q cavities. Today, almost all signal processing tasks needed in communication networks are performed in the electrical domain and therefore involve conversions between the electrical and optical domains, which limits the bandwidth as well as increasing the energy consumption. This is the motivation for exploring whether some of these signal processing functionalities could beneficially be realized in the optical domain.

All-optical switching has been demonstrated using many different materials and device configurations, such as semiconductor optical amplifiers [20–22], parametric processes in fibers [23], silicon waveguides [24], interferometers [25], microrings [26], and photonic crystal cavities [4, 27, 28]. Cavity-based switches rely on the dynamical change of a cavity resonance induced by a pump pulse, the control, which determines the transmission of the data signal to be processed. Photonic crystal switches, as demonstrated in References 4 and 27, are ultra compact structures with dimensions on the order of the wavelength, and switching with pulse energies in the few femtojoule range has been demonstrated by exploiting the field enhancement

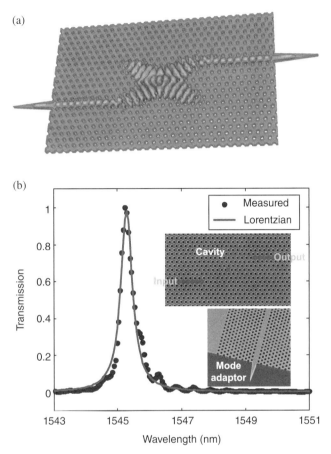

FIGURE 2.2 Photonic crystal cavity switch. (a) Schematic field profile of a signal propagating through the structure at resonance. (b) Measured transmission (normalized to 1 at peak) versus input wavelength for fabricated device. Insets show scanning electron microscope pictures of the cavity region as well as the mode adaptors used for coupling to external fibers. The device is fabricated in the InP material platform.

and filtering effect enabled by a high-Q cavity [4]. Below we review some of the important features of such switches.

An example of a nanocavity switch is shown in Figure 2.2. The structure is similar to the structures in References 27 and 29 and employs mode adaptors as suggested in [30] for reducing reflections at the facet interface as well as improving external coupling to the photonic crystal defect waveguide. The transmission of a signal from the input to the output waveguide depends on the spectrum of the signal relative to the cavity resonance. In the linear regime, where the cavity resonance is unaffected by the input signal, this can be used for simple filtering purposes [7]. For higher input signal intensities, where nonlinear optical effects lead to a change of the refractive index in the region of the cavity and thereby to a shift of the cavity resonance frequency, the transmission becomes nonlinear in the signal intensity and bistability may occur [3]. By keeping the signal intensity below the level of nonlinear effects, its transmission may also be controlled via the addition of a strong control signal. In this way one can realize an optically controlled switching action, for example, for demultiplexing a high bit rate optical signal. We notice, however, that a multimode cavity is required in order to separate the signal and control beams by using different frequencies and filtering at the output. Spatial as well as spectral separation may be achieved using a four-port structure as demonstrated in Reference 31.

For the analysis of cavity-based switching structures we use a temporal coupled mode theory (CMT) approach, as developed in References 7, 32, 33. In the CMT approach, the dynamical variations of the cavity field, $a(t)$, and the output field, $s_{\text{out}}(t)$, are governed by the equations

$$\frac{d}{dt}a(t) = -i\Delta\omega a(t) - \gamma_c a(t) + \sqrt{\gamma}\, s_{\text{in}}(t) \tag{2.5}$$

$$s_{\text{out}}(t) = \sqrt{\gamma}\, a(t), \tag{2.6}$$

where s_{in} is the input field in the waveguide, γ_c is the total (amplitude) decay rate of the cavity field, γ is the coupling rate between waveguide and cavity and $\Delta\omega$ is the dynamical shift of the cavity resonance induced by the optical field itself,

$$\Delta\omega = \omega_c - \omega_L + \delta_{\text{NL}}(t), \quad \delta_{\text{NL}}(t) = -F_{\text{Kerr}}|a(t)|^2 + F_{\text{car}} N(t), \tag{2.7}$$

in which ω_L is the carrier frequency of the input field, taken as the reference frequency for the cavity field as well, ω_c is the passive (cold) cavity resonance frequency, F_{Kerr} is the effective Kerr coefficient that accounts for the material nonlinearity as well as the field distribution in the cavity [34, 35], and F_{car} accounts for the free carrier dispersion effect induced by the carrier density $N(t)$. Equations (2.5) and (2.6) are formulated for the case of a so-called in-line coupled cavity, where the signal has to pass through the cavity to be transmitted from the input to the output waveguide. It can, however, easily be generalized to the case of a side-coupled cavity, including the effect of an additional scattering center in the waveguide [36].

We notice, that the proper calculation of the various coupling factors is complicated by the open nature of the cavity, which implies that even the definition of the proper cavity volume is nontrivial [13] and needs to be carried out within the framework of non-Hermitian differential equations [34, 37]. As a model for the carrier density, we assume that it is determined by linear and nonlinear absorption effects,

$$\frac{d}{dt}N(t) = -\frac{N(t)}{\tau_{\text{eff}}} + \alpha_{\text{LA}}(N)|a(t)|^2 + \alpha_{\text{TPA}}|a(t)|^4, \tag{2.8}$$

where $\alpha_{\text{LA}}(N)$ accounts for linear absorption and may depend on the carrier density due to bandfilling, α_{TPA} accounts for two-photon absorption (TPA), and τ_{eff} is the effective carrier lifetime. As discussed below, the relaxation of the carrier density in reality needs to be described by several different time constants, since both carrier diffusion, surface recombination and band-to-band recombination of carriers may be important.

In the linear regime, where the opticsal signal is sufficiently weak that it does not affect the cavity resonance, we may take $\delta_{NL}(t) \simeq 0$, and the spectrum in the output waveguide becomes

$$|s_{out}(\Omega)|^2 = \frac{\gamma^2}{\Omega^2 + \gamma_c^2} |s_{in}(\Omega)|^2, \qquad (2.9)$$

which clearly shows that the cavity acts as a filter of the bandpass type; only input signals with frequencies within the cavity spectrum are transmitted from input to output. For a white input spectrum, we recover the intrinsic line shape of the cavity with a FWHM of $\kappa = 2\gamma_c$. Dynamical switching of an input signal is obtained by changing the cavity resonance frequency via the local refractive index. This index change would typically be induced by an additional strong control signal. However, if the input signal itself becomes strong enough, it will affect its own transmission, leading to effects of bistability and self-switching. We shall first consider this case, which can be considered a specific example of a class of nonlinear Fabry–Perot interferometers [3, 38].

Figure 2.3 shows an example calculation of bistability in a nonlinear cavity [28]. This specific case uses a design adopted from Reference 39, in which the cavity is formed by high index rods in air, and the structure has two resonances and two sets of input–output waveguides. Furthermore, the Kerr nonlinearity is assumed to be dominating; for further details we refer to Reference 28. The figure shows the output energy (scaled by the input pulse width) as a function of the input peak

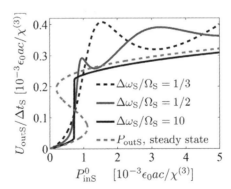

FIGURE 2.3 Calculated output pulse energy versus input pulse energy for different pulse durations, ranging from very long to comparable with the cavity lifetime. Power is measured in units of $10^{-3}\epsilon_0 ac/\chi^{(3)}$, where a is the lattice constant and $\chi^{(3)}$ is the third-order susceptibility. The curves are calculated for different pulse bandwidths Ω_S and a fixed cavity bandwidth $\Delta\omega_S = 2.34 \times 10^{-3} c/a$, and the signal is red-tuned by three cavity bandwidths. Reproduced from Reference 28 with the permission of the Optical Society of America.

power for square pulses of different widths. The steady-state solution corresponding to a CW input field is also plotted and shows the characteristic s-shaped curve, giving three solutions for a certain range of input fields, with the middle one being unstable [3, 38]. Pulses that are long compared to the cavity decay time are seen to follow the stable branches of this curve, while pulses which are comparable to or shorter than the cavity decay time lead to an oscillatory variation of the output power when the input power is varied. This behavior is well understood [28] and reflects a transient interference (beating) between the incoming pulse and the detuned cavity mode. Within the cavity, the period of the temporal oscillations is thus governed by the (inverse) detuning, and depending on the relation between the pulsewidth and the oscillation period, constructive or destructive interference may take place. Since the cavity resonance depends on the field intensity at the position of the cavity, the effective detuning and thereby the oscillation period varies with the input intensity, causing the local extrema of the curve. While the figure shows the case of a rectangular pulse, qualitatively similar behavior is found for Gaussian pulses [28]. The nonlinear dependence of the output pulse energy on the input energy may be used for improving the quality of a data signal—an operation known as amplitude regeneration. Intensity fluctuations between data pulses, which could lead to bit errors upon detection, would thus be suppressed in comparison with the stronger data pulses. When cascading many transmission spans, such nonlinear shaping effects can be used for regeneration [40]. It is to be noted, though, that transmission loss incurred upon the reshaping action, need to be compensated by additional amplification, therefore adding noise [40]. The bistability observed for long pulses may also be used for realizing a memory, as recently demonstrated in the photonic crystal platform [41].

If the data signal is kept at a power level low enough not to induce any index changes, the transmission of the data signal may be controlled via an injected control signal that changes the cavity resonance. An example of a measurement is shown in Figure 2.4 for the configuration shown in Figure 2.2 [42]. The transmission of a weak signal pulse is shown as a function of the time delay with respect to a stronger control pulse. For wavelengths around 1.5 μm, one-photon transitions are below the bandgap of InP and the refractive index changes are caused by bandfilling and dispersion of free carriers generated by TPA of the pump pulse. In this case the pump pulse leads to a reduction of the refractive index and thereby a blue-shift of the cavity resonance. Depending on whether the probe pulse is tuned to the blue or the red side of the cold cavity resonance wavelength, the control pulse therefore leads to either an increase or a reduction of the transmitted signal energy, as clearly seen in Figure 2.4. A qualitatively similar behavior was observed in the GaAs material platform [27]. The theoretical curves in Figure 2.4 are fits using the CMT model introduced earlier, but extended to allow for more than one characteristic relaxation time for the carrier density. Similarly to Reference 29, we find a slow relaxation component with a time constant on the order of a few hundred picoseconds, reflecting the recombination time of the free carriers generated by TPA and slow diffusion. In addition, however, there is a fast relaxation component with a characteristic time on the order of a few to several tens of picoseconds that may be attributed to carrier diffusion [4]. The spatial profile of the carriers generated by TPA is governed by the cavity mode excited by

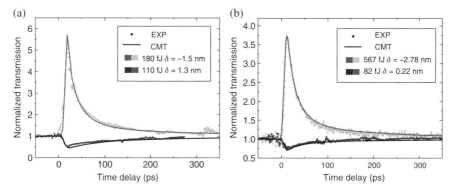

FIGURE 2.4 Example of probe pulse transmission versus temporal delay with respect to the control signal for Q factors of (a) 4100 and (b) 1200 and different pulse energies. Markers are experimental results and solid lines are fits using the CMT model, extended to include three relaxation time constants. The probe pulse width is ~9 ps, and the pump pulse width is ~5–6 ps. In (a), the sample has a resonance at 1545.2 nm and the pump pulse wavelength is 1545 nm. In (b), the sample has a resonance at 1541.8 nm and the pump pulse wavelength is 1541.5 nm. The transmission is normalized to 1 at large negative time delays when the probe arrives before the pump pulse. *(For a color version of this figure, see the color plate section.)*

the pump pulse. Since this distribution has an extent on the order of the wavelength its spatial gradients are large, thus implying fast initial redistribution of the carriers due to diffusion. This reduces the fraction of the carriers that overlap with the central part of the cavity mode and therefore leads to a fast relaxation of the cavity resonance frequency toward its value before excitation by the pump. The responses shown in Figure 2.4 also display a small, ultrafast, component due to Kerr effects, where the pump pulse leads to a positive change of the refractive index and therefore a red-shift of the cavity opposing the carrier-induced shift, as well as TPA of the signal induced by the pump, as seen in semiconductor waveguides [43]. These effects, however, remains small compared to the TPA-induced carrier dispersion for the regime of pulse energies and pulse widths investigated here.

The record-low switching energies reported in Reference 4, approaching the attojoule range, were obtained using a quaternary semiconductor compound made from InGaAsP with a composition corresponding to a bandgap at 1.47 μm, where TPA and (linear) one-photon absorption both contribute to free carrier generation for the considered wavelengths. In general, the relative role of the different processes that contribute to the shift of the cavity resonance frequency will depend on the input pulse energy as well as its width relative to the photon lifetime in the cavity and the characteristic relaxation times of the involved processes. We notice that although the response may be dominated by ultrafast carrier diffusion, the slow relaxation of the carrier density by recombination may be detrimental when operating at high bit rates due to the appearance of patterning effects [28]. This is similar to the case of semiconductor quantum dot amplifiers, where the slow recovery of a carrier reservoir imposes bit rate limitations [44]. In the case of semiconductor waveguide-based switches, such limitations have been overcome by the use of differential techniques [45, 46] and similar approaches might be implemented for cavity-based switches. Also, the use of a second pump beam to deplete the excited cavity by the use of coherent effects has been suggested [35], but has not yet been experimentally demonstrated. Finally, we mention the possibility of using the so-called Fano effects to reduce the switching energy by considering structures where the light may take two paths through the switch; one via the quasidiscrete cavity resonance and one path representing a continuum of waveguide modes [36, 47].

2.4 EMITTERS IN CAVITIES

If an excited emitter is embedded in a cavity, the qualitative behavior of the system will depend not only on the individual properties of the emitter and the cavity, but also on the emitter–cavity coupling coefficient g_c. This coupling coefficient is proportional to the emitter dipole moment as well as the field strength of the cavity mode at the position of the emitter. If light leaks from the cavity at a rate that is larger than the coupling, $\kappa \gg g_c$, the decay of the emitter is exponential and the system is said to be in the *weak coupling regime*. If, on the other hand, $\kappa \ll g_c$, the system is said to be in the *strong coupling regime*, where back-coupling effects from the cavity are so strong that photons may be reabsorbed before exiting the cavity and the emitter population consequently oscillates in time. In the following subsections, we shall quantify these conditions. We shall start by considering the regime of weak coupling, introducing the concept of local density of states and its relation to the emitter decay rate and move on to the regime of strong coupling and its possible realizations.

2.4.1 Weak Coupling: The Purcell Effect

As pointed out by Purcell as early as 1946 [48], a high-quality optical cavity may provide an increase in the spontaneous emission rate relative to that in a homogeneous medium by a certain factor, which is now named the Purcell factor. To derive the Purcell factor, we first note that for weak light–matter interaction, the radiative rate of spontaneous emission Γ_R from a narrow-band emitter, with resonance frequency ω_{eg} and situated at the position $\mathbf{r} = \mathbf{r}_c$, may be written as the product

$$\Gamma_R(\mathbf{r}_c, \omega_{eg}) = \frac{\pi \omega_{eg}}{\hbar \epsilon_0} d^2 \rho_L(\mathbf{r}_c, \omega_{eg}), \tag{2.10}$$

where $\hbar$ is the reduced Planck constant, ϵ_0 is the permittivity of free space, d is the magnitude of the dipole moment, and $\rho_L(\mathbf{r}, \omega)$ denotes the so-called (projected) local density of optical states (LDOS). The expression (2.10) for the spontaneous emission rate may be derived, for example, using Fermi's Golden Rule and requires the weak coupling condition that the LDOS is broad compared to the resulting emission linewidth. The LDOS is a convenient metric for the electromagnetic response at a given position and frequency and derives its name from being a measure of the density of available optical states into which a photon can be emitted. It is closely related to the electromagnetic Green's tensor and can be calculated through the relation

$$\rho_L(\mathbf{r}, \omega) = \frac{2\omega}{\pi c^2} \text{Im}\{\mathbf{e_d} \mathbf{G}(\mathbf{r}, \mathbf{r}, \omega) \mathbf{e_d}\}, \qquad (2.11)$$

in which $\mathbf{e_d}$ denotes a unit vector in the dipole moment direction and c is the speed of light. The Green's tensor $\mathbf{G}(\mathbf{r}, \mathbf{r}', \omega)$ may be interpreted as the electric field at the position $\mathbf{r}$ due to a point dipole source at the position $\mathbf{r}'$. In this way we can understand the LDOS as the electromagnetic response of the environment to light originating from the emitter position. In homogeneous media with refractive index n_B, the imaginary part of the Green's tensor in Eq. (2.11) is known analytically to be $\text{Im}\{\mathbf{e_d} \mathbf{G_B}(\mathbf{r}, \mathbf{r}, \omega) \mathbf{e_d}\} = n_B \omega / 6\pi c$. Therefore, from Eqs. (2.10) and (2.11), the decay rate of the emitter relative to the rate in a homogeneous medium Γ_B is

$$\frac{\Gamma_R(\mathbf{r_c}, \omega_{eg})}{\Gamma_B(\omega_{eg})} = \frac{6\pi c}{n_B \omega_{eg}} \text{Im}\{\mathbf{e_d} \mathbf{G}(\mathbf{r_c}, \mathbf{r_c}, \omega_{eg}) \mathbf{e_d}\}. \qquad (2.12)$$

Although the Green's tensor is known for certain simple material configurations, in general structures it must be calculated by numerical or other approximate means. A formulation in terms of the Green's tensor is convenient for the discussion of the Purcell factor, since for $\mathbf{r}$ and $\mathbf{r}'$ both inside the cavity the transverse part of the Green's tensor may be expanded in terms of the cavity modes $\tilde{\mathbf{f}}_\mu(\mathbf{r})$ as [12]

$$\mathbf{G}^T(\mathbf{r}, \mathbf{r}'; \omega) = c^2 \sum_\mu \frac{\tilde{\mathbf{f}}_\mu(\mathbf{r}) \tilde{\mathbf{f}}_\mu(\mathbf{r}')}{2\tilde{\omega}_\mu (\tilde{\omega}_\mu - \omega)}, \qquad (2.13)$$

in which the cavity modes are normalized by the inner product

$$\langle\langle \tilde{\mathbf{f}}_\mu | \tilde{\mathbf{f}}_\lambda \rangle\rangle = \lim_{V \to \infty} \left(\int_V \epsilon_r(\mathbf{r}) \tilde{\mathbf{f}}_\mu(\mathbf{r}) \cdot \tilde{\mathbf{f}}_\lambda(\mathbf{r}) d\mathbf{r} + i \frac{n_B c}{\tilde{\omega}_\mu + \tilde{\omega}_\lambda} \int_{\partial V} \tilde{\mathbf{f}}_\mu(\mathbf{r}) \cdot \tilde{\mathbf{f}}_\lambda(\mathbf{r}) dA \right), \qquad (2.14)$$

where $\epsilon_r(\mathbf{r})$ is the relative permittivity distribution defining the cavity, dA is an infinitesimal surface element, and $\tilde{\omega}_\mu$ is the complex eigenfrequency of the mode $\tilde{\mathbf{f}}_\mu$. Note that the second term in Eq. (2.14), which includes an integral over the surface of the volume V, is important for the normalization of the leaky cavity modes. Although both terms in Eq. (2.14) increase with increasing integration domain V, the sum converges remarkably fast to a finite value [13].

Close to a cavity resonance, one term ($\mu = c$) dominates the expansion of the Green's tensor in Eq. (2.13) and therefore the expansion can be well approximated by this term only. In this way, the Purcell factor F_P may be viewed as the single mode limit of the relative decay rate in Eq. (2.12), evaluated at the field maximum $\mathbf{r_c}$ and at the resonance frequency $\omega_{eg} = \omega_c$. Starting from Eqs. (2.12) and (2.13) with just a single term, and noting that $\text{Im}\{\mathbf{G}(\mathbf{r}, \mathbf{r}; \omega)\} = \text{Im}\{\mathbf{G}^T(\mathbf{r}, \mathbf{r}; \omega)\}$, we have

$$F_P = \frac{3\pi c^3}{n_c \omega_c} \text{Im}\left\{ i \frac{\tilde{\mathbf{f}}_c^2(\mathbf{r}_c)}{\omega_c \gamma_c} \right\}, \tag{2.15}$$

where $n_c = \sqrt{\epsilon_r(\mathbf{r}_c)}$, and we have discarded a small term $(\gamma_c)^2$. We define

$$\epsilon_r(\mathbf{r}_c) \tilde{\mathbf{f}}_c^2(\mathbf{r}_c) = \frac{\epsilon_r(\mathbf{r}_c) \tilde{\mathbf{f}}_c^2(\mathbf{r}_c)}{\langle\langle \tilde{\mathbf{f}}_c | \tilde{\mathbf{f}}_c \rangle\rangle} \equiv \frac{1}{v_Q}, \tag{2.16}$$

where v_Q is complex in general and has units of volume. From v_Q we can define an effective mode volume as [13]

$$\frac{1}{V_{\text{eff}}} = \text{Re}\left\{ \frac{1}{v_Q} \right\}, \tag{2.17}$$

and using $Q = \omega_c/2\gamma_c$ and $\lambda_c = 2\pi c/\omega_c$, we can write the Purcell factor in the well-known form due to Purcell [48] as

$$F_P = \frac{3}{4\pi^2} \left(\frac{\lambda_c}{n_c} \right)^3 \frac{Q}{V_{\text{eff}}}. \tag{2.18}$$

This prescription provides a rigorous and unambiguous way of calculating the effective mode volume for arbitrary leaky optical cavities. In the limit of infinite Q, that is, in cases where the cavity modes are strictly zero outside the cavity, the generalized effective mode volume v_Q is real and equals the effective mode volume commonly adopted in the literature.

The Purcell factor provides a convenient, dimensionless figure of merit for characterization of different cavities. The actual decay rate of an emitter in the cavity may be calculated in a similar way using Eqs. (2.10), (2.11) and (2.13) to be

$$\Gamma_R = \frac{2d^2}{\hbar \epsilon_0 \epsilon_r} \frac{Q}{V_{\text{eff}}}. \tag{2.19}$$

For quantum dots, the Purcell effect has been demonstrated experimentally by time-resolved measurements with both micropillar cavities [49] and photonic crystal cavities [50–52]. We emphasize that Eqs. (2.18) and (2.19) assume the emitter to be resonant with the cavity mode and to be located at an antinode of the field. If instead it is located at a node of the field, the spontaneous emission can be quenched, compared to emission in a homogeneous medium.

2.4.2 Strong Coupling: Vacuum Rabi Oscillations

In the previous section we considered the regime, where the emitter is weakly coupled to the cavity, resulting in an exponential decay of the excitation. When the emitter–cavity coupling is increased, the weak coupling condition may no longer be valid, and the system may enter the regime of strong coupling in which emitter and photon populations oscillate in time. In the limit of a single excited emitter and no externally injected field, this phenomenon is denoted vacuum Rabi oscillations since the field strength governing the oscillation frequency is given by the vacuum field. These fragile quantum phenomena have been observed even in solid-state systems, for example, for micropillar [53] and photonic crystal cavities [54]. The regime of single-photon excitation is of practical interest for the emerging field of quantum information technology [55], where sources of single photons are needed for the implementation of quantum cryptography systems as well as quantum computers. We shall not go into detail with these applications, but merely outline some of the most important dynamical effects in such a coupled cavity–emitter system in the single-photon regime. Figure 2.5 is a schematic illustration of an emitter with ground and excited states $|g\rangle$ and $|e\rangle$, which is coupled to the zero- and one-photon states $|n=0\rangle$ and $|n=1\rangle$ of an optical cavity. The emitter and cavity coupling is quantified by the coupling constant g_c.

We start by considering the case of an ideal optical cavity in which the optical modes vanish outside the cavity. From the discussion in Section 2.2, we know that this is an approximation, which corresponds to the limit of infinite Q factor of the cavity and which cannot be true in general. Nevertheless, this approach leads to Hermitian cavity modes and therefore enables a rigorous and systematic derivation of the quantum mechanical model for the coupled emitter cavity system. In this approach, we start with the assumption of infinite Q factor, and subsequently add additional losses to the model in a controlled manner. As in the previous section, we restrict the model to an optical cavity with a single dominant mode. By introducing operators for the creation and destruction of emitter states and optical states we may write the Hamiltonian for the system in the rotating wave and dipole approximation as

$$H = \hbar\omega_{eg} c_e^\dagger c_e + \hbar\omega_c a_c^\dagger a_c + \hbar g_c (a_c^\dagger c_g^\dagger c_e + c_e^\dagger c_g a_c), \tag{2.20}$$

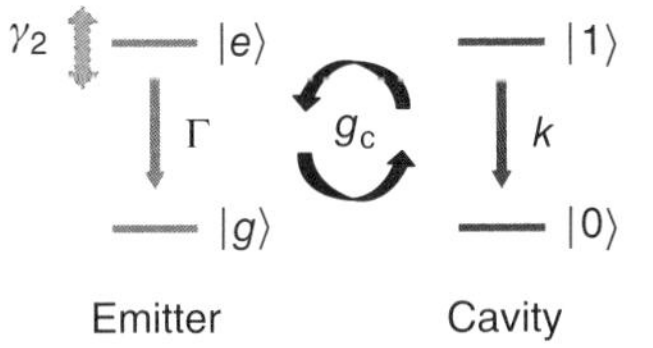

FIGURE 2.5 Schematic of the interaction between a quantum dot two-level system and a cavity in the one-photon limit. The emitter–cavity coupling is quantified by the coupling rate g_c, the cavity decay rate is κ, the emitter decay rate, due to nonradiative processes as well as coupling to other modes but the cavity mode, is Γ and finally the emitter is exposed to pure dephasing resulting from scattering with phonons, for example, with rate γ_2.

in which ω_c is the (real) frequency of the optical mode, and g_c is the coupling strength between the cavity mode and the emitter. The operator $c_e^\dagger$ (c_e) is the creation (annihilation) operator for the excited (e) state of the two-level system, and the corresponding ground (g) state operators are eliminated through the use of $c_e^\dagger c_e + c_g^\dagger c_g = 1$. The creation and annihilation operators for the cavity mode are $a_c^\dagger$ and a_c, respectively. Restricting attention to the case of a single quantum of excitation, the dynamics of the system can be captured by expansion in the subspace composed of the following basis:

$$\{|g\rangle = c_g^\dagger |0\rangle, \ |e\rangle = c_e^\dagger |0\rangle, \ |c\rangle = a_c^\dagger c_g^\dagger |0\rangle\}. \qquad (2.21)$$

By introducing the operator $\sigma_{\alpha\beta} = |\alpha\rangle\langle\beta|$ in this one-excitation subspace, the Hamiltonian becomes

$$H = \hbar\omega_{eg}\sigma_{ee} + \hbar\omega_c\sigma_{cc} + \hbar g_c(\sigma_{ce} + \sigma_{ec}). \qquad (2.22)$$

The equation of motion for the time-dependent density matrix of the system, $\rho(t)$, is obtained from

$$\partial_t \rho(t) = -\frac{i}{\hbar}[H, \rho(t)]. \qquad (2.23)$$

At this point we return to the question of losses. Because of the assumption of the cavity mode vanishing outside the cavity, Eq. (2.23) does not account for optical losses. However, we can add optical losses to the model by including the decay rate κ for the density matrix element ρ_{cc} corresponding to the cavity mode excitation. From the discussion of the complex cavity resonance frequency $\tilde{\omega}_c = \omega_c - i\gamma_c$ in Section 2.2, we realize that the decay rate is simply $\kappa = 2\gamma_c$. In addition to optical loss, the population of the excited level (e) of the two-level system may decay through nonradiative decay or through radiative decay to other modes than the dominating cavity mode. These processes are included in the model by the decay rate Γ for the density matrix element ρ_{ee} corresponding to the emitter excitation.

Population decay necessarily implies dephasing of the corresponding coherence between the two levels. In solid-state materials, such as a quantum dot embedded in a semiconductor, elastic scattering processes with phonons lead to additional loss of coherence without associated population decay, characterized by a so-called pure dephasing rate γ_2. The various scattering processes are also illustrated in Figure 2.5. The use of simple decay rates to model the different loss channels imply a Markovian approach where system excitations are lost by interaction with reservoirs and memory effects are neglected. Technically, the relaxation processes are included as Lindblad operators using a standard approach as described, for example, in References 56 and 57. A detailed investigation of the role of non-Markovian effects for the case of scattering by phonons may be found in References 58 and 59. With the inclusion

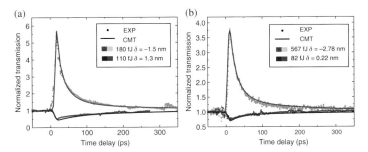

FIGURE 2.4 Example of probe pulse transmission versus temporal delay with respect to the control signal for Q factors of (a) 4100 and (b) 1200 and different pulse energies. Markers are experimental results and solid lines are fits using the CMT model, extended to include three relaxation time constants. The probe pulse width is ~9 ps, and the pump pulse width is ~5–6 ps. In (a), the sample has a resonance at 1545.2 nm and the pump pulse wavelength is 1545 nm. In (b), the sample has a resonance at 1541.8 nm and the pump pulse wavelength is 1541.5 nm. The transmission is normalized to 1 at large negative time delays when the probe arrives before the pump pulse.

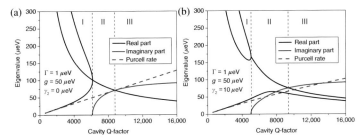

FIGURE 2.6 Illustration of strong and weak coupling regimes for a two-level system coupled to a cavity. Calculated eigenvalues as function of cavity quality factor for coupling factor $\hbar g_c = 50\ \mu eV$, emitter background decay rate $\hbar\Gamma = 1\ \mu eV$, and pure dephasing rates (a) $\hbar\gamma_2 = 0$, (b) $\hbar\gamma_2 = 10\ \mu eV$. Black lines: Real part of eigenvalues. Red lines: Imaginary part. Blue dashed line: Purcell-enhanced rate obtained by adiabatic elimination of coherence.

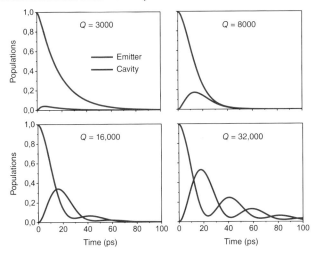

FIGURE 2.7 Temporal evolution of expectation value for emitter population (blue lines) and cavity population (red lines) for parameters corresponding to Figure 2.6b and different values of the quality factor: $Q = 3000$ (Regime I), $Q = 8000$ (Regime II), $Q = 16,000$ (Regime III), and $Q = 32,000$ (Regime III).

Photonics: Scientific Foundations, Technology and Applications, Volume II, First Edition.
Edited by David L. Andrews.
© 2015 John Wiley & Sons, Inc. Published 2015 by John Wiley & Sons, Inc.

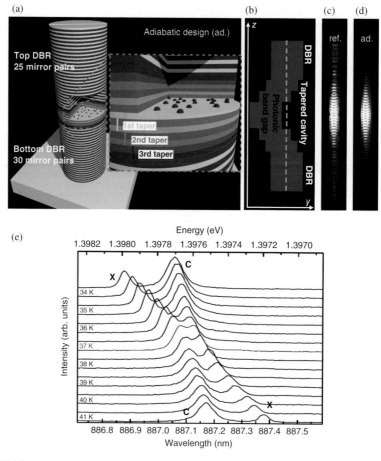

FIGURE 2.8 (a) Schematic of micropillar cavity with embedded layer of quantum dots. A normal and an adiabatic cavity design are illustrated. (b) Illustration of the variation of the photonic bandgap for the adiabatic design. Mode profiles for (c) reference and (d) adiabatic design. (e) Measured emission spectra showing anti-crossing between emitter (X) and cavity (C) peak, demonstrating that strong coupling is realized. The different spectra were obtained by tuning the temperature from 34 K to 41 K. Reprinted with permission from Reference 62 (Copyright 2012 by the American Physical Society).

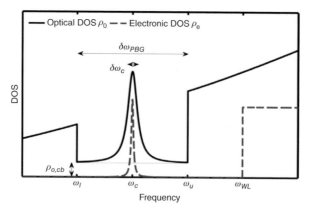

FIGURE 2.10 Schematic of the electronic (red dashed line) and optical (black line) density of states for an emitter embedded in a defect nanocavity. Reprinted with the permission from Reference 71. (Copyright 2012 by the American Physical Society.)

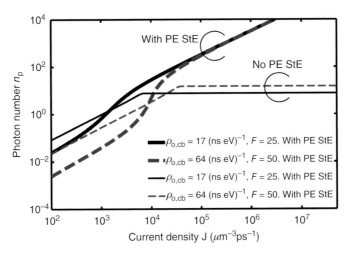

FIGURE 2.11 Photon number versus current density for a quantum dot nanocavity laser with the inclusion of Purcell-enhanced (PE) stimulated emission (StE) and without. Results for different values of the background density of states and the Purcell factor (F) are shown. Reprinted with the permission from Reference 71. (Copyright 2012 by the American Physical Society.)

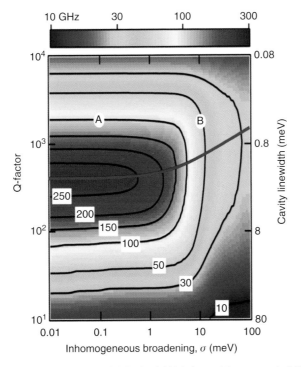

FIGURE 2.12 Calculated maximum 3 dB modulation bandwidth (color map) for a nanocavity LED versus degree of inhomogeneous broadening and cavity Q factor. Contour lines are drawn for constant modulation bandwidth (in GHz). Reprinted with the permission from Reference 73. (Copyright 2012 by the American Physical Society.)

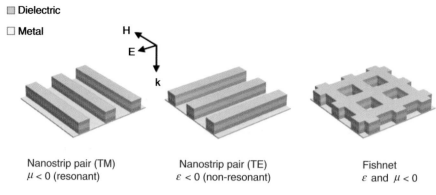

FIGURE 3.2 Basic idea behind fishnet NIM.

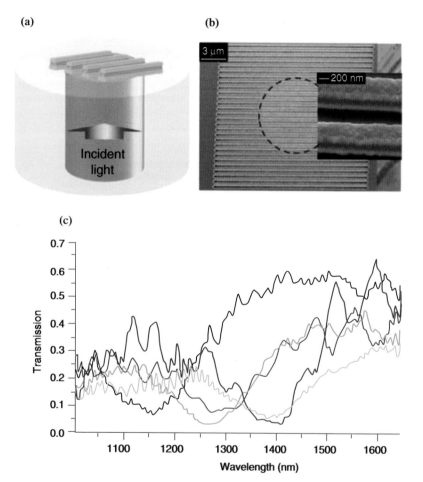

FIGURE 3.5 (a) Schematic of transverse cross section of optical fiber with a pattern in center; (b) SEM image of nanostrips covering core of fiber, which can be seen in center; (c) measured transmission for five fiber MMs samples illuminated with TM polarized wave.

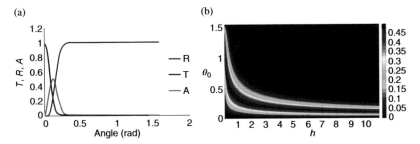

FIGURE 3.8 (a) Absorption, reflectivity, and transmittivity as function of incident angle θ_0. (b) Conversion efficiency at point $\varsigma = 1$ as a function of h and θ_0.

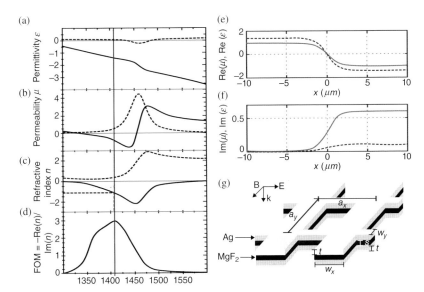

FIGURE 3.9 Real and imaginary parts of (a) ε, (b) μ, and (c) n for experimentally fabricated fishnet MM [13]. (d) Figure of merit for fishnet MM. (e) Real and (f) imaginary parts of ε (blue dashed line) and μ (solid green line) for the simulated transition layer with $h = 2\lambda$ and $\theta_0 = \pi/17$. (g) Diagram of experimentally fabricated fishnet MM [71].

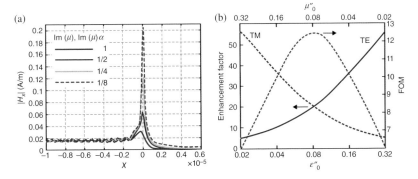

FIGURE 3.10 (a) $|H_x|$ for TM incident wave at $\theta_0 = \pi/17$ and $h = 2\lambda$ as a function of material loss. (b) Enhancement factor and FOM for TE and TM polarizations as a function of ε_0'' and μ_0''.

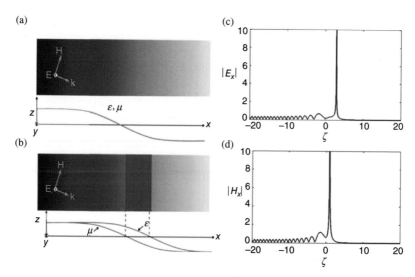

FIGURE 3.11 Transition layer showing index variation along x for profile with zeros of ε and μ (a) coinciding and (b) spatially separated, (c) absolute value of electric field component E_x for TM wave as functions of ζ for $l = 2$, (d) absolute value of magnetic field component H_x for TE wave as functions of ζ for $l = -4$.

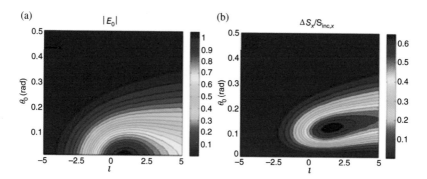

FIGURE 3.12 (a) $|E_0|$ and (b) $\Delta S_x/S_{0x}$ as functions of l and θ_0 for TE polarization.

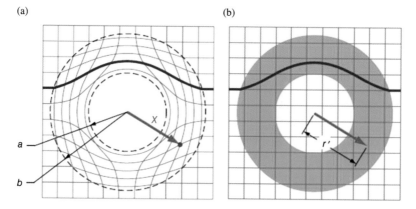

FIGURE 3.13 Transmission of a ray (blue line) through a medium under the (a) topological interpretation, and (b) material interpretation [72].

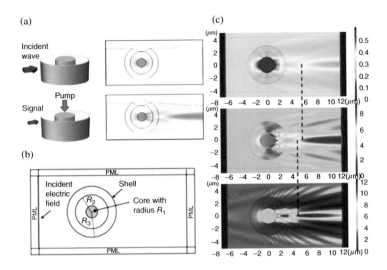

FIGURE 3.14 (a) Schematic of a wave concentrator based on self-action of single high intensity beam (top left), nonlinear concentrator for a weak signal propagating in nonlinear medium with refractive index modified by a strong pump (bottom left), and power flow in a linear (top right) and nonlinear concentrator based on self-action (bottom right). (b) Schematic of a lens. (c) Time averaged power flow illustrating the predicted lensing effect in the case of focusing nonlinearity. Focal line shifts toward the lens as incident field increases.

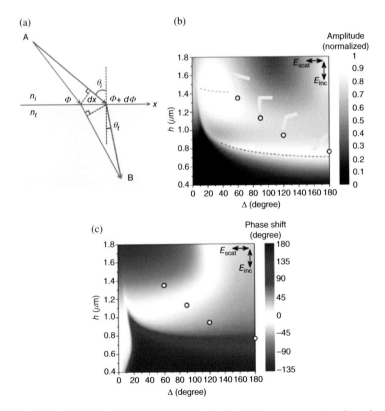

FIGURE 3.15 (a) Schematics of interface between two artificially structured media introducing abrupt phase shift in light path that varies as function of position along the interface. (b)–(c) Calculated amplitude and phase shift of the cross-polarized scattered light for V-antennas consisting of gold rods as functions of length h and angle between the rods Δ [96].

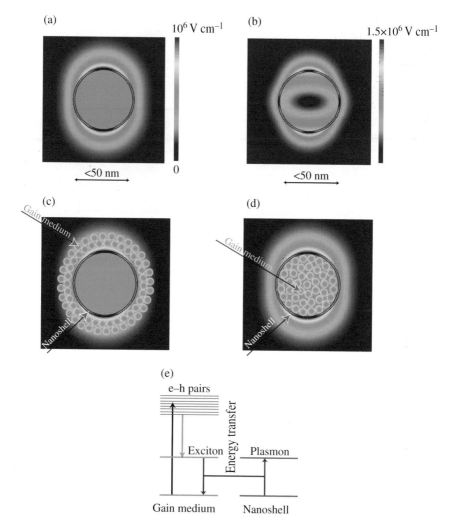

FIGURE 4.2 Schematic of spaser geometry, local fields, and fundamental processes leading to spasing. Adapted from Reference 17. (a) Nanoshell geometry and the local optical field distribution for one SP in an axially symmetric dipole mode. The nanoshell has the aspect ratio $\eta = 0.95$. The local field magnitude is color-coded by the scale bar on the right-hand side of the panel. (b) The same as (a) but for a quadrupole mode. (c) Schematic of a nanoshell spaser where the gain medium is outside of the shell, on the background of the dipole-mode field. (d) The same as (c) but for the gain medium inside the shell. (e) Schematic of the spasing process. The gain medium is excited and population-inverted by an external source, as depicted by the black arrow, which produces electron–hole pairs in it. These pairs relax, as shown by the green arrow, to form the excitons. The excitons undergo decay to the ground state emitting SPs into the nanoshell. The plasmonic oscillations of the nanoshell stimulates this emission, supplying the feedback for the spaser action.

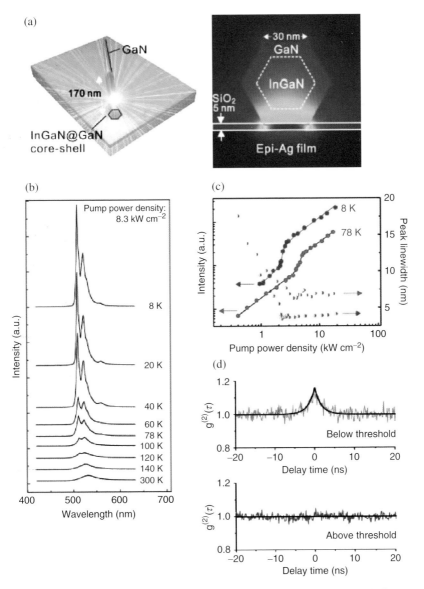

FIGURE 4.3 InGaN nanospaser and its properties. (a) Schematic of geometry of InGaN/GaN core-shell nanospaser (left) and theoretical intensity of its spasing eigenmode. (b) Series of emission spectra: temperature-dependent spasing behavior from 8 to 300 K. The spasing threshold at 140 K is clearly visible. (c) The L–L (light–light) plots at the main lasing peak (510 nm) are shown with the corresponding linewidth-narrowing behavior when the spaser is measured at 8 K (red) and 78 K (blue), with lasing thresholds of 2.1 and 3.7 kW cm^{-2}, respectively. (d) Second-order photon correlation function $g^{(2)}(\tau)$ measured at 8 K. The upper curve is recorded below the spasing threshold, and the lower above the threshold. Adapted from Reference 32.

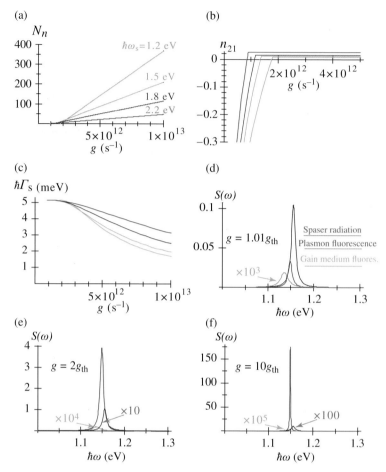

FIGURE 4.5 Spaser SP population and spectral characteristics in the stationary state. The computations are done for a silver nanoshell with the external radius $R_2 = 12$ nm; the detuning of the gain medium from the spasing SP mode is $\hbar(\omega_{21} - \omega_n) = -0.02$ eV. The other parameters are indicated in Section 4.2.4. (a) Number N_n of plasmons per spasing mode as a function of the excitation rate g (per one chromophore of the gain medium). Computations are done for the dipole eigenmode with the spasing frequencies ω_s as indicated, which were chosen by the corresponding adjustment of the nanoshell aspect ratio. (b) Population inversion n_{12} as a function of the pumping rate g. The color coding of the lines is the same as in panel (a). (c) The spectral width Γ_s of the spasing line (expressed as $\hbar\Gamma_s$ in meV) as a function of the pumping rate g. The color coding of the lines is the same as in panel (a). (d)–(f) Spectra of the spaser for the pumping rates g expressed in the units of the threshold rate g_{th}, as indicated in the panels. The curves are color coded and scaled as indicated. Adapted from Reference 17.

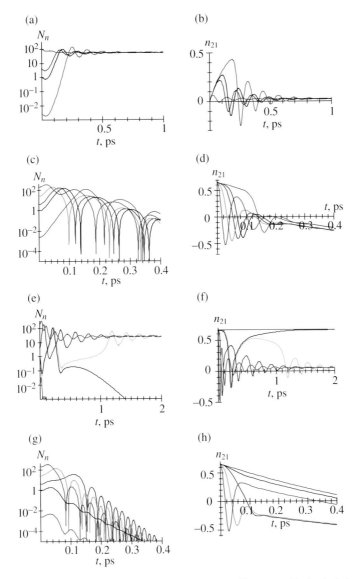

FIGURE 4.6 Ultrafast dynamics of spaser. (a) For monostable spaser (without a saturable absorber), dependence of SP population in the spasing mode N_n on time t. The spaser is stationary pumped at a rate of $g = 5 \times 10^{12}$ s^{-1}. The color-coded curves correspond to the initial conditions with the different initial SP populations, as shown in the graphs. (b) The same as (a) but for the temporal behavior of the population inversion n_{21}. (c) Dynamics of a monostable spaser (no saturable absorber) with the pulse pumping described as the initial inversion $n_{21} = 0.65$. Coherent SP population N_n is displayed as a function of time t. Different initial populations are indicated by color-coded curves. (d) The same as (c) but for the corresponding population inversion n_{21}. (e) The same as (a) but for bistable spaser with the saturable absorber in concentration $n_a = 0.66 n_c$. (f) The same as (b) but for the bistable spaser. (g) The same as (e) but for the pulse pumping with the initial inversion $n_{21} = 0.65$. (h) The same as (g) but for the corresponding population inversion n_{21}. Adapted from Reference 17.

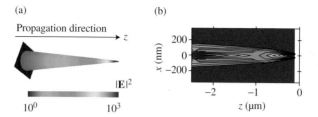

FIGURE 4.7 (a) Geometry of the nanoplasmonic tapered waveguide. The propagation direction of the SPPs is indicated by the arrow. Intensity $I(\mathbf{r}) = |\mathbf{E}(\mathbf{r})|^2$ of the local fields relative to the excitation field is shown by color. The scale of the intensities is indicated by the color bar in the center. (b) Local electric field intensity $I(\mathbf{r})$ is shown in the longitudinal cross section of the system. The radius of the waveguide gradually, adiabatically decreases from 50 to 2 nm. Adapted from Reference 93.

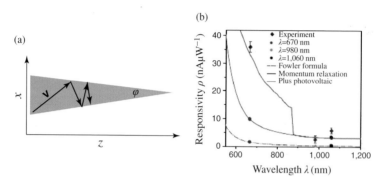

FIGURE 4.11 (a) Schematic of electron momentum relaxation in a tapered waveguide. With each collision with the wall, electron velocity direction (indicated by the arrows) rotates clockwise by a 2φ angle, where φ is the apex angle of the waveguide cone. (b) Experimental device responsivity ρ (blue diamonds) and its theoretical curve as a function of wavelength and considering Fowler equation (blue dotted line) or modified equation (blue continuous line). The red line shows the same calculations, adding the photovoltaic (PV) contribution obtained from known literature.

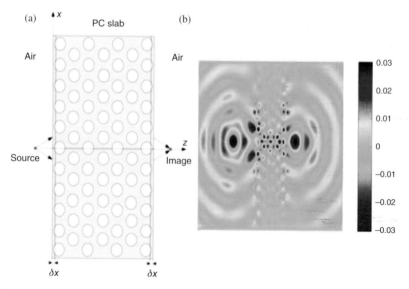

FIGURE 5.7 (a) Schematic diagram of an imaging system formed from a photonic slab utilizing negative refraction. (b) Field distribution for a point source and its image across a photonic slab.

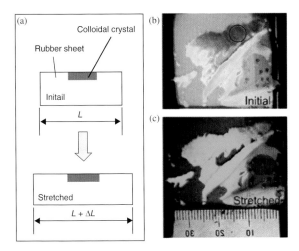

FIGURE 5.9 Changes in the structural color of the colloidal crystal film covering a silicone rubber sheet. (a) Elastic deformation when the rubber is stretched from L to $L + \Delta L$. (b) Photographic image of the initial sheet (L). (c) Photographic image of the stretched sheet ($L + \Delta L$). Reprinted with permission from Fudouzi and Sawada [109]. Copyright 2006 American Chemical Society.

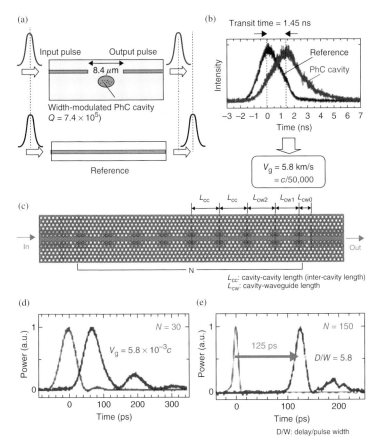

FIGURE 5.11 Slow light in a cavity. (a), (b) Slow light propagation in a single ultrahigh Q nanocavity; (c) schematic of coupled nanocavities; (d), (e) slow light propagation in coupled nanocavities [118].

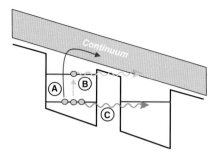

FIGURE 6.5 Dark current mechanisms in QWIPs and QDIPs. (a) Thermionic emission; (b) thermally assisted tunneling; (c) sequential tunneling.

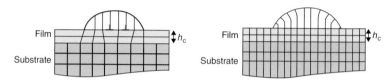

FIGURE 6.9 (Left) Schematic illustrating an SK dot that has relaxed via defect formation and (Right) schematic of a dot that has coherent relaxed without defects.

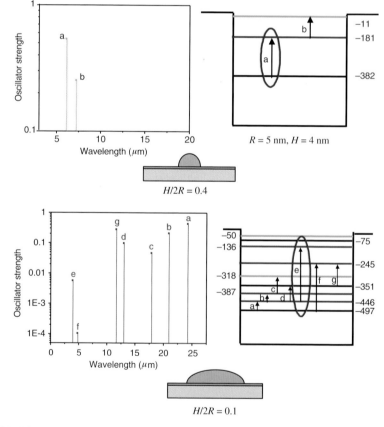

FIGURE 6.10 Calculation of energy levels and oscillator strengths for InGaAs dots in InGaP barriers. The dots have different aspect ratios and fixed height.

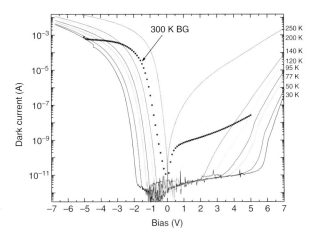

FIGURE 6.16 Dark current measured as a function of bias for InGaAs/InGaP QDIP at different temperatures.

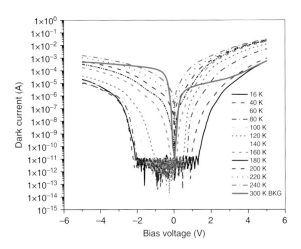

FIGURE 6.26 Dark current at various temperatures and 300 K background photocurrent.

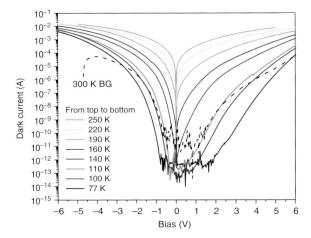

FIGURE 6.28 Dark current measured as a function of bias for InAs/GaAs/AlInAs/InP QDIP at different temperatures. Also shown is the 300 K background photocurrent with a 150° field of view (dashed line).

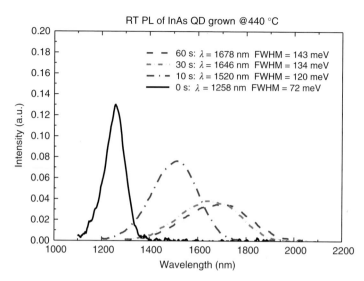

FIGURE 6.31 Room temperature PL from a single layer of InAs QDs grown and capped with InP at 440°C with different ripening times.

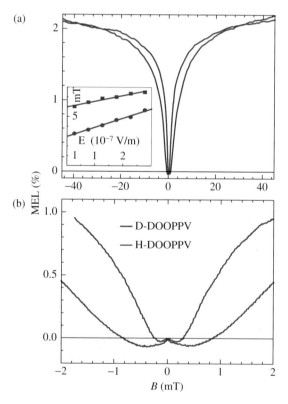

FIGURE 7.6 Room temperature MEL response of H- and D-DOOPPV (red and blue lines, respectively) measured at bias voltage $V = 2.5$ V, plotted on large (**a**) and small (**b**) magnetic field scales, where the respective regular and ultra-small-field MEL responses are separated. **Inset to a**: The field, $B_{1/2}$, at half the MEL maximum for the two polymers as a function of the applied bias voltage, V, given in terms of the internal electric field in the polymer layer; the lines are linear fits to guide the eye.

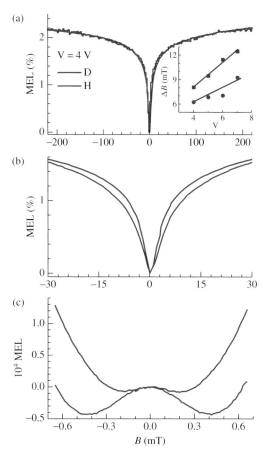

FIGURE 7.8 (a)–(c) MEL(B) response of OLEDs based on $H_{18}Alq_3$ (red line) and $D_{18}Alq_3$ (blue line) measured at room temperature and bias $V = 4$ V, plotted at three *different B scales*. The $D_{18}Alq_3$ response was normalized to that of $H_{18}Alq_3$ at $B \sim 250$ mT. Inset in (a): The full width, ΔB at MEL = 0.8% plotted versus V for $H_{18}Alq_3$ (red) and $D_{18}Alq_3$ (blue).

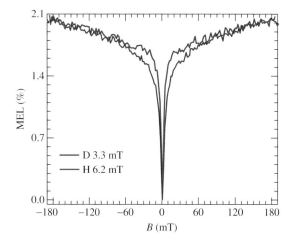

FIGURE 7.9 MEL(B) response of OLEDs based on $H_{18}Alq_3$ and $D_{18}Alq_3$ saturate-exposed to oxygen, measured at $V = 4$ V and room temperature. The full width, ΔB, at MEL = 1% is 12.4 mT (6.6 mT) for the $H_{18}Alq_3$ ($D_{18}Alq_3$) device. The response of $D_{18}Alq_3$ was normalized to that of $H_{18}Alq_3$ at $B = 200$ mT.

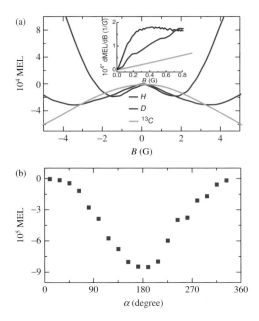

FIGURE 7.11 MEL(B) response of OLEDs based on D-, H-, and C13-DOOPPV (red, blue, green, respectively). (a) Up to 0.5 mT when B_E is shielded; it shows a maximum at $B_{coil} = 0$ and minima at $B_{coil} = \pm B_m$. Inset: The first derivative $|d(MEL)/dB|$ of the response in (a) up to 0.08 mT; $|d(MEL)/dB|$ is largest for the D-DOOPPV-based OLED. (b) MEL(α) response for H-DOOPPV at $B_{coil} = 0.05$ mT. The line through the data points is a fit using the data in Figure 7.11a and a procedure described in the text.

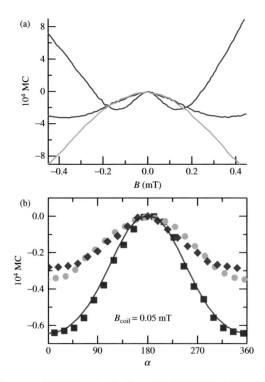

FIGURE 7.12 MC(B) response of OLEDs based on D-, H-, and C13-DOOPPV (red, blue, green, respectively). (a) Up to 0.5 mT, when the Earth's magnetic field, B_E, is shielded, showing maximum at $B_{coil} = 0$ and minima at $B_{coil} = \pm B_m$. The inset shows the chemical backbone of D-DOO-PPV. (b) MC(α) response for the three isotopes using $B_{coil} = 0.05$ mT. The line through the data points is a fit using the data in Figure 7.12a and the procedure described in the text.

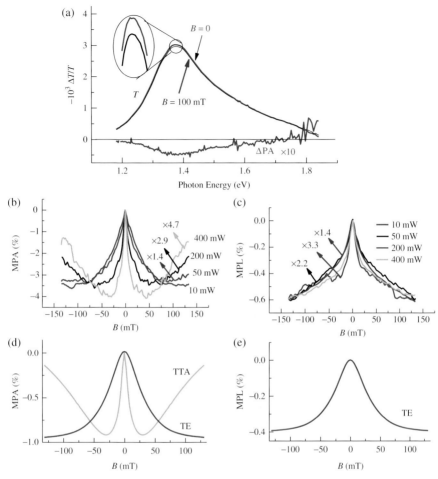

FIGURE 7.14 Excited state spectra and magnetic field effects in pristine MEH-PPV films. (a) The triplet PA band, PA_T at $B = 0$ and 100 mT (black and red lines, respectively), generated by $h\nu_L = 2.54$ eV at $I_L = 0.2$ W, and their difference $\Delta PA_T = [PA_T(100\ mT) - PA_T(0)]$ (blue line). The region near the peak is magnified (circle). (b) $MPA_T(B)$ measured at 1.37 eV for various laser excitation intensities (normalized). (c) $MPL(B)$ measured at 2.05 eV for various laser excitation intensities (normalized). (d) Model calculations of $MPA_T(B)$ using the TE (blue line, corresponds to the 10 mW data in (b)) and TTA (green line, corresponds to the 400 mW data in (b)) mechanisms (see text, Section 7.5.3). (e) Model calculation of $MPL(B)$ using the SE quenching model (see text, Section 7.5.2).

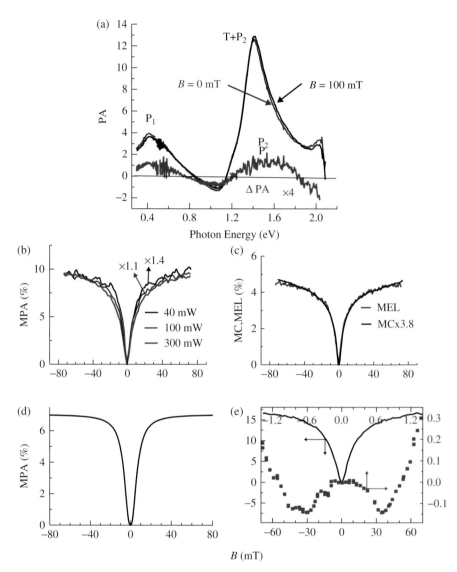

FIGURE 7.15 Excited state spectra and magnetic field effects in UV irradiated MEH-PPV film compared to MFE in OLED device. (a) PA spectrum at $I_L = 0.1$ W for $B = 0$ (black line) and $B = 100$ mT (red line) and their difference $\Delta PA = [PA(100\ mT)-PA(0)]$ (blue line) in MEH-PPV film. (b) MPA(B) measured at 1.4 eV for various laser excitation intensities (normalized). (c) MEL(B) and MC(B) in MEH-PPV OLED device. (d) Model calculations of MPA$_P$(B) in films using the PP mechanism (see text, Section 7.5.1). (e) MPA(B) at 1.1 eV up to $B = 1.5$ mT (filled squares) and up to $B = 60$ mT (blue line, inset).

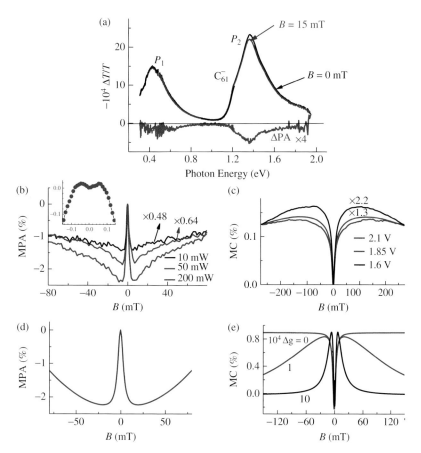

FIGURE 7.16 Excited state spectra and magnetic field effects in MEH-PPV/PCBM film and OLED device. (a) PA spectrum of MEH-PPV film at $I_L = 0.2$ W for $B = 0$ (black line) and $B = 15$ mT (red line) and their difference $\Delta PA = PA(15\text{ mT})-PA(0)$ (blue line). (b) MPA(B) measured at 1.37 eV for various laser excitation intensities (normalized). Inset: High resolution data, showing USMPA peaks at $|B| \sim 0.1$ mT. This data was measured under conditions of complete apparatus shielding from the earth magnetic field and any stray field. (c) MC(B) in OLED at various bias voltages, V. (d) and (e) Model calculations of MPA$_{pp}$(B) and MC(B), respectively, using the "Δg + HFI" mechanism (see text, Section 7.5.1).

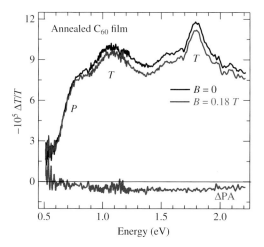

FIGURE 7.17 Photomodulation spectra of annealed C_{60} film at $T = 80$ K and $I_L = 0.2$ W cm^{-2} for $B = 0$ (black line) and $B = 180$ mT (red line). The blue negative line is the difference spectrum $\Delta PA = PA(B = 180\text{ mT})-PA(0)$.

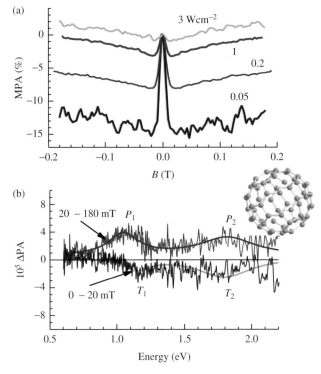

FIGURE 7.18 (a) MPA(B) response of an annealed C_{60} film at various pump excitation intensities, measured at photon energy $E = 1.8$ eV and $T = 80$ K. (b) The spectra $\Delta PA(B_1, B_2, E)$ for $B_1 = 0$, $B_2 = 20$ mT (black line, lower curves), and $B_1 = 20$ mT, $B_2 = 180$ mT (blue line, upper curves) for $I_L = 1.5$ W cm^{-2}. The smooth green and red lines through the data are to guide the eye, and show the TE- and polaron-related MPA bands, respectively.

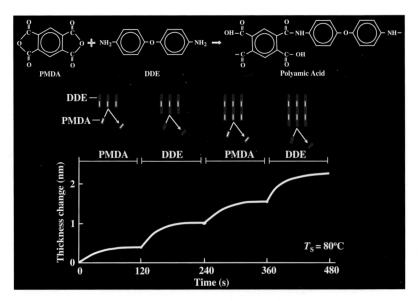

FIGURE 8.5 Experimental demonstration of MLD using PMDA and DDE. From Reference 6.

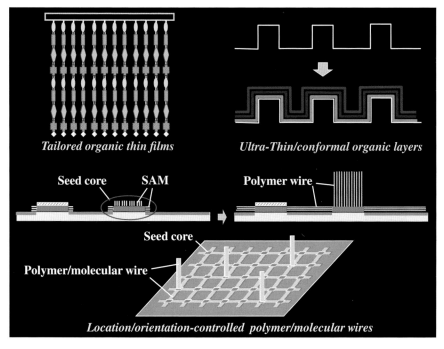

FIGURE 8.6 Expected structures grown by MLD.

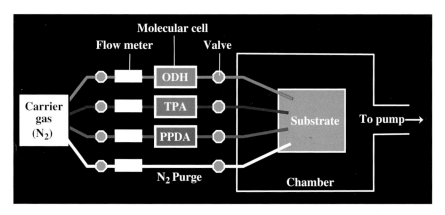

FIGURE 8.7 Schematic illustration of the carrier-gas type MLD. From Reference 32.

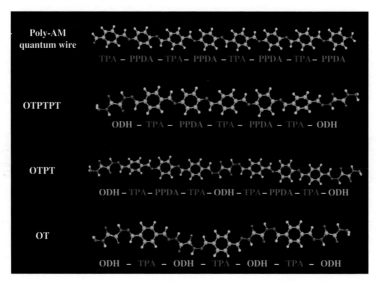

FIGURE 8.13 Molecular structures of a poly-AM quantum wire and polymer MQDs obtained by the MO method. From Reference 32.

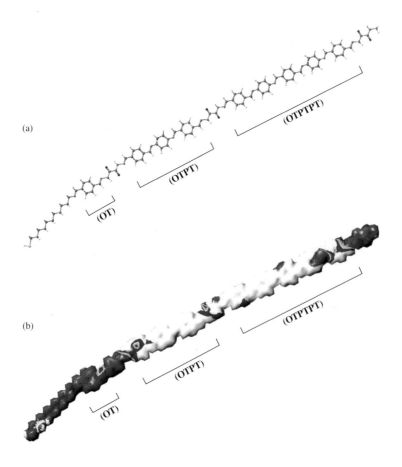

FIGURE 8.15 (a) Molecular structure and (b) electron density for 3QD. From Reference 32.

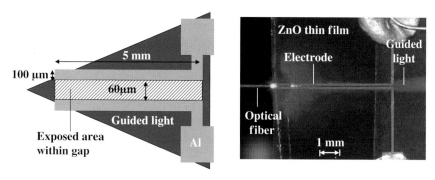

FIGURE 8.26 Setup for photocurrent measurements by the "guided light" configuration.

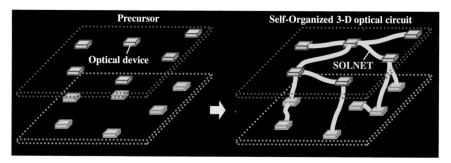

FIGURE 8.27 Concept of a self-organized three-dimensional integrated optical circuit consisting of SOLNET.

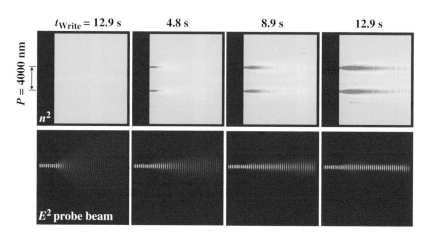

FIGURE 8.31a Simulation for parallel waveguides of SOLNET with no luminescent targets.

(i) SOLNET with no luminescent targets

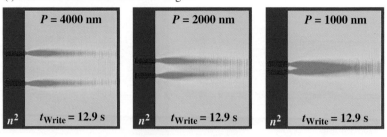

(ii) R-SOLNET *with luminescent targets*

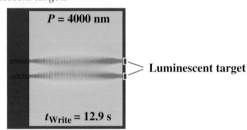

FIGURE 8.31b Simulation for parallel waveguides of (i) SOLNET with no luminescent targets and (ii) R-SOLNET *with luminescent targets*.

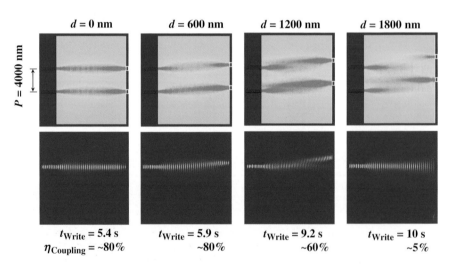

FIGURE 8.31c Simulation for parallel waveguides of R-SOLNET *with luminescent targets* for various lateral misalignments.

(a) During write beam exposure

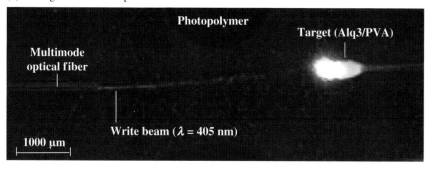

(b) After write beam exposure for 2 min

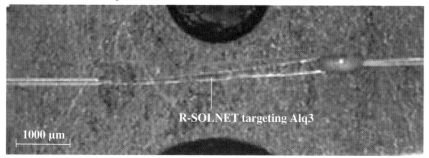

FIGURE 8.32 Experimental demonstration of R-SOLNET targeting Alq3 dispersed in PVA. From Reference 54.

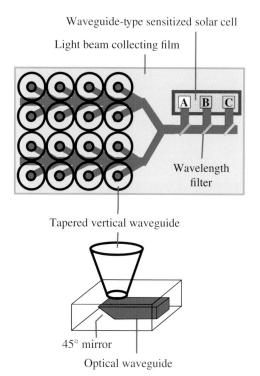

FIGURE 8.39 Concept of the film-based integrated solar cell.

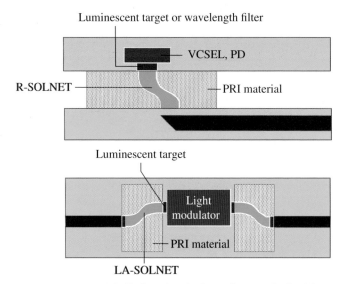

FIGURE 8.41 Concept of the three-dimensional integrated optical interconnects within computers.

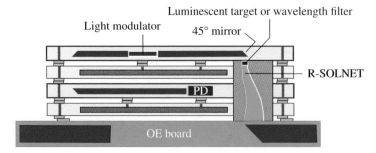

FIGURE 8.42 (a) Optical solder of R-SOLNET and LA-SOLNET for self-aligned optical couplings and (b) self-aligned vertical waveguides of R-SOLNET.

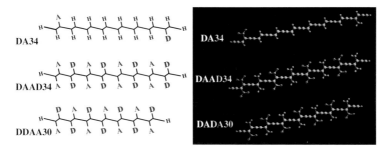

FIGURE 8.44 Models of quantum dots with poly-diacetylene backbones. From Reference 13.

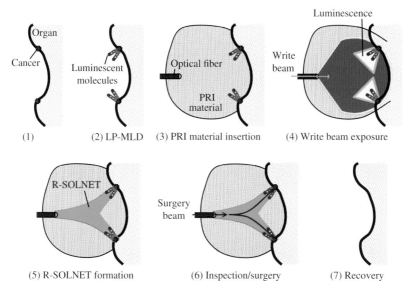

FIGURE 8.47 Concept of SOLNET-assisted laser surgery. From Reference 33.

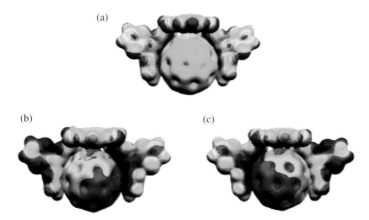

FIGURE 9.7 Electrostatic potential (B3LYP/6-31G**) calculated for 7–C_{60} in (a) the ground electronic state and (b) the charge-separated HOMO→LUMO+4, and (c) HOMO−1→LUMO+4 excited states. Blue: positive potential, red: negative potential (color scale for $\delta^+ \to \delta^-$: blue→green→yellow→orange→red). Reprinted from Reference 21 with permission from Wiley-VCH.

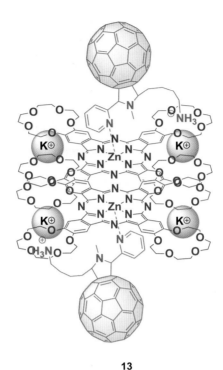

13

FIGURE 9.11 $(ZnPc)_2$–$(C_{60})_2$ supramolecular tetrad (**13**).

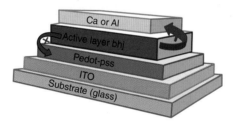

FIGURE 9.16 Typical sandwich-like architecture for a (OPV) solar cell. Polymers and fullerene derivatives are bended on the active layer.

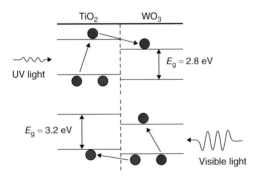

FIGURE 10.3 Band diagram showing the band edge positions of TiO_2 and WO_3 and the movement of electrons (red) and holes (blue) upon illumination with UV light (top) and visible light (bottom) [34]. The movement of charge carriers explains the superior water splitting performance of the WO_3/TiO_2 core/shell nanorod arrays.

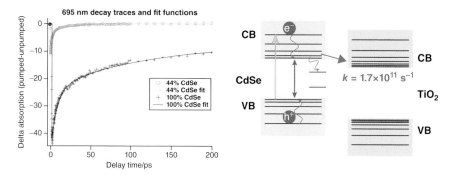

FIGURE 10.4 Time traces from Reference 52 showing the recovery of the CdSe transient bleach feature and a band diagram showing the relative band positions of CdSe and TiO_2. In the presence of TiO_2 the bleach feature recovers much faster. The band diagram to the right shows the excitation of CdSe by visible light, electrons and holes relaxing to the band edge, and electrons being injected from CdSe into TiO_2. The electron transfer rate, k_{ET}, is very fast due to the strong electronic coupling between the two materials.

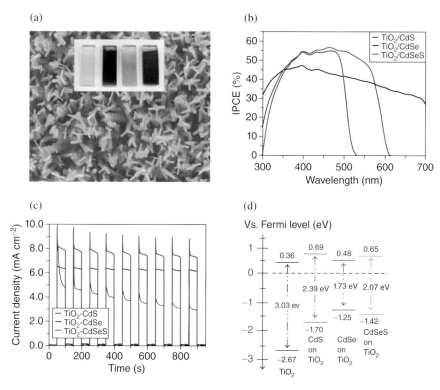

FIGURE 10.5 (a) SEM image of CdSeS rods grown on TiO_2 nanowires using CVD deposition, inset showing from left to right films of TiO_2, CdS nanorods on TiO_2, CdSe nanorods on TiO_2, and CdSeS nanorods on TiO_2. (b) IPCE plots for the sensitized samples showing that CdS has high efficiency but can only absorb wavelengths less than 520 nm, CdSe can absorb all of the visible spectrum, and CdSeS sensitized has high efficiency and intermediate visible light absorption extending to 600 nm. (c) White light on/off scans showing the photocurrent density from the TiO_2 films with different sensitizers. (d) Diagram showing the band positions of TiO_2, CdS, CdSe, and CdSeS alloy quantum dots. The CdSeS QDs were shown to have the optimal band positions for absorbing solar light and injecting charge into TiO_2. From Reference 59.

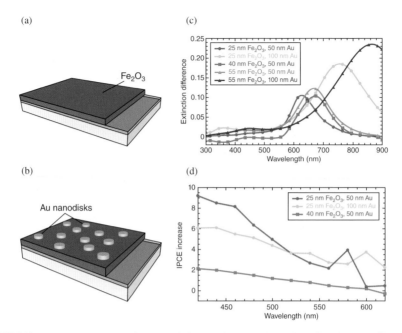

FIGURE 10.6 (a) Schematic drawing of hematite thin film deposited on conductive glass. Film thicknesses of 25, 40, and 55 nm were compared. (b) Schematic of lithographically deposited Au nanodisks on hematite thin films. Nanodisks were made with diameters of 50 and 100 nm. (c) Plots of the difference in extinction spectrum (with gold nanodisks–without gold nanodisks) for films of different thicknesses with different diameter gold nanodisks. (d) IPCE action difference spectra (IPCE with gold nanodisk–IPCE of hematite film) for three of the films showing enhancement of IPCE at the SPR wavelength of the gold nanodisks. From Reference 75.

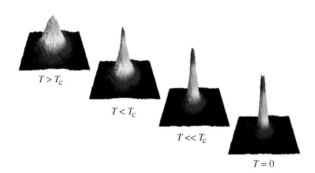

FIGURE 11.1 The onset of BEC is seen as a sharp peak in the density in the center of the trap. In the figures, the temperature is lowered from left to right. To the far right, an almost pure condensate with a negligible thermal component can be seen. Pictures courtesy of A. Arnold and are from the BEC experiment at University of Strathclyde, Glasgow, UK [43].

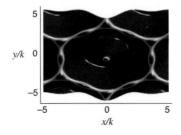

FIGURE 11.3 The intensity profile obtained from the six lasers in Eqs. (11.36)–(11.38) which creates an edge-centered honeycomb lattice for the cold atoms [56]. Red indicates high intensity and blue low intensity.

of the various loss channels through the Lindblad formalism we end up with the following equations for the elements of the density matrix:

$$\partial_t \rho_{ee} = -\Gamma \rho_{ee}(t) - ig_c[\rho_{ce}(t) - \rho_{ec}(t)] \tag{2.24}$$

$$\partial_t \rho_{cc} = -\kappa \rho_{cc}(t) - ig_c[\rho_{ec}(t) - \rho_{ce}(t)] \tag{2.25}$$

$$\partial_t \rho_{ce} = [i\Delta_c - \gamma_{ce}]\rho_{ce}(t) - ig_c[\rho_{ee}(t) - \rho_{cc}(t)] \tag{2.26}$$

$$\partial_t \rho_{ec} = [-i\Delta_c - \gamma_{ce}]\rho_{ec}(t) + ig_c[\rho_{ee}(t) - \rho_{cc}(t)], \tag{2.27}$$

where we defined the detuning $\Delta_c = \omega_{eg} - \omega_c$ and

$$\gamma_{ce} = \gamma_2 + \frac{1}{2}(\Gamma + \kappa), \tag{2.28}$$

which is the effective dephasing rate of the off-diagonal coherence terms, including contributions from population decay as well as pure dephasing. Equations (2.24)–(2.27) constitute the famous Jaynes–Cummings model, restricted to the case of a single excitation, and describes the dynamics of the coupled emitter–cavity system. Depending on the relative sizes of the parameters, this model can describe a number of qualitatively very different regimes, including the weak coupling regime with Purcell-enhanced exponential decay of the emitter, and the strong coupling regime with vacuum Rabi oscillations. For simplicity, we shall limit the following analysis to the case of zero detuning where the strongest coupling between emitter and cavity is achieved. Experimentally, this condition is often achieved by varying the temperature of the system. For a quantum dot emitter embedded in a photonic crystal defect cavity, the emitter transition frequency thus red-shifts with increasing temperature while the cavity resonance hardly changes.

In order to connect to the results of the previous section, we first consider the case where the dephasing and cavity decay rates are much larger than the light–matter coupling strength, $\gamma_{ce}, \kappa \gg g_c$, such that the off-diagonal terms of the density matrix, that is, the coherences ρ_{ce} and ρ_{ec} can be adiabatically eliminated and the cavity population can be assumed to be zero. In this case, the time derivative for the coherence variables is approximated by zero, and the coherences adiabatically follow the emitter population and the equation for the emitter population reduces to

$$\partial_t \rho_{ee} = -\Gamma_P \rho_{ee}; \quad \Gamma_P = \Gamma + \frac{4g_c^2}{\Gamma + \kappa + 2\gamma_2}. \tag{2.29}$$

Here, Γ_P is the Purcell-enhanced emission rate, which is seen to scale with cavity Q factor for $\Gamma + 2\gamma_2 \ll \kappa$. Comparing to Eq. (2.19), which was derived in the same regime of weak coupling, we may express the coupling strength as

$$g_c^2 = \frac{d^2 \omega_{eg}}{2\hbar \epsilon_0 \epsilon_r V_{\text{eff}}}. \tag{2.30}$$

If the Q factor is so large that the cavity decay rate becomes comparable to the light–matter coupling strength, the adiabatic approximation breaks down. In that case the emitted photon may be reabsorbed and the decay is no longer exponential. The general case may be analyzed by help of the eigenvalues characterizing the dynamical evolution of the density matrix elements in Eqs. (2.24)–(2.27). Introducing the column vector $\mathbf{x} = (\rho_{ee}, \rho_{cc}, \rho_{ce} - \rho_{ec}, \rho_{ce} + \rho_{ec})^T$, the eigenvalue equation is expressed as

$$\mathbf{A}\mathbf{x} = \lambda \mathbf{x}, \qquad (2.31)$$

where the matrix $\mathbf{A}$ is identified by comparison with Eqs. (2.24)–(2.27), λ is the eigenvalue and $\mathbf{x}$ is the eigenvector. The eigenvalues are determined by the roots of the characteristic polynomium

$$D(\lambda) = (\gamma_{ce} + \lambda)\{(\Gamma + \lambda)(\kappa + \lambda)(\gamma_{ce} + \lambda) + 2g_c^2(\kappa + \Gamma + 2\lambda)\}, \qquad (2.32)$$

from which it is clear that $\lambda = -\gamma_{ce}$ is always an eigenvalue. This reflects the characteristic decay rate of the real part of the emitter–photon coherence, i.e., $(\rho_{ce} + \rho_{ec})/2$. The corresponding imaginary part, on the other hand, is affected by the emitter and cavity occupation. If the pure dephasing rate is negligible, that is, $\gamma_2 = 0$, the other three roots of Eq. (2.32) can be written in closed form as

$$\lambda = -\frac{1}{2}(\kappa + \Gamma), \quad \lambda = -\frac{1}{2}(\kappa + \Gamma) \pm \frac{1}{2}\sqrt{(\kappa - \Gamma)^2 - 16g_c^2}. \qquad (2.33)$$

Figure 2.6a shows an example of the eigenvalues as a function of cavity Q factor. The emitter decay rate derived using the adiabatic approximation is shown as a dashed line. All eigenvalues remain real and negative for $g_c < |\kappa - \Gamma|/4$, implying that

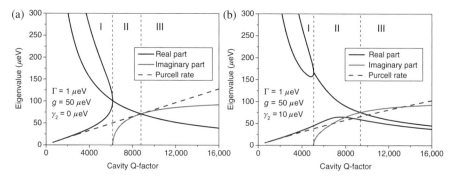

FIGURE 2.6 Illustration of strong and weak coupling regimes for a two-level system coupled to a cavity. Calculated eigenvalues as function of cavity quality factor for coupling factor $\hbar g_c = 50\,\mu\text{eV}$, emitter background decay rate $\hbar\Gamma = 1\,\mu\text{eV}$, and pure dephasing rates (a) $\hbar\gamma_2 = 0$, (b) $\hbar\gamma_2 = 10\,\mu\text{eV}$. Black lines: Real part of eigenvalues. Red lines: Imaginary part. Blue dashed line: Purcell-enhanced rate obtained by adiabatic elimination of coherence. *(For a color version of this figure, see the color plate section.)*

the decay of an initially excited emitter is monotonic. This is denoted by the weak coupling regime. On the other hand, for $g_c > |\kappa\Gamma|/4$ two of the eigenvalues acquire nonzero imaginary parts $\pm i\Omega_R$ with the vacuum Rabi frequency Ω_R given by

$$\Omega_R = \frac{1}{2}\sqrt{16g_c^2 - (\kappa - \Gamma)^2}. \tag{2.34}$$

This signifies the appearance of oscillatory variations of the emitter and photon occupation with period $T_R = 2\pi/\Omega_R$. Such oscillations, however, can only be observed if the Rabi frequency is larger than all the characteristic decay rates, given by the real parts of the eigenvalues. In the present case of zero pure dephasing rate, this is fulfilled for $g_c > \sqrt{(\kappa^2 + \Gamma^2)/8}$, which may be denoted by the vacuum Rabi splitting regime, since in this case the spectrum of the light emitted from the cavity will show two distinctive peaks. The spectrum of the emitted light can be calculated using Laplace transform techniques, leading to spectral peaks centered at the imaginary parts of the eigenvalues and having spectral widths given by the real part of the eigenvalues. In Figure 2.6a the three qualitatively different regimes are indicated; weak coupling (I), strong coupling (II and III) and vacuum Rabi oscillations (III). The intermediate regime II is neither characterized by an exponential decay, nor is the coupling of the cavity and emitter strong enough to induce oscillations.

When the pure dephasing rate is nonzero, there is no simple, exact expression for the eigenvalues, but approximate results have been stated [60]. Nevertheless, one can still calculate the eigenvalues, and Figure 2.6b shows an example of the calculated eigenvalues when taking into account a finite pure dephasing rate. Pure dephasing is seen to lower the required Q factor for the onset of strong coupling, defined as the point at which a set of eigenvalues with finite imaginary part appears. This is similar to the way a finite emitter decay rate lowers the strong coupling threshold in the absence of pure dephasing. However, it is important to point out that since pure dephasing adversely affects the linewidth, the required cavity Q factor to observe vacuum Rabi splitting actually increases with the rate of pure dephasing, as expected physically. Figure 2.7 shows examples of the temporal evolution of the expectation value of the emitter and cavity populations in the three qualitatively different coupling regimes. The emitter is initially in the excited state and parameters were chosen corresponding to Figure 2.6b. The decay of the emitter population is monotonous in Regime I ($Q = 3000$), where the system is in the weak coupling regime, and in regime II ($Q = 8000$), where the system is in the strong coupling regime but is overdamped. The decrease of the emitter decay time when the Q factor is increased from $Q = 3000$ to $Q = 8000$ is clearly observable. Oscillatory variations of the populations appear in Regime III ($Q = 16,000$ and $Q = 32,000$). The observable number of oscillations scale as the ratio of the Rabi splitting to the largest real part of the eigenvalues. Therefore, very high Q factors are required in order to observe several oscillations. The onset of vacuum Rabi oscillations depends on the ratio g_c/κ, and from Eqs. (2.29) and (2.30) this ratio is seen to scale as $Q/\sqrt{V_{eff}}$. Therefore, for an emitter with a given dipole moment, one can reach the regime of vacuum Rabi oscillations by increasing

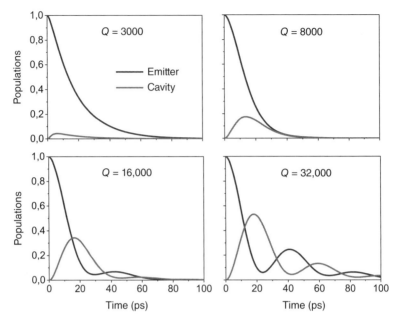

FIGURE 2.7 Temporal evolution of expectation value for emitter population (blue lines) and cavity population (red lines) for parameters corresponding to Figure 2.6b and different values of the quality factor: $Q = 3000$ (Regime I), $Q = 8000$ (Regime II), $Q = 16,000$ (Regime III), and $Q = 32,000$ (Regime III). *(For a color version of this figure, see the color plate section.)*

the Q factor and lowering the effective mode volume. We note, however, that these are typically conflicting demands.

In the presence of pure dephasing, the Q factor required to observe vacuum Rabi oscillations increases, as discussed above. In the micropillar geometry, the reduction of mode volume leads to the consideration of structures with ultrasmall diameters. For standard micropillar designs, however, the cavity mode becomes poorly matched to the modes of the mirror regions for submicron diameters, leading to scattering losses and Q factors on the order of or smaller than 2000 [61, 62]. In addition, substantial variations of the Q factor are observed, depending on the detailed matching with the Bloch modes of the periodic Bragg region [15]. One solution is to gradually change, or taper, the thickness of the layers composing the Bragg grating from the cavity region to the periodic mirror region as illustrated in Figure 2.8 for the case of a micropillar cavity [62]. A similar approach has been successfully applied also for the design of photonic crystal cavities, where it is known as a multistep heterostructure [16]. The working principle of the technique can be explained by invoking the theory of Bloch modes, so that it is a matter of engineering those modes rather than the usual transverse waveguide modes considered in standard tapers. In this way it was possible to achieve Q factors larger than 10,000 for micropillar cavities with diameters smaller than 1 µm [62], enabling the observation of a vacuum Rabi splitting in the frequency spectrum, cf. Figure 2.8. In comparison with other works, such as Reference 53, the quantum dots in this experiment have modest oscillator strengths on the order of 10

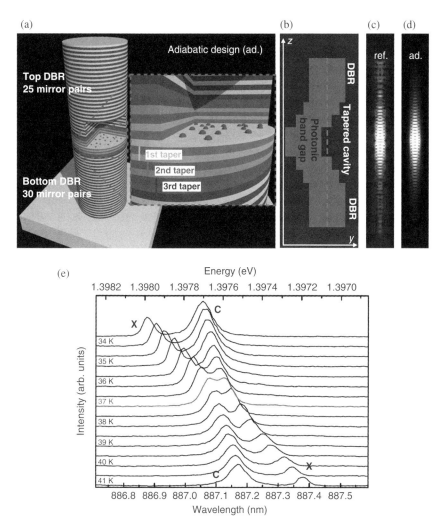

FIGURE 2.8 (a) Schematic of micropillar cavity with embedded layer of quantum dots. A normal and an adiabatic cavity design are illustrated. (b) Illustration of the variation of the photonic bandgap for the adiabatic design. Mode profiles for (c) reference and (d) adiabatic design. (e) Measured emission spectra showing anti-crossing between emitter (X) and cavity (C) peak, demonstrating that strong coupling is realized. The different spectra were obtained by tuning the temperature from 34 K to 41 K. Reprinted with permission from Reference 62 (Copyright 2012 by the American Physical Society). *(For a color version of this figure, see the color plate section.)*

and the key to the large vacuum Rabi splitting is the low modal volume obtained simultaneously with a high quality factor by Bloch-wave engineering of the cavity region. The same cavity design principle was recently used to demonstrate lasers with spontaneous emission factors on the order of 0.5 [63]

The properties of coupled quantum dot cavity systems remains an active field of research because of the rich dynamics of the systems. For applications in quantum computation schemes, the coherence of subsequently emitted photons, as quantified by the single-photon indistinguishability [64], is very important since it quantifies the ability of the single photons to interfere with each other. The degree of indistinguishability is, however, adversely affected by dephasing and its evaluation requires the consideration of two-time correlation functions [65]. In this case the use of a Lindblad approach for modeling the coupling to the phonon reservoir has been shown to be insufficient and non-Markovian effects become important [59].

2.5 NANOCAVITY LASERS AND LEDs

Light sources may operate in qualitatively different regimes, as illustrated in Figure 2.9. In the simplest case, an excited emitter decays into free space by spontaneous emission, Figure 2.9a. If the emitter is embedded in a cavity, a photon originally generated by spontaneous emission may be back-reflected by one of the cavity mirrors and stimulate emission of another, identical, photon, as shown in Figure 2.9b. This, however, requires that the emitter has been re-excited or that the stimulated emission occurs from another emitter. If the stimulated emission rate can be made large enough to balance the cavity loss rate, the cavity photon population builds up to the point where laser oscillation eventually takes place. Besides providing feedback for laser oscillation, the presence of the cavity also fundamentally changes the emission rate of the emitter, spontaneous as well as stimulated, via the Purcell effect discussed in Section 2.4.1. This effect is not important in present-day commercial lasers, which rely on large cavity volumes and/or spectrally broad emitters, but it becomes significant for ultrasmall cavities, where it may be used to lower the laser threshold. Below-threshold operation in a regime where the spontaneous emission rate is Purcell enhanced may also be of practical interest. Indeed, such a cavity-enhanced LED has been predicted to enable the realization of large modulation bandwidths not restricted by the usual damping mechanisms in lasers [66].

While the Purcell effect so far was discussed for the case of spontaneous emission, it also implies an enhancement of the rate of stimulated emission. Purcell-enhanced stimulated emission appears naturally within microscopic approaches, such as for example in Reference 67, where quantization of the electromagnetic field leads to a photon emission rate proportional to $F_P(n_p + 1)$, where F_P is the Purcell factor, n_p is the photon number, and the "1" signifies spontaneous emission, which in a

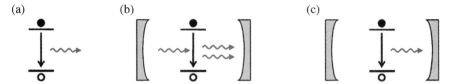

FIGURE 2.9 Illustration of different light sources: (a) light-emitting diode (LED), (b) laser, (c) cavity-enhanced LED. Reproduced from Reference 66 with the permission of the Optical Society of America.

quantum optics treatment appears as emission stimulated by vacuum fluctuations. Whether or not to regard spontaneous emission itself as a quantum phenomenon, however, is purely a matter of interpretation [68, 69]. Using the same argument, Purcell enhancement was phenomenologically included in the rate equation models commonly employed by the semiconductor laser community [70], assuming the necessary conditions to be fulfilled and without a treatment of the electronic density of states. Several recent works, however, include Purcell enhancement of the spontaneous emission rate, while the stimulated emission rate is unaffected. A rate equation model treating spontaneous and stimulated emission on equal footing and including details of the optical end electronic density of states has been reported in Reference 71 and will be briefly summarized below.

The dynamics of an ensemble of quantum dots in an optical cavity is modeled using the following rate equations for the carrier density N and photon density N_p in the cavity [71]:

$$\frac{dN}{dt} = J - R_c - R_b, \tag{2.35}$$

$$\frac{dN_p}{dt} = GR_c - \frac{N_p}{\tau_p}, \tag{2.36}$$

where J is the injection current density, $R_c = R_{sp} + R_{st}$ is the total emission rate into the cavity, including both spontaneous and stimulated contributions, R_b is the radiative decay to other modes than the dominating cavity mode, G is the confinement factor, which describes the overlap of the active material with the optical field, and $\tau_p = Q/\omega_c$ is the photon lifetime. For simplicity, we neglected nonradiative decay in Eq. (2.35). The radiative decay rates are determined through integration over all optical energies $\hbar\omega$ and electron energies E as

$$R_{sp} = \int\int \rho_o(\hbar\omega)\rho_e(E)f_2(E)(1-f_1(E))B_{21}\hbar\omega L(\hbar\omega - E)\hbar d\omega dE, \tag{2.37}$$

$$R_{st} = \int\int \rho_o(\hbar\omega)\rho_e(E)N_P V_{eff}(f_2(E)-f_1(E))B_{21}\hbar\omega L(\hbar\omega - E)\hbar d\omega dE, \tag{2.38}$$

in which ρ_e and ρ_o denote electronic and optical densities of state (DOS), respectively. These functions are illustrated in Figure 2.10. The electronic DOS may include inhomogeneous broadening of the emitter due to size fluctuations of quantum dots as well as the presence of a wetting layer. Homogeneous broadening, on the other hand, is included via the convolution with the Lorentzian lineshape function $L(\hbar\omega)$. The optical DOS accounts for the presence of a cavity mode, which appears as a localized defect mode within the photonic bandgap of the surrounding crystal. The background spontaneous emission rate, R_b, is obtained from Eq. (2.37) by replacing the cavity optical DOS with the background optical DOS, $\rho_{o,b}$. For further details on the representation of the DOS we refer to [71]. Other parameters appearing in the expressions for the decay rates are the recombination coefficient B_{21}, valence and

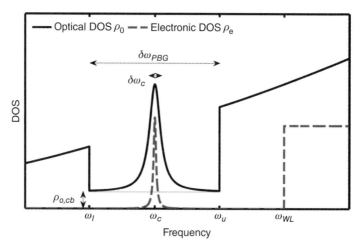

FIGURE 2.10 Schematic of the electronic (red dashed line) and optical (black line) density of states for an emitter embedded in a defect nanocavity. Reprinted with permission from Reference 71. (Copyright 2012 by the American Physical Society.) *(For a color version of this figure, see the color plate section.)*

conduction band Fermi functions f_1 and f_2, and photon density N_p, defined such that the photon number is $n_p = N_p V_{\text{eff}}$.

Figure 2.11 shows examples of calculated light-current characteristics for different parameter sets, comparing the case where Purcell enhancement of the stimulated emission is included with the case where it is not. For this choice of parameters, the laser stays below threshold if Purcell enhancement is not included, while the proper inclusion of Purcell enhancement increases the stimulated emission rate sufficiently to balance cavity losses and a transition to lasing is observed. The apparent clamping of the photon number observed in Figure 2.11 in the absence of Purcell enhancement takes place because the quantum dot states are completely filled and the emission rate is limited by the recombination time. For these calculations, parameters were chosen such that the total linewidth of the emitter is narrow compared to the cavity, enabling a high Purcell enhancement to be achieved. The predictions of the rate equation model presented above were shown to agree very well with the results of a microscopic approach [71]. The microscopic models usually assume a fixed value for the fraction of spontaneous emission into the lasing mode, which is typically denoted the spontaneous emission β-factor and defined as $\beta = R_{\text{sp}}/(R_{\text{sp}} + R_{\text{b}})$. The rate equation model, on the other hand, enables calculations in which the β-factor varies with injection level [71]. Due to the variation with Fermi level of the overlap between the Fermi functions and the cavity and background optical DOS, the spontaneous emission factor is found to vary significantly with carrier density.

The possibility of using the Purcell effect to increase the emission rate was proposed as a means to realize sources with modulation bandwidths greatly exceeding that of conventional semiconductor devices [66]. In particular, operation of the device

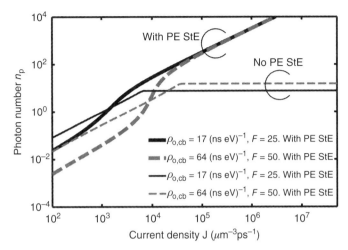

FIGURE 2.11 Photon number versus current density for a quantum dot nanocavity laser with the inclusion of Purcell-enhanced (PE) stimulated emission (StE) and without. Results for different values of the background density of states and the Purcell factor (F) are shown. Reprinted with permission from Reference 71. (Copyright 2012 by the American Physical Society.) *(For a color version of this figure, see the color plate section.)*

as a cavity-enhanced LED appears to be very attractive. In Reference 66, modulation bandwidths on the order of several hundred GHz were predicted for a quantum well structure with modal volumes realizable using photonic crystal cavities. Subsequent work, however, showed that the broadband nature of quantum well transitions severely limits the possibility of Purcell enhancing the modulation bandwidth [72] in agreement with Eq. (2.29) for large emitter broadening, $\gamma_2 \gg \kappa$. Figure 2.12 shows the calculated modulation bandwidth for a nanocavity LED, i.e., operated below the threshold current, versus degree of inhomogeneous broadening and cavity Q factor for a small homogeneous broadening of 100 µeV as appropriate for low temperature and low density operation. The results are taken from Reference 73, to which we refer for further details of the model and parameter values. The conclusion that we draw from this figure is that Purcell enhancement may indeed be used to increase the modulation bandwidth to the range of several hundred GHz. This, however, requires that the inhomogeneous broadening of the quantum dot transition energy due to size fluctuations is below $\simeq 1$ meV, which is difficult to realize using present-day self-assembled growth techniques. Also, the Q factor needs to be appropriately chosen; in this case there is a maximum around $Q = 500$. For smaller Q factors, the Purcell enhancement drops off, while for larger Q factors the photon lifetime becomes so large that it prohibits fast operation. These trends are well accounted for by approximate expressions for the modulation bandwidth [73]. We notice that the reduction in speed due to the finite carrier capture rate from the wetting layer into the quantum dots is not included in Figure 2.12, but can be a bottleneck at very high modulation speeds.

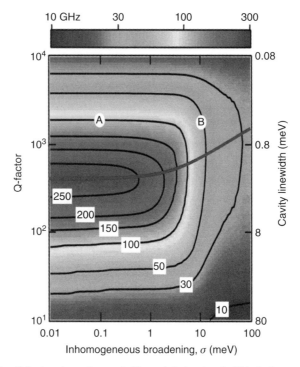

FIGURE 2.12 Calculated maximum 3 dB modulation bandwidth (color map) for a nanocavity LED versus degree of inhomogeneous broadening and cavity Q factor. Contour lines are drawn for constant modulation bandwidth (in GHz). Reprinted with permission from Reference 73. (Copyright 2012 by the American Physical Society.) *(For a color version of this figure, see the color plate section.)*

The calculations reported above are based on simplified representations of the optical density of states for the cavity and background, and it remains an important task to explore the possibilities offered by different cavity configurations for increasing device speed and controlling the spontaneous emission factor. One of the problems with the ultrasmall cavity designs is that the output power will be very limited. By coupling several cavities in arrays this problem may in principle be overcome [74], but it places strict demands on the fabrication tolerance for all the cavities in the array to stay coherently coupled [75].

2.6 SUMMARY

The possibility of fabricating ultrasmall optical cavities that embed emitters such as semiconductor quantum dots or nonlinear optical materials has revitalized the field of cavity photonics. Engineering the cavity offers possibilities for control of the emitter decay rate and may be used to improve and control the properties of lasers and LEDs. Exploiting the optical energy build-up in a cavity, optical switching may be realized for ultra-low excitation energies and in very compact structures with possible applications in optical signal processing. A high-quality cavity that is strongly coupled to an emitter may enter a regime of Rabi oscillations even in the single-photon regime. The exploration of this regime is important for future applications within quantum information technology.

ACKNOWLEDGMENTS

We acknowledge helpful discussions with Bjarne Tromborg and Kresten Yvind from DTU Fotonik, and with Alfredo de Rossi and Sylvain Combrie from Thales Research and Technology. This work was partly financed by Villum Foundation through the NATEC Centre of Excellence (Grant 8692), the European Commission through the COPERNICUS project and the Danish Council for Independent Research (FTP 10-093651).

REFERENCES

[1] K. J. Vahala, "Optical microcavities," *Nature* **424**, 839–846 (2003).
[2] T. J. Kippenberg and K. J. Vahala, "Cavity optomechanics: back-action at the mesoscale," *Science* **321**, 1172–1176 (2008).
[3] H. M. Gibbs, *Optical Bistability: Controlling Light with Light*. (Academic Press, Orlando, 1985).
[4] K. Nozaki, T. Tanabe, A. Shinya, S. Matsuo, T. Sato, H. Taniyama, and M. Notomi, "Sub-femtojoule all-optical switching using a photonic-crystal nanocavity," *Nat. Photonics* **4**, 477–483 (2010).
[5] B. W. Shore and P. L. Knight, "The Jaynes-Cummings Model," *J. Mod. Opt.* **40**, 1195–1238 (1993).
[6] S. Reitzenstein and A. Forchel, "Quantum dot micropillars," *J. Phys. D. Appl. Phys.* **43**, 033001-1-25 (2010).
[7] S. D. Joannopoulos, J. D., Johnson, S.G. Meade, *Photonic Crystals: Molding the Flow of Light*. (Princeton University Press, New Jersey, 2008).
[8] Y. Akahane, T. Asano, B.-S. Song and S. Noda, "High-Q photonic nanocavity in a two-dimensional photonic crystal," *Nature* **425**, 944–947 (2003).
[9] C. Sauvan, P. Lalanne, and J.-P. Hugonin, "Tuning holes in photonic-crystal nanocavities," *Nature* **429**, 2003 (2004).
[10] S. Reitzenstein, C. Hofmann, A. Gorbunov, M. Strauß, S. H. Kwon, C. Schneider, A. Loffler, S. Hofling, M. Kamp, and A. Forchel, "AlAs-GaAs micropillar cavities with quality factors exceeding 150.000," *Appl. Phys. Lett.* **90** (25), 251109-1-3 (2007).
[11] Y. Takahashi, H. Hagino, Y. Tanaka, B.-S. Song, T. Asano, and S. Noda, "High-Q nanocavity with a 2-ns photon lifetime," *Opt. Express* **15**, 17206–17213 (2007).
[12] K. Lee, P. Leung, and K. Pang, "Dyadic formulation of morphology-dependent resonances. I. Completeness relation," *J. Opt. Soc. Am. B* **16** (9), 1409–1417 (1999).

[13] P. T. Kristensen, C. Van Vlack, and S. Hughes, "Generalized effective mode volume for leaky optical cavities," *Opt. Lett.* **37** (10), 1649–1651 (2012).

[14] P. Lalanne, C. Sauvan, and J. Hugonin, "Photon confinement in photonic crystal nanocavities," *Laser Photonics Rev.* **2**, 514–526 (2008).

[15] N. Gregersen, S. Reitzenstein, C. Kistner, M. Strauss, C. Schneider, S. Hofling, L. Worschech, A. Forchel, T. R. Nielsen, J. Mork et al., "Numerical and experimental study of the Q-factor of high-Q micropillars," *IEEE J. Quantum Electron.* **46**, 1470–1483 (2010).

[16] B.-S. Song, S. Noda, T. Asano, and Y. Akahane, "Ultra-high-Q photonic double-heterostructure nanocavity," *Nat. Mater.* **4**, 207–210 (2005).

[17] Y. Tanaka, T. Asano, and S. Noda, "Design of photonic crystal nanocavity with Q-factor of 1 billion," *J. Light. Technol.* **26** (11), 1532–1539 (2008).

[18] T. Baba, "Slow light in photonic crystals," *Nat. Photonics* **2**, 465–473 (2008).

[19] M. Okano, T. Yamada, J. Sugisaka, N. Yamamoto, M. Itoh, T. Sugaya, K. Komori, and M. Mori, "Analysis of two-dimensional photonic crystal L-type cavities with low-refractive-index material cladding," *J. Opt.* **12**, 075101-1-10 (2010).

[20] R. J. Manning, A. D. Ellis, A. J. Poustie, and K. J. Blow, "Semiconductor laser amplifiers for ultrafast all-optical signal processing," *J. Opt. Soc. Am. B* **14**, 3204–3216 (1997).

[21] K. Stubkjaer, "Semiconductor optical amplifier-based all-optical gates for high-speed optical processing," *IEEE J. Sel. Top. Quantum Electron.* **6**, 1428–1435 (2000).

[22] M. Nielsen and J. Mork, "Bandwidth enhancement of SOA-based switches using optical filtering: theory and experimental verification," *Opt. Express* **14**, 1260–1265 (2006).

[23] P. A. Andrekson, H. Sunnerud, S. Oda, T. Nishitani, and J. Yang, "Ultra-fast, atto-Joule switch using fiber-optic parametric amplifier operated in saturation," *Opt. Express* **16**, 10956–10961 (2008).

[24] J. Y. Lee, L. Yin, G. P. Agrawal, and P. M. Fauchet, "Ultrafast optical switching based on nonlinear polarization rotation in silicon waveguides," *Opt. Express* **18**, 11514–11523 (2010).

[25] L. O'Faolain, S. A. Schulz, D. M. Beggs, T. P. White, M. Spasenovic, L. Kuipers, F. Morichetti, A. Melloni, S. Mazoyer, J. P. Hugonin, et al., "Loss engineered slow light waveguides," *Opt. Express* **18**, 27627–27638 (2010).

[26] V. R. Almeida, C. A. Barrios, R. R. Panepucci, and M. Lipson, "All-optical control of light on a silicon chip," *Nature* **431**, 1081–1084 (2004).

[27] C. Husko, A. De Rossi, S. Combrie, Q. V. Tran, F. Raineri, and C. W. Wong, "Ultrafast all-optical modulation in GaAs photonic crystal cavities," *Appl. Phys. Lett.* **94** (2), 021111-1-3 (2009).

[28] M. Heuck, P. T. Kristensen, and J. Mork, "Energy-bandwidth trade-off in all-optical photonic crystal microcavity switches," *Opt. Express* **19**, 18410–18422 (2011).

[29] K. Nozaki, T. Tanabe, A. Shinya, S. Matsuo, T. Sato, H. Taniyama, and M. Notomi, "Sub-femtojoule all-optical switching using a photonic-crystal nanocavity," *Nat. Photonics* **4**, 477–483 (2010).

[30] Q. V. Tran, S. Combrie, P. Colman, and A. De Rossi, "Photonic crystal membrane waveguides with low insertion losses," *Appl. Phys. Lett.* **95** (6), 061105-1-3 (2009).

[31] Y. Yu, M. Heuck, S. Ek, N. Kuznetsova, K. Yvind, and J. Mork, "Experimental demonstration of a four-port photonic crystal cross-waveguide structure," *Appl. Phys. Lett.* **101**, 251113-1-3 (2012).

[32] H. A. Haus, *Waves and Fields in Optoelectronics* (Prentice Hall, 1984).

[33] A. de Rossi, M. Lauritano, S. Combrié, Q. Tran, and C. Husko, "Interplay of plasma-induced and fast thermal nonlinearities in a GaAs-based photonic crystal nanocavity," *Phys. Rev. A* **79**, 043818–1-9 (2009).

[34] M. Heuck, J. Mork, and P. T. Kristensen, "A non-hermitian approach to non-linear switching dynamics in coupled cavity-waveguide systems," in *CLEO Tech. Dig.*, (San Jose, CA), p. Paper JW4A.6 (Optical Society of America, 2012).

[35] P. T. Kristensen, M. Heuck, and J. Mork, "Optimal switching using coherent control," *Appl. Phys. Lett.* **102** (4), 041107–1-4 (2013).

[36] M. Heuck, P. T. Kristensen, Y. Elesin, and J. Mork, "Improved switching using Fano resonances in photonic crystal structures," *Opt. Lett.* **38** (14), 2466–2468 (2013).

[37] H. M. Lai, P. T. Leung, K. Young, P. W. Barber, and S. C. Hill, "Time-independent perturbation for leaking electromagnetic modes in open systems with application to resonances in microdroplets," *Phys. Rev. A* **41** (9), 5187–5198 (1990).

[38] G. Khitrova and H. M. Gibbs, "Nonlinear optics of normal-mode-coupling semiconductor microcavities," *Rev. Mod. Phys.* **71**, 1591–1639 (1999).

[39] M. F. Yanik, S. Fan, M. Soljačić, and J. D. Joannopoulos, "All-optical transistor action with bistable switching in a photonic crystal cross-waveguide geometry," *Opt. Lett.* **28**, 2506–2508 (2003).

[40] J. Mork, F. Ohman, and S. Bischoff, "Analytical expression for the bit error rate of cascaded all-optical regenerators," *IEEE Photonics Technol. Lett.* **15**, 1479–1481 (2003).

[41] K. Nozaki, A. Shinya, S. Matsuo, Y. Suzaki, T. Segawa, T. Sato, Y. Kawaguchi, R. Takahashi, and M. Notomi, "Ultralow-power all-optical RAM based on nanocavities," *Nat. Photonics* **6**, 248–252 (2012).

[42] Y. Yu, Private communication (2013).

[43] J. Mork and J. Mark, "Carrier heating in InGaAsP laser amplifiers absorption," *Appl. Phys. Lett.* **64**, 2206–2208 (1994).

[44] T. Berg, S. Bischof, I. Magnusdottir, and J. Mork, "Ultrafast gain recovery and modulation limitations in self-assembled quantum-dot devices," *IEEE Photonics Technol. Lett.* **13**, 541–543 (2001).

[45] S. Nakamura, Y. Ueno, and K. Tajima, "Femtosecond switching with semiconductor-optical-amplifier-based symmetric MachZehnder-type all-optical switch," *Appl. Phys. Lett.* **78** (25), 3929–3931 (2001).

[46] M. Nielsen and J. Mork, "Increasing the modulation bandwidth of semiconductor-optical-amplifier-based switches by using optical filtering," *J. Opt. Soc. Am. B* **21**, 1606–1619 2004).

[47] K. Nozaki, A. Shinya, S. Matsuo, T. Sato, E. Kuramochi, and M. Notomi, "Ultralow-energy and high-contrast all-optical switch involving Fano resonance based on coupled photonic crystal nanocavities," *Opt. Express* **21**, 11877–11888 (2013).

[48] E. Purcell, "Spontanoeus emission probability at radio frequencies," *Proc. Am. Phys. Soc.* **69** (11), 681 (1946).

[49] J. Gérard, B. Sermage, and B. Gayral, "Enhanced spontaneous emission by quantum boxes in a monolithic optical microcavity," *Phys. Rev. Lett.* **81**, 1110–1113 (1998).

[50] A. Badolato, K. Hennessy, M. Atatüre, J. Dreiser, E. Hu, P. M. Petroff, and A. Imamoglu, "Deterministic coupling of single quantum dots to single nanocavity modes," *Science* **308**, 1158–1161 (2005).

[51] A. Kress, F. Hofbauer, N. Reinelt, M. Kaniber, H. Krenner, R. Meyer, G. Böhm, and J. Finley, "Manipulation of the spontaneous emission dynamics of quantum dots in two-dimensional photonic crystals," *Phys. Rev. B* **71**, 241304–1-4 (2005).

[52] D. Englund, D. Fattal, E. Waks, G. Solomon, B. Zhang, T. Nakaoka, Y. Arakawa, Y. Yamamoto, and J. Vučkovič, "Controlling the spontaneous emission rate of single quantum dots in a two-dimensional photonic crystal," *Phys. Rev. Lett.* **95**, 200–203 (2005).

[53] J. P. Reithmaier, G. Sek, A. Loftier, C. Hofmann, S. Kuhn, S. Reitzenstein, L. V. Keldysh, V. D. Kulakovskii, T. L. Reinecke, and A. Forchel, "Strong coupling in a single quantum dot-semiconductor microcavity system," *Nature* **432**, 197–200 (2004).

[54] T. Yoshie, M. Loncar, A. Scherer, and Y. Qiu, "High frequency oscillation in photonic crystal nanolasers," *Appi. Phys. Lett.* **84**, 3543–3545 (2004).

[55] J. Claudon, J. Bleuse, N. S. Malik, M. Bazin, P. Jaffrennou, N. Gregersen, C. Sauvan, P. Lalanne, and J.-M. Gérard, "A highly efficient single-photon source based on a quantum dot in a photonic nanowire," *Nat. Photonics* **4**, 174–177 (2010).

[56] H. J. Carmichael, *Statistical Methods in Quantum Optics 1 - Master Equations* (Springer, 1999).

[57] P. Kaer, T. R. Nielsen, P. Lodahl, A.-P. Jauho, and J. Mork, "Microscopic theory of phonon-induced effects on semiconductor quantum dot decay dynamics in cavity QED," *Phys. Rev. B* **86**, 085302–1-20 (2012).

[58] P. Kaer, T. R. Nielsen, P. Lodahl, A.-P. Jauho, and J. Mork, "Non-Markovian model of photon-assisted dephasing by electron-phonon interactions in a coupled quantum-dot cavity system," *Phys. Rev. Lett.* **104**, 157401–1-4 (2010).

[59] P. Kaer, P. Lodahl, A.-P. Jauho, and J. Mork, "Microscopic theory of indistinguishable single-photon emission from a quantum dot coupled to a cavity: The role of non-Markovian phonon-induced decoherence," *Phys. Rev. B* **87**, 081308(R)–1-5 (2013).

[60] A. Auffèves, D. Gerace, J.-M. Gérard, M. F. Santos, L. C. Andreani, and J.-P. Poizat, "Controlling the dynamics of a coupled atom-cavity system by pure dephasing," *Phys. Rev. B* **81**, 245419 (2010).

[61] P. Lalanne, J. P. Hugonin, and J. M. Gerard, "Electromagnetic study of the quality factor of pillar microcavities in the small diameter limit," *Appl. Phys. Lett.* **84** (23), 4726–4728 (2004).

[62] M. Lermer, N. Gregersen, F. Dunzer, S. Reitzenstein, S. Höfling, J. Mork, L. Worschech, M. Kamp, and A. Forchel, "Bloch-wave engineering of quantum dot micropillars for cavity quantum electrodynamics experiments," *Phys. Rev. Lett.* **108**, 057402–1-4 (2012).

[63] M. Lermer, N. Gregersen, M. Lorke, E. Schild, P. Gold, J. Mork, C. Schneider, A. Forchel, S. Reitzenstein, S. Hofling et al. "High beta lasing in micropillar cavities with adiabatic layer design," *Appl. Phys. Lett.* **102** (5), 052114–1-4 (2013).

[64] A. Kiraz, M. Atatüre, and A. Imamoglu, "Quantum-dot single-photon sources: prospects for applications in linear optics quantum-information processing," *Phys. Rev. A* **69**, 032305–1-10 (2004).

[65] J. Bylander, I. Robert-Philip, and I. Abram, "Interference and correlation of two independent photons," *Eur. Phys. J. D* **22**, 295–301 (2003).

[66] E. K. Lau, A. Lakhani, R. S. Tucker, and M. C. Wu, "Enhanced modulation bandwidth of nanocavity light emitting devices," *Opt. Express* **17**, 7790–7799 (2009).

[67] C. Gies, J. Wiersig, M. Lorke, and F. Jahnke, "Semiconductor model for quantum-dot-based microcavity lasers," *Phys. Rev. A* **75**, 0138031–11 (2007).

[68] P. Milonni, J. Ackerhalt, and W. Smith, "Interpretation of radiative corrections in spontaneous emission," *Phys. Rev. Lett.* **31**, 958–960 (1973).

[69] P. Milonni, "Why spontaneous emission?," *Am. J. Phys.* **52** (4), 340 (1984).

[70] H. Yokoyama and S. D. Brorson, "Rate equation analysis of microcavity lasers," *J. Appl. Phys.* **66**, 4801–4805 (1989).

[71] N. Gregersen, T. Suhr, M. Lorke, and J. Mork, "Quantum-dot nano-cavity lasers with Purcell-enhanced stimulated emission," *Appl. Phys. Lett.* **100** (13), 131107 (2012).

[72] T. Suhr, N. Gregersen, K. Yvind, and J. Mork, "Modulation response of nanoLEDs and nanolasers exploiting Purcell enhanced spontaneous emission," *Opt. Express* **18**, 11230–11241 (2010).

[73] T. Suhr, N. Gregersen, M. Lorke, and J. Mork, "Modulation response of quantum dot nanolight-emitting-diodes exploiting purcell-enhanced spontaneous emission," *Appl. Phys. Lett.* **98** (21), 211109-1-3 (2011).

[74] H. Altug and J. Vuckovic, "Photonic crystal nanocavity array laser," *Opt. Express* **13**, 8820–8828 (2005).

[75] T. Suhr, P. T. Kristensen, and J. Mork, "Phase-locking regimes of photonic crystal nanocavity laser arrays," *Appl. Phys. Lett.* **99**, 251104-1-3 (2011).

3

METAMATERIALS: STATE-OF-THE ART AND FUTURE DIRECTIONS

NATALIA M. LITCHINITSER[1] AND VLADIMIR M. SHALAEV[2]
[1] University at Buffalo, The State University of New York, Buffalo, NY, USA
[2] Purdue University, West Lafayette, IN, USA

3.1 INTRODUCTION

Despite the impressive progress in the field of modern photonics, further breakthroughs, such as scaling photonic components down in size and imaging with subwavelength resolution, are still needed. However, this is often challenging to achieve due to the constraints imposed by the optical properties of naturally existing materials. The emergence of metamaterials (MMs) provides new ways to overcome the limitations of modern optics, materials science, and their applications. Metamaterials are artificial structures that are engineered to have properties that cannot be found in nature. These artificial materials are made of carefully designed building blocks, or "meta-atoms," which are typically much larger than conventional atoms but much smaller than the wavelength of incident light, allowing the material to act as an effective medium with the desired optical properties. The meta-atoms are designed so that when combined together they enable new degrees of freedom for manipulating light waves. In particular, the value of the refractive index, or more precisely the values of the dielectric permittivity and magnetic permeability of the material, can be tailored to steer the waves in a prescribed trajectory or to impose certain amplitude, phase, or polarization properties onto the transmitted or reflected beams [1–5].

Since the field of MMs is still relatively new, we start with a brief historical overview and discuss its current status and future directions. Most of the known transparent optical materials are characterized by a positive real part of the dielectric permittivity, ε', and a positive real part of the magnetic permeability, μ', and therefore they are sometimes referred to as "double-positive" materials. In these materials, the real part of the refractive index of the medium, n', is positive as well and electromagnetic waves can propagate through such media. All known naturally existing optical materials possess a relative magnetic permeability that is very close or equal to 1, that is, they are nonmagnetic at optical frequencies. As a result, light–matter interactions in optics are limited to the electric field component coupling to materials. On the other hand, coupling magnetic field component to materials at optical frequencies can be enabled using meta-atoms, revealing one of the unique properties enabled by MMs—magnetism in optics [2, 6–9]. Another group of materials with one of the material parameters being positive and the other being negative, the so-called "single-negative" materials, can only support evanescent waves and thus are, optically "opaque" when considered separately. However, when combined together, a periodic stack of ε-negative and μ-negative MMs gives rise to yet a new type of photonic bandgap structures, such as omnidirectional and polarization-insensitive zero-phase gap materials [10]. Finally, another possibility is a so-called "double-negative" material with both dielectric permittivity and magnetic permeability being negative in the same frequency range. Once again, such materials have not been found in nature and it was not even obvious if such materials could exist from a fundamental physics standpoint until a detailed theoretical study was conducted by Victor Veselago in 1967 [11]. In particular, it was predicted that media containing both negative ε' and μ' not only support propagating waves, but a real part of the refractive index in such media is negative. In the early years, the term "metamaterials" was largely associated with these new and unusual materials with a negative index of refraction. Nowadays, it is used in a much broader context and signifies any engineered structure with the feature size much smaller than the wavelength of light that enable new light–matter interactions, which cannot be realized using naturally existing materials.

3.2 NEGATIVE-INDEX MATERIALS

Since a theoretical analysis [11] and an experimental demonstration [12] of negative-index materials (NIMs) were the starting points for the entire MM field, we begin with a review of their unusual properties and fascinating potential applications. When considering Maxwell's equations and the Poynting vector definition for two different cases of media with the same absolute $|\varepsilon|$ and $|\mu|$ values, the wave vector $\vec{k}$, electric field $\vec{E}$, and magnetic field $\vec{H}$ form the right-handed triplet when ε' and μ' are both positive. In this case, wave vector $\vec{k}$ and the Poynting vector are parallel, signifying that the phase and energy propagate in the same direction. These materials are referred to as positive-index materials (PIMs) or right-handed materials (RHMs). In the case of an NIM, vectors $\vec{k}$, $\vec{E}$ and $\vec{H}$ form a left-handed triplet; thus, NIMs are also referred to as left-handed materials (LHM). In this case, wave vector $\vec{k}$ and the Poynting vector are antiparallel, that is, pointing in opposite directions. This property is often considered as the general definition of an NIM.

The first optical NIMs were fabricated and tested in 2005 by two research groups [13, 14]. One of these first optical NIMs was based on pairs of nanorods made of

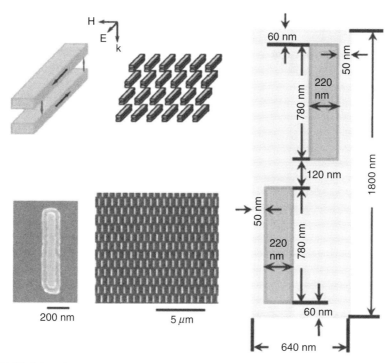

FIGURE 3.1 Diagram and experimentally fabricated images of gold nanorod pair MM structures [13].

gold, as shown in Figure 3.1, which were placed in a two-dimensional array and illuminated with light at a wavelength of 1.5 μm. The origin of a negative refractive index in such paired nanorods can be understood as follows. The electric resonances of individual nanorods originate from the excitation of the surface plasmon polaritons (SPPs) on the metal–air interface. In a paired nanorod configuration, two types of SPP waves can be supported: symmetric and antisymmetric. The electric field, oriented parallel to the nanorods, induces parallel currents (symmetric plasmon mode) in both nanorods, leading to the excitation of a dipole moment. The magnetic field, oriented perpendicular to the plane of the nanorods, excites antiparallel currents (antisymmetric plasmon mode) in the pair of nanorods. Combined with the displacement currents between the nanorods, they induce a resonant magnetic dipole moment. The excited moments are co-directed with the incident field when the wavelength of the incident light is above the resonance, and they are counter-directed to the incident fields at wavelengths below the resonance. The excitation of such plasmon resonances for both the electric and magnetic field components results in the resonant response of the refractive index. In particular, the refractive index can become negative at wavelengths below the resonance. However, it is important that the electric and magnetic resonances occur at similar wavelengths, which is quite difficult to achieve. It should be noted that the condition of simultaneously negative ε and μ is a sufficient but not necessary condition for having negative real part of the index of refraction [15]. It was shown that the materials would possess a negative refractive index, the permittivity $\varepsilon = \varepsilon' + i\varepsilon''$ and the permeability $\mu = \mu' + i\mu''$ obey the condition $\varepsilon'\mu'' + \mu'\varepsilon'' < 0$ suggesting that negative refractive index can be realized even if $\mu' > 0$ (still, the excitation of the magnetic response is essential so that μ has to be different from 1). In fact, both of the first optical NIMs demonstrated in 2005 belong to this category. The challenge with this approach to negative index of refraction is that the figure of merit (FOM), defined as

$$\text{FOM} = \frac{-\text{Re}(n)}{\text{Im}(n)}, \tag{3.1}$$

is usually very small in this case. Therefore, later, a superior design known as, a fishnet structure was proposed where under proper conditions a higher FOM can be obtained.

It is noteworthy that an electric resonance is not really required in order to obtain a negative ε'. Therefore, it was suggested to use a resonant magnetic structure along with a nonresonant metallic structure that provides negative permittivity in a range of wavelengths including those corresponding to the magnetic resonance. For example, such background negative permittivity can be achieved using pairs of continuous silver or gold nanostrips that do not have an electrical resonance at the wavelength of interest [16], while a magnetic resonance with a negative permeability is obtained using another set of carefully designed pairs of metallic nanostrips. The resulting structure, known as the "fishnet" NIM, is shown in Figure 3.2. In the fishnet structure, the pairs of broader metal strips provide negative permeability originating from asymmetric currents and the pairs of narrower metal strips act as a diluted metal. The highest reported FOM for fishnet structures without any loss compensation was about 3 [16].

An important property of NIMs is the fact that when light passes from a positive index medium into a negative index medium it undergoes negative refraction at the interface [11]. After passing through the interface, the refracted light will lie on the

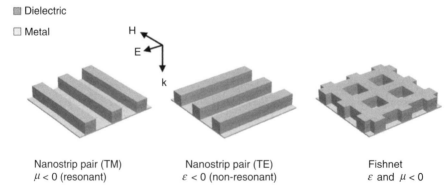

FIGURE 3.2 Basic idea behind fishnet NIM. *(For a color version of this figure, see the color plate section.)*

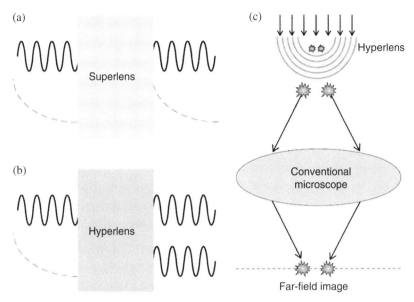

FIGURE 3.3 Schematics of (a) superlens, (b) hyperlens, and (c) imaging system that uses a hyperlens. Solid lines in (a) and (b) correspond to propagating waves; dashed lines correspond to evanescent waves [20].

same side of the normal as the incident light in a NIM and the $\vec{k}$ vector points in a direction toward the interface while the energy flow ($\vec{S}$ vector) points away from the interface. An especially interesting effect occurs in the case of light passing through an NIM, slab sandwiched between the two PIMs. The slab acts similar to a lens in a sense that light incident on its surface is negatively refracted inside the PIM, gets focused inside the slab, then, it is positively refracted upon exiting the slab and is finally refocused outside the slab, forming a real image of the object [11]. Besides this lens-like effect, an important unique phenomenon first predicted by Pendry [17], is an "amplification" of evanescent components of the incident wave in an NIM slab, as shown in Figure 3.3a. These evanescent components carry information about subwavelength features of the object; consequently, the ability to carry them from an object plane to the image plane results in imaging below the diffraction limit or "superlensing" [17]. In contrast, in conventional lenses, only the propagating components contribute to formation of the image since the evanescent components decay quickly in the lens. Therefore, the resolution of a conventional optical lens is diffraction limited, which means that the minimum feature size, d_{min}, that can be resolved is

$$d_{min} \approx \frac{2\pi}{k_{max}} = \lambda, \qquad (3.2)$$

where k_{max} is a maximum wave vector and λ is a wavelength of light.

The ingenious prediction of the superlensing effect by Pendry [17] inspired the enormous progress in the field of MMs that we witness today. While it was theoretically predicted that an ideal, lossless, impedance-matched NIM with a refractive index of −1 would provide ultrahigh resolution, the performance of the superlens was found to be extremely sensitive to light attenuation inside the material and the material's quality so that in practice it is limited to the case when both an object and an image are in proximity to the NIM slab. Thus, several studies were devoted to addressing the limitations of the original superlens. One of the major developments in this direction was the invention of a so-called hyperlens (shown in Fig. 3.3b) [18–21]. Instead of amplifying evanescent wave components, a hyperlens, which is made of a highly anisotropic material with hyperbolic dispersion curves, converts them into propagating waves that can be focused by a conventional lens in the far field, as illustrated in Figure 3.3c.

While the explosion of interest in NIMs for imaging applications has received a lot of attention in the literature, other interesting effects of NIMs predicted by Veselago such as backward Cerenkov radiation, reversed Doppler effect as well as backward phase-matching conditions enabling a plethora of new regimes of nonlinear wave interactions have also been discovered [22–54].

It is well known that many nonlinear phenomena rely on phase matching either between waves at different frequencies or those at the same frequency [22, 23]. It is worth mentioning that MM's properties are strongly frequency dependent, suggesting that the same material can serve as a PIM in one range of frequencies and as an NIM in a different range. Therefore, one could attempt phase matching two or more waves, some of which propagate in the PIM and the others in the NIM. On the other hand, we could combine two different materials such that the frequency of the wave is fixed; however, one of the materials appears as a PIM at this fixed frequency, while another one functions as an NIM. These two possibilities correspond to combining PIM and NIM structures in the spectral domain and in the spatial domain, respectively. Here, we discuss in some detail a novel regime of the second harmonic generation (SHG) process enabled by the unique properties of NIMs. Other studies have predicted a number of novel phenomena in PIMs and NIMs combined either in spatial or in the spectral domains, such as backward phase-matched nonlinear couplers or optical parametric amplifiers, respectively.

One of the most remarkable new features of SHG in metamaterials is that the direction of second harmonic (SH) propagation is reversed in a $\chi^{(2)}$ nonlinear medium if the fundamental frequency (FF) propagates in the NIM regime, while the SH wave propagates in the PIM. This is in contrast to conventional PIMs, where the FF wave and the SH wave propagate in the same direction and exchange energy upon propagation. Both k-vectors and Poynting vectors are co-directional. In the case when the FF at ω_1 falls into the NIM regime of propagation and the SH at ω_2 propagates in the PIM, the vectors $\vec{k}_1$ and $\vec{S}_1$ point in opposite directions, while $\vec{k}_2$ and $\vec{S}_2$ point in the same direction. As a result, in order to satisfy the phase-matching condition, $\vec{k}_1$ and $\vec{k}_2$ are parallel and co-directional; $\vec{S}_1$ and $\vec{S}_2$ are antiparallel, such that the energy at ω_1 propagates to the right and is converted to ω_2, whose energy propagates to the left [25–27, 51] The result of such "backward" phase matching is the modified

Manley–Rowe relation for the PIM–NIM structure. Indeed, a Manley–Rowe relation in a PIM–NIM system requires that the difference (instead of the sum as in conventional materials) of squared amplitudes of the FF and SH remains invariant along the propagation direction [27]. It is noteworthy that such a modified Manley–Rowe relation is well known in the context of distributed feedback structures. However, an important difference is that in a PIM–NIM structure, "effective feedback" results from the inherent property of the NIM–counter-directional wave and Poynting vectors.

Such effective feedback is directly related to the fact that the boundary conditions for the FF and SH waves in the NIM are specified at opposite interfaces of the slab of a finite length, while in the conventional PIM, both conditions are specified at the front interface. As a result, the conversion at any point within the NIM slab depends on the total thickness of the slab. Moreover, in the limit of a semi-infinite NIM, both FF and SH waves disappear at infinity; thus, a 100% conversion efficiency of the incoming wave at the FF to the SH frequency propagating in the opposite direction is expected. An NIM slab of this nature works as a nonlinear mirror.

Recently, the first experimental demonstration of the nonlinear optical mirror effect at microwave frequencies in a metallic waveguide loaded with a magnetic nonlinear MM made of varactor-loaded split-ring resonator (shown in Fig. 3.4) was reported [50]. SHG was studied in three configurations including reflected SH phase matching in a negative-index spectral range, transmitted SH quasi-phase matching, and simultaneous quasi-phase matching of both the reflected and transmitted SH waves near a zero-index spectral range.

3.3 MAGNETIC METAMATERIALS

As discussed above, magnetism is particularly weak at optical frequencies and $\mu \approx 1$. On the other hand, MMs were shown to enable an artificial magnetic response over a broad range of optical wavelengths [6–9]. Indeed, for example a pair of nanostrips consisting of a dielectric layer sandwiched between two metal layers can support a symmetric resonance that leads to artificial permittivity and asymmetric resonance, which enables an artificial permeability [9].

First demonstrations of such nanostrip-based magnetic optical MMs were primarily done in the form of thin films with thicknesses of the order of or less than the optical wavelength. In order to use them in applications such as remote sensing or imaging, it is important to address light in- and out-coupling issues. From this viewpoint, waveguide-coupled MMs would be desirable. Recently, a fiber-coupled magnetic MM structure was proposed and demonstrated (see Fig. 3.5) [55].

Fiber optics is a mature technology that enables long-distance, low-loss light delivery [56]. Moreover, many devices, such as fiber Bragg gratings, fiber tapers, and directional couplers, can be easily realized and placed in-line with the light transmission channel. Fibers are routinely used in sensing applications [57], for example, to deliver light to and from the sample, or in near-field optics, for example, near-field optical microscopy (NSOM) [58]. Therefore, merging this well-developed fiber optics technology with the novel unique properties of MMs would likely enable

(a)

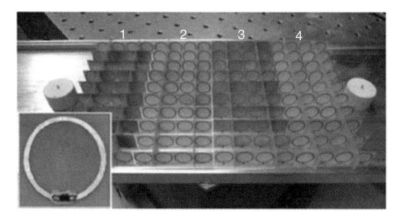

(b)

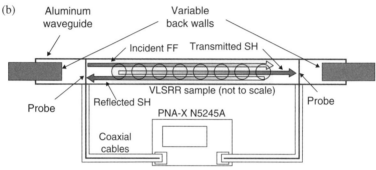

FIGURE 3.4 (a) Image of waveguide loaded with four identical sections of varactor-loaded split-ring resonator. Inset shows an enlarged view of unit cell. (b) Diagram of experimental setup used to measure the second harmonic spectra [50].

a plethora of advanced applications. Recently, first steps toward combining fiber technology with plasmonics have been reported [59, 60].

To demonstrate that MM structures can be integrated on a fiber, a standard single mode fiber (SMF-28) with a 9 µm core and a 125 µm diameter was used. Metamaterials structures were designed to have magnetic resonance in the range between 1100 and 1600 nm. Figures 3.4a and 3.4b show a schematic of the fiber MM and a nanostrip array, each composed of 40 nm silver (Ag) layer, followed by a 40 nm lithium fluoride (LiF) layer, and by another 40 nm Ag layer.

Let us briefly review the basic physics enabling unusual optical properties of nanostrip-based MMs shown in Figure 3.5. The electric field, oriented parallel to the nanostrips induces parallel currents in both nanostrips, leading to the excitation of an electric dipole moment. The magnetic field, oriented perpendicular to the plane of the nanostrips, excites antiparallel currents in the pair of nanostrips. Combined with the displacement currents between the nanostrips, they induce a resonant magnetic

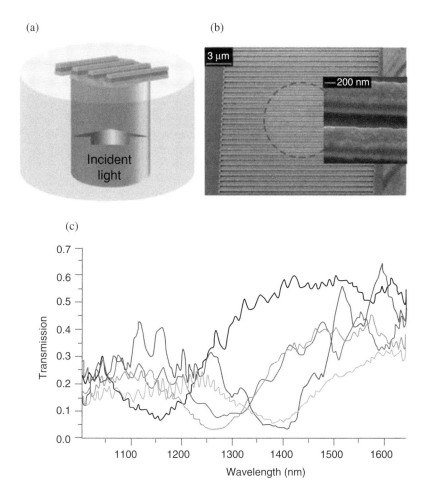

FIGURE 3.5 (a) Schematic of transverse cross section of optical fiber with a pattern in center; (b) SEM image of nanostrips covering core of fiber, which can be seen in center; (c) measured transmission for five fiber MMs samples illuminated with TM polarized wave. *(For a color version of this figure, see the color plate section.)*

dipole moment as described above. The excited moments are co-directed with the incident field when the wavelength of an incident light is above the resonance; they are counter-directed to the incident fields at wavelengths below the resonance, leading to positive and negative magnetic permeability, respectively.

Finite element method-based numerical simulation used for designing and modeling the actual experimental results took into account the feedback from fabrication and measurements, including the trapezoidal configuration of the fabricated nanostrips and the removal of approximately 80 nm of fiber core in the process of patterning using a focused ion beam. Therefore, the structure used in the numerical simulations consisted of four layers: two layers of Ag with LiF in between and a layer of silica glass. The dielectric permittivity of silver was taken into account using the Drude–Lorentz model with five oscillators:

$$\varepsilon = \varepsilon_1 - \frac{\omega_p^2}{\omega^2 + i\Gamma_p\omega} + \sum_{m=1}^{n} \frac{f_m \omega_m^2}{\omega_m^2 - \omega^2 - i\Gamma\omega'} \qquad (3.3)$$

where in the Drude term, $\varepsilon_1 = 2.1485$ is the static dielectric constant, $\omega_p = 9.1821$ eV is the plasma frequency, and ω_p (0.0210 eV) is the damping rate for the Drude term. Note that the contribution of the Drude term dominates in the metal at optical frequencies. The Lorentz terms mainly contribute to the loss at wavelengths shorter than 400 nm. Experimentally measured transmission plots for all the samples are shown in Figure 3.4c.

In summary, in this section we reviewed one of the unique properties enabled by the MMs approach—magnetism in optics. Moreover, we discussed how such novel magnetic MMs could be integrated with mature optical fiber technology. Fiber–MMs integration described here may provide fundamentally new solutions for photonic-on-a-chip systems for sensing, biomedical applications, such as fiber-based endoscopy with subwavelength resolution enabled by MM technology and advanced image processing.

3.4 GRADED-INDEX TRANSITION METAMATERIALS

In this section, we discuss recent progress in the field of light propagation in transition metamaterials, a class of artificial graded-index materials with dielectric permittivity and magnetic permeability gradually changing between positive, zero, and negative values. The initial studies predicted strong, polarization-sensitive, anomalous field enhancement near the zero refractive index point under oblique incidence of the plane wave on a realistic, lossy transition metamaterial layer, potentially enabling a variety of applications in microwave, terahertz, and optical metamaterials, including subwavelength transmission and low intensity nonlinear optical devices [61–67]. Later, the phenomenon of resonant field enhancement was investigated in the cases of Gaussian and Hermite–Gaussian beam propagation in transition metamaterials, indicating that in these cases resonant enhancement occurs for both normal and oblique incidence due to the fact that the beams are composed of a combination of plane waves, each with a different angular vector. We also studied various graded-index structures and identified a possibility of field enhancement in both positive–zero– index structures and positive–zero–positive index profile cases. Moreover, more complex refractive index distributions with, for example, two zero-index crossings, may result in a formation of tunable resonant cavities.

A theoretical model describing the wave propagation in transition MMs can be derived from Maxwell's equations. Using Maxwell's equations, we have

$$\nabla \times \vec{H} = \frac{i\omega}{c}\varepsilon\vec{E} \qquad (3.4)$$

$$\nabla \times \vec{E} = -\frac{i\omega}{c}\mu\vec{H} \qquad (3.5)$$

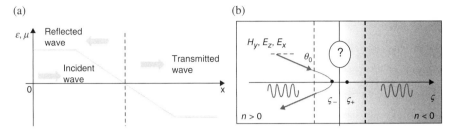

FIGURE 3.6 (a) Schematic of a transition layer between PIM (left) and NIM (right). (b) Diagram of wave behavior in transition layer with linear graded profile.

Considering TM or TE polarized waves such that either magnetic or electric field is polarized along the y-axis and both μ and ε profiles are inhomogeneous with respect to x. Wave propagation is governed by the following equation:

$$\frac{\partial^2 H_y}{\partial x^2} - \frac{1}{\varepsilon}\frac{\partial \varepsilon}{\partial x}\frac{\partial H_y}{\partial x} + \frac{\omega^2}{c^2}\left(\varepsilon\mu - \sin^2\theta_0\right)H_y = 0 \qquad (3.6)$$

$$\frac{\partial^2 E_y}{\partial x^2} - \frac{1}{\mu}\frac{\partial \mu}{\partial x}\frac{\partial E_y}{\partial x} + \frac{\omega^2}{c^2}\left(\varepsilon\mu - \sin^2\theta_0\right)E_y = 0. \qquad (3.7)$$

Figure 3.6a shows schematic of a transition layer between the PIM (left) and the NIM (right). The profiles have regions where both material parameters linearly decrease from a positive index region to a negative index region and two regions where the parameters are constant. Since there is a negative index region now beyond the point where the refractive index is zero, the wave that is transmitted beyond this point is a propagating wave (Fig. 3.6b).

The dimensionless wave equation can then be written as

$$\frac{\partial^2 \Phi}{\partial \varsigma^2} - \frac{1}{\upsilon}\frac{\partial \upsilon}{\partial \varsigma}\frac{\partial \Phi}{\partial \varsigma} + \left(h^2 k_0^2\right)\left(\varepsilon\mu - \sin^2\theta_0\right)\Phi = 0, \qquad (3.8)$$

where $k_0 = (\omega/c_0)\sqrt{\varepsilon_0\mu_0}$, $\varsigma = x/h$, $\upsilon = \varepsilon$ or μ for a TM or TE polarized wave, respectively. The incident field can be written in the form $H_y(E_y) = \Phi(x)\exp(ik_0\sin\theta_0 z)$, and the explicit forms of the profiles are

$$\varepsilon(x) = \varepsilon_0\left(1 - x/h\right), \quad \mu(x) = \mu_0\left(1 - x/h\right), \qquad (3.9)$$

where the materials' parameter h is a measure of the transition gradient.

Consider the case of a TE polarized wave. In the case of normal incidence, that is, when $\sin\theta_0 = 0$, the solution is simply given by

$$\Phi = C_1 \exp\left(\frac{-ik_0(1-\varsigma)^2}{2}\right) + C_2 \exp\left(\frac{ik_0(1-\varsigma)^2}{2}\right). \qquad (3.10)$$

From Eq. (3.10) it is clear that no unusual behavior of the electric and magnetic field components is expected at normal incidence; however, it was found that the wave propagation at oblique incidence is far more surprising.

First of all, from Eq. (3.8), it is clear that there are two points possible at which the wave can be totally internally reflected. These so-called turning points are given by

$$\varsigma_{\pm} = 1 \pm \sin\theta_0. \tag{3.11}$$

The analytical solution to (3.8) can be written in terms of a confluent hypergeometric function, U, that describes the behavior of the magnetic field within the entire layer as

$$\Phi(\varsigma) = C_1 \exp\left(\frac{-ihk_0(1-\varsigma)^2}{2}\right) ihk_0(1-\varsigma)^2 U\left(1 - \frac{ih^2 k_0^2 \sin^2\theta_0}{4hk_0}, 2, ihk_0(1-\varsigma)^2\right)$$

$$+ C_2 \exp\left(\frac{ihk_0(1-\varsigma)^2}{2}\right) ihk_0(1-\varsigma)^2 U\left(1 + \frac{ih^2 k_0^2 \sin^2\theta_0}{4hk_0}, 2, -ihk_0(1-\varsigma)^2\right). \tag{3.12}$$

Near the singular point the solution can be approximated as

$$\Phi(\varsigma) \approx (C_1 - C_2)\left(1 + \frac{h^2 k_0^2 \sin^2\theta_0 (1-\varsigma)^2}{2} \ln(hk_0(1-\varsigma))\right),$$

$$- \frac{i}{2}(C_1 + C_2)\left(1 - \frac{\pi h^2 k_0^2 \sin^2\theta_0}{4hk_0}\right) hk_0(1-\varsigma)^2 \tag{3.13}$$

where the values of the coefficients C_1 and C_2 are obtained from boundary conditions. The first term in this expression has a logarithmic singularity. The second term corresponds to a regular solution of Eq. (3.8). The logarithmic term is well defined for $\varsigma < 1$; however, for $\varsigma > 1$, the definition of the logarithm is not unique and depends on the choice of the path around.

Following the approach used in References 68 and 69 and Maxwell's equations, we obtain

$$E_x = \frac{ic}{\omega\varepsilon}\frac{\partial H_y}{\partial z} = -\frac{c}{\omega\varepsilon} k_0 \sin\theta_0 \Phi \exp(ik_0 \sin\theta_0 z - \omega t)$$

$$= -\sqrt{\frac{\varepsilon_0}{\mu_0}} H_0 (1-\varsigma)^{-1} \sin\theta_0 \exp(ik_0 \sin\theta_0 z - \omega t) + O((1-\varsigma)^2), \tag{3.14}$$

$$E_z = -\frac{ic}{\omega\varepsilon}\frac{\partial H_y}{\partial x} = -\frac{ic}{\omega\varepsilon}\frac{1}{h}\frac{\partial \Phi}{\partial \varsigma} \exp(ik_0 \sin\theta_0 z)$$

$$= -ik_0 h \mu_0 H_0 \sin^2\theta_0 \ln(hk_0(1-\varsigma)) \exp(ik_0 \sin\theta_0 z - \omega t) + O((1-\varsigma)^2), \tag{3.15}$$

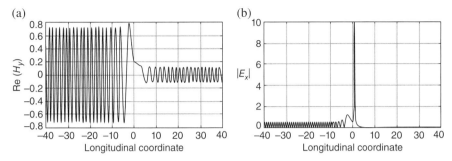

FIGURE 3.7 (a) Real part of the magnetic field component H_y as a function of longitudinal coordinate. (b) Absolute value of E_x as a function of longitudinal coordinate.

where the value of the constant H_0 is the value of the field at $\varsigma = 1$. While the y-component of the magnetic field H_y is continuous, the x-component of the magnetic field E_x is singular at $\varsigma = 1$; z-component E_z experiences a jump when the value of ε changes sign from positive to negative.

Figure 3.7 shows the results of full numerical simulations confirming the above predictions. The magnetic field forms a standing wave pattern in the PIM region due to the interference of incident and reflected waves. Beyond the turning point, the wave decreases as it approaches the singular point at $\varsigma = 1$. As it passes the singular point it converts back into a propagating wave once again. The magnitude of the electric field component E_x shows the predicted resonant enhancement at the point $\varsigma = 1$, where it is predicted to be singular.

The numerical algorithm for calculating the field solution within the medium is based on the sweep method, which was previously implemented in Reference 70 to solve the equation that describes wave propagation in inhomogenous plasmas. The algorithm uses a finite-difference scheme that generates an $(N \times 1) \times (N \times 1)$ tridiagonal matrix of coefficients based on the coefficients and boundary conditions of a differential equation of the form

$$p(x)\frac{d^2 y}{dx^2} + q(x)\frac{dy}{dx} + r(x)y = f(x). \quad (3.16)$$

The LU decomposition is then performed on that tridiagonal matrix using first a backward recursion, and then a forward recursion to find all of the y values. The forward and backward recursive steps give this algorithm the name "sweep method."

Figure 3.8a shows the coefficients R, T, and A as a function of incident angle θ_0. As the incident angle increases, the reflectivity reaches 1 and the absorption and transmittivity both go to zero.

Contour plots in Figure 3.8b show the absorption as both parameters θ_0 and h are swept over a range of possible values. This figure indicates that for a wide range of h values high absorption is maintained if the angle of incidence stays small. When the transition between PIM and NIM regions gets sharper, there is a wider range of

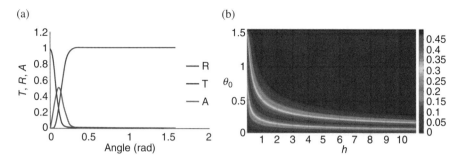

FIGURE 3.8 (a) Absorption, reflectivity, and transmittivity as function of incident angle θ_0. (b) Conversion efficiency at point $\varsigma = 1$ as a function of h and θ_0. *(For a color version of this figure, see the color plate section.)*

incidence angles that provide high absorption; however, as the width of transition layer becomes smaller than the wavelength of incident light, strictly speaking, the model has to be modified to include the effects of spatial dispersion.

Finally, it is noteworthy to find that the difference of the longitudinal components of the Poynting vector $S_x = (c/4\pi) E_y H_z$ (averaged over rapid field oscillations) before and after the transition of the wave through the point $n = 0$ is given by

$$\Delta S_x = c\pi E_0^2 (h/\lambda)\, \varepsilon_0 \sin^2 \theta_0. \qquad (3.17)$$

Note that while the model discussed so far assumed losses (imaginary parts of ε and μ) to be infinitesimally small, at the point $\varsigma = 1$ ($n = 0$) the real parts of ε and μ are zero; therefore, their contribution of the small imaginary parts become significant and can no longer be neglected. Consequently, the dissipation of energy Q occurs due to these losses at $\varsigma = 1$ and $Q = S_x$. As a result, transition through the point where ε and μ change sign even with infinitesimal loss is accompanied by a finite dissipation of incident wave energy. The specific mechanism of dissipation though would be determined by a physical model of an MM.

These results may have different implications for the design of particular applications of graded-index MMs. As MM perfect absorber research has gained great interest [66–68], a particular design could make use of a transition layer to enhance its absorptive properties. There are cases, however, where graded-index interfaces can make unwanted appearances on structures, such as semiconductors and superlenses. In these structures, the possibility of resonant absorption if the index goes through zero could be decreased by properly designing the MM structure.

Next, we discuss the effect of more realistic losses on the resonant enhancement phenomena. The values of the parameters for this study were based on experimentally obtained data from Wegener's group [71]. The fishnet MM was designed to be operated at $\lambda = 1.41$ μm, and was made out of Ag layers with MgF_2 in between. The actual experimental results are shown in Figure 3.9. This material had the best FOM that was reported at the time of the study.

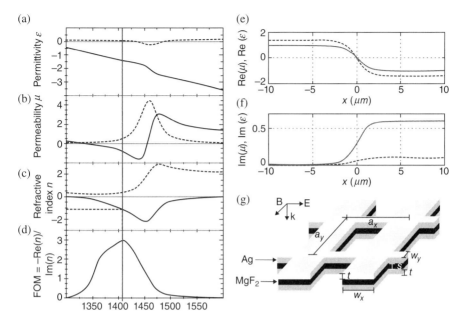

FIGURE 3.9 Real and imaginary parts of (a) ε, (b) μ, and (c) n for experimentally fabricated fishnet MM [13]. (d) Figure of merit for fishnet MM. (e) Real and (f) imaginary parts of ε (blue dashed line) and μ (solid green line) for the simulated transition layer with $h = 2\lambda$ and $\theta_0 = \pi/17$. (g) Diagram of experimentally fabricated fishnet MM [71]. *(For a color version of this figure, see the color plate section.)*

Figure 3.10 shows the absolute value of the longitudinal component of the magnetic field for the case of losses corresponding to the measured values of imaginary parts of ε and μ and those reduced by a certain factor α. The results show that the effect of resonant field enhancement takes place even in the case of relatively high losses and can be significantly increased by decreasing the losses by a factor of ~10.

After analyzing the enhancement of the fields, the amount of enhancement can be quantified by defining the enhancement factors for each input polarization as

$$\eta_{\text{TE}} \approx 2 \frac{|E_y|_{x=0}}{\langle|E_y(x)|\rangle_{x:\text{PIM}}} M \eta_{\text{TM}} \approx 2 \frac{|H_y|_{x=0}}{\langle|H_y(x)|\rangle_{x:\text{PIM}}} E$$

$$M = \frac{\mu_0'}{\mu_0''} E = \frac{\varepsilon_0'}{\varepsilon_0''},$$

(3.18)

where the average values of the fields shown in (3.17) are taken in the PIM region. The FOM of the transition layer must also be considered to get a picture of how the

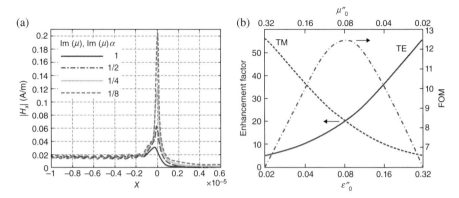

FIGURE 3.10 (a) $|H_x|$ for TM incident wave at $\theta_0 = \pi/17$ and $h = 2\lambda$ as a function of material loss. (b) Enhancement factor and FOM for TE and TM polarizations as a function of ε_0'' and μ_0''. (*For a color version of this figure, see the color plate section.*)

MM performs with its associated losses. The relation for the FOM has been shown to be

$$\text{FOM} = \frac{|\text{Re}(n)|}{\text{Im}(n)} = \frac{(E+M)}{\left(EM - 1 + \sqrt{(1+E^2)(1+M^2)}\right)}. \quad (3.19)$$

A comparison of both the enhancement factor and the FOM in the material for each polarization is shown in Figure 3.10b. Given a particular value for $\varepsilon_0''(\mu_0'')$ at one point, the value of other can be found to lie on a vertical line drawn to the axis directly above (below) it. The blue lines show the enhancement factors for the TE (solid line) and TM (dashed line) polarizations, and as expected, the enhancement for the TE (TM) polarization is maximized when μ_0'' (ε_0'') is minimized. The FOM shown by the red curve has its peak value when both ε_0'' and μ_0'' are equal. In fact, the minima of the FOM curve correspond to the maximum enhancement factor for both the TE and TM case. While fabricated MMs are typically designed to have the highest FOM possible, the results in Figure 3.9b show that it is the enhancement factor that should be optimized instead of the FOM when transition MM structures are optimized.

Up to this point, we have considered the case of the ε and μ profiles of the transition layers such that the zeros of the real parts of each profile were at the same location (Fig. 3.11a). When MMs are fabricated, it is often difficult to ensure that both ε and μ overlap exactly, and in many cases the electric and magnetic resonances are attributed to different structures/components of the same unit cell. It is worthwhile, therefore, to investigate how the resonant enhancement changes when the zeros of each profile are separated by a certain distance, l (Fig. 3.11b). The material parameters' profiles can be expressed as [66]

$$\varepsilon(x) = \varepsilon_0 \left(\frac{x-l}{h}\right), \mu(x) = \mu_0 \left(\frac{x}{h}\right). \quad (3.20)$$

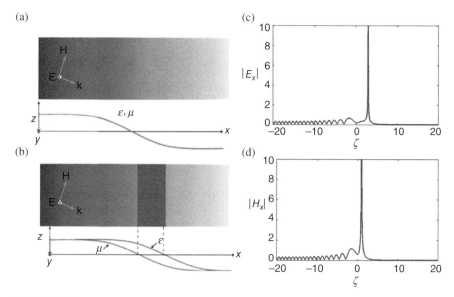

FIGURE 3.11 Transition layer showing index variation along x for profile with zeros of ε and μ (a) coinciding and (b) spatially separated, (c) absolute value of electric field component E_x for TM wave as functions of ζ for $l = 2$, (d) absolute value of magnetic field component H_x for TE wave as functions of ζ for $l = -4$. *(For a color version of this figure, see the color plate section.)*

The main steps of the analysis of wave propagation in a MM with spatially separated linearly graded profiles of ε and μ that are similar to those outlined above the locations of the turning points are now given by

$$\zeta_{1,2} = \frac{l}{2h} \mp \sqrt{\frac{l^2}{4h^2} + \sin^2 \theta_0}, \qquad (3.21)$$

where the distance between the zeros of the profiles l plays an important role in determining their exact position. In addition, as the distance between the zeros increases, the distance between the turning points does so as well.

Figures 3.11c and 3.11d show that the location of the peak of the longitudinal component of the magnetic field (for TE) and the electric field (for TM) is strongly polarization dependent. The results shown in Figures 3.11c and 3.11d can be easily realized in laboratory experiments by simply turning the polarizer in order to switch from TE to TM polarization.

As follows from Eq. (3.17), the change in Poynting vector ΔS_x is determined by the value of E_0, which stands for the electric field value at the point $\varepsilon, \mu = 0$ and by the angle of incidence θ. The value of E_0 depends on l and θ. Figure 3.12 shows the results of a parametric study of E_0 as a function of l and θ. In this study, the parameter l was varied from positive to negative values corresponding to the cases of different

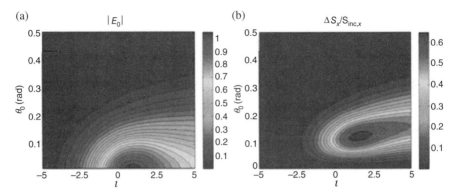

FIGURE 3.12 (a) $|E_0|$ and (b) $\Delta S_x/S_{0x}$ as functions of l and θ_0 for TE polarization. *(For a color version of this figure, see the color plate section.)*

positions of zeroes of ε and μ with respect to each other. Figure 3.12a shows a clear maximum point corresponding to a positive value l for a certain θ.

Finally, the amount of the energy dissipated ΔS_x was studied as function of two parameters: the separation between the zero points of ε and μ given by the parameter l and the angle of incidence θ. Figure 3.12b shows a contour plot of $\Delta S_x(l,\theta)/S_{0x}$, where S_{0x} is the amount of incident energy. It clearly shows that the amount of dissipated energy is increasing as l increases, assuming $l > 0$ and decreases as l decreases and becomes less than zero. Finally, it is noteworthy that an incident angle corresponding to the maximum absorption is independent of the value of the distance between $\varepsilon = 0$ and $\mu = 0$ points.

In summary, transition MMs may be of considerable interest for a variety of metamaterials-based applications. Depending on the application, the amount of resonant absorption may be minimized or maximized by changing the parameters of the transition layer and more generally the spatial profiles of the material parameters. For example, resonant absorption may affect the performance of a superlens with diffused boundaries or other structures based on, for example, doped semiconductors, and therefore needs to be minimized. On the other hand, applications such as "perfect" absorbers or nanodetectors may benefit from the resonant absorption effect that should be maximized in this case. Resonant absorption phenomenon near the zero index point could be of considerable importance in the context of transformation optics as it could significantly change spatial field distributions and lead to undesired absorption.

3.5 TRANSFORMATION OPTICS

A majority of earlier studies of MMs was focused on materials (or effective media) that were both homogeneous and made use of a negative index of refraction. Currently, however, it has become obvious that there are many fascinating potential

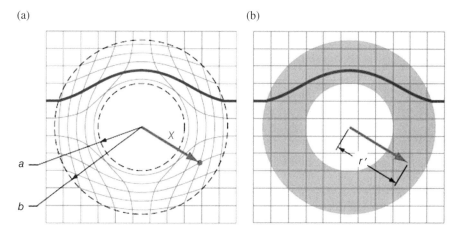

FIGURE 3.13 Transmission of a ray (blue line) through a medium under the (a) topological interpretation, and (b) material interpretation [72]. *(For a color version of this figure, see the color plate section.)*

applications of MMs that do not require a negative index of refraction. The emerging field of transformation optics (TO) enabling manipulation of light propagation by tailoring the profiles of ε and μ is one of the examples of a broader class of MMs. The basic idea of the TO approach is illustrated in Figure 3.13, which shows two equivalent approaches to guiding light around an object. In Figure 3.13a, the material parameters are unchanged, but it is the space itself that is "stretched" to force light flow in a direction shown by thick solid line. This is referred to as topological interpretation. However, since Maxwell's equations are form invariant, the same effect of light propagation can be achieved by changing the values of ε and μ as functions of coordinates but not altering the space (material interpretation). The equivalence of both interpretations is illustrated in Figures 3.13a and 3.13b [72]. It is the realization of this property of Maxwell's equations that led to the design of cloaking structures. These materials function as shells that cloak objects at their core by steering light around them, eliminating reflections off of their surface. In the first experimentally demonstrated cloaking structure [73], the radial index variation was achieved by making 10 concentric rings of split-ring resonator arrays and varying two of the geometrical dimensions of the SRRs at each radial distance. Since MMs operate as an effective medium and are dispersive if constructed with resonant elements, cloaking structures typically work within a narrow band centered near the resonance frequency of constituent unit cells. Despite these limitations, there have been a number of advancements (some featuring broader operating bands) in cloaking, including nonmagnetic cloaks, carpet, three dimensional (3D), and nonmetallic cloaking schemes [74–80].

Recently, various functionalities enabled by a TO approach have been proposed, including electromagnetic wave concentrators, lenses, graded-index waveguides and bends, black holes, and even illusion devices [81–90]. Here, we discuss in some detail wave concentrator functionality enabled by a combination of TO and nonlinear optics [91].

An electromagnetic (EM) concentrator based on the transformation optics approach was first reported by Rahm et al. [81,82], followed by a paper by Wang et al. [83] that proposed reduced material parameters for designing the EM concentrator that are easier to realize in practice. The purpose of this device is to focus incident EM waves, enhancing the EM energy density in a particular area. Naturally, this enhanced energy density can be exploited in a number of nonlinear devices. Recently, a number of such nonlinearly controlled devices have been proposed, including nonlinear cloaking devices and light-tunable reflection, shaping, and focusing of electromagnetic waves in MMs [92–94]. As an example let us consider the behavior of an EM concentrator with a core comprised of focusing Kerr nonlinearity.

A transformation optics approach can be used to design a linear graded-index outer shell of the device that enables channeling of the incident plane wave into the linear inner core region. Next, we replace linear core material with nonlinear medium. Figure 3.14 shows two possible configurations of the nonlinear concentrator.

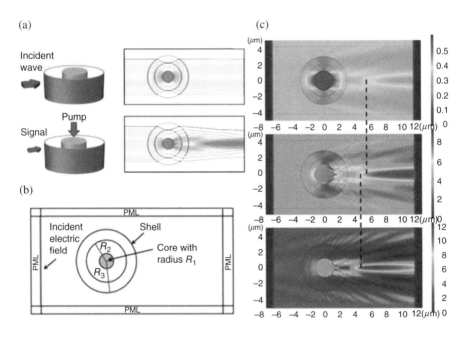

FIGURE 3.14 (a) Schematic of a wave concentrator based on self-action of single high intensity beam (top left), nonlinear concentrator for a weak signal propagating in nonlinear medium with refractive index modified by a strong pump (bottom left), and power flow in a linear (top right) and nonlinear concentrator based on self-action (bottom right). (b) Schematic of a lens. (c) Time averaged power flow illustrating the predicted lensing effect in the case of focusing nonlinearity. Focal line shifts toward the lens as incident field increases. *(For a color version of this figure, see the color plate section.)*

In the first implementation schematically shown in Figure 3.14a (top left), high intensity incident light modifies the refractive index of the core of the concentrator and modifies its own propagation through either self-focusing or self-defocusing. The second possible implementation, shown in Figure 3.14a (bottom left), utilizes a strong pump beam that propagates along the core of the device and modifies the refractive index of the core through the Kerr effect; a weak signal that propagates in the direction perpendicular to the core is focused or defocused as it crosses the core with a modified (increased or decreased) refractive index. The device configuration shown in Figure 3.14a (bottom left) can also be realized using other mechanisms of modulation of the refractive index of the core (e.g., electrical or thermal). A simple physical picture of performance of the device shown in Figure 3.14a (top left)—an axicon-like lensing with a variable focus line, can be gained from a simple ray picture of light refraction at the shell–core interface or by considering a power flow inside and outside of the structure, as shown in Figure 3.14a (right). Figure 3.14a (top right) shows the power flow in a linear concentrator. The basic principle of the device is based on a combination of transformation optics design of the outer shell of the cylindrical concentrator made of linear graded-index MM, and nonlinear changes in the refractive index of the inner region that results from nonlinear correction to the linear refractive index due to Kerr nonlinearity. As the intensity of an incoming field increases, the intensity concentrated in the inner region of the concentrator increases, resulting in changes of the refractive index in that region due to the nonlinear part of the refractive index that is proportional to the intensity. This change in refractive index, in turn, changes how light is refracted at the inner region boundary as it enters the inner region and as it leaves it; this leads to the changes in the position of the extended focal spot that moves from infinity (corresponding to the linear case) toward the concentrator (as incident intensity increases) as shown in Figure 3.14a (bottom right). In the case of defocusing Kerr nonlinearity, the refractive index of the core would decrease with increasing intensity of an incoming beam, leading to the beam defocusing/splitting effect.

A schematic diagram along with the simulation domain of the concentrator is as shown in Figure 3.14b. This transformation is mapped to material tensors as follows

$$\varepsilon_r^{i,j} = \mu_r^{i,j} = \begin{cases} \begin{bmatrix} 1 & 0 & 0 \\ 0 & 1 & 0 \\ 0 & 0 & (R_2/R_1)^2 \end{bmatrix} & 0 \leq r \leq R1 \\ \begin{bmatrix} (r+M)/r & 0 & 0 \\ 0 & r/(r+M) & 0 \\ 0 & 0 & N(r+M)/r \end{bmatrix} & R_1 < r \leq R_3 \end{cases} \quad (3.22)$$

where $M = (R_2 - R_1)R_3/(R_3 - R_2)$, $N = [(R_3 - R_2)/(R_3 - R_1)]^2$. The computational domain is terminated by perfectly matched layers (PMLs) to absorb the

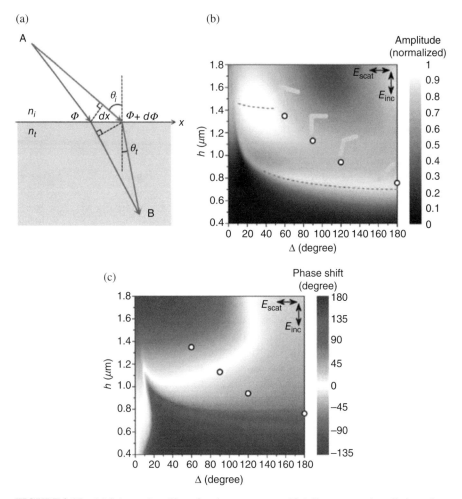

FIGURE 3.15 (a) Schematics of interface between two artificially structured media introducing abrupt phase shift in light path that varies as function of position along the interface. (b)–(c) Calculated amplitude and phase shift of the cross-polarized scattered light for V-antennas consisting of gold rods as functions of length h and angle between the rods Δ [96]. *(For a color version of this figure, see the color plate section.)*

scattered field. The structure is illuminated by a transverse electric (TE) polarized plane wave incident from the left having a wavelength of 780 nm. In summary, the above lens illustrates just one of the many novel functionalities enabled by a combination of transformation and nonlinear optical approaches. Such a lens has a variable focus, which is tunable based on the intensity of the incident electric field, by incorporating Kerr-type focusing nonlinearity in the core of the device. Potential applications for this device include the development of ultra-compact optical components for all optical circuits.

3.6 METASURFACES

In this last part of the chapter we discuss an entirely new direction of MM research focused on reduced dimensionality structures emerging as a promising practical approach to the development of ultra-compact on-chip optical components. Metasurfaces have an extensive range of potential applications in a wide frequency range from low microwave to optical frequencies, including controllable "smart" surfaces, ultra-compact cavity resonators, novel wave-guiding structures, electromagnetic wave absorbers, biomedical and microfluidic tunable devices, and terahertz switches [95–106].

At optical frequencies, this approach gave rise to the new field of "flat photonics" that utilizes subwavelength-thick metasurfaces to overcome current limitations of conventional 3D MM designs, such as significant losses, limited bandwidth, and difficulties of large-scale fabrication and integration. Metasurfaces enable unprecedented control of spatial, spectral, topological, and polarization properties of light by introducing abrupt phase shifts over the scale of the wavelength along the optical path. Such abrupt phase shifts can be introduced in the optical path by properly engineering the interface. This approach, the control of the wavefront no longer relies on the phase accumulated during the propagation of light, but is achieved via the phase shifts experienced by light as it scatters off the optically thin array of subwavelength-spaced nanoantennas. Linear gradients of phase discontinuities lead to planar reflected and refracted wavefronts, while nonlinear phase gradients lead to the formation of complex wavefronts, such as helical wavefronts of vortex beams.

For example, in one of the first studies in this area [96], it was shown that the introduction of an abrupt phase discontinuity at the interface between two media with a constant phase gradient along the interface, the generalized Snell's law can be written as follows:

$$\sin(\theta_t)n_t - \sin(\theta_i)n_i = k_0^{-1}\frac{d\Phi}{dx}, \qquad (3.23)$$

where θ_t is the angle of refraction; Φ and $\Phi + d\Phi$ are the phase discontinuities at the locations where the two paths cross the interface; dx is the distance between the crossing points; n_i and n_t are the refractive indices of the two media; and $k_0 = 2\pi/\lambda_0$, where λ_0 is the vacuum wavelength (see Fig. 3.15).

One of the most impressive examples of an entirely new range of opportunities for light manipulation opened by interfacial phase discontinuities is shown in Figure 3.16. Here, a plasmonic interface was designed and fabricated to produce a vortex beam when illuminated by normally incident linearly polarized light. A vortex beam has a spiral equal-phase wave front and carries an orbital angular momentum, or a topological charge that is the number of turns of the wavefront within one wavelength [96].

Figure 3.16c shows that the vortex beam has an annular intensity distribution in the cross-section, where the dark region at the center corresponds to a phase singularity [96]. Figure 3.16d demonstrates the result of interfering the vortex beam with a

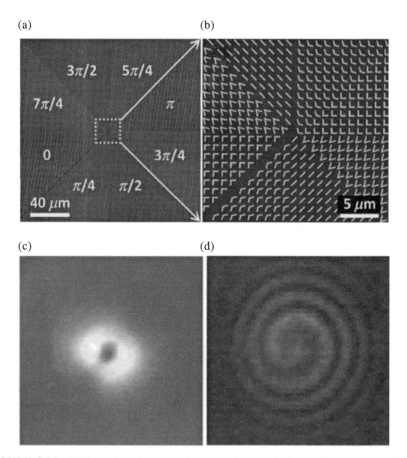

FIGURE 3.16 (a) Scanning electron microscope image of plasmonic pattern consisting of eight regions, arranged so as to generate phase shift that varies azimuthally from 0 to 2π. (b) Zoom-in view of the center part of (a). (c) Measured far-field intensity distribution of optical vortex generated by the pattern in (a). (d) Measured spiral patterns created by interference of vortex beam and co-propagating Gaussian beam [96].

co-propagating Gaussian beam, producing a spiral interference pattern that proves the presence of the helical wavefront. Vortices are usually generated by employing a spiral phase plate or a computer-generated hologram [107–109]. However, future ultra-compact optoelectronic signal processing systems necessitate the development of new ways of generating and controlling such complex light beams using chip-scale components. The first step toward manipulating light on a chip was a recent demonstration of metasurfaces operating at visible and near-infrared wavelengths shown in Figure 3.17 [98].

Metasurfaces is a rapidly developing sub-field of metamaterials research with a broad range of potential applications. Recently, a graded-index metasurface able to convert propagating waves to surface waves with nearly 100% efficiency was

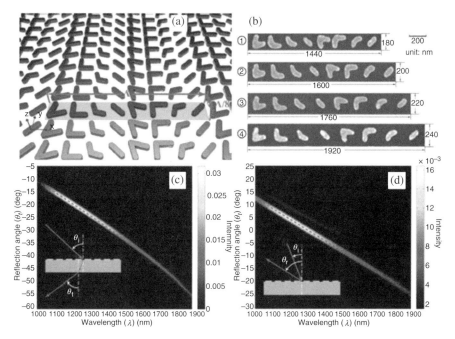

FIGURE 3.17 (a) A schematic view of an optical antenna array with the unit cell shown in blue. (b) Scanning electron microscope images of the unit cells of four antenna arrays with different periodicities. (c) and (d) Measured relative intensity profiles for the cases of the refraction (c) and reflection (d) of the cross-polarized light [98].

proposed and demonstrated [104]. As opposed to conventional prism or grating couplers [110–113], the momentum mismatch between propagating and surface waves was compensated by the reflection-phase gradient of the metasurface. These structures open entirely new design opportunities for efficient surface plasmon couplers, antireflection surfaces, and light absorbers.

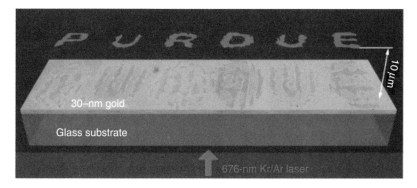

FIGURE 3.18 A phase holography image obtained with a plasmonic metasurface [106].

By now, optical metasurfaces have been used to implement such functionalities as light bending, flat lenses, half-wave plates, and quarter-wave plates. Recently, the first experimental demonstrations of a phase holography image obtained with a plasmonic metasurface were reported [105, 106]. A holographic image of the word "PURDUE" produced by such a metasurface is shown in Figure 3.18 [106]. The sample was illuminated by a Kr/Ar laser at 676 nm from the glass substrate side. The images were obtained at a plane 10 μm above the metasurface.

In summary, the initial studies of metasurfaces briefly reviewed in this section show a significant potential for ultrathin plasmonic metasurfaces to control and manipulate amplitude, phase, and polarization properties of light potentially enabling a plethora of ultra-compact devices and novel functionalities on a subwavelength scale.

REFERENCES

[1] J. B. Pendry and D. R. Smith, "Metamaterials and negative refractive index," *Science* **305**, 788–792 (2004).

[2] V. M. Shalaev, "Optical negative-index metamaterials," *Nat. Photon.* **1**, 41–48 (2007).

[3] V. M. Shalaev, "Transforming light," *Science* **322**, 384–386 (2008).

[4] N. M. Litchinitser, I. R. Gabitov A. I. Maimistov, and V. M. Shalaev, *Negative Refractive Index Metamaterials in Optics, for Progress in Optics*, edited by E. Wolf, Vol. 51, Chap. 1, Elsevier, pp. 1–68 (2008).

[5] C. M. Soukoulis and M. Wegener, "Past achievements and future challenges in the development of three-dimensional photonic metamaterials," *Nat. Photon.* **5**, 523–530 (2011).

[6] C. Enkrich, M. Wegener, S. Linden, S. Burger, L. Zschiedrich, F. Schmidt, J. Zhou, T. Koschny, and C. M. Soukoulis, "Magnetic metamaterials at telecommunication and visible frequencies," *Phys. Rev. Lett.* **95**, 203901 (2005).

[7] G. Dolling, C. Enkrich, M. Wegener, J. Zhou, C. M. Soukoulis, and S. Linden, "Cut-wire pairs and plate pairs as magnetic atoms for optical metamaterials," *Opt. Lett.* **30**, 3198–3200 (2005).

[8] N. Feth, C. Enkrich, M. Wegener, and S. Linden, "Large-area magnetic metamaterials via compact interference lithography," *Opt. Express* **15**, 501–507 (2007).

[9] W. Cai, U. K. Chettiar, H.-K. Yuan, V. C. de Silva, A. V. Kildishev, V. P. Drachev, and V. M. Shalaev, "Metamagnetics with rainbow colors," *Opt. Express* **15**, 3333–3341 (2007).

[10] R. A. Depine, M. L. Martínez-Riccia, J. A. Monsoriub, E. Silvestrec, and P. Andrésc, "Zero permeability and zero permittivity band gaps in 1D metamaterial photonic crystals," *Phys. Lett. A* **364**, 352–355 (2007).

[11] V. G. Veselago, "The electrodynamics of substances with simultaneously negative values of ε and μ," *Soviet Physics Uspekhi* **10** (4), 509–514 (1968).

[12] R. A. Shelby, D. R. Smith, and S. Shultz, "Experimental verification of a negative index of refraction," *Science* **292** (5514), 77–79 (2001).

[13] V. M. Shalaev, W. Cai, U. K. Chettiar, H.-K. Yuan, A. K. Sarychev, V. P. Drachev, and A. V. Kildishev, "Negative index of refraction in optical metamaterials," *Opt. Lett.* **30**, 3356–3358 (2005).

[14] S. Zhang, W. Fan, N. C. Panoiu, K. J. Malloy, R. M. Osgood, and S. R. Brueck, "Experimental demonstration of near-infrared negative-index metamaterials," *Phys. Rev. Lett.* **95**, 137404 (2005).

[15] R. A. Depine and A. Lakhtakia, "A new condition to identify isotropic dielectric-magnetic materials displaying negative phase velocity," *Microwave Opt. Tech. Lett.* **41**, 315–316 (2004).

[16] G. Dolling, M. Wegener, C. M. Soukoulis, and S. Linden, "Negative-index metamaterial at 780 nm wavelength," *Opt. Lett.* **32**, 53–55 (2007).

[17] J. B. Pendry, "Negative refraction makes a perfect lens," *Phys. Rev. Lett.* **85**, 3966–3969 (2000).

[18] Z. Jacob, L. V. Alekseyev, and E. Narimanov, "Optical hyperlens: far-field imaging beyond the diffraction limit," *Opt. Express* **14**, 8247–8256 (2006).

[19] A. Salandrino and N. Engheta, "Far-field subdiffraction optical microscopy using metamaterial crystals: theory and simulations," *Phy. Rev. B* **74**, 075103 (2006).

[20] N. M. Litchinitser and V. M. Shalaev, "Negative refraction," in *The McGraw-Hill 2008 Yearbook of Science & Technology*, McGraw-Hill Professional, pp. 230–233 (2008).

[21] Z. Liu, H. Lee, Y. Xiong, C. Sun, and X. Zhang, "Optical hyperlens magnifying sub-diffraction-limited objects," *Science* **315**, 1686 (2007).

[22] R. Boyd, *Nonlinear Optics*, 3rd ed. (Academic Press, 2008).

[23] Y.-R. Shen, *The Principles of Nonlinear Optics* (Wiley-Interscience, 2002).

[24] N. Mattiucci, G. D'Aguanno, M.J. Bloemer, and M. Scalora, "Second-harmonic generation from a positive-negative index material heterostructure," *Phys. Rev. E* **72**, 066612 (2005).

[25] I. V. Shadrivov, A. A. Zharov, and Y. S. Kivshar, "Second-harmonic generation in nonlinear left-handed metamaterials," *J. Opt. Soc. Am. B* **23**, 529–534 (2006).

[26] A. K. Popov, V. V. Slabko, and V. M. Shalaev, "Second harmonic generation in left-handed metamaterials," *Laser Phys. Lett.* **3**, 293–297 (2006).

[27] A. K. Popov and V. M. Shalaev, "Negative-index metamaterials: second-harmonic generation, Manley–Rowe relations and parametric amplification," *Appl. Phys. B* **84**, 131–137 (2006).

[28] A. K. Popov and V. M. Shalaev, "Compensating losses in negative-index metamaterials by optical parametric amplification," *Opt. Lett.* **31**, 2169 (2006).

[29] M. Scalora, G. D'Aguanno, M. Bloemer, M. Centini, D. de Ceglia, N. Mattiucci, and Y. S. Kivshar, "Dynamics of short pulses and phase matched second harmonic generation in negative index materials," *Opt. Express* **14**, 4746–4756 (2006).

[30] G. D'Aguanno, N. Mattiucci, M. Scalora, and M. J. Bloemer, "Second-harmonic generation at angular incidence in a negative-positive index photonic band-gap structure," *Phys. Rev. E* **74**, 026608 (2006).

[31] I. V. Shadrivov, A. A. Zharov, and Y. S. Kivshar, "Second harmonic generation in nonlinear left-handed materials," *J. Opt. Soc. Am. B* **23**, 529 (2006).

[32] M. V. Gorkunov, I. V. Shradivov, and Y. S. Kivshar, "Enhanced parametric processes in binary metamaterials," *Appl. Phys. Lett.* **88**, 071912 (2006).

[33] R. Hegde and H. Winful, "Optical bistability in periodic nonlinear structures containing left handed materials," *Microw. Opt. Technol. Lett.* **46**, 6 (2006).

[34] N. M. Litchinitser, I. R. Gabitov, and A. I. Maimstov, "Optical bistability in a nonlinear optical coupler with a negative index channel," *Phys. Rev. Lett.* **99**, 113902 (2007).

[35] M. Scalora, D. de Ceglia, G. D'Aguanno, N. Mattiucci, N. Akozbek, M. Centini, and M. J. Bloemer, "Gap solitons in a nonlinear quadratic negative-index cavity," *Phys. Rev. E* **75**, 066606 (2007).

[36] V. Roppo, M. Centini, C. Sibilia, M. Bertolotti, D. de Ceglia, M. Scalora, N.t Akozbek, M. J. Bloemer, J. W. Haus, O. G. Kosareva et al., "Role of phase matching in pulsed second-harmonic generation: walk-off and phase-locked twin pulses in negative-index media," *Phys. Rev. A* **76**, 033829(2007).

[37] A. K. Popov, S. A. Myslivets, T. F. George, and V. M. Shalaev, "Four-wave mixing, quantum control, and compensating losses in doped negative-index photonic metamaterials," *Opt. Lett.* **32**, 3044 (2007).

[38] D. de Ceglia, A. D'Orazio, M. De Sario, V. Petruzzelli, F. Prudenzano, M. Centini, M. G. Cappeddu, M. J. Bloemer, and M. Scalora, "Enhancement and inhibition of second-harmonic generation and absorption in a negative index cavity," *Opt. Lett.* **32**, 265–267 (2007).

[39] S. Gao and S. He, "Four-wave mixing in left-handed materials," *J. Nonlinear Optic. Phys. Mat.* **16**, 485 (2007).

[40] A. I. Maimistov and I. R. Gabitov, "Nonlinear optical effects in artificial materials," *Eur. Phys. J. Spec. Topics* **147**, 265–286(2007).

[41] I. V. Shadrivov, A. B. Kozyrev, D. W. van der Weide, and Y. S. Kivshar, "Nonlinear magnetic metamaterials," *Opt. Express* **16**, 20266(2008).

[42] V. Roppo, M. Centini, D. de Ceglia, M. A. Vicenti, J. W. Haus, N. Akozbek, M. J. Bloemer, and M. Scalora, "Anomalous momentum states, non-specular reflections, and negative refraction of phase-locked, second-harmonic pulses," *Metamaterials'2007 Congress* **2**, 135–144 (2008).

[43] E. Centeno and C. Ciracì "Theory of backward second-harmonic localization in nonlinear left-handed media," *Phys. Rev. B* **78**, 235101 (2008).

[44] A.V. Kildishev, Y. Sivan, N.M. Litchinitser, and V.M. Shalaev, "Frequency-domain modeling of TM wave propagation in optical nanostructures with a third-order nonlinear response," *Opt. Lett.* **34**, 3364 (2009).

[45] E. Poutrina, D. Huang, and D. R. Smith, "Analysis of nonlinear electromagnetic metamaterials," *New J. Phys.* **12**, 093010 (2010).

[46] A.V. Kildishev, "Modeling nonlinear effects in 2D optical metamagnetics," *Metamaterials* **4**, 77 (2010).

[47] A.V. Kildishev and N.M. Litchinitser, "Efficient simulation of non-linear effects in 2D optical nanostructures to TM waves," *Opt. Commun.* **283**, 1628 (2010).

[48] A. D. Boardman, R. C. Mitchell-Thomas, N. King, and Yu. Rapoport, "Bright spatial solitons in metamaterials," *Opt. Commun.* **283**, 1585 (2010).

[49] V. Roppo, C. Ciracì, C. Cojocaru, and M. Scalora, "Second harmonic generation in a generic negative index medium," *J. Opt. Soc. Am. B* **27**, 1671–1679 (2010).

[50] A. Rose, D. Huang, and D. R. Smith, "Controlling the second harmonic in a phase-matched negative-index metamaterial bulk NLMMs," *Phys. Rev. Lett.* **107**, 063902 (2011).

[51] Z. Kudyshev, I. R. Gabitov, and A. I. Maimistov, "The effect of phase mismatch on second harmonic generation in negative index materials," arXiv:1102.0538 (2011).

[52] I. V. Shadrivov, P. V. Kapitanova, S. I. Maslovski, and Y. S. Kivshar, "Metamaterials controlled with light," *Phys. Rev. Lett.* **109**, 083902 (2012).

[53] G. Venugopal, Zh. Kudyshev, and N. Litchinitser, "Asymmetric positive-negative index nonlinear waveguide couplers," *IEEE J. Sel. Top. Quantum Electron.* **18**, 2 (2012).

[54] Zh. Kudyshev, G. Venugopal, and N. M. Litchinitser, "Generalized analytical solutions for nonlinear positive-negative index couplers," *Phys. Res. Internat.*, Article ID 945807, 4 pages (2012).

[55] X. Wang, G. Venugopal, J. Zeng, Y. Chen, D. Ho Lee, N. M. Litchinitser, and A. N. Cartwright, "Optical fiber metamagnetics," *Opt. Express* **19**, 19813–19821 (2011).

[56] G. P. Agrawal, *Fiber Optics Communications*, 3rd ed. (Wiley-Interscience, 2002).

[57] E. Udd, *Fiber Optic Sensors: An Introduction for Engineers and Scientists*, 1st ed. (Wiley-Interscience, 1991).

[58] L. Novotny and B. Hecht, *Principles of Nano-Optics* (Cambridge University Press, 2006).

[59] T. Thio, H. J. Lezec, and T. W. Ebbesen, "Strongly enhanced optical transmission through subwavelength holes in metal films," *Physica B* **279**(1–3), 90–93 (2000).

[60] E. Cubukcu, N. Yu, E. J. Smythe, L. Diehl, K. B. Crozier, and F. Capasso, "Plasmonic laser antennas and related devices," *IEEE J. Sel. Top. Quantum Electron.* **14**(6), 1448–1461 (2008).

[61] N. M. Litchinitser, A. I. Maimistov, I. R. Gabitov, R. Z. Sagdeev, and V. M. Shalaev, "Metamaterials: electromagnetic enhancement at zero-index transition," *Opt. Lett.* **33**(20), 2350–2352 (2008).

[62] K. Kim, D.-H. Lee, and H. Lim, "Resonant absorption and mode conversion in a transition layer between positive index and negative-index media," *Opt. Express* **16**, 18505–18513 (2008).

[63] M. Dalarsson and P. Tassin, "Analytical solution for wave propagation through a graded index interface between a right-handed and a left-handed material," *Opt. Express* **17**, 6747–6752 (2009).

[64] I. Mozjerin, E. A. Gibson, E. P. Furlani, I. R. Gabitov, and N. M. Litchinitser, "Electromagnetic enhancement in lossy optical transition metamaterials," *Opt. Lett.* **35**, 3240–3242 (2010).

[65] E. A. Gibson, M. Pennybacker, A. I. Maimistov, I. R. Gabitov, and N. M. Litchinitser, "Resonant absorption in transition metamaterials: parametric study," *J. Opt.* **13**(5), 024013 (2011).

[66] E. A. Gibson, I. R. Gabitov, A. I. Maimistov, and N. M. Litchinitser, "Transition metamaterials with spatially separated zeros," *Opt. Lett.* **36**, 3624–3626 (2011).

[67] F. Alali and N. M. Litchinitser, "Gaussian beams in near-zero transition metamaterials," *Opt. Commun.* **291**, 179–183 (2013).

[68] V. L. Ginzburg, *The Propagation of Electromagnetic Waves in Plasma* (Pergamon, 1970).

[69] L. D. Landau, E. M. Lifshitz, and L. P. Pitaevskii, *Electrodynamics of Continuous Media*, 2nd ed. (Butterworth-Heinemann, 1984), Vol. **8**.

[70] A. O. Korotkevich, A. C. Newell, and V. E. Zakharov, "Communication through plasma sheaths," *J. Appl. Phys.* **102**, 083305 (2007).

[71] G. Dolling, C. Enkrich, M. Wegener, C. M. Soukoulis, and S. Linden, "Low-loss negative-index metamaterial at telecommunication wavelengths," *Opt. Lett.* **31**, 1800–1802 (2006).

[72] J. B. Pendry, D. Schurig, and D. R. Smith, "Controlling electromagnetic fields," *Science* **312**, 1780–1782 (2006).

[73] D. Schurig, J. J. Mock, B. J. Justice, S. A. Cummer, J. B. Pendry, A. F. Starr, and D. R. Smith, "Metamaterial electromagnetic cloak at microwave frequencies," *Science* **314**, 977–980 (2006).

[74] W. Cai, U. K. Chettiar, A. V. Kildishev, and V. M. Shalaev, "Optical cloaking with metamaterials," *Nat. Photon.* **1**, 224 (2007).

[75] A. V. Kildishev, W. Cai, U. K. Chettiar, and V. M. Shalaev, "Transformation optics: approaching broadband electromagnetic cloaking," *New J. Phys.* **10**, 115029 (2008).

[76] J. Li and J. B. Pendry, "Hiding under the carpet: a new strategy for cloaking," *Phys. Rev. Lett.* **101**, 203901 (2008).

[77] R. Liu, C. Ji, J. J. Mock, J. Y. Chin, T. J. Cui, and D. R. Smith, "Broadband ground-plane cloak," *Science* **323**, 366–369 (2009).

[78] J. Valentine, J. Li, T. Zentgraf, G. Bartal, and X. Zhang, "An optical cloak made of dielectrics," *Nat. Mater.* **8**, 568–571 (2009).

[79] T. Ergin, N. Stenger, P. Brenner, J. B. Pendry, and M. Wegener, "Three-Dimensional invisibility cloak at optical wavelengths," *Science* **328**, 337–339 (2010).

[80] L. H. Gabrielli, J. Cardenas, C. B. Poitras, and M. Lipson, "Silicon nanostructure cloak operating at optical frequencies," *Nat. Photon.* **3**, 461–463 (2009).

[81] M. Rahm, D. Schurig, D. A. Roberts, S. A. Cummer, D. R. Smith, and J. B. Pendry, "Design of electromagnetic cloaks and concentrators using form-invariant coordinate transformations of Maxwell's equations," *Photon. Nanostruct.: Fundam. Applic.* **6**, 87–95 (2008).

[82] A. V. Kildishev and V. M. Shalaev, "Engineering space for light via transformation optics," *Opt. Lett.* **33**, 43 (2008).

[83] W. Wang, L. Lin, J. Ma, C. Wang, J. Cui, C. Du, and X. Luo, "Electromagnetic concentrators with reduced material parameters based on coordinate transformation," *Opt. Express* **16**, 11431–11437 (2008).

[84] N. I. Landy, S. Sajuyigbe, J. J. Mock, D. R. Smith, and W. J. Padilla, "Perfect metamaterial absorber," *Phys. Rev. Lett.* **100**, 207402 (2008).

[85] E. E. Narimanov and A. V. Kildishev, "Optical black hole: broadband omnidirectional light absorber," *Appl. Phys. Lett.* **95**, 041106 (2009).

[86] A. V. Kildishev, L. J. Prokopeva, and E. E. Narimanov, "Cylinder light concentrator and absorber: theoretical description," *Opt. Express* **18**, 16646–16662 (2010).

[87] U. Leonhardt and T. G. Philbin, "Transformation optics and the geometry of light, progress in optics," *Elsevier* **53**, 69–152 (2009).

[88] N. M. Litchinitser and V. M. Shalaev, "Metamaterials: transforming theory into reality," *J. Opt. Soc. Am. B* **26**, B161–B169 (2009).

[89] E. E. Narimanov and A. V. Kildishev, "Optical black hole: broadband omnidirectional light absorber," *Appl. Phys. Lett.* **95**(3), 041106 (2009).

[90] Y. Lai, H. Y. Chen, D. Z. Han, J. J. Xiao, Z.-Q. Zhang, and C. T. Chan, "Illusion optics: the optical transformation of an object into another object," *Phys. Rev. Lett.* **102**, 253902–253905 (2009).

[91] A. Pandey and N. M. Litchinitser, "Nonlinear light concentrators," *Opt. Lett.* **37**, 5238–5240 (2012).

[92] N. A. Zharova, I. V. Shadrivov, A. A. Zharov, and Y. S. Kivshar, "Nonlinear control of invisibility cloaking," *Opt. Express* **20**, 14954–14959 (2012).

[93] I. V. Shadrivov, P. V. Kapitanova, S. I. Maslovski, and Y. S. Kivshar, "Metamaterials controlled with light," *Phys. Rev. Lett.* **109**, 083902(4) (2012).

[94] H. Odabasi, F. L. Teixeira, and W. C. Chew, "Impedance-matched absorbers and optical pseudo black holes," *J. Opt. Soc. Am. B* **28**, 1317–1323 (2011).

[95] C. L. Holloway, E. F. Kuester, J. A. Gordon, J. O'Hara, J. Booth, and D. R. Smith, "An overview of the theory and applications of metasurfaces: the two-dimensional equivalents of metamaterials, antennas and propagation magazine," *IEEE* **54**, 10–35 (2012).

[96] N. Yu, P. Genevet, M. A. Kats, F. Aieta, J. P. Tetienne, F. Capasso, and Z. Gaburro, "Light propagation with phase discontinuities: generalized laws of reflection and refraction," *Science* **334**, 333–337 (2011).

[97] P. Genevet, N. Yu, F. Aieta, J. Lin, M. A. Kats, R. Blanchard, M. O. Scully, Z. Gaburro, and F. Capasso, "Ultra-thin plasmonic optical vortex plate based on phase discontinuities," *Appl. Phys. Lett.* **100**, 013101 (2012).

[98] X. Ni, N. K. Emani, A. V. Kildishev, A. Boltasseva, and V. M. Shalaev, "Broadband light bending with plasmonic nanoantennas," *Science* **335**, 427 (2012).

[99] F. Aieta, P. Genevet, M. A. Kats, N. Yu, R. Blanchard, Z. Gaburro, and F. Capasso, "Aberration-free ultra-thin flat lenses and axicons at telecom wavelengths based on plasmonic metasurfaces," *Nano Lett.* **12**, 4932–4936 (2012).

[100] J. Hao, Y. Yuan, L. Ran, T. Jiang, J. A. Kong, C. Chan, and L. Zhou, "Manipulating electromagnetic wave polarizations by anisotropic metamaterials," *Phys. Rev. Lett.* **99**, 63908 (2007).

[101] Z. Zhu, C. Guo, K. Liu, W. Ye, X. Yuan, B. Yang, and T. Ma, "Metallic nanofilm half-wave plate based on magnetic plasmon resonance," *Opt. Lett.* **37**, 698–700 (2012).

[102] A. Drezet, C. Genet, and T. W. Ebbesen, "Miniature plasmonic wave plates," *Phys. Rev. Lett.* **101**, 43902 (2008).

[103] Y. Zhao and A. Alù, "Manipulating light polarization with ultrathin plasmonic metasurfaces," *Phys. Rev. B* **84**, 205428 (2011).

[104] S. Sun, Q. He, S. Xiao, Q. Xu, X. Li, and L. Zhou, "Gradient-index meta-surfaces as a bridge linking propagating waves and surface waves," *Nat. Mater.* **11**, 426–431(2012).

[105] F. Zhou, Y. Liu, and W. Cai, "Plasmonic holographic imaging with V-shaped nanoantenna array," *Opt. Express* **21**, 4348–4354 (2013).

[106] A. V. Kildishev, X. Ni, A. Emadeldin, and V. Shalaev, "Plasmonic metasurfaces for phase holography: bianisotropic characterization," in The 43nd Winter Colloquium on the Physics of Quantum Electronics, January 6–10, 2013 – Snowbird, Utah, USA.

[107] M. W. Beijersbergen, R. P. C. Coerwinkel, M. Kristensen, and J. P. Woerdman, "Helical-wavefront laser beams produced with a spiral phaseplate," *Opt. Commun.* **112**, 321 (1994).

[108] N. R. Heckenberg, R. McDuff, C. P. Smith, and A. G. White, "Generation of optical phase singularities by computer-generated holograms," *Opt. Lett.* **17**(3), 221–233 (1992).

[109] V. Yu. Bazhenov, M. V. Vasnetsov, and M. S. Soskin, "Laser beams with screw dislocations in their wavefronts," *JETP Lett.* **52**, 429 (1990).

[110] E. Kretschmann and H. Raether, "Radiative decay of nonradiative surface plasmons excited by light," *Z. Naturforsch.* **23A**, 2135–2136 (1968).

[111] H. Raether, *Surface Plasmons on Smooth and Rough Surfaces and on Gratings* (Springer, 1988).

[112] M. Neviere, R. Petit, and M. Cadilhac, "About the theory of optical grating coupler-waveguide systems," *Opt. Commun.* **8**, 113–117 (1973).

[113] Y. B. Tang, Z. C. Wang, L. Wosinski, U. Westergren, and S. L. He, "Highly efficient nonuniform grating coupler for silicon-on-insulator nanophotonic circuits," *Opt. Lett.* **35**, 1290–1292 (2010).

4

QUANTUM NANOPLASMONICS

MARK I. STOCKMAN

Ludwig Maximilian University, Munich, Germany
Max Plank Institute for Quantum Optics, Garching at Munich, Germany
Georgia State University, Atlanta, GA, USA

4.1 INTRODUCTION

Nanoplasmonics is a branch of optical condensed matter science devoted to optical phenomena on the nanoscale in nanostructured metal systems; see, for example, References 1 and 2. A remarkable property of such systems is their ability to keep the optical energy concentrated on the nanoscale due to modes called surface plasmons (SPs). It is well known [3] that the existence of SPs depends entirely on the property that dielectric function ε_m has a negative real part, $\operatorname{Re}\varepsilon_m < 0$. The SPs are well pronounced as resonances when the losses are small enough, that is, $\operatorname{Im}\varepsilon_m \ll -\operatorname{Re}\varepsilon_m$. This is a known property of a good plasmonic metal, valid, for example, for silver in the most of the visible region. We will call a substance a good plasmonic metal if these two properties

$$\operatorname{Re}\varepsilon_m < 0, \quad \operatorname{Im}\varepsilon_m \ll -\operatorname{Re}\varepsilon_m \qquad (4.1)$$

are satisfied simultaneously.

Not just a promise anymore [4], nanoplasmonics has delivered a number of important applications: ultrasensing [5], scanning near-field optical microscopy [1,6], SP-enhanced photodetectors [7], thermally assisted magnetic recording [8], generation of extreme uv [9], biomedical tests [5,10], SP-assisted thermal cancer treatment [11], plasmonic enhanced generation of extreme ultraviolet (EUV) pulses [12] and extreme

---

Photonics: Scientific Foundations, Technology and Applications, Volume II, First Edition.
Edited by David L. Andrews.
© 2015 John Wiley & Sons, Inc. Published 2015 by John Wiley & Sons, Inc.

ultraviolet to soft x-ray (XUV) pulses [9], and many others—see also Reference 13. While the dielectric permittivity, ε_m, can only be described quantum mechanically, plasmonics is conventionally formulated in terms of classical optical fields. There are different ways to define what is quantum plasmonics: either the underlying metal is described quantum mechanically, usually in the framework of time-dependent density functional theory [14, 15], or plasmons are considered as quantum fields [16, 17]. We note that, traditionally, quantum mechanical description of matter and materials is referred to as quantum chemistry. Hence, we will reserve the term "quantum plasmonics" for theories based on quantized plasmonic fields.

There are two developed areas of quantum plasmonics where phenomena can be fully understood only on the basis of quantized plasmonic fields. These are surface plasmon amplification by stimulated emission of radiation (spaser and gain in nanoplasmonics) described in Section 4.2 and plasmonic generation of hot electrons presented in Section 4.3 in conjunction with nanoscopy.

4.2 SPASER AND NANOPLASMONICS WITH GAIN

4.2.1 Introduction to Spasers and Spasing

To continue its vigorous development, nanoplasmonics needs an active device—near-field generator and amplifier of nanolocalized optical fields, which has until recently been absent. A nanoscale amplifier in microelectronics is the metal-oxide-semiconductor field effect transistor (MOSFET) [18, 19], which has enabled all contemporary digital electronics, including computers and communications, and enabled the present-day technology as we know it. However, the MOSFET is limited by frequency and bandwidth to $\lesssim 100$ GHz, which is already a limiting factor in further technological development. Another limitation of the MOSFET is its high sensitivity to temperature, electric fields, and ionizing radiation, which limits its use in extreme environmental conditions, nuclear technology and warfare.

An active element of nanoplasmonics is the spaser (surface plasmon amplification by stimulated emission of radiation), which was proposed [16, 20] as a nanoscale quantum generator of nanolocalized coherent and intense optical fields.

Spaser is a nanoplasmonic counterpart of laser: it is a quantum generator and nanoamplifier where photons as the generated quanta are replaced by SPs. Spaser consists of a metal nanoparticle, which plays the role of the laser cavity (resonator), and the gain medium. Figure 4.1 schematically illustrates geometry of a spaser introduced in the original article [16], which contains a V-shaped metal nanoparticle surrounded by a layer of semiconductor nanocrystal quantum dots.

The idea of spaser has been further developed theoretically [17, 21–29]. Nanospasers have been observed experimentally [30–32]. Also a number of surface plasmon polariton (SPP) spasers (also called nanolasers) have been experimentally observed [33–40].

One of the most significant ideas generalizing the spaser was the concept of lasing spaser, which has been introduced as a nanofilm that is a periodic plasmonic nanostrucrure (a two-dimensional (2D) plasmonic crystal) containing gain components. The

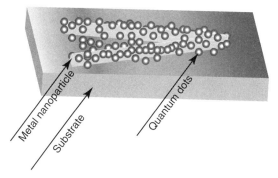

FIGURE 4.1 A schematic of the spaser as originally proposed in Reference 16. The resonator of the spaser is a metal nanoparticle shown as a golden V-shape. It is covered by the gain medium depicted as nanocrystal quantum dots. This active medium is supported by a neutral substrate.

individual metal unit cells of this plasmonic crystal are spasers that work in unison (with equal phases) generating a light wave in the direction normal to the surface. The lasing spaser has originally been introduced theoretically [28, 41, 42]. Recently, lasing spasers have been experimentally demonstrated, which consist of a 2D plasmonic crystal nanofilm and adjacent gain medium [43, 44]; see also Reference 45.

4.2.2 Spaser Fundamentals

As we have already mentioned, the spaser is a nanoplasmonic counterpart of the laser [16, 21]. The laser has two principal elements: resonator (or cavity) that supports photonic mode(s) and the gain (or active) medium that is population-inverted and supplies energy to the lasing mode(s). An inherent limitation of the laser is that the size of the laser cavity in the propagation direction is at least half the wavelength and practically more than that even for the smallest lasers developed [33, 34, 46]. In the spaser [16], this limitation is overcome. The spasing modes are surface plasmons (SPs) whose localization length is on the nanoscale [47] and is only limited by the minimum inhomogeneity scale of the plasmonic metal and the nonlocality radius [48] $l_{nl} \sim 1$ nm. So, the spaser is truly nanoscopic—its minimum total size can be just a few nanometers.

The resonator of a nanospaser can be any plasmonic metal nanoparticle whose total size R is much less than the wavelength λ and whose metal thickness is between l_{nl} and skin depth $l_s \approx 25$nm, which supports an SP mode with required frequency ω_n. This metal nanoparticle should be surrounded by the gain medium that overlaps with the spasing SP eigenmode spatially and whose emission line overlaps with this eigenmode spectrally [16]. As an example, we consider a model of a nanoshell spaser [17, 21, 49], which is illustrated in Figure 4.2. Panel (a) shows a silver nanoshell carrying a single SP (plasmon population number $N_n = 1$) in the dipole eigenmode. It is characterized by a uniform field inside the core and hot spots at the poles outside the shell with the maximum field reaching $\sim 10^6$ V cm^{-1}. Similarly, Figure 4.2b shows the quadrupole mode in the same nanoshell. In this case, the mode electric field is

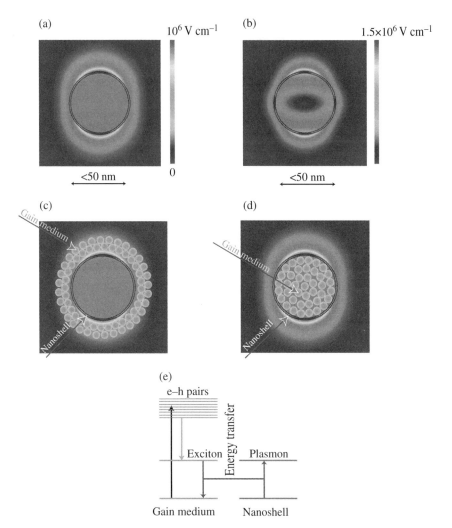

FIGURE 4.2 Schematic of spaser geometry, local fields, and fundamental processes leading to spasing. Adapted from Reference 17. (a) Nanoshell geometry and the local optical field distribution for one SP in an axially symmetric dipole mode. The nanoshell has the aspect ratio $\eta = 0.95$. The local field magnitude is color-coded by the scale bar on the right-hand side of the panel. (b) The same as (a) but for a quadrupole mode. (c) Schematic of a nanoshell spaser where the gain medium is outside of the shell, on the background of the dipole-mode field. (d) The same as (c) but for the gain medium inside the shell. (e) Schematic of the spasing process. The gain medium is excited and population-inverted by an external source, as depicted by the black arrow, which produces electron–hole pairs in it. These pairs relax, as shown by the green arrow, to form the excitons. The excitons undergo decay to the ground state emitting SPs into the nanoshell. The plasmonic oscillations of the nanoshell stimulates this emission, supplying the feedback for the spaser action. (*For a color version of this figure, see the color plate section.*)

nonuniform, exhibiting hot spots of ~1.5×10^6 V cm^{-1} of the modal electric field at the poles. These fields are high because the SP modal volume is so small, nanoscopic.

These high values of the modal fields is the underlying physical reason for a very strong feedback in the spaser. Under our conditions, the electromagnetic retardation within the spaser volume can be safely neglected. Also, the radiation of such a spaser is a weak effect: the decay rate of plasmonic eigenmodes is dominated by the internal loss in the metal. Therefore, it is sufficient to consider only quasistatic eigenmodes [3, 47] and not their full electrodynamic counterparts [50].

For the sake of numerical illustrations of our theory, we will use the dipole eigenmode (Fig. 4.2a). There are two basic ways to place the gain medium: (i) outside the nanoshell, as shown in panel (c), and (ii) in the core, as in panel (d), which was originally proposed in Reference 49. As we have verified, these two designs lead to comparable characteristics of the spaser. However, the placement of the gain medium inside the core illustrated in Figure 4.2d has a significant advantage because the hot spots of the local field are not covered by the gain medium and are sterically available for applications.

Note that any l-multipole mode of a spherical particle is indeed $2l + 1$-times degenerate. This may make the spasing mode to be polarization unstable, like in lasers without polarizing elements. In reality, the polarization may be clamped and become stable due to deviations from the perfect spherical symmetry, which exist naturally or can be introduced deliberately. More practical shape for a spaser may be a nanorod, which has a mode with the stable polarization along the major axis. However, a nanorod is a more complicated geometry for theoretical treatment, and we will consider it elsewhere.

The level diagram of the spaser gain medium and the plasmonic metal nanoparticle is displayed in Figure 4.2e along with a schematic of the relevant energy transitions in the system. The gain medium chromophores may be semiconductor nanocrystal quantum dots [16, 42], dye molecules [51, 52], rare-earth ions [49], or electron–hole excitations of an unstructured semiconductor [33, 46]. For certainty, we will use a semiconductor-science language of electrons and holes in quantum dots.

The pump excites electron–hole pairs in the chromophores (Figure 4.2e), as indicated by the vertical black arrow, which relax to form excitons. The excitons constitute the two-level systems that are the donors of energy for the SP emission into the spasing mode. In vacuum, the excitons would recombine emitting photons. However, in the spaser geometry, the photoemission is strongly quenched due to the resonance energy transfer to the SP modes, as indicated by the red arrows in the panel. The probability of the radiativeless energy transfer to the SPs relative to that of the radiative decay (photon emission) is given by the so-called Purcell factor

$$\sim \frac{\bar{\lambda}^3 Q}{R^3} \gg 1, \tag{4.2}$$

where R is a characteristic size of the spaser metal core. Thus this radiativeless energy transfer to the spaser mode is the dominant process whose probability is by orders of magnitude greater than that of the free-space (far-field) emission.

The SPs already in the spaser mode create high local fields that excite the gain medium and stimulate more emissions to this mode, which is the feedback mechanism. If this feedback is strong enough, and the lifetime of the spaser SP mode is long enough, then an instability develops leading to an avalanche of SP emissions in the spasing mode and spontaneous symmetry breaking, establishing the phase coherence of the spasing state. Thus, the establishment of spasing is a nonequilibrium phase transition, as in the physics of lasers.

4.2.3 Brief Overview of Latest Progress in Spasers

After the original theoretical proposal and prediction of the spaser [16], there has been an active development in this field, both theoretical and experimental. There has also been a US patent issued on spaser [20].

Among theoretical developments, a nanolens spaser has been proposed [53], which possesses a nanofocus ("the hottest spot") of the local fields. In References 16 and 53, the necessary condition of spasing has been established on the basis of the perturbation theory.

There have been theories published describing the SPP spasers (or, "nanolasers" as sometimes they are called) phenomenologically, on the basis of classic linear electrodynamics by considering the gain medium as a dielectric with a negative imaginary part of the permittivity; for example, see Reference 49. Very close fundamentally and technically are works on the loss compensation in metamaterials [54–57]. Such linear-response approaches do not take into account the nature of the spasing as a nonequilibrium phase transition, at the foundation of which is spontaneous symmetry breaking: establishing coherence with an arbitrary but sustained phase of the SP quanta in the system [17]. Spaser is necessarily a deeply nonlinear (nonperturbative) phenomenon where the coherent SP field always saturates the gain medium, which eventually brings about establishment of the stationary (or, continuous wave, CW) regime of the spasing [17]. This leads to principal differences of the linear-response results from the microscopic quantum-mechanical theory in the region of spasing, as we discuss here in conjunction with Figure 4.5.

There has also been a theoretical publication on a bowtie spaser (nanolaser) with electrical pumping [58]. It is based on balance equations and only the CW spasing generation intensity is described. Yet another theoretical development has been a proposal of the lasing spaser [41], which is made of a plane array of spasers.

There have also been a theoretical proposal of a spaser ("nanolaser") consisting of a metal nanoparticle coupled to a single chromophore [59]. In this paper, a dipole–dipole interaction is illegitimately used at very small distances r where it has a singularity (diverging for $r \to 0$), leading to a dramatically overestimated coupling with the SP mode. As a result, a completely unphysical prediction of CW spasing due to single chromophore has been obtained [59]. In contrast, our theory [17] is based on the full (exact) field of the spasing SP mode without the dipole (or, any multipole) approximation. As our results of Section 4.2.5 show, hundreds of chromophores per metal nanoparticle are realistically required for the spasing even under the most favorable conditions.

There has been a vigorous experimental investigation of the spaser and the concepts of spaser. Stimulated emission of SPPs has been observed in a proof-of-principle experiment using pumped dye molecules as an active (gain) medium [51]. There have also been later experiments that demonstrated strong stimulated emission compensating a significant part of the SPP loss [52, 60–63]. As a step toward the lasing spaser, the first experimental demonstration has been reported of a partial compensation of the Joule losses in a metallic photonic metamaterial using optically pumped PbS semiconductor quantum dots [42]. There have also been experimental investigations reporting the stimulated emission effects of SPs in plasmonic metal nanoparticles surrounded by gain media with dye molecules [64, 65].

The full loss compensation and amplification of the long-range SPPs at $\lambda = 882$ nm in a gold nanostrip waveguide with a dyes solution as a gain medium has been observed [66]. Another example of full loss compensation has recently been obtained for thin (∼20 nm thickness) gold stripes (width ∼1 μm) surrounded by a gain medium containing donor–acceptor with a Förster energy transfer to increase the Stokes shift and decrease absorption at the probe frequency.

At the present time, there have been a number of the successful experimental observations of the spaser and SPP spasers (the so-called nanolasers). An electrically pumped nanolaser with semiconductor gain medium has been demonstrated [33] where the lasing modes are SPPs with a one-dimensional (1D) confinement to a ∼50 nm size. Other electrically pumped nanolasers (SPP spasers) have recently been fabricated and their lasing observed based on a diode with an intrinsic InGaAs gain media and silver nanocavities as plasmonic cores [37, 67, 68]. The latest of these nanolasers [68] operates at room temperature and has a relatively small cavity volume $V_c \approx 0.67 \lambda^3$, where vacuum wavelength $\lambda = 1591$ nm. This volume is still much larger than the modal volumes of the spasers with tighter confinement, especially SP-mode spasers—see below. A nanolaser with an optically pumped semiconductor gain medium and a hybrid semiconductor/metal (CdS/Ag) SPP waveguide has been demonstrated with an extremely tight transverse (2D) mode confinement to ∼10 nm size [34]. This has been followed by the development of CdS/Ag nanolasers generating a visible single mode at room temperature with a tight 1D confinement (∼20 nm) and a 2D confinement in the plane of the structure to an area ∼1 μm^2 [35]. A highly efficient SPP spaser in the communication range ($\lambda = 1.46$ μm) with an optical pumping based on a gold film and an InGaAs semiconductor quantum-well gain medium has recently been reported [36].

Another class of spasers observed are random spasers composed of a rough metal nanofilm as a plasmonic component and a dye-doped polymeric film as a gain medium [39]. The spasing in such systems competes with loss compensation for SPPs propagating at the interface—see also Section 4.2.7.

Historically, the first spaser observed was a nanoparticle spaser [30]. This spaser is a chemically synthesized gold nanosphere of radius 7 nm surrounded by a dielectric shell of a 21 nm outer radius containing immobilized dye molecules. Under nanosecond optical pumping in the absorption band of the dye, this spaser develops a relatively narrow-spectrum and intense visible emission that exhibits a pronounced threshold in pumping intensity. The observed characteristics of this spaser are in an excellent qualitative agreement and can be fully understood on the basis of the corresponding theoretical results described in Section 4.2.5.

4.2.3.1 Nanospaser with Semiconductor Gain Media It is of both fundamental and applied importance to develop nanoscale-size spasers (nanospasers) with semiconductor gain media. The photochemical and electrochemical stability of the semiconductor gain media is the main attraction of such a design. Belonging to this class, spasers have recently been fabricated and their operation observed, comprised of a InGaN-core/InN-shell semiconductor-nanorod gain medium and silver film as a plasmonic component [31, 32]. They generate on localized SP modes. One of these [32] is a nanospaser with a deeply sub-wavelength mode size based on an epitaxial silver nanofilm [32]. Such a design bears a promise of practical applications due to its stability and small modal volume leading to high operational speed—see Section 4.2.6.

In Figure 4.3, we display geometry of this InGaN-core/InN-shell nanorod spaser and properties of its spasing mode. The active region of the spaser (Fig. 4.3a, left panel) is a core-shell nanocylinder with a 30-nm diameter core of InGaN surrounded by thick shell of GaN. The latter is a wide bandgap semiconductor that plays the role of an insulator. The active nanorod is separated by the metal by a 5-nm layer of silica. The plasmonic component of this spaser is a flat layer of epitaxial silver. The high monocrystalline quality of the silver film is instrumental in reducing the threshold of the spaser and increasing its output. The calculated intensity for the spasing eigenmode is shown in the right panel of Figure 4.3a. Similar to the gap modes introduced in Reference 69, this eigenmode is concentrated in the thin layer of a low-permittivity dielectric (silica) between the two high-permittivity media: GaN and silver. The modal fields do penetrate sufficiently into the gain medium providing the feedback necessary for the spaser functioning.

Under 8.3 kWcm^{-2} optical pumping with frequency above the bandgap of InGaN, a series of the emission spectra of a single spaser is displayed in Figure 4.3b. At room temperature, $T = 300$ K, the emission is a spontaneous fluorescence in a wide yellow-green spectral band near the bandgap of InGaN. The first evidence of the spasing appears at $T = 120$ K as a small notch at the green side of the spectrum. As the temperature decreases to $T = 8$ K, the narrow line at $\lambda \approx 500$ nm becomes dominant and narrow. This change of the spectrum over the threshold is in qualitative agreement with theory—see Section 4.2.5 and, in particular, Figures 4.5d–4.5f.

The L–L line is the dependence of the light intensity out (the intensity of the radiation emitted by the spaser within the linewidth spectral range) versus the intensity of the pumping radiation. The theoretical prediction for the spaser is that after reaching the spasing threshold, the L–L line becomes linear with universally unit slope—see Figure 4.5a and its discussion in Section 4.2.5.

The experimentally obtained L–L line of the nanorod spaser shown in Figure 4.3c is in excellent agreement with this prediction. Note that this figure is presented in the double-logarithmic scale. There are two curves in this figure taken at different temperatures, which are similar, though at a lower temperature the intensity out is higher and the threshold is lower. The parts of the curves at lower pumping intensities

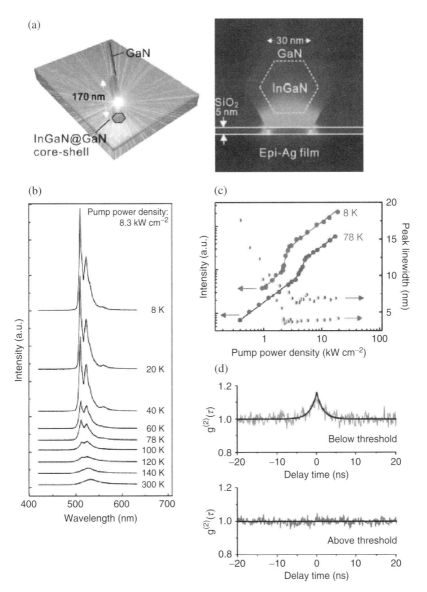

FIGURE 4.3 InGaN nanospaser and its properties. (a) Schematic of geometry of InGaN/GaN core-shell nanospaser (left) and theoretical intensity of its spasing eigenmode. (b) Series of emission spectra: temperature-dependent spasing behavior from 8 to 300 K. The spasing threshold at 140 K is clearly visible. (c) The L–L (light–light) plots at the main lasing peak (510 nm) are shown with the corresponding linewidth-narrowing behavior when the spaser is measured at 8 K (red) and 78 K (blue), with lasing thresholds of 2.1 and 3.7 kW cm^{-2}, respectively. (d) Second-order photon correlation function $g^{(2)}(\tau)$ measured at 8 K. The upper curve is recorded below the spasing threshold, and the lower above the threshold. Adapted from Reference 32. (*For a color version of this figure, see the color plate section.*)

are also unit-slope straight lines corresponding to spontaneous fluorescence. With the increased intensity, the curves enter a transitional regime of amplified spontaneous emission where the slopes are greater than one. The regime of developed spasing takes place at high intensities where the L–L curves become unit-slope straight lines without a saturation. As has already been mentioned above, this is a universal behavior.

This universal unsaturable behavior can be very simply understood qualitatively—cf. Reference 70. The excitation rate $\dot{N}_e$ of the upper spasing level is linearly proportional to pumping intensity I_p, $\dot{N}_e = \sigma_e I_p$, where σ_e is the total excitation cross section into the conduction band of the semiconductor gain medium. In the developed spasing regime, plasmon population N_n of the spasing eigenmode becomes large, asymptotically $N_n \to \infty$. Correspondingly, the stimulated decay rate, which is $\propto N_n$, becomes large and dominates any spontaneous decay rate. Thus, all the excitation events to the conduction band end up with the emission of an SP into the spasing mode whose SP population becomes $N_n = \dot{N}_e/\gamma_n$, where γ_n is the SP decay rate is given below by Eq. (4.7)—see also References 16 and 71.

Finally, radiation rate $\dot{N}_r$ for a spaser becomes

$$\dot{N}_r = \sigma_e \gamma^{(r)}/\gamma_n, \qquad (4.3)$$

where $\gamma^{(r)}$ is the SP radiative decay rate. Of course. in reality the straight-line, unsaturable L–L curves will end when the pumping intensities become so high that the nonlinearity in the spaser metal develops (including, but not limited to, thermal nonlinearity), or optical breakdown occurs, or heat production will physically damage the spaser.

As theory shows (see Section 4.2.6 and Figure 4.6a), under steady pumping, a generating spaser reaches its stationary regime within ~100 fs. Correspondingly, we expect that any fluctuation in the emission radiated by the generating spaser relaxes back to the mean level within the same time. A measure of the fluctuations of the spaser-radiation intensity $I(t)$ with time t is the second-order autocorrelation function

$$g^{(2)}(\tau) = \frac{\langle I(t+\tau)I(t)\rangle}{\langle I(t)\rangle^2}, \qquad (4.4)$$

where τ is the delay time, and $\langle \cdots \rangle$ denotes quantum-mechanical (theory) or temporal (experiment) averaging.

Experimentally, $g^{(2)}(\tau)$ has been measured for a single spaser in Reference 32. The result is reproduced in Figure 4.3d. The upper curve is recorded below the spasing threshold; at the zero delay, it shows a peak, which is characteristic of incoherent radiation. If such radiation is produced by many independent emitters, it has Gaussian statistics, and the peak value should be $g^{(2)}(0) = 2$—this effect was introduced by Hanbury Brown and Twiss and used by them for stellar interferometry [72]. For the upper curve of Figure 4.3d, $g^{(2)}(0)$ is significantly less. This may be due to various reasons, in particular, insufficient temporal resolution of the photodetection or partial coherence between the individual emitters of the gain medium induced by their interaction via plasmonic fields.

In sharp contrast, above the spasing threshold, the autocorrelation function in Figure 4.3d is a constant at all delays. As we have already pointed out, this is due to the fact that after an emission of a photon, the number of plasmons in the spaser is restored within ∼100 fs, while the temporal resolution of the photodetection in Reference 32 is $\Delta\tau \gtrsim 100$ ps, that is, three orders of magnitude coarser. The physical reason for $g^{(2)}(\tau) = $ const is that the spaser under steady-state pumping tends to keep a constant plasmon population. After the emission of a photon, this population is decreased by one. However, very rapidly, within ∼100 fs, it restores to the pre-emission level. This transitional restoration process is too fast and the photodetectors of Reference 32 miss it, producing $g^{(2)}(\tau) = $ const.

4.2.4 Equations of Spaser

4.2.4.1 Quantum Density Matrix Equations (Optical Bloch Equations) for Spaser

The SP eigenmodes $\varphi_n(\mathbf{r})$ are described by a wave equation given in References 16 and 47. The electric field operator of the quantized SPs is an operator [16]

$$\hat{\mathbf{E}}(\mathbf{r}) = -\sum_n A_n \nabla\varphi_n(\mathbf{r})(\hat{a}_n + \hat{a}_n^\dagger), \quad A_n = \left(\frac{4\pi\hbar s_n}{\varepsilon_d s_n'}\right)^{1/2}, \tag{4.5}$$

where $\hat{a}_n^\dagger$ and $\hat{a}_n$ are the SP creation and annihilation operators, $-\nabla\varphi_n(\mathbf{r}) = \mathbf{E}_n(\mathbf{r})$ is the modal field of an nth eigenmode, s_n is its eigenvalue ($1 \geq s_n \geq 0$),

$$s(\omega) = \frac{\varepsilon_d}{\varepsilon_d - \varepsilon_m(\omega)} \tag{4.6}$$

is Bergman's spectral parameter, ω_n is eigenmode's frequency, ε_d is permittivity of the embedding dielectric, and $s_n' = \text{Re}[ds(\omega_n)/d\omega_n]$. Note that we have corrected a misprint in Reference 16 by replacing the coefficient 2π by 4π.

For the sake of reference, we give here an explicit expression for the plasmon decay rate γ_n, which can be calculated (see Reference 16) in terms of the spectral parameter as

$$\gamma_n = \frac{\text{Im}[s(\omega_n)]}{s_n'}, \quad s_n' \equiv \left.\frac{\partial\text{Re}[s(\omega)]}{\partial\omega}\right|_{\omega=\omega_n}. \tag{4.7}$$

In terms of the dielectric permittivity as a function of frequency, one can express

$$s'(\omega) = \frac{\varepsilon_d}{|\varepsilon_d - \varepsilon(\omega)|^2}\text{Re}\frac{\partial\varepsilon_m(\omega)}{\partial\omega}, \quad \gamma_n = \frac{\text{Im}\varepsilon_m(\omega_n)}{\text{Re}\frac{\partial\varepsilon_m(\omega_n)}{\partial\omega_n}}. \tag{4.8}$$

Importantly, the spectral width γ_n is a universal function of frequency.

The spaser Hamiltonian has the form

$$\hat{H} = \hat{H}_g + \hbar \sum_n \omega_n \hat{a}_n^\dagger \hat{a}_n - \sum_p \hat{\mathbf{E}}(\mathbf{r}_p)\hat{\mathbf{d}}^{(p)}, \tag{4.9}$$

where $\hat{H}_g$ is the Hamiltonian of the gain medium, p is a number (label) of a gain medium chromophore, $\mathbf{r}_p$ is its coordinate vector, and $\hat{\mathbf{d}}^{(p)}$ is its dipole moment operator. In this theory, we treat the gain medium quantum mechanically but the SPs quasiclassically, considering $\hat{a}_n$ as a classical quantity (c-number) a_n with time dependence as $a_n = a_{0n} \exp(-i\omega t)$, where a_{0n} is a slowly-varying amplitude. The number of coherent SPs per spasing mode is then given by $N_p = |a_{0n}|^2$. This approximation neglects the quantum fluctuations of the SP amplitudes. However, when necessary, we will take into account these quantum fluctuations, in particular, to describe the spectrum of the spaser.

Introducing $\rho^{(p)}$ as the density matrix of a pth chromophore, we can find its equation of motion in a conventional way by commuting it with the Hamiltonian (4.9) as

$$i\hbar \dot{\rho}^{(p)} = [\rho^{(p)}, \hat{H}], \tag{4.10}$$

where the dot denotes temporal derivative. We use the standard rotating wave approximation (RWA), which only takes into account the resonant interaction between the optical field and chromophores. We denote $|1\rangle$ and $|2\rangle$ as the ground and excited states of a chromophore, with the transition $|2\rangle \rightleftharpoons |1\rangle$ resonant to the spasing plasmon mode n. In this approximation, the time dependence of the nondiagonal elements of the density matrix is $(\rho^{(p)})_{12} = \bar{\rho}_{12}^{(p)} \exp(i\omega t)$, and $(\rho^{(p)})_{21} = \bar{\rho}_{12}^{(p)*} \exp(-i\omega t)$, where $\bar{\rho}_{12}^{(p)}$ is an amplitude slowly varying in time, which defines the coherence (polarization) for the $|2\rangle \rightleftharpoons |1\rangle$ spasing transition in a pth chromophore of the gain medium.

Introducing a rate constant Γ_{12} to describe the polarization relaxation and a difference $n_{21}^{(p)} = \rho_{22}^{(p)} - \rho_{11}^{(p)}$ as the population inversion for this spasing transition, we derive an equation of motion for the nondiagonal element of the density matrix as

$$\dot{\bar{\rho}}_{12}^{(p)} = -[i(\omega - \omega_{12}) + \Gamma_{12}]\bar{\rho}_{12}^{(p)} + i a_{0n} n_{21}^{(p)} \tilde{\Omega}_{12}^{(p)*}, \tag{4.11}$$

where

$$\tilde{\Omega}_{12}^{(p)} = -A_n \mathbf{d}_{12}^{(p)} \nabla \varphi_n(\mathbf{r}_p)/\hbar \tag{4.12}$$

is the one-plasmon Rabi frequency for the spasing transition in a pth chromophore, and $\mathbf{d}_{12}^{(p)}$ is the corresponding transitional dipole element. Note that always $\mathbf{d}_{12}^{(p)}$ is either real or can be made real by a proper choice of the quantum state phases, making the Rabi frequency $\tilde{\Omega}_{12}^{(p)}$ also a real quantity.

An equation of motion for n_{21}^p can be found in a standard way by commuting it with $\hat{H}$. To provide conditions for the population inversion ($n_{21}^p > 0$), we imply existence of a third level. For simplicity, we assume that it very rapidly decays into the excited state $|2\rangle$ of the chromophore, so its own populations is negligible. It is pumped by an external source from the ground state (optically or electrically) with some rate that we will denote g. In this way, we obtain the following equation of motion:

$$\dot{n}_{21}^{(p)} = -4\text{Im}\left[a_{0n}\bar{\rho}_{12}^{(p)}\tilde{\Omega}_{21}^{(p)}\right] - \gamma_2\left(1+n_{21}^{(p)}\right) + g\left(1-n_{21}^{(p)}\right), \quad (4.13)$$

where γ_2 is the decay rate $|2\rangle \to |1\rangle$.

The stimulated emission of the SPs is described as their excitation by the coherent polarization of the gain medium. The corresponding equation of motion can be obtained using Hamiltonian (4.9) and adding the SP relaxation with a rate of γ_n as

$$\dot{a}_{0n} = [i(\omega - \omega_n) - \gamma_n]a_{0n} + ia_{0n}\sum_p \rho_{12}^{(p)*}\tilde{\Omega}_{12}^{(p)}. \quad (4.14)$$

As an important general remark, the system of Eqs. (4.11), (4.13), and (4.14) is highly nonlinear; each of these equations contains a quadratic nonlinearity: a product of the plasmon-field amplitude a_{0n} by the density matrix element ρ_{12} or population inversion n_{21}. Altogether, this is a six-order nonlinearity. This nonlinearity is a fundamental property of the spaser equations, which makes the spaser generation always an essentially nonlinear process that involves a noneqilibrium phase transition and a spontaneous symmetry breaking: establishment of an arbitrary but sustained phase of the coherent SP oscillations.

A relevant process is spontaneous emission of SPs by a chromophore into a spasing SP mode. The corresponding rate $\gamma_2^{(p)}$ for a chromophore at a point $\mathbf{r}_p$ can be found in a standard way using the quantized field (4.5) as

$$\gamma_2^{(p)} = 2\frac{A_n^2}{\hbar\gamma_n}\left|\mathbf{d}_{12}\nabla\varphi_n(\mathbf{r}_p)\right|^2 \frac{(\Gamma_{12}+\gamma_n)^2}{(\omega_{12}-\omega_n)^2 + (\Gamma_{12}+\gamma_n)^2}. \quad (4.15)$$

As in Schawlow–Towns theory of laser-line width [73], this spontaneous emission of SPs leads to the diffusion of the phase of the spasing state. This defines width γ_s of the spasing line as

$$\gamma_s = \frac{\sum_p\left(1+n_{21}^{(p)}\right)\gamma_2^{(p)}}{2(2N_p+1)}. \quad (4.16)$$

This width is small for a case of developed spasing when $N_p \gg 1$. However, for $N_p \sim 1$, the predicted width may be too high because the spectral diffusion theory assumes that $\gamma_s \lesssim \gamma_n$. To take into account this limitation in a simplified way, we will interpolate to find the resulting spectral width Γ_s of the spasing line as $\Gamma_s = \left(\gamma_n^{-2} + \gamma_s^{-2}\right)^{-1/2}$.

We will also examine the spaser as a bistable (logical) amplifier. One of the ways to set the spaser in such a mode is to add a saturable absorber. This is described by the same Eqs. (4.11)–(4.14) where the chromophores belonging to the absorber are not pumped by the external source directly, that is, for them in Eq. (4.13) one has to set $g = 0$.

Numerical examples are given for a silver nanoshell where the core and the external dielectric have the same permittivity of $\varepsilon_d = 2$; the permittivity of silver is adopted from Reference 74. The following realistic parameters of the gain medium are used (unless indicated otherwise): $d_{12} = 1.5 \times 10^{-17}$ esu, $\hbar\Gamma_{12} = 10$ meV, $\gamma_2 = 4 \times 10^{12} \text{s}^{-1}$ (this value takes into account the spontaneous decay into SPs), and density of the gain medium chromophores is $n_c = 2.4 \times 10^{20} \text{cm}^{-3}$, which is realistic for dye molecules but may be somewhat high for semiconductor quantum dots that were proposed as the chromophores [16] and used in experiments [42]. We will assume a dipole SP mode and chromophores situated in the core of the nanoshell as shown in Figure 4.2d. This configuration is of advantage both functionally (because the region of the high local fields outside the shell is accessible for various applications) and computationally (the uniformity of the modal fields makes the summation of the chromophores trivial, thus greatly facilitating numerical procedures).

4.2.4.2 Equations for CW Regime
Physically, the spaser action is a result of spontaneous symmetry breaking when the phase of the coherent SP field is established from the spontaneous noise. Mathematically, the spaser is described by homogeneous differential Eqs. (4.11)–(4.14). These equations become homogeneous algebraic equations for the CW case. They always have a trivial, zero solution. However, they may also possess a nontrivial solution describing spasing. An existence condition of such a nontrivial solution is

$$(\omega_s - \omega_n + i\gamma_n)^{-1} \times \qquad (4.17)$$
$$(\omega_s - \omega_{21} + i\Gamma_{12})^{-1} \sum_p \left|\tilde{\Omega}_{12}^{(p)}\right|^2 n_{21}^{(p)} = -1.$$

The population inversion of a pth chromophore $n_{21}^{(p)}$ is explicitly expressed as

$$n_{21}^{(p)} = (g - \gamma_2) \times \qquad (4.18)$$
$$\left\{ g + \gamma_2 + 4N_n \left|\tilde{\Omega}_{12}^{(p)}\right|^2 \Big/ \left[(\omega_s - \omega_{21})^2 + \Gamma_{12}^2\right] \right\}^{-1}.$$

From the imaginary part of Eq. (4.18) we immediately find the spasing frequency ω_s,

$$\omega_s = (\gamma_n \omega_{21} + \Gamma_{12}\omega_n) / (\gamma_n + \Gamma_{12}), \qquad (4.19)$$

which generally does not coincide with either the gain transition frequency ω_{21} or the SP frequency ω_n, but is between them (this is a frequency walk-off phenomenon similar to that of laser physics). Substituting Eq. (4.19) back into Eqs. (4.18)–(4.19), we obtain a system of equations

$$\frac{(\gamma_n + \Gamma_{12})^2}{\gamma_n \Gamma_{12} \left[(\omega_{21} - \omega_n)^2 + (\Gamma_{12} + \gamma_n)^2 \right]}$$

$$\times \sum_p \left| \tilde{\Omega}_{12}^{(p)} \right|^2 n_{21}^{(p)} = 1, \quad (4.20)$$

$$n_{21}^{(p)} = (g - \gamma_2)$$

$$\times \left[g + \gamma_2 + \frac{4 N_n \left| \tilde{\Omega}_{12}^{(p)} \right|^2 (\Gamma_{12} + \gamma_n)}{(\omega_{12} - \omega_n)^2 + (\Gamma_{12} + \gamma_n)^2} \right]^{-1}. \quad (4.21)$$

This system defines the stationary (CW-generation) number of SPs per spasing mode, N_n.

Since $n_{21}^{(p)} \leq 1$, from Eqs. (4.20) and (4.21), we immediately obtain a necessary condition of the existence of spasing,

$$\frac{(\gamma_n + \Gamma_{12})^2}{\gamma_n \Gamma_{12} \left[(\omega_{21} - \omega_n)^2 + (\Gamma_{12} + \gamma_n)^2 \right]} \sum_p \left| \tilde{\Omega}_{12}^{(p)} \right|^2 \geq 1. \quad (4.22)$$

This expression is fully consistent with Reference 16. The following order of magnitude estimate of this spasing condition has a transparent physical meaning and is of heuristic value,

$$\frac{d_{12}^2 Q N_c}{\hbar \Gamma_{12} V_n} \gtrsim 1, \quad (4.23)$$

where $Q = \omega / \gamma_n$ is the quality factor of SPs, V_n is the volume of the spasing SP mode, and N_c is the of number of the gain medium chromophores within this volume. Deriving this estimate, we have neglected the detuning, that is, set $\omega_{21} - \omega_n = 0$. We also used the definitions of A_n of Eq. (4.5) and $\tilde{\Omega}_{12}^{(p)}$ given by Eq. (4.12), and the estimate $|\nabla \varphi_n(\mathbf{r})|^2 \sim 1/V$ following from the normalization of the SP eigenmodes $\int |\nabla \varphi_n(\mathbf{r})|^2 d^3 r = 1$ of Reference 47. The result of Eq. (4.23) is, indeed, in agreement with Reference 16 where it was obtained in different notations.

It follows from Eq. (4.23) that for the existence of spasing, it is beneficial to have a high quality factor Q, a high density of the chromophores, and a large transition dipole (oscillator strength) of the chromophore transition. The small modal volume V_n (at a given number of the chromophores N_c) is beneficial for this spasing condition: physically, it implies strong feedback in the spaser. Note that for the given density of the chromophores $n_c = N_c/V_n$, this spasing condition does not explicitly depend on the spaser size, which opens up a possibility of spasers of a very small size limited from the bottom by only the nonlocality radius $l_{nl} \sim 1$ nm. Another important property of Eq. (4.23) is that it implies the quantum-mechanical nature of spasing and spaser amplification: this condition essentially contains the Planck constant $\hbar$ and, thus, does not have a classical counterpart. Note that in contrast to lasers, the spaser theory and Eqs. (4.22) and (4.23) in particular do not contain speed of light, that is, they are quasistatic.

Now we will examine the spasing condition and reduce it to a requirement for the gain medium. First, we substitute all the definitions and assume the perfect resonance between the generating SP mode and the gain medium, that is, $\omega_n = \omega_{21}$. As a result, we obtain from Eq. (4.22),

$$\frac{4\pi}{3} \frac{s_n |\mathbf{d}_{12}|^2}{\hbar \gamma_n \Gamma_{12} \varepsilon_d s'_n} \int_V [1 - \Theta(\mathbf{r})] |\mathbf{E}_n(\mathbf{r})|^2 d^3r \geq 1, \tag{4.24}$$

where the integral is extended over the volume V of the system, and Θ is a characteristic function equal one inside the metal and zero insides the dielectric; it takes into account a simplifying realistic assumption that the gain medium occupies the entire space free from the core's metal. We also assume that the orientations of the transition dipoles $\mathbf{d}_{12}^{(p)}$ are random and average over them, which results in the factor of 3 in the denominator in Eq. (4.24).

One can show that (see Reference 71)

$$\int_V [1 - \Theta(\mathbf{r})] |\mathbf{E}_n(\mathbf{r})|^2 d^3r = 1 - s_n. \tag{4.25}$$

Next, we give approximate expressions for the spectral parameter (4.6), which are very accurate for the realistic case of $Q \gg 1$,

$$\operatorname{Im} s(\omega) = \frac{s_n^2}{\varepsilon_d} \operatorname{Im} \varepsilon_m(\omega) = \frac{1}{Q} s_n (1 - s_n). \tag{4.26}$$

Taking into account Eqs. (4.25) and (4.26), we obtain from Eq. (4.24) a necessary condition of spasing at a frequency ω as

$$\frac{4\pi}{3} \frac{|\mathbf{d}_{12}|^2 n_c [1 - \operatorname{Re} s(\omega)]}{\hbar \Gamma_{12} \operatorname{Re} s(\omega) \operatorname{Im} \varepsilon_m(\omega)} \geq 1, \tag{4.27}$$

For the sake of comparison, consider a continuous gain medium composed of the same chromophores as the gain shell of the spaser. Its gain G (whose dimensionality is cm^{-1}) is given by a standard expression

$$G = \frac{4\pi}{3} \frac{\omega}{c} \frac{\sqrt{\varepsilon_d} |\mathbf{d}_{12}|^2 n_c}{\hbar \Gamma_{12}}. \tag{4.28}$$

Substituting it into Eq. (4.27), we obtain the spasing criterion in terms of the gain as

$$G \geq G_{\text{th}}, \quad G_{\text{th}} = \frac{\omega}{c\sqrt{\varepsilon_d}} \frac{\operatorname{Re} s(\omega)}{1 - \operatorname{Re} s(\omega)} \operatorname{Im} \varepsilon_m(\omega), \quad (4.29)$$

where G_{th} has a meaning of the threshold gain needed for spasing. Importantly, this gain depends only on the dielectric properties of the system and spasing frequency but not on the geometry of the system or the distribution of the local fields of the spasing mode (hot spots, etc.) explicitly. However, note that the system's geometry (along with the permittivities) does define the spasing frequencies.

In Figure 4.4, we illustrate the analytical expression (4.29) for gold, silver, and aluminum embedded in a dielectric with $\varepsilon_d = 2$ (simulating a light glass). These curves are computed from Eq. (4.29) assuming that the metal core is embedded into the gain medium with the real part of the dielectric function equal to ε_d. As we see from Figure 4.4, the spasing is possible for silver in the near-ir communication range and the adjacent red portion of the visible spectrum for a gain $G < 10^4$ cm^{-1} (regions below the dashed line in Figure 4.4), which is realistically achievable with direct bandgap semiconductors (DBGSs). In contrast, spasing for aluminum requires a significantly higher gain, $G \gtrsim 5 \times 10^4$ cm^{-1}, which may pose a problem to achieve, especially with electrical pumping.

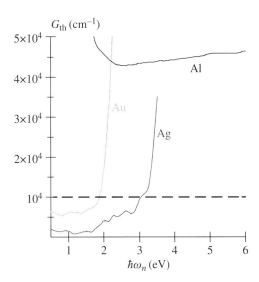

FIGURE 4.4 Threshold gain for spasing G_{th} for silver, gold, and aluminum, as indicated in the graphs, as a function of the spasing frequency ω. The dashed line separates the area $g_{\text{th}} < 10^4$cm^{-1}, where spasing can relatively easily be achieved with direct bandgap semiconductors (DBGSs). The real part of the gain medium permittivity is set as $\varepsilon_d = 2$. Adapted from Reference 24.

4.2.5 Spaser in CW Regime

The "spasing curve" (a counterpart of the L–L curve, for lasers), which is the dependence of the coherent SP population N_n on the excitation rate g, obtained by solving Eqs. (4.20) and (4.21), is shown in Figure 4.5a for four types of the silver nanoshells with the frequencies of the spasing dipole modes as indicated, which are in the range from near-ir ($\hbar\omega_s = 1.2$ eV) to mid-visible ($\hbar\omega_s = 2.2$ eV). In all cases, there is a pronounced threshold of the spasing at an excitation rate $g_{\text{th}} \sim 10^{12} \text{s}^{-1}$. Soon after the threshold, the dependence $N_n(g)$ becomes linear, which means that every quantum of excitation added to the gain medium with a high probability is stimulated to be emitted as an SP, adding to the coherent SP population.

While this is similar to conventional lasers, still there is a dramatic difference for the spaser. In lasers, a similar relative rate of the stimulated emission is achieved at a photon population of $\sim 10^{18} - 10^{20}$, while in the spaser, the SP population is $N_n \lesssim 100$. This is due to the much stronger feedback in spasers because of the much smaller modal volume V_n—see discussion of Eq. (4.23). The shape of the spasing curves of Figure 4.5a (a threshold with the linear dependence almost immediately above the threshold) is in qualitative agreement with the experiment [30].

Note that if the gain-medium emission spectral line is within the plasmon-mode spectral contour, and the Purcell factor (4.2) is high enough, which can be a realistic situation, than all transitions in the gain medium, both spontaneous and stimulated, will result in the excitation of SP quanta. In such a case, the total SP population of the mode, $N_n^{(\text{tot})}$, will, obviously, be determined by the gain-medium excitation rate, g, as

$$N_n^{(\text{tot})} = \frac{g}{\gamma_n}. \tag{4.30}$$

In such a case, there will be no appreciable threshold in the mode's *total* SP population and, consequently, in the emitted radiation. This would look like the "thresholdless nanolasing" observed in recent experiments [38]. In reality, the *coherent* SP population and radiation will always have a threshold. This threshold is always evident in the second-order correlation function, $g^{(2)}$, as observed [32] and described in Section 4.2.3—see Figure 4.3 and the corresponding discussion.

The population inversion number n_{21} as a function of the excitation rate g is displayed in Figure 4.5b for the same set of frequencies (and with the same color coding) as in panel (a). Before the spasing threshold, n_{21} increases with g to become positive with the onset of the population inversion just before the spasing threshold. For higher g, after the spasing threshold is exceeded, the inversion n_{21} becomes constant (inversion clamping). The clamped levels of the inversion are very low, $n_{21} \sim 0.01$, which again is due to the very strong feedback in the spaser.

The spectral width Γ_s of the spaser generation is due to the phase diffusion of the quantum SP state caused by the noise of the spontaneous emission of the SPs into the spasing mode, as described by Eq. (4.16). This width is displayed in Figure 4.5c as a function of the pumping rate g. At the threshold, Γ_s is that of the SP line γ_n but for

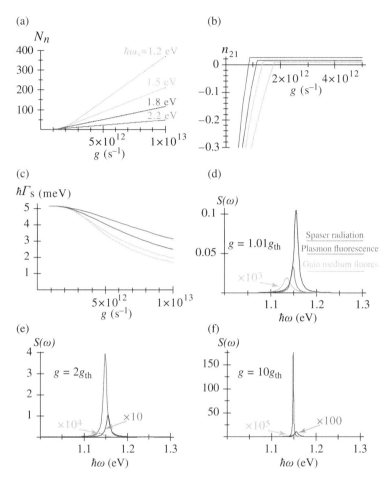

FIGURE 4.5 Spaser SP population and spectral characteristics in the stationary state. The computations are done for a silver nanoshell with the external radius $R_2 = 12$ nm; the detuning of the gain medium from the spasing SP mode is $\hbar(\omega_{21} - \omega_n) = -0.02$ eV. The other parameters are indicated in Section 4.2.4. (a) Number N_n of plasmons per spasing mode as a function of the excitation rate g (per one chromophore of the gain medium). Computations are done for the dipole eigenmode with the spasing frequencies ω_s as indicated, which were chosen by the corresponding adjustment of the nanoshell aspect ratio. (b) Population inversion n_{12} as a function of the pumping rate g. The color coding of the lines is the same as in panel (a). (c) The spectral width Γ_s of the spasing line (expressed as $\hbar\Gamma_s$ in meV) as a function of the pumping rate g. The color coding of the lines is the same as in panel (a). (d)–(f) Spectra of the spaser for the pumping rates g expressed in the units of the threshold rate g_{th}, as indicated in the panels. The curves are color coded and scaled as indicated. Adapted from Reference 17. (*For a color version of this figure, see the color plate section.*)

stronger pumping, as the SPs accumulate in the spasing mode, it decreases $\propto N_n^{-1}$, as given by Eq. (4.16). This decrease of Γ_s reflects the higher coherence of the spasing state with the increased number of SP quanta and, correspondingly, lower quantum fluctuations. As we have already mentioned, this is similar to the lasers as described by the Schawlow–Townes theory [73].

The developed spasing in a dipole SP mode will show itself in the far field as an anomalously narrow and intense radiation line. The shape and intensity of this line in relation to the lines of the spontaneous fluorescence of the isolated gain medium and its SP-enhanced fluorescence line in the spaser is illustrated in Figures 4.5d–4.5f. Note that for the system under consideration, there is a 20 meV red shift of the gain medium fluorescence with respect to the SP line center. It is chosen so to illustrate the spectral walk-off of the spaser line. For one percent in the excitation rate above the threshold of the spasing (panel (d)), a broad spasing line (red color) appears comparable in intensity to the SP-enhanced spontaneous fluorescence line (blue color). The width of this spasing line is approximately the same as that of the fluorescence, but its position is shifted appreciably (spectral walk-off) toward the isolated gain medium line (green color). For the pumping twice more intense (panel (e)), the spaser-line radiation dominates, but its width is still close to that of the SP line due to significant quantum fluctuations of the spasing state phase. Only when the pumping rate is an order of magnitude above the threshold, the spaser line strongly narrows [panel (f)], and it also completely dominates the spectrum of the radiation. This is a regime of small quantum fluctuations, which is desired in applications.

These results in the spasing region are different in the most dramatic way from previous phenomenological models, which are based on linear electrodynamics where the gain medium that has negative imaginary part of its permittivity plus lossy metal nanosystem, described purely electrodynamically [49, 55]. For instance, in a "toy model" [55], the width of the resonance line tends to zero at the threshold of spasing and then broadens up again. This distinction of the present theory is due the nature of the spasing as a spontaneous symmetry breaking (nonequilibrium phase transition with a randomly established but sustained phase) leading to the establishment of a coherent SP state. This nonequilibrium phase transition to spasing and the spasing itself are contained in the present theory due to the fact that the fundamental equations of the spasing (4.11), (4.13), and (4.14) are nonlinear, as we have already discussed in conjunction with these equations—see the text after Eq. (4.14). The previous publications on gain compensation by loss [49,55,57] based on linear electrodynamic equations do not contain spasing. Therefore, they are not applicable in the region of the complete loss compensation and spasing, though their results are presented for that region.

4.2.6 Spaser as Ultrafast Quantum Nanoamplifier

4.2.6.1 Problem of Setting Spaser as an Amplifier As we have already mentioned in Section 4.2.1, a fundamental and formidable problem is that, in contrast to the conventional lasers and amplifiers in quantum electronics, the spaser has an inherent feedback that typically cannot be removed. Such a spaser will develop generation and accumulation of the macroscopic number of coherent SPs in the spasing mode. This leads to the the population inversion clamping in the CW regime at a very low level—cf. Figure 4.5b. This CW regime corresponds to the net amplification equal zero, which means that the gain exactly compensates the loss, which condition is expressed by Eq. (4.20). This is a consequence of the nonlinear gain saturation. This holds for any stable CW generator (including any spaser or laser) and precludes using them as amplifiers.

There are several ways to set a spaser as a quantum amplifier. One of them is to reduce the feedback, that is, to allow some or most of the SP energy in the spaser to escape from the active region, so the spaser will not generate in the region of amplification. Such a root has successfully been employed to build a SPP plasmonic amplifier on the long-range plasmon polaritons [66]. A similar root for the SP spasers would be to allow some optical energy to escape either by a near-field coupling or by a radiative coupling to far-field radiation. The near-field coupling approach is promising for building integrated active circuits out of the spasers. Another root has been used in Reference 75, which employed symmetric SPP modes in a thin gold strip. Such modes have much lower loss than the antisymmetric modes at the expense of much weaker confinement (transverse modal area $\sim \lambda^2$). The lower loss allows one to use the correspondingly lower gain and, therefore, avoids both spasing at localized SP modes and random lasing due to back-scattering from gold imperfections.

Following Reference 17, we consider here two distinct approaches for setting the spasers as quantum nanoamplifiers. The first is a transient regime based on the fact that the establishment of the CW regime and the consequent inversion clamping and the total gain vanishing require some time that is determined mainly by the rate of the quantum feedback and it depends also on the relaxation rates of the SPs and the gain medium. After the population inversion is created by the onset of pumping and before the spasing spontaneously develops, as we show later in this Section, there is a time interval of approximately 250 fs, during which the spaser provides usable (and as predicted, quite high) amplification—see Section 4.2.6.

The second approach to set the spaser as a logical quantum nanoamplifier is a bistable regime that is achieved by introducing a saturable absorber into the active region, which prevents the spontaneous spasing. Then injection of a certain above-threshold amount of SP quanta will saturate the absorber and initiate the spasing. Such a bistable quantum amplifier will be considered in Section 4.2.6.

The temporal behavior of the spaser has been found by direct numerical solution of Eqs. (4.11)–(4.14). This solution is facilitated by the fact that in the model under consideration, all the chromophores experience the same local field inside the nanoshell, and there are only two types of such chromophores: belonging to the gain medium and the saturable absorber, if it is present.

4.2.6.2 Monostable Spaser as a Nanoamplifier in Transient Regime Here we consider a monostable spaser in a transient regime. This implies that no saturable absorber is present. We will consider two pumping regimes: stationary and pulse.

Starting with the stationary regime, we assume that the pumping at a rate (per one chromophore) of $g = 5 \times 10^{12}$ s^{-1} starts at a moment of time $t = 0$ and stays constant after that. Immediately at $t = 0$, a certain number of SPs are injected into the spaser. We are interested in its temporal dynamics from this moment on.

The dynamical behavior of the spaser under this pumping regime is illustrated in Figures 4.6a and 4.6b. As we see, the spaser, which starts from an arbitrary initial population N_n, rather rapidly, within a few hundred femtoseconds approaches the same stationary ("logical") level. At this level, an SP population of $N_n = 67$ is established, while the inversion is clamped at a low level of $n_{21} = 0.02$. On the way to this stationary state, the spaser experiences relaxation oscillations in both the SP numbers and inversion, which have a trend to oscillate out of phase (compare panels (a) and (b)). This temporal dynamics of the spaser is quite complicated and highly nonlinear (unharmonic). It is controlled not by a single relaxation time but by a set of the relaxation rates. Clearly, among these are the energy transfer rate from the gain medium to the SPs and the relaxation rates of the SPs and the chromophores.

In this mode, the main effect of the initial injection of the SPs (described theoretically as different initial values of N_n) is in the interval of time it is required for the spaser to reach the final (CW) state. For very small N_n, which in practice can be supplied by the noise of the spontaneous SP emission into the mode, this time is approximately 250 fs (cf.: the corresponding SP relaxation time is less then 50 fs). In contrast, for the initial values of $N_n = 1 - 5$, this time shortens to 150 fs.

Now consider the second regime: pulse pumping. The gain-medium population of the spaser is inverted at $t = 0$ to saturation with a short (much shorter than 100 fs) pump pulse. Simultaneously, at $t = 0$, some number of plasmons are injected (say, by an external nanoplasmonic circuitry). In response, the spaser should produce an amplified pulse of the SP excitation. Such a function of the spaser is illustrated in Figures 4.6c and 4.6d.

As we see from panel (c), independently from the initial number of SPs, the spaser always generates a series of SP pulses, of which only the first pulse is large (at or above the logical level of $N_n \sim 100$) (an exception is a case of little practical importance when the initial $N_n = 120$ exceeds this logical level, when two large pulses are produced). The underlying mechanism of such a response is the rapid depletion of the inversion seen in panel (d), where energy is dissipated in the metal of the spaser. The characteristic duration of the SP pulse ~ 100 fs is defined by this depletion and controlled by the energy transfer and SP relaxation rates. This time is much shorter than the spontaneous decay time of the gain medium. This acceleration is due to the stimulated emission of the SPs into the spasing mode (which can be called a "stimulated Purcell effect"). There is also a pronounced trend: the lower is initial SP population N_n, the later the spaser produces the amplified pulse. In a sense, this spaser functions as a pulse-amplitude to time-delay converter.

4.2.6.3 Bistable Spaser with Saturable Absorber as an Ultrafast Nanoamplifier
Now let us consider a bistable spaser as a quantum threshold (or, logical) nanoamplifier. Such a spaser contains a saturable absorber mixed with the gain medium with parameters indicated at the end of Section 4.2.4 and the concentration of the saturable absorber $n_a = 0.66 n_c$. This case of a bistable spaser amplifier is of a particular interest because in this regime, the spaser comes as close as possible in its functioning

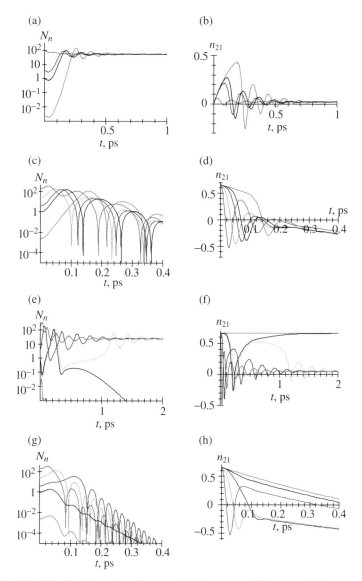

FIGURE 4.6 Ultrafast dynamics of spaser. (a) For monostable spaser (without a saturable absorber), dependence of SP population in the spasing mode N_n on time t. The spaser is stationary pumped at a rate of $g = 5 \times 10^{12}$ s^{-1}. The color-coded curves correspond to the initial conditions with the different initial SP populations, as shown in the graphs. (b) The same as (a) but for the temporal behavior of the population inversion n_{21}. (c) Dynamics of a monostable spaser (no saturable absorber) with the pulse pumping described as the initial inversion $n_{21} = 0.65$. Coherent SP population N_n is displayed as a function of time t. Different initial populations are indicated by color-coded curves. (d) The same as (c) but for the corresponding population inversion n_{21}. (e) The same as (a) but for bistable spaser with the saturable absorber in concentration $n_a = 0.66 n_c$. (f) The same as (b) but for the bistable spaser. (g) The same as (e) but for the pulse pumping with the initial inversion $n_{21} = 0.65$. (h) The same as (g) but for the corresponding population inversion n_{21}. Adapted from Reference 17. (*For a color version of this figure, see the color plate section.*)

to the semiconductor-based (mostly, MOSFET-based) digital nanoamplifiers. As in the previous subsection, we will consider two cases: the stationary and short-pulse pumping.

We again start with the case of the stationary pumping at a rate of $g = 5 \times 10^{12}$ s^{-1}. We show in Figures 4.6e and 4.6f the dynamics of such a spaser. For a small initial population $N_n = 5 \times 10^{-3}$ simulating the spontaneous noise, the spaser is rapidly (faster than in 50 fs) relaxing to the zero population [panel (e)], while its gain-medium population is equally rapidly approaching a high level [panel (f)] $n_{21} = 0.65$ that is defined by the competition of the pumping and the enhanced decay into the SP mode (the purple curves). This level is so high because the spasing SP mode population vanishes and the stimulated emission is absent. After reaching this stable state (which one can call, say, "logical zero"), the spaser stays in it indefinitely long despite the continuing pumping.

In contrast, for initial values N_n of the SP population large enough (for instance, for $N_n = 5$, as shown by the blue curves in Figures 4.6e and 4.6f), the spaser tends to the "logical one" state where the stationary SP population reaches the value of $N_n \approx 60$. Due to the relaxation oscillations, it actually exceeds this level within a short time of $\lesssim 100$ fs after the seeding with the initial SPs. As the SP population N_n reaches its stationary (CW) level, the gain medium inversion n_{21} is clamped down at a low level of a few percent, as typical for the CW regime of the spaser. This "logical one" state salso persists indefinitely, as long as the inversion is supported by the pumping.

There is a critical curve (separatrix) that divides the two stable dynamics types (leading to the logical levels of zero and one). For the present set of parameters, this separatrix starts with the initial population of $N_n \approx 1$. For a value of the initial N_n slightly below 1, the SP population N_n experiences a slow (hundreds fs in time) relaxation oscillation but eventually relaxes to zero (Fig. 4.6e, black curve), while the corresponding chromophore population inversion n_{21} relaxes to the high value $n_{21} = 0.65$ (panel (f), black curve). In contrast, for a value of N_n slightly higher than 1 (light blue curves in panels (e) and (f)), the dynamics is initially close to the separatrix but eventually the initial slow dynamics tends to the high SP population and low chromophore inversion through a series of the relaxation oscillations. The dynamics close to the separatrix is characterized by a wide range of oscillation times due to its highly nonlinear character. The initial dynamics is slowest (the "decision stage" of the bistable spaser that lasts $\gtrsim 1$ ps). The "decision time" is diverging infinitesimally close to the separatrix, as is the characteristic of any threshold (logical) amplifier.

The gain (amplification coefficient) of the spaser as a logical amplifier is the ratio of the high CW level to the threshold level of the SP population N_n. For this specific spaser with the chosen set of parameters, this gain is ≈ 60, which is more than sufficient for the digital information processing. Thus this spaser can make a high-gain, ~ 10 THz-bandwidth logical amplifier or dynamical memory cell with excellent prospects of applications.

The last but not the least regime to consider is that of the pulse pumping in the bistable spaser. In this case, the population inversion ($n_{21} = 0.65$) is created by a short pulse at $t = 0$ and simultaneously initial SP population N_n is created. Both are simulated as the initial conditions in Eqs. (4.11)–(4.14). The corresponding results are displayed in Figures 4.6g and 4.6h.

When the initial SP population exceeds the critical one of $N_n = 1$ (the blue, green, and red curves), the spaser responds with generating a short (duration less than 100 fs) pulse of the SP population (and the corresponding local fields) within a time $\lesssim 100$ fs (panel (g)). Simultaneously, the inversion is rapidly (within $\sim$100 fs) exhausted (panel (h)).

In contrast, when the initial SP population N_n is less than the critical one (that is, $N_n < 1$ in this specific case), the spaser rapidly (within a time $\lesssim 100$ fs) relaxes as $N_n \to 0$ through a series of realaxation oscillations—see the black and magenta curves in Figure 4.6g. The corresponding inversion decays in this case almost exponentially with a characteristic time $\sim$1 ps determined by the enhanced energy transfer to the SP mode in the metal—see the corresponding curves in panel (h). Note that the SP population decays faster when the spaser is above the generation threshold due to the stimulated SP emission leading to the higher local fields and enhanced relaxation.

4.2.7 Compensation of Loss by Gain and Spasing

4.2.7.1 Introduction to Loss Compensation by Gain A problem for many applications of plasmonics and metamaterials is posed by losses inherent in the interaction of light with metals. There are several ways to bypass, mitigate, or overcome the detrimental effects of these losses, which we briefly discuss below:

(i) The most common approach consists in employing effects where the losses are not fundamentally important such as SPP propagation used in sensing [13], ultramicroscopy [76, 77], and solar energy conversion [78]. For realistic losses, there are other effects and applications that are not prohibitively suppressed by the losses and useful, in particular, sensing based on SP resonances and surface-enhanced Raman scattering (SERS) [5, 13, 79–81].

(ii) Another promising idea is to use superconducting plasmonics to dramatically reduce losses [82–85]. However, this is only applicable for frequencies below the superconducting gap widths, that is, in the terahertz region.

(iii) Yet another proposed direction is using highly doped semiconductors where the Ohmic losses can be significantly lower due to much lower free carrier concentrations [86]. However, a problem with this approach may lie in the fact that the usefulness of plasmonic modes depends not on the loss *per se* but on the quality factor Q, which for doped semiconductors may not be higher than that for the plasmonic metals.

(iv) One of the alternative approaches to low-loss plasmonic metamaterials is based on our idea of the spaser: it is using a gain to compensate the dielectric (Ohmic) losses [87, 88]. In this case, the gain medium is included into the metamaterials. It surrounds the metal plasmonic component in the same manner as in the spasers. The idea is that the gain will provide quantum amplification compensating the loss in the metamaterials quite analogously to the spasers.

We will consider theory of the loss compensation in the plasmonic metamaterials using gain [22, 23]. Below we show that the full compensation or overcompensation of the optical loss in a dense resonant gain metamaterial leads to an instability that is resolved by its spasing (i.e., by becoming a generating spaser). We further show analytically that the conditions of the complete loss compensation by gain and the threshold condition of spasing—see Eqs. (4.27) and (4.29)—are identical. Thus the full compensation (overcompensation) of the loss by gain in such a metamaterial will cause spasing. This spasing limits (clamps) the gain—see Section 4.2.5—and, consequently, inhibits the complete loss compensation (overcompensation) at any frequency.

4.2.7.2 Permittivity of Nanoplasmonic Metamaterial
We will consider, for certainty, an isotropic and uniform metamaterial that, by definition, in a range of frequencies ω, can be described by the effective permittivity $\bar{\varepsilon}(\omega)$ and permeability $\bar{\mu}(\omega)$. We will concentrate on the loss compensation for the optical electric responses; similar consideration with identical conclusions for the optical magnetic responses is straightforward. Our theory is applicable for the true three-dimensional (3D) metamaterials whose size is much greater than the wavelength λ (ideally, an infinite metamaterial).

Consider a small piece of such a metamaterial with sizes much greater than the unit cell but much smaller than λ. Such a piece is a metamaterial itself. Let us subject this metamaterial to a uniform electric field $\mathbf{E}(\omega) = -\nabla \phi(\mathbf{r}, \omega)$ oscillating with frequency ω. Note that $\mathbf{E}(\omega)$ is the amplitude of the macroscopic electric field inside the metamaterial. We will denote the local field at a point $\mathbf{r}$ inside this metamaterial as $\mathbf{e}(\mathbf{r}, \omega) = -\nabla \varphi(\mathbf{r}, \omega)$. We assume standard boundary conditions

$$\varphi(\mathbf{r}, \omega) = \phi(\mathbf{r}, \omega), \qquad (4.31)$$

for $\mathbf{r}$ belonging to the surface S of the volume under consideration.

To present our results in a closed form, we first derive a homogenization formula used in Reference 89 (see also references cited therein). By definition, the electric displacement in the volume V of the metamaterial is given by a formula

$$\mathbf{D}(\mathbf{r}, \omega) = \frac{1}{V} \int_V \varepsilon(\mathbf{r}, \omega) \mathbf{e}(\mathbf{r}, \omega) dV, \qquad (4.32)$$

where $\varepsilon(\mathbf{r}, \omega)$ is a position-dependent permittivity. This can be identically expressed (by multiplying and dividing by the conjugate of the macroscopic field E^*) and, using the Gauss theorem, transformed to a surface integral as

$$\begin{aligned} D &= \frac{1}{VE^*(\omega)} \int_V \mathbf{E}^*(\omega) \varepsilon(\mathbf{r}, \omega) \mathbf{e}(\mathbf{r}, \omega) dV \\ &= \frac{1}{VE^*(\omega)} \int_S \phi^*(\mathbf{r}, \omega) \varepsilon(\mathbf{r}, \omega) \mathbf{e}(\mathbf{r}, \omega) d\mathbf{S}, \end{aligned} \qquad (4.33)$$

where we took into account the Maxwell continuity equation $\nabla [\varepsilon(\mathbf{r},\omega)\mathbf{e}(\mathbf{r},\omega)] = 0$. Now, using the boundary conditions of Eq. (4.31), we can transform it back to the volume integral as

$$D = \frac{1}{VE^*(\omega)} \int_S \varphi^*(\mathbf{r})\varepsilon(\mathbf{r},\omega)\mathbf{e}(\mathbf{r},\omega)d\mathbf{S}$$

$$= \frac{1}{VE^*(\omega)} \int_V \varepsilon(\mathbf{r},\omega) |\mathbf{e}(\mathbf{r},\omega)|^2 dV. \quad (4.34)$$

From the last equality, we obtain the required homogenization formula as an expression for the effective permittivity of the metamaterial

$$\bar{\varepsilon}(\omega) = \frac{1}{V|E(\omega)|^2} \int_V \varepsilon(\mathbf{r},\omega) |\mathbf{e}(\mathbf{r},\omega)|^2 dV. \quad (4.35)$$

4.2.7.3 Plasmonic Eigenmodes and Effective Resonant Permittivity of Metamaterials

This piece of the metamaterial with the total size $R \ll \lambda$ can be treated in the quasistatic approximation. The local field inside the nanostructured volume V of the metamaterial is given by the eigenmode expansion [47, 90, 91]

$$\mathbf{e}(\mathbf{r},\omega) = \mathbf{E}(\omega) - \sum_n \frac{a_n}{s(\omega) - s_n} \mathbf{E}_n(\mathbf{r}), \quad (4.36)$$

$$a_n = \mathbf{E}(\omega) \int_V \theta(\mathbf{r})\mathbf{E}_n(\mathbf{r})dV,$$

where we remind that $\mathbf{E}(\omega)$ is the macroscopic field. In the resonance, $\omega = \omega_n$, only one pole term in Eq. (4.36) dominates, and it becomes

$$\mathbf{e}(\mathbf{r},\omega) = \mathbf{E}(\omega) + i\frac{a_n}{\operatorname{Im} s(\omega_n)} \mathbf{E}_n(\mathbf{r}). \quad (4.37)$$

The first term in this equation corresponds to the mean (macroscopic) field and the second one describes the deviations of the local field from the mean field containing contributions of the hot spots [92]. The mean root square ratio of the second term (local field) to the first (mean field) is estimated as

$$\sim \frac{f}{\operatorname{Im} s(\omega_n)} = \frac{fQ}{s_n(1-s_n)}, \quad (4.38)$$

where we took into account that $E_n \sim V^{-1/2}$, and

$$f = \frac{1}{V} \int_V \theta(\mathbf{r})dV, \quad (4.39)$$

where f is the metal fill factor of the system, and Q is the plasmonic quality factor. Deriving expression (4.38), we have also taken into account an equality $\operatorname{Im} s(\omega_n) = s_n(1 - s_n)/Q$, which is valid in the assumed limit of the high quality factor, $Q \gg 1$ (see the next paragraph).

For a good plasmonic metal, $Q \gg 1$—see Reference 71. For most metal-containing metamaterials, the metal fill factor is not small, typically $f \gtrsim 0.5$. Thus, it is very realistic to assume that

$$\frac{fQ}{s_n(1 - s_n)} \gg 1. \tag{4.40}$$

If so, the second (local) term of the field (4.37) dominates and, with good precision, the local field is approximately the eigenmode's field

$$\mathbf{e}(\mathbf{r}, \omega) = i \frac{a_n}{\operatorname{Im} s(\omega_n)} \mathbf{E}_n(\mathbf{r}). \tag{4.41}$$

Substituting this into Eq. (4.35), we obtain a homogenization formula

$$\bar{\varepsilon}(\omega) = b_n \int_V \varepsilon(\mathbf{r}, \omega) \left[\mathbf{E}_n(\mathbf{r})\right]^2 dV, \tag{4.42}$$

where $b_n > 0$ is a real positive coefficient whose specific value is

$$b_n = \frac{1}{3V} \left(\frac{Q \int_V \theta(\mathbf{r}) \mathbf{E}_n(\mathbf{r}) dV}{s_n(1 - s_n)} \right)^2 \tag{4.43}$$

Using Eq. (4.42), it is possible to show that the effective permittivity (4.42) simplifies exactly to

$$\bar{\varepsilon}(\omega) = b_n \left[s_n \varepsilon_m(\omega) + (1 - s_n)\varepsilon_h(\omega)\right], \tag{4.44}$$

where $\varepsilon_h(\omega)$ is the permittivity of the host gain medium—see Eq. (4.45).

4.2.8 Conditions of Loss Compensation by Gain and Spasing

In the case of the full inversion (maximum gain) and in the exact resonance, the host medium permittivity acquires the imaginary part describing the stimulated emission as given by the standard expression

$$\varepsilon_h(\omega) = \varepsilon_d - i \frac{4\pi}{3} \frac{|\mathbf{d}_{12}|^2 n_c}{\hbar \Gamma_{12}}, \tag{4.45}$$

where $\varepsilon_d = \operatorname{Re} \varepsilon_h$, $\mathbf{d}_{12}$ is a dipole matrix element of the gain transition in a chromophore center of the gain medium, Γ_{12} is a spectral width of this transition, and n_c

is the concentration of these centers (these notations are consistent with those used in Sections 4.2.4–4.2.6). Note that if the inversion is not maximum, then this and subsequent equations are still applicable if one sets as the chromophore concentration n_c the inversion density: $n_c = n_2 - n_1$, where n_2 and n_1 are the concentrations of the chromophore centers of the gain medium in the upper and lower states of the gain transition, respectively.

The condition for the full electric loss compensation in the metamaterial and amplification (overcompensation) at the resonant frequency $\omega = \omega_n$ is

$$\operatorname{Im} \bar{\varepsilon}(\omega) \leq 0. \tag{4.46}$$

Taking Eq. (4.44) into account, this reduces to

$$s_n \operatorname{Im} \varepsilon_m(\omega) - \frac{4\pi}{3} \frac{|\mathbf{d}_{12}|^2 n_c (1 - s_n)}{\hbar \Gamma_{12}} \leq 0. \tag{4.47}$$

Finally, taking into account that $\operatorname{Im} \varepsilon_m(\omega) > 0$, we obtain from Eq. (4.47) the condition of the loss (over)compensation as

$$\frac{4\pi}{3} \frac{|\mathbf{d}_{12}|^2 n_c [1 - \operatorname{Re} s(\omega)]}{\hbar \Gamma_{12} \operatorname{Re} s(\omega) \operatorname{Im} \varepsilon_m(\omega)} \geq 1, \tag{4.48}$$

where the strict inequality corresponds to the overcompensation and net amplification. In Eq. (4.45), we have assumed nonpolarized gain transitions. If these transitions are all polarized along the excitation electric field, the concentration n_c should be multiplied by a factor of 3.

Equation (4.48) is a fundamental condition, which is precise (assuming that the requirement (4.40) is satisfied, which is very realistic for metamaterials) and general. Moreover, it is fully analytical and, actually, very simple. Remarkably, it depends only on the material characteristics and does not explicitly contain any geometric properties of the metamaterial system or the local fields. (Note that the system's geometry does affect the eigenmode frequencies and thus enters the problem implicitly.) In particular, the hot spots, which are prominent in the local fields of nanostructures [47, 92], are completely averaged out due to the integrations in Eqs. (4.35) and (4.42).

The condition (4.48) is completely nonrelativistic (quasistatic)—it does not contain speed of light c, which is characteristic of also the spaser. It is useful to express this condition also in terms of the total stimulated emission cross section $\sigma_e(\omega)$ (where ω is the central resonance frequency) of a chromophore of the gain medium as

$$\frac{c \sigma_e(\omega) \sqrt{\varepsilon_d} n_c [1 - \operatorname{Re} s(\omega)]}{\omega \operatorname{Re} s(\omega) \operatorname{Im} \varepsilon_m(\omega)} \geq 1. \tag{4.49}$$

We see that Eq. (4.48) *exactly* coincides with a spasing condition expressed by Eq. (4.27). This brings us to an important conclusion: the full compensation (overcompensation) of the optical losses in a metamaterial (which is resonant and dense enough to satisfy condition (4.40)) and the spasing occur under precisely the same conditions.

We have considered above in Section 4.2.4 the conditions of spasing, which are equivalent to (4.49). These are given by one of equivalent conditions of Eqs. (4.27), (4.29), and (4.48). It is also illustrated in Figure 4.4. We stress that exactly the same conditions are for the full loss compensation (overcompensation) of a dense resonant plasmonic metamaterial with gain.

We would like also to point out that the criterion given by the equivalent conditions of Eqs. (4.27), (4.29), (4.48), or (4.49) is derived for localized SPs, which are describable in the quasistatic approximation, and is not directly applicable to the propagating plasmonic modes (SPPs). However, we expect that very localized SPPs, whose wave vector $k \lesssim l_s$, can be described by these conditions because they are, basically, quasistatic. For instance, the SPPs on a thin metal wire of a radius $R \lesssim l_s$ are described by a dispersion relation [93]

$$k_x \approx \frac{1}{R} \left[-\frac{\varepsilon_m}{2\varepsilon_d} \left(\ln \sqrt{-\frac{4\varepsilon_m}{\varepsilon_d}} - \gamma \right) \right]^{-1/2}. \quad (4.50)$$

where $\gamma \approx 0.57721$ is the Euler constant. This relation is obviously quasistatic because it does not contain speed of light c.

4.2.8.1 Discussion of Spasing and Loss Compensation by Gain
This fact of the equivalence of the full loss compensation and spasing is intimately related to the general criteria of the thermodynamic stability with respect to small fluctuations of electric and magnetic fields—see Chapter 9 of Reference 94,

$$\operatorname{Im} \bar{\varepsilon}(\omega) > 0, \quad \operatorname{Im} \bar{\mu}(\omega) > 0, \quad (4.51)$$

which must be *strict* inequalities for all frequencies for electromagnetically stable systems. For systems in thermodynamic equilibrium, these conditions are automatically satisfied.

However, for the systems with gain, the conditions (4.51) can be violated, which means that such systems can be electromagnetically unstable. The first of conditions (4.51) is opposite to Eqs. (4.46) and (4.48). This has a transparent meaning: the electrical instability of the system is resolved by its spasing.

The significance of these stability conditions for gain systems can be elucidated by the following *gedanken* experiment. Take a small isolated piece of such a metamaterial (which is a metamaterial itself). Consider that it is excited at an optical frequency ω either by a weak external optical field **E** or acquires such a field due to fluctuations (thermal or quantum). The energy density $\mathcal{E}$ of such a system is given by the Brillouin formula [94]

$$\mathcal{E} = \frac{1}{16\pi} \frac{\partial \omega \operatorname{Re} \bar{\varepsilon}}{\partial \omega} |\mathbf{E}|^2. \qquad (4.52)$$

Note that for the energy of the system to be definite, it is necessary to assume that the loss is not too large, $|\operatorname{Re} \bar{\varepsilon}| \gg \operatorname{Im} \bar{\varepsilon}$. This condition is realistic for many metamaterials, including all potentially useful ones.

The internal optical energy-density loss per unit time Q (i.e., the rate of the heat-density production in the system) is [94]

$$Q = \frac{\omega}{8\pi} \operatorname{Im} \bar{\varepsilon} |\mathbf{E}|^2. \qquad (4.53)$$

Assume that the internal (Ohmic) loss dominates over other loss mechanisms such as the radiative loss, which is also a realistic assumption since the Ohmic loss is very large for the experimentally studied systems and the system itself is very small (the radiative loss rate is proportional to the volume of the system). In such a case of the dominating Ohmic losses, we have $d\mathcal{E}/dt = Q$. Then Eqs. (4.52) and (4.53) can be resolved together yielding the energy $\mathcal{E}$ and electric field $|\mathbf{E}|$ of this system to evolve with time t exponentially as

$$|\mathbf{E}| \propto \sqrt{\mathcal{E}} \propto e^{-\Gamma t}, \quad \Gamma = \omega \operatorname{Im} \bar{\varepsilon} \Big/ \frac{\partial(\omega \operatorname{Re} \bar{\varepsilon})}{\partial \omega}. \qquad (4.54)$$

We are interested in a resonant case when the metamaterial possesses a resonance at some eigenfrequency $\omega_n \approx \omega$. For this to be true, the system's behavior must be plasmonic, that is, $\operatorname{Re} \bar{\varepsilon}(\omega) < 0$. Then the dominating contribution to $\bar{\varepsilon}$ comes from a resonant SP eigenmode n with a frequency $\omega_n \approx \omega$. In such a case, the dielectric function [47] $\bar{\varepsilon}(\omega)$ has a simple pole at $\omega = \omega_n$. As a result, $\partial(\omega \operatorname{Re} \bar{\varepsilon})/\partial \omega \approx \omega \partial \operatorname{Re} \bar{\varepsilon}/\partial \omega$ and, consequently, $\Gamma = \gamma_n$, where γ_n is the SP decay rate given in Reference 71, and the metal dielectric function ε_m is replaced by the effective permittivity $\bar{\varepsilon}$ of the metamaterial. Thus, Eq. (4.54) is fully consistent with the spectral theory of SPs.

If the losses are not very large so that energy of the system is meaningful, the Kramers-Kronig causality requires [94] that $\partial(\omega \operatorname{Re} \bar{\varepsilon})/\partial \omega > 0$. Thus, $\operatorname{Im} \bar{\varepsilon} < 0$ in Eq. (4.54) would lead to a negative decrement,

$$\Gamma < 0, \qquad (4.55)$$

implying that the initial small fluctuation starts exponentially grow in time in its field and energy, which is an instability. Such an instability is indeed not impossible: it will result in spasing that will eventually stabilize $|\mathbf{E}|$ and $\mathcal{E}$ at finite stationary (CW) levels of the spaser generation.

Note that the spasing limits (clamps) the gain and population inversion making *the net gain to be precisely zero* [17] in the stationary (continuous wave or CW) regime see Section 4.2.6 and Figure 4.5b. Above the threshold of the spasing, the population inversion of the gain medium is clamped at a rather low level $n_{21} \sim 1\%$. The corresponding net amplification in the CW spasing regime is exactly zero, which is a condition for the CW regime. This makes the complete loss compensation and its overcompensation impossible in a dense resonant metamaterial where the feedback is created by the internal inhomogeneities (including its periodic structure) and the facets of the system.

4.2.8.2 Discussion of Published Research on Spasing and Loss Compensations
In an experimental study of the lasing spaser [42], a nanofilm of PbS quantum dots (QDs) was positioned over a 2D metamaterial consisting of an array of negative split ring resonators. When the QDs were optically pumped, the system exhibited an increase of the transmitted light intensity on the background of a strong luminescence of the QDs but apparently did not reach the lasing threshold. The polarization-dependent loss compensation was only ~ 1 %. Similarly, for an array of split ring resonators over a resonant quantum well, where the inverted electron-hole population was excited optically [95], the loss compensation did not exceed ~ 8 %. The relatively low loss compensation in these papers may be due either to random spasing and/or spontaneous or amplified spontaneous emission enhanced by this plasmonic array, which reduces the population inversion.

A dramatic example of possible random spasing is presented in Reference 52. The system studied was a Kretschmann-geometry SPP setup [96] with an added $\sim 1 - \mu m$ polymer film containing Rodamine 6G dye in the $n_c = 1.2 \times 10^{19} cm^{-3}$ concentration. When the dye was pumped, there was outcoupling of radiation in a range of angles. This was a threshold phenomenon with the threshold increasing with the Kretschmann angle. At the maximum of the pumping intensity, the widest range of the outcoupling angles was observed, and the frequency spectrum at every angle narrowed to a peak near a single frequency $\hbar\omega \approx 2.1$ eV.

These observations of Reference 52 can be explained by the spasing where the feedback is provided by roughness of the metal. At the high pumping, the localized SPs (hots spots), which possess the highest threshold, start to spase in a narrow frequency range around the maximum of the spasing criterion—the left-hand side of Eq. (4.48). Because of the sub-wavelength size of these hot spots, the Kretschmann phase-matching condition is relaxed, and the radiation is outcoupled into a wide range of angles.

The SPPs of Reference 52 excited by the Kretschmann coupling are short-range SPPs, very close to the antisymmetric SPPs. They are localized at subwavelength distances from the surface, and their wave length in the plane is much shorter the ω/c. Thus they can be well described by the quasistatic approximation and the present theory is applicable to them. Substituting the above-given parameters of the dye and the extinction cross section $\sigma_e = 4 \times 10^{-16}$ cm^2 into Eq. (4.49), we find that the conditions of Reference 52 are above the threshold, supporting our assertion of the spasing. Likewise, the amplified spontaneous emission and, possibly spasing, appear to have prevented the full loss compensation in an SPP system of Reference 63. Note that recently, random spasing for rough surfaces surrounded by dye gain media was shown experimentally in two independent observations [39, 97].

Note that the long-range SPPs of Reference 66 are localized significantly weaker (at distances $\sim \lambda$) than those excited in Kretschmann geometry. Thus the long-range SPPs experience a much weaker feedback, and the amplification instead of the spasing can be achieved. Generally, the long-range SPPs are fully electromagnetic (non-quasistatic) and are not describable in the present theory. Similarly, relatively weakly confined, full electromagnetic are symmetric SPP modes on thin gold strips in Reference 75 where the amplification has been demonstrated.

As we have already discussed in conjunction with Figure 4.4, the spasing is readily achievable with the gain medium containing common DBGSs or dyes. There have been numerous experimental observations of the spaser. Among them is a report of an SP spaser with a 7-nm gold nanosphere as its core and a laser dye in the gain medium [30], observations of the SPP spasers (also known as nanolasers) with silver as a plasmonic-core metal and DBGS as the gain medium with a 1D confinement [33, 36], a tight 2D confinement [34], and a 3D confinement [35]. There also has been a report on observation of an SPP microcylinder spaser [98]. A high efficiency room-temperature semiconductor spaser with a DBGS InGaAS gain medium operating near 1.5 μm (i.e., in the communication near-ir range) has been reported [36].

The research and development in the area of spasers (nanolasers) as quantum nano-generators is very active and will undoubtedly lead to further rapid advances. The next in line is the spaser as an ultrafast nanoamplifier, which is one of the most important tasks in nanotechnology.

In periodic metamaterials, plasmonic modes generally are propagating waves (SPPs) that satisfy Bloch theorem [99] and are characterized by quasi-wavevector **k**. These are propagating waves except for the band edges where $\mathbf{ka} = \pm\pi$, where **a** is the lattice vector. At the band edges, the group velocity v_g of these modes is zero, and these modes are localized, that is, they are SPs. Their wave function is periodic with period $2a$, which may be understood as a result of the Bragg reflection from the crystallographic planes. Within this $2a$ period, these band-edge modes can indeed be treated quasistatically because $2a \ll l_s, \bar{\lambda}$. If any of the band-edge frequencies is within the range of compensation (where the condition (4.27)—or, (4.29)—is satisfied), the system will spase. In fact, at the band edge, this metamaterial with gain is similar to a distributed feedback (DFB) laser [100]. It actually is a DFB spaser, which, as all the DFB lasers, generates in a band-edge mode.

In fact, there have recently been two observations of lasing spasers with optical pumping generating on the band-edge modes [43, 44], see also our Research Highlight in Reference 45. In Reference 43, the metal component of the lasing spaser was a periodic nanohole array in a silver nanofilm, and the gain component was semiconductor diode of InGaAs/InP. In Reference 44, the plasmonic metal component was a periodic planar array of either gold or silver nanoparticle while the gain medium was a polymer nanolayer composed of polyurethane and IR-140 dye.

These spasers [43, 44] generated well-collimated radiation beam normally to the surfaces of the nanofilms. This suggests that they generated on modes at the boundary of the second Brillouin zone (in the extended-zone representation). This follows from the fact that the phases of the underlying mode in the plane is the same at the adjacent crystallographic planes. Note that the the lowest DFB modes have these phases differing by π and, correspondingly, are dark, not emitting in the far field. These modes, when generating, will still clamp the inversion and limit the loss compensation. Similar situation was, probably, the case in Reference 101 where a limited loss compensation in extraordinary transmission was observed.

Moreover, not only the SPPs, which are exactly at the band edge, will be localized and be able to supply the feedback for spasing. Due to unavoidable disorder caused by fabrication defects in metamaterials, there will be scattering of the SPPs from these defects. Close to the band edge, the group velocity becomes small, $v_g \to 0$. Because the scattering cross section of any wave is $\propto v_g^{-2}$, the corresponding SPPs experience Anderson localization [102]. Also, there always will be SPs nanolocalized at the defects of the metamaterial, whose local fields are hot spots. Each of such hot spots within the bandwidth of conditions (4.27) or (4.29) will be a generating spaser, which clamps the inversion and precludes the full loss compensation. Such spasing at random defects has been observed experimentally [39].

4.3 ADIABATIC HOT-ELECTRON NANOSCOPY

4.3.1 Introduction to Adiabatic Hot-Electron Nanoscopy

Near-field microscopy with nanoscopic resolution (nanoscopy) has become a widely used plasmonic-based technology, see, for example, References 1, 103–107. It is based on scanning of an object by nanoscopic nanolocalized near-field of SPs. There are several approaches to the plasmonic nanoscopy. In one of them—called aperture NSOM (near-field scanning optical microscope) a metallized tapered optical fiber is used to create a nanoscopic near-field optical source used for scanning, see, for example, References 103 and 108. The other type—apertureless NSOM—is based on enhanced optical field at the end of tapered plasmonic-metal probe, see, for example, References 106 and 109.

There are two known interrelated problems with the apretureless NSOMs: (i) the fraction of energy of the laser sources delivered to the nanotip is very small, $\sim 10^{-5}$, due to its small cross section, and (ii) there is a significant background unrelated to the tip coming from the nanoscale region illuminated by the excitation source on the object. A radical solution to both these problems has been introduced [93] by an approach of adiabatic concentration (see also References 110 and 111).

In what follows, in Section 4.3.2, we will describe the principles of adiabatic concentration and briefly review experimental progress achieved in this area. In Section 4.3.3, we will describe adiabatic concentration as a nanoscale source of hot electrons and nanoscopy based on it.

4.3.2 Adiabatic Concentration of Optical Energy and Hot Electrons

The central problem of the nanooptics is the delivery and concentration (nanofocusing) of the optical radiation energy on the nanoscale, which is formidable because the wavelength of light is on the microscale, many orders of magnitude too large. Coupling laser radiation to the nanoscale through, for example, tapered optical fibers [112] or by focusing on metal tips [113] leads to an enormous loss: only a miniscule part of the excitation energy is transferred to the nanoscale.

Following Reference 93, we show that it is possible to focus and concentrate in three dimensions the optical radiation energy on the nanoscale without major losses. This can be done by exciting the SPPs propagating toward a tip of a tapered metal-nanowire surface-plasmonic waveguide. This propagation of SPPs causes their rapid adiabatic slowing down and asymptotic stopping. This phenomenon leads to giant concentration of energy on the nanoscale. The SPPs are adiabatically transformed into localized SPs that are purely electric oscillations that can and do nano-localize [47] leading to the 3D nanofocusing.

When the radius of the plasmonic taper becomes significantly less than the smallest electromagnetic scale, which is skin depth, $R \ll l_s \approx 25$ nm for most metals in the optical range, then the only parameter of the dimensionality of length that determines SP wave vector, k, becomes R. This implies that the SP wavelength, λ_{SP}, its longitudinal concentration length, and its extension in space outside of the taper are all $\sim R$. This leads to very simple scaling of the local electric field components normal E_x and longitudinal E_z, local optical intensity I, SP wave vector $k_z = 2\pi/\lambda_{SP}$, and its decay length L_{SP} as

$$\lambda_{SP} = \frac{2\pi}{k_z} \sim R, I \sim I^{(0)} \left(\frac{R}{l_s}\right)^{-3}, E_z \sim E_x \propto E^{(0)} \left(\frac{R}{l_s}\right)^{-3}, L_{SP} \sim QR, \quad (4.56)$$

where the upper index $\ldots^{(0)}$ refers to quantities at $R \sim l_s$.

This scaling suggests that the SP undergo *3D adiabatic concentration* in a region of volume $\sim R^3$. Because a realistic minimum radius of the adiabatic taper is $R_{min} \sim$ 3 nm, an expected increase of the local intensity is by a factor $I/I_0 \sim (l_s/R_{min})^3 \sim 10^3$. Note that the scaling of Eq. 4.56 is obtained neglecting the propagation loss of the SPs. As the quantitative results to be discussed in this section show, this assumption is realistic: the fraction of optical energy delivered to the tip of the taper is $\sim 50\%$. Thus the adiabatic compression is, indisputably, the most efficient method of delivering optical energy to the nanoscale.

In Figure 4.7, we display the local optical-field intensity at the surface of a silver obtained [94] from quantitative eikonal approximation also called Wentzel-Kramers-Brillouin (WKB) or quasiclassical approximation in quantum mechanics [114]. These intensity distributions clearly show the adiabatic concentration by a factor of $I/I_0 \approx 10^3$, in line with an estimate given in the previous paragraph. The adiabatic concentration effect causes nanofocusing: formation of a nanoscopic hot spot of the local fields at the end of the taper, which is clearly seen at the surface of the metal in Figure 4.7a and in the volume of the tip in Figure 4.7b. The latter is due to the fact that the adiabatic nanofocusing occurs for $R \ll l_s$ forcing the optical fields into the metal.

Such high efficiency of the adiabatic concentration is due to the fact that the adiabaticity condition is satisfied very well for the example considered. This eliminates reflection of the SPP waves. Another fundamental reason for this efficiency is that

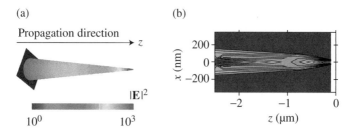

FIGURE 4.7 (a) Geometry of the nanoplasmonic tapered waveguide. The propagation direction of the SPPs is indicated by the arrow. Intensity $I(\mathbf{r}) = |\mathbf{E}(\mathbf{r})|^2$ of the local fields relative to the excitation field is shown by color. The scale of the intensities is indicated by the color bar in the center. (b) Local electric field intensity $I(\mathbf{r})$ is shown in the longitudinal cross section of the system. The radius of the waveguide gradually, adiabatically decreases from 50 to 2 nm. Adapted from Reference 93. (*For a color version of this figure, see the color plate section.*)

the SPPs on a straight adiabatic cone are dark waves, because $k_S P \gg \omega/c$. Thus the adiabatic compression does not suffer radiative loss in contrast to curved adiabatic waveguides considered in transformation optics approach [115]. Similar high efficiency and absence of back-reflected SPP waves in nanoplasmonic tapers were confirmed in direct experiments [116].

Another fundamentally important property of adiabatic concentration is that it is a wide-band effect: it exists in the entire plasmonic region, which for silver extends from near-UV to mid-IR. Consequently, the adiabatic nanofocusing is potentially an ultrafast effect: the spectral-width-limited duration of a pulse at the tip can be as short as several hundred attoseconds. Experimentally, adiabatic nanofocusing of ultrashort pulses has recently been demonstrated experimentally [117,118]. In these experiments, the adiabatic nanofocusing was achieved where ultrashort pulses of duration ~10 fs were obtained in the taper-apex hot spots with size ~10 nm.

Returning to theory of adiabatic nanofocusing, the underlying cause of this effect is universal scaling of SPP phase velocity v_p and group velocity v_g (cf. Eq. 4.56)

$$v_p \sim v_g \sim \omega R. \tag{4.57}$$

This scaling also follows from the fact that the only quantity with the dimensionality of distance relevant to the system is the instantaneous radius, R, of the taper. This scaling is in an excellent agreement with calculation results shown in Figure 4.8a where the linear dependence on z and, consequently on $R \propto z$ is clearly seen. Note that it can well be described by asymptotic Eq. (4.50). The slowing down of SPPs at the tip of a taper has been quantitatively confirmed in direct time-resolved experiments [119].

Thus the SPP propagation velocities, both v_g and v_p, tend to zero at the apex (this is adiabatic slowing-down and asymptotic stopping of the SPPs, or effect of "slow light"). As a result of this SPP slowing down, the optical energy accumulates at the apex and the fields oscillate in space with spatial frequency infinitely increasing toward the apex of the tip, which is a point of essential singularity – see Figure 4.8b.

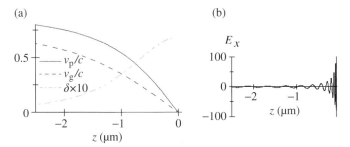

FIGURE 4.8 (a) Phase velocity v_p, group velocity v_g, and adiabatic parameter δ (scaled by a factor of 10) are shown as functions of the plasmonic taper radius. (b) Radial optical electric field at the surface of the metal taper in the units of the excitation field against the coordinate. Calculations are for silver with parameters from Reference 74. Adapted from Reference 93.

Properties of the adiabatic concentration can be traced in Figure 4.9 where normal component E_x [panel (a)] and tangential component E_z of the SPP electric field is displayed as a snapshot distribution in space in the xz plane. As one can see, the adiabatic concentration mostly occurs for $R \lesssim l_s$ when nanofocusing in all three dimensions is evident concurrent with dramatically faster oscillations of the local field in the propagation direction (z).

4.3.3 Adiabatic Hot-Electron Nanoscope

Due to its ability to deliver a significant fraction of the far-field energy to the near field, the adiabatic concentration is an unparalleled approach to nanoscopy. In fact, it has been used as a foundation of nanoscopy and ultrasensing in a number of

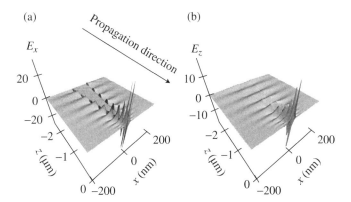

FIGURE 4.9 Snapshot of instantaneous fields. Normal component E_x (a) and tangential component E_z (b) of the local optical electric field are shown in the longitudinal-radial cross-section (xz) plane of the system. The fields are plotted in the units of the far-zone (excitation) field. Adapted from Reference 93.

works [76, 77, 117, 118, 120–125]. Here we will present a recent article [121] that has introduced a new approach to nanoscopy based on hot electrons generated in the process of adiabatic nanofocusing.

Decay of surface plasmons has always been a parasitic effect that reduces plasmonic quality factor Q, decreases local fields, limits sensitivity of nanoplasmonic detection, negatively affects plasmonic-enhanced photovoltaic devices, etc. There are two major mechanisms of plasmonic decay: radiative loss and radiationless decay. As we have already pointed out in Section 4.3.2, the radiative decay of SPPs in the process of adiabatic concentration is negligible if the adiabaticity condition is well satisfied, which is the case in known experimental implementations. The primary act of the radiationless decay occurs mainly via dephasing (Landau damping), which leads to decay of an SPP into electron–hole pair. It leaves energy in the electron system but destroys coherence of the collective electron oscillations, which eliminates enhanced local fields characteristic of nanoplasmonic phenomena. This is a rapid process that limits plasmonic lifetimes to $\sim 10 - 100$ fs depending on metal and spectral range.

Recently a positive side of the plasmonic dephasing has been revealed by utilizing the hot electrons created in this process to create electric current in Schottky junctions [126, 127]. The Schottky (metal-semiconductor) junction approach provides an advantage with respect to the conventional photocurrent generation based on interband transitions in semiconductors because Schottky barrier can be significantly lower than the bandgap required to sustain satisfactory energy efficiency in semiconductor junctions.

However, both energy and quantum efficiency of the Schottky-junction-based processes are quite low (in a $\lesssim 1\%$ range). The main reason for that is the limitation imposed by linear momentum conservation at a smooth interface. It was shown that a rough interface in the Schottky contact increases photocurrent-generation efficiency by two orders of magnitude with respect to that with a smooth interface [128].

We have shown [121] that the adiabatic nanoconcentration (also known as nanofocusing or superfocusing) of SPPs with Schottky-contact photocurrent generation resolves the efficiency problem even for a smooth-surface taper yielding a high quantum efficiency and photocurrent responsivity—see the later part of this section. A schematic of the experiment [121] is shown in Figure 4.10a. Nanotip of a gold taper (see its geometry in Figure 4.10b) contains a grating coupling light to SPPs. These propagate toward the apex that is in contact with a semiconductor (n-doped GaAs) forming a Schottky diode working at a weak reverse bias. The tip is mounted at a cantilever of an atomic force microscope (AFM), which scans the object. The output of this nanoscope is both AFM object height and the Schottky current.

AFM-obtained topography of one of the test objects is shown in Figure 4.10c. It is microscopic area of a GaAs surface where a zig-zag of oxide lines of a nanoscopic width is fabricated by chemical oxidation. Figure 4.10d displays distribution of the Schottky photocurrent. One can see that the oxide lines indeed produce zero current while the clean semiconductor surface produces a uniform high current distribution. The resolution is excellent and the signal-to-noise (S/N) ratio is very high.

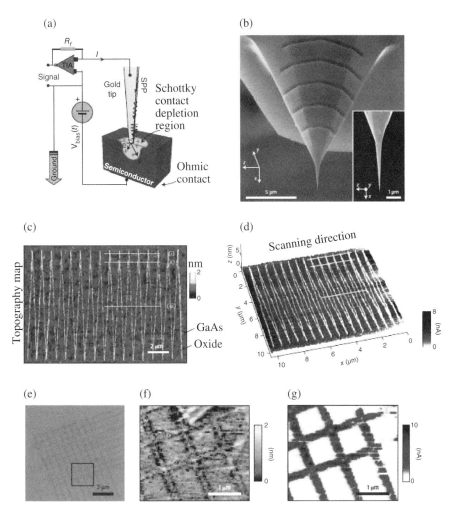

FIGURE 4.10 Schematic of experimental setup and 3D hot-electron maps of specific custom-realized locally patterned samples. (a) Electric and plasmonic schematic of experiment. (b) Scanning electron microscopy (SEM) images of grating and adiabatic cone used in experiments. (c) High resolution atomic force microscope (AFM) topography and height profiles of a continuous oxide pattern deposited on GaAs made by a top-down fabrication technique. (d) Photocurrent imaging overlaid on 3D topography, showing simultaneously the achieved current and topographic resolution. (e) SEM image of Ga ion-implanted GaAs sample. (f) and (g) Topography and plasmonic hot-electron maps, generated at 980-nm laser excitation, acquired in the region indicated by a black rectangle in panel (e). Adapted from Reference 121.

Another studied object is depicted in Figure 4.10e by scanning electron microscopy (SEM). This is a rectangular microscopic grid of nanoscopic-width lines obtained by implantation of Ga ions. As Figure 4.10f shows, AFM contrast of this surface is very poor. In contrast, the adiabatic SPP Schottky nanoscopy, shown in Figure 4.10g, produces a very high contrast, low noise picture. This result shows this type of nanoscopy to be a true "chemical nanoscale vision", because, as inherent in the Schottky effect, it is very sensitive to the position of the Fermi level of the semiconductor and, correspondingly, to its doping with a nanoscale spatial resolution.

To understand the origin of a high photocurrent responsivity in experiments, consider the Fowler's formula [129] for probability η of over-the-barrier passing of a Schottky-junction barrier,

$$\eta = \frac{(\hbar\omega - e\Phi_B)^2}{4E_F\hbar\omega}, \tag{4.58}$$

where e is electron charge, E_F is the Fermi energy of the semiconductor, and Φ_B is the height of the Schottky barrier between the metal and the dielectric. An especially small factor in this expression for η is $(\hbar\omega - e\Phi_B)/E_F$ which can realistically be as small as $\sim 10^{-2}$. This factor is, on the order of magnitude, a steric angle at the normal to the surface within which electrons must fly to pass over the Schottky barrier when the tangential linear momentum is conserved.

When the conservation of the tangential linear momentum is conserved by collisions with metal surfaces, this expression for η changes to a different one, as derived in Reference 121,

$$\eta = \frac{\hbar\omega - e\Phi_B}{2\hbar\omega}, \tag{4.59}$$

which is by a factor $\sim E_F/(\hbar\omega - e\Phi_B) \gg 1$ greater than that predicted by Fowler's formula (4.58).

Figure 4.11a illustrates how collisions with the surfaces of the plasmonic taper relax the linear momentum conservation. Analyzing it, it is useful to take into account that the mean free path of an electron in metal is typically ~ 40 nm and is much greater than the taper radius at the apex. Therefore an electron experiences many collisions on its free path, as depicted in schematic of Figure 4.11a. Each such collision rotates its velocity by double the apex angle, 2π. As a result, electron gets into the escape cone, leaves the metal, and enters the semiconductor creating the photocurrent. Figure 4.11b displays experimental results superimposed on theoretical curves. As one can see, Fowler's formula (4.58) dramatically underestimates current while Eq. (4.59) is in excellent agreement with it.

This described effect, due to its high efficiency (on the order of tens per cent) is perfectly suited to be in the foundation of a new chemical vision nanoscopy. This one can find application in different areas, in particular in microelectronics industry due to its demonstrated high sensitivity to semiconductor doping. Another plausible application may be in molecular and cell biology since many biological molecules

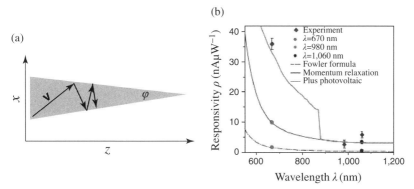

FIGURE 4.11 (a) Schematic of electron momentum relaxation in a tapered waveguide. With each collision with the wall, electron velocity direction (indicated by the arrows) rotates clockwise by a 2φ angle, where φ is the apex angle of the waveguide cone. (b) Experimental device responsivity ρ (blue diamonds) and its theoretical curve as a function of wavelength and considering Fowler equation (blue dotted line) or modified equation (blue continuous line). The red line shows the same calculations, adding the photovoltaic (PV) contribution obtained from known literature. (*For a color version of this figure, see the color plate section.*)

are organic semiconductors. This type of nanoscopy can provide information on electronic transitions at the surface as the red curve in Figure 4.11b demonstrates. Finally, due to its high efficiency, applications of this approach to solar energy conversion appears to be promising.

ACKNOWLEDGMENTS

The primary support for this work was provided by MURI Grant No. N00014-13-1-0649 from the U.S. Office of Navy Research. Additional support was provided by Grant No. DE-FG02-11ER46789 from the Materials Sciences and Engineering Division, Office of the Basic Energy Sciences, Office of Science, U.S. Department of Energy, and Grant No. DE-FG02-01ER15213 from the Chemical Sciences, Biosciences and Geosciences Division, Office of the Basic Energy Sciences, Office of Science, U.S. Department of Energy.

REFERENCES

[1] L. Novotny and B. Hecht, *Principles of Nano-Optics* (Cambridge University Press, Cambridge, New York, 2006).

[2] T. V. Shahbazyan and M. I. Stockman (eds.), *Plasmonics: Theory and Applications* (Springer, Dordrecht, Heidelberg, New York, London, 2013).

[3] D. J. Bergman and D. Stroud, "Properties of macroscopically inhomogeneous media," in *Solid State Physics*, edited by H. Ehrenreich and D. Turnbull (Academic Press, Boston, 1992), Vol. 46, pp. 148–270.

[4] H. A. Atwater, "The promise of plasmonics," *Sci. Am.* **296**, 56–63 (2007).

[5] J. N. Anker, W. P. Hall, O. Lyandres, N. C. Shah, J. Zhao, and R. P. V. Duyne, "Biosensing with plasmonic nanosensors," *Nature Mater.* **7**, 442–453 (2008).

[6] A. Israel, M. Mrejen, Y. Lovsky, M. Polhan, S. Maier, and A. Lewis, "Near-field imaging probes electromagnetic waves," *Laser Focus World* **43**, 99–102 (2007).

[7] L. Tang, S. E. Kocabas, S. Latif, A. K. Okyay, D. S. Ly-Gagnon, K. C. Saraswat, and D. A. B. Miller, "Nanometre-scale germanium photodetector enhanced by a near-infrared dipole antenna," *Nat. Phot.* **2**, 226–229 (2008).

[8] W. A. Challener, C. Peng, A. V. Itagi, D. Karns, W. Peng, Y. Peng, X. Yang, X. Zhu, N. J. Gokemeijer, Y. T. Hsia, et al., "Heat-assisted magnetic recording by a near-field transducer with efficient optical energy transfer," *Nat. Phot.* **3**, 220–224 (2009).

[9] I.-Y. Park, S. Kim, J. Choi, D.-H. Lee, Y.-J. Kim, M. F. Kling, M. I. Stockman, and S.-W. Kim, "Plasmonic generation of ultrashort extreme-ultraviolet light pulses," *Nat. Phot.* **5**, 677–681 (2011).

[10] N. Nagatani, R. Tanaka, T. Yuhi, T. Endo, K. Kerman, Y. Takamura, and E. Tamiya, "Gold nanoparticle-based novel enhancement method for the development of highly sensitive immunochromatographic test strips," *Sci. Technol. Adv. Mater.* **7**, 270–275 (2006).

[11] L. R. Hirsch, R. J. Stafford, J. A. Bankson, S. R. Sershen, B. Rivera, R. E. Price, J. D. Hazle, N. J. Halas, and J. L. West, "Nanoshell-mediated near-infrared thermal therapy of tumors under magnetic resonance guidance," *Proc. Natl. Acad. Sci. USA* **100**, 13549–13554 (2003).

[12] S. Kim, J. H. Jin, Y. J. Kim, I. Y. Park, Y. Kim, and S. W. Kim, "High-harmonic generation by resonant plasmon field enhancement," *Nature* **453**, 757–760 (2008).

[13] M. I. Stockman, "Nanoplasmonics: the physics behind the applications," *Phys. Today* **64**, 39–44 (2011).

[14] D. C. Marinica, A. K. Kazansky, P. Nordlander, J. Aizpurua, and A. G. Borisov, "Quantum plasmonics: nonlinear effects in the field enhancement of a plasmonic nanoparticle dimer," *Nano Lett.* **12**, 1333–1339 (2012).

[15] S. Thongrattanasiri, A. Manjavacas, P. Nordlander, and F. J. G. de Abajo, "Quantum junction plasmons in graphene dimers," *Laser Photonics Rev.* (2013), doi:10.1002/lpor. 201200101

[16] D. J. Bergman and M. I. Stockman, "Surface plasmon amplification by stimulated emission of radiation: quantum generation of coherent surface plasmons in nanosystems," *Phys. Rev. Lett.* **90**, 027402–1–4 (2003).

[17] M. I. Stockman, "The spaser as a nanoscale quantum generator and ultrafast amplifier," *J. Opt.* **12**, 024004–1–13 (2010).

[18] D. Kahng, inventor, "Electric field controlled semiconductor device," US Patent No. 3,102,230 (1963).

[19] Y. Tsividis, *Operation and Modeling of the MOS Transistor* (McGraw-Hill, New York, 1999).

[20] M. I. Stockman and D. J. Bergman, inventors, "Surface plasmon amplification by stimulated emission of radiation (spaser)," US Patent No. 7,569,188 (2009).

[21] M. I. Stockman, "Spasers explained," *Nat. Phot.* **2**, 327–329 (2008).

[22] M. I. Stockman, "Spaser action, loss compensation, and stability in plasmonic systems with gain," *Phys. Rev. Lett.* **106**, 156802–1–4 (2011).

[23] M. I. Stockman, "Loss compensation by gain and spasing," *Phil. Trans. R. Soc. A* **369**, 3510–3524 (2011).
[24] D. Li and M. I. Stockman, "Electric spaser in the extreme quantum limit," *Phys. Rev. Lett.* **110**, 106803–1–5 (2013).
[25] A. A. E. Saleh and J. A. Dionne, "Waveguides with a silver lining: low threshold gain and giant modal gain in active cylindrical and coaxial plasmonic devices," *Phys. Rev. B* **85**, 045407–1–12 (2012).
[26] D. G. Baranov, A. P. Vinogradov, A. A. Lisyansky, Y. M. Strelniker, and D. J. Bergman, "Magneto-optical spaser," *Opt. Lett.* **38**, 2002–2004 (2013).
[27] O. L. Berman, R. Y. Kezerashvili, and Y. E. Lozovik, "Graphene nanoribbon based spaser," *Phys. Rev. B* **88**, 235424–1–7 (2013).
[28] Y.-W. Huang, W. T. Chen, P. C. Wu, V. A. Fedotov, N. I. Zheludev, and D. P. Tsai, "Toroidal lasing spaser," *Sci. Rep.* **3**, 1237–1–4 (2013).
[29] X. Meng, U. Guler, A. V. Kildishev, K. Fujita, K. Tanaka, and V. M. Shalaev, "Unidirectional spaser in symmetry-broken plasmonic core-shell nanocavity," *Sci. Rep.* **3**, 1241–1–5 (2013).
[30] M. A. Noginov, G. Zhu, A. M. Belgrave, R. Bakker, V. M. Shalaev, E. E. Narimanov, S. Stout, E. Herz, T. Suteewong, and U. Wiesner, "Demonstration of a spaser-based nanolaser," *Nature* **460**, 1110–1112 (2009).
[31] C. Y. Wu, C. T. Kuo, C. Y. Wang, C. L. He, M. H. Lin, H. Ahn, and S. Gwo, "Plasmonic green nanolaser based on a metal-oxide-semiconductor structure," *Nano Lett.* **11**, 4256–4260 (2011).
[32] Y.-J. Lu, J. Kim, H.-Y. Chen, C. Wu, N. Dabidian, C. E. Sanders, C.-Y. Wang, M.-Y. Lu, B.-H. Li, X. Qiu, et al., "Plasmonic nanolaser using epitaxially grown silver film," *Science* **337**, 450–453 (2012).
[33] M. T. Hill, M. Marell, E. S. P. Leong, B. Smalbrugge, Y. Zhu, M. Sun, P. J. van Veldhoven, E. J. Geluk, F. Karouta, Y.-S. Oei, et al., "Lasing in metal-insulator-metal sub-wavelength plasmonic waveguides," *Opt. Express* **17**, 11107–11112 (2009).
[34] R. F. Oulton, V. J. Sorger, T. Zentgraf, R.-M. Ma, C. Gladden, L. Dai, G. Bartal, and X. Zhang, "Plasmon lasers at deep subwavelength scale," *Nature* **461**, 629–632 (2009).
[35] R.-M. Ma, R. F. Oulton, V. J. Sorger, G. Bartal, and X. Zhang, "Room-temperature sub-diffraction-limited plasmon laser by total internal reflection," *Nat. Mater.* **10**, 110–113 (2010).
[36] R. A. Flynn, C. S. Kim, I. Vurgaftman, M. Kim, J. R. Meyer, A. J. Mäkinen, K. Bussmann, L. Cheng, F. S. Choa, and J. P. Long, "A room-temperature semiconductor spaser operating near 1.5 micron," *Opt. Express* **19**, 8954–8961 (2011).
[37] K. Ding, Z. C. Liu, L. J. Yin, M. T. Hill, M. J. H. Marell, P. J. van Veldhoven, R. Nöetzel, and C. Z. Ning, "Room-temperature continuous wave lasing in deep-subwavelength metallic cavities under electrical injection," *Phys. Rev. B* **85**, 041301–1–5 (2012).
[38] M. Khajavikhan, A. Simic, M. Katz, J. H. Lee, B. Slutsky, A. Mizrahi, V. Lomakin, and Y. Fainman, "Thresholdless nanoscale coaxial lasers," *Nature* **482**, 204–207 (2012).
[39] J. K. Kitur, G. Zhu, A. B. Yu, and M. A. Noginov, "Stimulated emission of surface plasmon polaritons on smooth and corrugated silver surfaces," *J. Optics* **14**, 114015–1–8 (2012).

[40] R. M. Ma, R. F. Oulton, V. J. Sorger, and X. Zhang, "Plasmon lasers: coherent light source at molecular scales," *Laser Photonics Rev.* 1–21 (2012), doi: 10.1002/lpor.201100040

[41] N. I. Zheludev, S. L. Prosvirnin, N. Papasimakis, and V. A. Fedotov, "Lasing spaser," *Nat. Phot.* **2**, 351–354 (2008).

[42] E. Plum, V. A. Fedotov, P. Kuo, D. P. Tsai, and N. I. Zheludev, "Towards the lasing spaser: controlling metamaterial optical response with semiconductor quantum dots," *Opt. Express* **17**, 8548–8551 (2009).

[43] F. V. Beijnum, P. J. V. Veldhoven, E. J. Geluk, M. J. A. D. Dood, G. W. T. Hooft, and M. P. V. Exter, "Surface plasmon lasing observed in metal hole arrays," *Phys. Rev. Lett.* **110**, 206802–1–5 (2013).

[44] W. Zhou, M. Dridi, J. Y. Suh, C. H. Kim, D. T. Co, M. R. Wasielewski, G. C. Schatz, and T. W. Odom, "Lasing action in strongly coupled plasmonic nanocavity arrays," *Nature Nano* **8**, 506–511 (2013).

[45] M. I. Stockman, "Lasing spaser in two-dimensional plasmonic crystals," *NPG Asia Mater.* **5**, e71 (2013).

[46] M. T. Hill, Y.-S. Oei, B. Smalbrugge, Y. Zhu, T. de Vries, P. J. van Veldhoven, F. W. M. van Otten, T. J. Eijkemans, J. P. Turkiewicz, H. de Waardt, et al., "Lasing in metallic-coated nanocavities," *Nat. Phot.* **1**, 589–594 (2007).

[47] M. I. Stockman, S. V. Faleev, and D. J. Bergman, "Localization versus delocalization of surface plasmons in nanosystems: can one state have both characteristics?," *Phys. Rev. Lett.* **87**, 167401–1–4 (2001).

[48] I. A. Larkin and M. I. Stockman, "Imperfect perfect lens," *Nano Lett.* **5**, 339–343 (2005).

[49] J. A. Gordon and R. W. Ziolkowski, "The design and simulated performance of a coated nano-particle laser," *Opt. Express* **15**, 2622–2653 (2007).

[50] D. J. Bergman and D. Stroud, "Theory of resonances in the electromagnetic scattering by macroscopic bodies," *Phys. Rev. B* **22**, 3527–3539 (1980).

[51] J. Seidel, S. Grafstroem, and L. Eng, "Stimulated emission of surface plasmons at the interface between a silver film and an optically pumped dye solution," *Phys. Rev. Lett.* **94**, 177401–1–4 (2005).

[52] M. A. Noginov, G. Zhu, M. Mayy, B. A. Ritzo, N. Noginova, and V. A. Podolskiy, "Stimulated emission of surface plasmon polaritons," *Phys. Rev. Lett.* **101**, 226806–1–4 (2008).

[53] K. Li, X. Li, M. I. Stockman, and D. J. Bergman, "Surface plasmon amplification by stimulated emission in nanolenses," *Phys. Rev. B* **71**, 115409–1–4 (2005).

[54] Z. G. Dong, H. Liu, T. Li, Z. H. Zhu, S. M. Wang, J. X. Cao, S. N. Zhu, and X. Zhang, "Resonance amplification of left-handed transmission at optical frequencies by stimulated emission of radiation in active metamaterials," *Opt. Express* **16**, 20974–20980 (2008).

[55] M. Wegener, J. L. Garcia-Pomar, C. M. Soukoulis, N. Meinzer, M. Ruther, and S. Linden, "Toy model for plasmonic metamaterial resonances coupled to two-level system gain," *Opt. Express* **16**, 19785–19798 (2008).

[56] A. Fang, T. Koschny, M. Wegener, and C. M. Soukoulis, "Self-consistent calculation of metamaterials with gain," *Phys. Rev. B (Rapid Communications)* **79**, 241104(R)–1–4 (2009).

[57] S. Wuestner, A. Pusch, K. L. Tsakmakidis, J. M. Hamm, and O. Hess, "Overcoming losses with gain in a negative refractive index metamaterial," *Phys. Rev. Lett.* **105**, 127401–1–4 (2010).

[58] S. W. Chang, C. Y. A. Ni, and S. L. Chuang, "Theory for bowtie plasmonic nanolasers," *Opt. Express* **16**, 10580–10595 (2008).

[59] I. E. Protsenko, A. V. Uskov, O. A. Zaimidoroga, V. N. Samoilov, and E. P. O'Reilly, "Dipole nanolaser," *Phys. Rev. A* **71**, 063812 (2005).

[60] M. Ambati, S. H. Nam, E. Ulin-Avila, D. A. Genov, G. Bartal, and X. Zhang, "Observation of stimulated emission of surface plasmon polaritons," *Nano Lett.* **8**, 3998–4001 (2008).

[61] Z. K. Zhou, X. R. Su, X. N. Peng, and L. Zhou, "Sublinear and superlinear photoluminescence from Nd doped anodic aluminum oxide templates loaded with Ag nanowires," *Opt. Express* **16**, 18028–18033 (2008).

[62] M. A. Noginov, V. A. Podolskiy, G. Zhu, M. Mayy, M. Bahoura, J. A. Adegoke, B. A. Ritzo, and K. Reynolds, "Compensation of loss in propagating surface plasmon polariton by gain in adjacent dielectric medium," *Opt. Express* **16**, 1385–1392 (2008).

[63] P. M. Bolger, W. Dickson, A. V. Krasavin, L. Liebscher, S. G. Hickey, D. V. Skryabin, and A. V. Zayats, "Amplified spontaneous emission of surface plasmon polaritons and limitations on the increase of their propagation length," *Opt. Lett.* **35**, 1197–1199 (2010).

[64] M. A. Noginov, G. Zhu, M. Bahoura, J. Adegoke, C. Small, B. A. Ritzo, V. P. Drachev, and V. M. Shalaev, "The effect of gain and absorption on surface plasmons in metal nanoparticles," *Appl. Phys. B* **86**, 455–460 (2007).

[65] M. A. Noginov, "Compensation of surface plasmon loss by gain in dielectric medium," *J. Nanophotonics* **2**, 021855–1–17 (2008).

[66] I. D. Leon and P. Berini, "Amplification of long-range surface plasmons by a dipolar gain medium," *Nat. Phot.* **4**, 382–387 (2010).

[67] K. Ding, L. Yin, M. T. Hill, Z. Liu, P. J. V. Veldhoven, and C. Z. Ning, "An electrical injection metallic cavity nanolaser with azimuthal polarization," *Appl. Phys. Lett.* **102**, 041110–1–4 (2013).

[68] K. Ding, M. T. Hill, Z. C. Liu, L. J. Yin, P. J. van Veldhoven, and C. Z. Ning, "Record performance of electrical injection sub-wavelength metallic-cavity semiconductor lasers at room temperature," *Opt. Express* **21**, 4728–4733 (2013).

[69] R. F. Oulton, V. J. Sorger, D. A. Genov, D. F. P. Pile, and X. Zhang, "A hybrid plasmonic waveguide for subwavelength confinement and long-range propagation," *Nat. Phot.* **2**, 496–500 (2008).

[70] D. Li and M. I. Stockman, "Electric spaser in the extreme quantum limit," *Phys. Rev. Lett.* **110**, 106803–1–5 (2013).

[71] M. I. Stockman, "Nanoplasmonics: past, present, and glimpse into future," *Opt. Express* **19**, 22029–22106 (2011).

[72] R. H. Brown and R. Q. Twiss, "A test of a new type of stellar interferometer on sirius," *Nature* **178**, 1046–1048 (1956).

[73] A. L. Schawlow and C. H. Townes, "Infrared and optical masers," *Phys. Rev.* **112**, 1940 (1958).

[74] P. B. Johnson and R. W. Christy, "Optical constants of noble metals," *Phys. Rev. B* **6**, 4370–4379 (1972).

[75] S. Kéna-Cohen, P. N. Stavrinou, D. D. C. Bradley, and S. A. Maier, "Confined surface plasmon-polariton amplifiers," *Nano Lett.* **13**, 1323–1329 (2013).

[76] F. De Angelis, G. Das, P. Candeloro, M. Patrini, M. Galli, A. Bek, M. Lazzarino, I. Maksymov, C. Liberale, L. C. Andreani, and E. Di Fabrizio, "Nanoscale chemical mapping using three-dimensional adiabatic compression of surface plasmon polaritons," *Nat. Nanotechnol.* **5**, 67–72 (2009).

[77] C. C. Neacsu, S. Berweger, R. L. Olmon, L. V. Saraf, C. Ropers, and M. B. Raschke, "Near-field localization in plasmonic superfocusing: a nanoemitter on a tip," *Nano Lett.* **10**, 592–596 (2010).

[78] H. A. Atwater and A. Polman, "Plasmonics for improved photovoltaic devices," *Nat. Mater.* **9**, 205–213 (2010).

[79] K. Kneipp, M. Moskovits, and H. Kneipp (eds.), *Surface Enhanced Raman Scattering: Physics and Applications* (Springer, Heidelberg, New York, Tokyo, 2006).

[80] J. Kneipp, H. Kneipp, B. Wittig, and K. Kneipp, "Novel optical nanosensors for probing and imaging live cells," *Nanomed. Nanotechnol. Biol. Med.* **6**, 214–226 (2010).

[81] M. I. Stockman, "Electromagnetic theory of SERS," in *Surface Enhanced Raman Scattering*, edited by M. M. K. Kneipp and H. Kneipp (Springer, Heidelberg, 2006), Vol. 103, pp. 47–66.

[82] F. J. Dunmore, D. Z. Liu, H. D. Drew, S. Dassarma, Q. Li, and D. B. Fenner, "Observation of below-gap plasmon excitations in superconducting $YBa_2Cu_3O_7$ films," *Phys. Rev. B* **52**, R731–R734 (1995).

[83] D. Schumacher, C. Rea, D. Heitmann, and K. Scharnberg, "Surface plasmons and Sommerfeld-Zenneck waves on corrugated surfaces: application to high-T_c superconductors," *Surf. Sci.* **408**, 203–211 (1998).

[84] V. A. Fedotov, A. Tsiatmas, J. H. Shi, R. Buckingham, P. de Groot, Y. Chen, S. Wang, and N. I. Zheludev, "Temperature control of Fano resonances and transmission in superconducting metamaterials," *Opt. Express* **18**, 9015–9019 (2010).

[85] A. Tsiatmas, A. R. Buckingham, V. A. Fedotov, S. Wang, Y. Chen, P. A. J. de Groot, and N. I. Zheludev, "Superconducting plasmonics and extraordinary transmission," *Appl. Phys. Lett.* **97**, 111106-1–3 (2010).

[86] A. Boltasseva and H. A. Atwater, "Low-loss plasmonic metamaterials," *Science* **331**, 290–291 (2011).

[87] V. M. Shalaev, "Optical negative-index metamaterials," *Nat. Photonics* **1**, 41–48 (2007).

[88] N. I. Zheludev, "A roadmap for metamaterials," *Opt. Photonics News* **22**, 30–35 (2011).

[89] M. I. Stockman, K. B. Kurlayev, and T. F. George, "Linear and nonlinear optical susceptibilities of Maxwell Garnett composites: dipolar spectral theory," *Phys. Rev. B* **60**, 17071–17083 (1999).

[90] M. I. Stockman, D. J. Bergman, and T. Kobayashi, "Coherent control of nanoscale localization of ultrafast optical excitation in nanosystems," *Phys. Rev. B* **69**, 054202-1–10 (2004).

[91] X. Li and M. I. Stockman, "Highly efficient spatiotemporal coherent control in nanoplasmonics on a nanometer-femtosecond scale by time reversal," *Phys. Rev. B* **77**, 195109-1–10 (2008).

[92] M. I. Stockman, L. N. Pandey, and T. F. George, "Inhomogeneous localization of polar eigenmodes in fractals," *Phys. Rev. B* **53**, 2183–2186 (1996).

[93] M. I. Stockman, "Nanofocusing of optical energy in tapered plasmonic waveguides," *Phys. Rev. Lett.* **93**, 137404–1–4 (2004).

[94] L. D. Landau and E. M. Lifshitz, *Electrodynamics of Continuous Media* (Pergamon, Oxford and New York, 1984).

[95] N. Meinzer, M. Ruther, S. Linden, C. M. Soukoulis, G. Khitrova, J. Hendrickson, J. D. Olitzky, H. M. Gibbs, and M. Wegener, "Arrays of Ag split-ring resonators coupled to InGaAs single-quantum-well gain," *Opt. Express* **18**, 24140–24151 (2010).

[96] E. Kretschmann and H. Raether, "Radiative decay of nonradiative surface plasmons excited by light," *Z. Naturforsch. A* **23**, 2135–2136 (1968).

[97] E. Heydari, R. Flehr, and J. Stumpe, "Influence of spacer layer on enhancement of nanoplasmon-assisted random lasing," *Appl. Phys. Lett.* **102**, 133110–133114 (2013).

[98] J. K. Kitur, V. A. Podolskiy, and M. A. Noginov, "Stimulated emission of surface plasmon polaritons in a microcylinder cavity," *Phys. Rev. Lett.* **106**, 183903–1–4 (2011).

[99] F. Bloch, "Über die Quantenmechanik der Elektronen in Kristallgittern," *Z. Phys. A* **52**, 555–600 (1929).

[100] H. Ghafouri-Shiraz, *Distributed Feedback Laser Diodes and Optical Tunable Filters* (John Wiley & Sons, West Sussex, England; Hoboken, NJ, 2003).

[101] F. V. Beijnum, P. J. van Veldhoven, E. J. Geluk, G. W. Hooft, apos, and M. P. van Exter, "Loss compensation of extraordinary optical transmission," *Appl. Phys. Lett.* **104**, (2014).

[102] P. W. Anderson, "Absence of diffusion in certain random lattices," *Phys. Rev.* **109**, 1492–1505 (1958).

[103] D. W. Pohl, W. Denk, and M. Lanz, "Optical stethoscopy: image recording with resolution lambda/20," *Appl. Phys. Lett.* **44**, 651–653 (1984).

[104] A. Lewis, A. Radko, N. B. Ami, D. Palanker, and K. Lieberman, "Near-field scanning optical microscopy in cell biology," *Trends Cell Biol.* **9**, 70–73 (1999).

[105] B. Knoll and F. Keilmann, "Near-field probing of vibrational absorption for chemical microscopy," *Nature* **399**, 134–137 (1999).

[106] L. Novotny and S. J. Stranick, "Near-field optical microscopy and spectroscopy with pointed probes," *Ann. Rev. Phys. Chem.* **57**, 303–331 (2006).

[107] L. Novotny and N. van Hulst, "Antennas for light," *Nat. Phot.* **5**, 83–90 (2011).

[108] M. Burresi, D. van Oosten, T. Kampfrath, H. Schoenmaker, R. Heideman, A. Leinse, and L. Kuipers, "Probing the magnetic field of light at optical frequencies," *Science* **326**, 550–553 (2009).

[109] A. Bouhelier, J. Renger, M. R. Beversluis, and L. Novotny, "Plasmon-coupled tip-enhanced near-field optical microscopy," *J. Microsc. - Oxford* **210**, 220–224 (2003).

[110] F. Keilmann, "Scanning tip for optical radiation," US Patent 4,994,818, Max Plank Geselschaft, Germany, USA (1991), pp. 1–7.

[111] A. J. Babajanyan, N. L. Margaryan, and K. V. Nerkararyan, "Superfocusing of surface polaritons in the conical structure," *J. Appl. Phys.* **87**, 3785–3788 (2000).

[112] A. A. Mikhailovsky, M. A. Petruska, M. I. Stockman, and V. I. Klimov, "Broadband near-field interference spectroscopy of metal nanoparticles using a femtosecond white-light continuum," *Opt. Lett.* **28**, 1686–1688 (2003).

[113] A. Hartschuh, H. N. Pedrosa, L. Novotny, and T. D. Krauss, "Simultaneous fluorescence and raman scattering from single carbon nanotubes," *Science* **301**, 1354–1356 (2003).

[114] C. Cohen-Tannoudji, B. Diu, and F. Laloe, *Quantum Mechanics* (John Wiley & Sons, New York, 1977).

[115] J. B. Pendry, A. Aubry, D. R. Smith, and S. A. Maier, "Transformation optics and subwavelength control of light," *Science* **337**, 549–552 (2012).

[116] E. Verhagen, M. Spasenovic, A. Polman, and L. Kuipers, "Nanowire plasmon excitation by adiabatic mode transformation," *Phys. Rev. Lett.* **102**, 203904–1–4 (2009).

[117] S. Berweger, J. M. Atkin, X. G. Xu, R. L. Olmon, and M. B. Raschke, "Femtosecond nanofocusing with full optical waveform control," *Nano Lett.* **11**, 4309–4313 (2011).

[118] S. Schmidt, B. Piglosiewicz, D. Sadiq, J. Shirdel, J. S. Lee, P. Vasa, N. Park, D.-S. Kim, and C. Lienau, "Adiabatic nanofocusing on ultrasmooth single-crystalline gold tapers creates a 10-nanometer-sized light source with few-cycle time resolution," *Acs Nano* **6**, 6040–6048 (2012).

[119] V. Kravtsov, J. M. Atkin, and M. B. Raschke, "Group delay and dispersion in adiabatic plasmonic nanofocusing," *Opt. Lett.* **38**, 1322–1324 (2013).

[120] F. D. Angelis, F. Gentile, F. M. G. Das, M. Moretti, P. Candeloro, M. L. Coluccio, G. Cojoc, A. Accardo, C. Liberale, R. P. Zaccaria, et al., "Breaking the diffusion limit with super-hydrophobic delivery of molecules to plasmonic nanofocusing sers structures," *Nat. Phot.* **5**, 682–687 (2011).

[121] A. Giugni, B. Torre, A. Toma, M. Francardi, M. Malerba, A. Alabastri, R. P. Zaccaria, M. I. Stockman, and E. D. Fabrizio, "Hot-electron nanoscopy using adiabatic compression of surface plasmons," *Nat. Nano* **8**, 845–852 (2013).

[122] M. B. Raschke, S. Berweger, J. M. Atkin, and R. L. Olmon, "Adiabatic tip-plasmon focusing for nano-Raman spectroscopy," *J. Phys. Chem. Lett.* **1**, 3427–3432 (2010).

[123] S. Berweger, J. M. Atkin, R. L. Olmon, and M. B. Raschke, "Light on the tip of a needle: plasmonic nanofocusing for spectroscopy on the nanoscale," *J. Phys. Chem. Lett.* **3**, 945–952 (2012).

[124] C. Ropers, C. C. Neacsu, T. Elsaesser, M. Albrecht, M. B. Raschke, and C. Lienau, "Grating-coupling of surface plasmons onto metallic tips: a nano-confined light source," *Nano Lett.* **7**, 2784–2788 (2007).

[125] D. Sadiq, J. Shirdel, J. S. Lee, E. Selishcheva, N. Park, and C. Lienau, "Adiabatic nanofocusing scattering-type optical nanoscopy of individual gold nanoparticles," *Nano Lett.* **11**, 1609–1613 (2011).

[126] I. Goykhman, B. Desiatov, J. Khurgin, J. Shappir, and U. Levy, "Locally oxidized silicon surface-plasmon schottky detector for telecom regime," *Nano Lett.* **11**, 2219–2224 (2011).

[127] M. W. Knight, H. Sobhani, P. Nordlander, and N. J. Halas, "Photodetection with active optical antennas," *Science* **332**, 702–704 (2011).

[128] I. Goykhman, B. Desiatov, J. Khurgin, J. Shappir, and U. Levy, "Waveguide based compact silicon schottky photodetector with enhanced responsivity in the telecom spectral band," *Opt. Express* **20**, 28594–28602 (2012).

[129] R. H. Fowler, "The analysis of photoelectric sensitivity curves for clean metals at various temperatures," *Phys. Rev.* **38**, 45–56 (1931).

5

DIELECTRIC PHOTONIC CRYSTALS

ROBERT H. LIPSON
Department of Chemistry, University of Victoria, Victoria, BC Canada

5.1 INTRODUCTION

Photonic crystals (PC) or photonic band gap (PBG) materials are structures having a dielectric constant profile that changes periodically on a distance scale approximately equal to a specific wavelength of light, λ. In principle, PCs can be fabricated with such spatial periodicities in one dimension (1D), two dimensions (2D), or three dimensions (3D) (Fig. 5.1). The effect on light passing through such a structure is dramatically different from propagation through a homogenous medium in that destructive interference via Bragg scattering within the periodic structure maximizes the overall reflection of the light at the PC/air interface. Put another way, destructive interference via Bragg scattering within a PC opens up PBGs which correspond to wavelength regions where light propagation is not supported.

Although various aspects of the principles behind PCs were established before 1979 [1] that date is often seen as the time when the concepts behind PCs were firmly established [2]. Subsequent papers by Yablonovitch [3] and John [4] in 1987 sparked an explosion of activity which continues to this day. The field has been reviewed several times over the last 20 years [5–10] and several books [11–13] are now available which detail the theory of PC–light interactions, PC fabrication, and their applications.

Well-known phenomena such as x-ray diffraction and grating operation are also based on Bragg scattering as are 1D multilayer dielectric structures first discussed by Lord Rayleigh in 1887 [14]. Furthermore, many examples of photonic crystals can be found in nature [15, 16]. In addition to minerals such as opals [17] a large variety of

---

Photonics: Scientific Foundations, Technology and Applications, Volume II, First Edition.
Edited by David L. Andrews.
© 2015 John Wiley & Sons, Inc. Published 2015 by John Wiley & Sons, Inc.

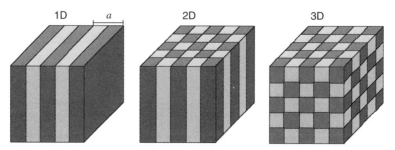

FIGURE 5.1 Schematics of a 1D, 2D, and 3D photonic crystal. The different shades represent materials with different dielectric indices. The spatial period of the material is called the lattice constant, a. (http://ab-initio.mit.edu/photons/)

aquatic species [18] and terrestrial flora [19, 20] and fauna [21–23] exhibit iridescent colorations which can be attributed to PC structures that have evolved in step with and for the development of visual systems, and for thermal regulation and signaling. Interestingly, the dielectric contrast of materials comprising these systems, defined as $\varepsilon_{high}/\varepsilon_{low}$, is usually low. Indeed it has been suggested that only those structures with a dielectric contrast ≥ 2 should be labeled as PCs [24]. An emphasis will therefore be placed here on PCs which have a large dielectric contrast because this condition leads to the appearance of PBGs in 2D and 3D, and allow many applications to be realized that depend on the modification of light dispersion within these structures.

5.2 FUNDAMENTALS

PBGs can be deduced by solving Maxwell's equations in a source-free periodic dielectric medium. The derivation of the band structure in photonic crystals begins by seeking time-dependent solutions for the magnetic field vector of the form $\vec{H}(\vec{r})e^{-i\omega t}$. In this case Maxwell's equations can be cast as an eigenvalue problem:

$$\vec{\nabla} \times \frac{1}{\varepsilon(\vec{r})} \vec{\nabla} \times \vec{H}(\vec{r}) = \left(\frac{\omega^2}{c^2}\right) \vec{H}(\vec{r}), \tag{5.1}$$

where $\varepsilon(\vec{r})$ is a function describing the dielectric constant at each position $\vec{r}$ in the lattice, ω is the frequency of the mode, and c is the speed of light. Finding magnetic field solutions based on Eq. (5.1) is preferred because the operator on the left side is both linear and Hermitian when the dielectric constant is everywhere nonabsorbing; that is, real and positive. Thus the solutions form a complete orthonormal set with real eigenfrequencies, ω. The electric field solutions can subsequently be derived from the magnetic field using

$$\vec{E}(\vec{r}) = \left(\frac{-ic}{\omega \varepsilon(\vec{r})}\right) \vec{\nabla} \times \vec{H}(\vec{r}). \tag{5.2}$$

The periodic dielectric function of a PC can be written as

$$\varepsilon(\vec{r}) = \varepsilon(\vec{r} + \vec{R}_i), \tag{5.3}$$

where $\vec{R}_i$, $i = 1, 2, 3$, are primitive lattice vectors for the crystal in three dimensions. The Bloch–Floquet theorem states that for spatially periodic systems the field eigenfunctions derived can be written as

$$\vec{H}(\vec{r}) = e^{i(\vec{k}\cdot\vec{r})} \vec{H}_{n,\vec{k}} \tag{5.4}$$

with eigenvalues $\omega_n(\vec{k})$. Substitution of Eq. (5.4) into Eq. (5.1) shows that $\vec{H}_{n,\vec{k}}$ satisfies

$$(\vec{\nabla} + i\vec{k}) \times \frac{1}{\varepsilon}(\vec{\nabla} + i\vec{k}) \times \vec{H}_{n,\vec{k}} = \left(\frac{\omega_n(\vec{k})}{c}\right)^2 \vec{H}_{n,\vec{k}}. \tag{5.5}$$

The solutions are also periodic functions of the wave vector $\vec{k}$ in that the same solution exists for $\vec{k}$ and $\vec{k} + \vec{G}$, where $\vec{G}$ is the primitive reciprocal lattice vector defined by

$$\vec{R}_i \cdot \vec{G}_i = 2\pi \delta_{ij}. \tag{5.6}$$

In 1D systems it can be shown that eigenstates with $\vec{k}$ and $\vec{k} + \frac{2\pi}{a}$ are the same. Hence one need only to find solutions over the first Brillouin zone between $-\frac{\pi}{a} \cdots +\frac{\pi}{a}$. Furthermore, time reversal symmetry shows that the eigenstates for $-\vec{k}$ are the same as for $\vec{k}$ and therefore unique solutions exist only over the range of wave vectors between 0 and $\frac{\pi}{a}$ corresponding to the irreducible Brillouin zone. In 2D and 3D systems the Brillouin zones form polygons and polyhedral, respectively. Since the solutions to Eq. (5.5) are over a finite domain, the eigenvalues $\omega_n(\vec{k})$ form a discrete set of continuous functions of $\vec{k}$ for $n = 1, 2, 3, \ldots$. These discrete frequencies plotted as a function of $\vec{k}$ form the band structure of a PC.

One approach to numerically solving Eq. (5.1) is the use of block-iterative frequency-domain methods where N plane waves combined with effective dielectric tensors, to solve for the n photonic bands ($n \ll N$) [25]. A shareware computer program, the MIT Photonic Bands (MPB) package that uses this methodology, is available on-line [26]. Some of the figures in this work were created using the MPB software.

5.2.1 Analogies

5.2.1.1 Quantum Mechanics Equation (5.1) can be compared to the time-independent Schrödinger wave equation

$$\left(-\frac{h}{2m}\nabla^2 V(\vec{r})\right)\Psi(\vec{r}) = E\Psi(\vec{r}), \tag{5.7}$$

where the operator in brackets on the left-hand side is the Hamiltonian operator $\hat{H}$, $\Psi(\vec{r})$ is the wave function (eigenfunction), and E is the energy eigenvalue $= \hbar\omega$. Like the operator in Eq. (5.1) $\hat{H}$ is Hermitian and the solutions of Eq. (5.7) form a discrete set of eigenfunctions, each characterized by an index or a set of indices (quantum numbers). The dielectric function $\varepsilon(\vec{r})$ of a PC plays a role of the potential function $V(\vec{r})$.

There are however also several very important differences. The first is that the $\vec{H}$-field and associated $\vec{E}$-field solutions of Eq. (5.1) are vectors while quantum mechanical wave functions are scalars. A quantum mechanical potential function is typically separable which in Cartesian coordinates means $V(\vec{r}) = V(x) + V(y) + V(z)$. Thus, the wave function can also be expressed (in Cartesian coordinates) as $\Psi(\vec{r}) = \Psi(x) \cdot \Psi(y) \cdot \Psi(z)$, and the total energy for the system as $E = E_x + E_y + E_z$. However, the curl operator in Eq. (5.1) can couple the solutions for different directions even if the dielectric function $\varepsilon(\vec{r})$ is separable.

Quantum mechanical results are also only valid for small sizes, typically nanometers or less. Conversely, if the ratio of the wavelength to lattice periodicity, λ/a, is maintained the electromagnetic solutions for a PC at one wavelength will be the same as the second system at the scaled wavelength. There is also no fundamental length scale for the dielectric constant. If two PCs have dielectric constants which are related by $\varepsilon'(\vec{r}) = \varepsilon(\vec{r})/s^2$ where s is a scaling factor, then the light frequency ω' scales as $s\omega$ [11].

Additional differences can be understood by examining the equation for an electromagnetic wave of frequency ω propagating in an inhomogeneous dielectric constant of the form: $\varepsilon(\vec{r}) = \varepsilon_0 + \varepsilon_{\text{fluc}}(\vec{r})$ where ε_0 and $\varepsilon_{\text{fluc}}(\vec{r})$ are the average dielectric constant and the spatially fluctuating portion, respectively [27]:

$$-\nabla^2 E - \frac{\omega^2}{c^2}\varepsilon_{\text{fluc}}(\vec{r})E = \varepsilon_0 \frac{\omega^2}{c^2}E. \tag{5.8}$$

While again reminiscent of the Schrödinger wave equation, quantum mechanical energy eigenvalues can be negative while the quantity $\varepsilon_0 \frac{\omega^2}{c^2}$ in Eq. (5.8) is always positive. This means that states do not exist in a PC where light is bound in negative regions of the potential. In addition, the scattering dielectric function $\varepsilon_{\text{fluc}}$ is multiplied explicitly by ω while $V(\vec{r})$ for quantum systems does not depend on the energy eigenvalues. Thus, the scattering phenomenon in a PC becomes less important as the photon energy is decreased. This is opposite to the situation quantum mechanically where lower energies typically lead to more electron localization. Similarly, at very high energies where geometric ray optics becomes more valid interference effects also become less important. Furthermore since $\varepsilon(\vec{r}) = \varepsilon_0 + \varepsilon_{\text{fluc}}(\vec{r})$ is everywhere positive, the energy eigenvalues $\varepsilon_0 \frac{\omega^2}{c^2}$ therefore will always be $> \frac{\omega^2}{c^2} \varepsilon_{\text{fluc}}(\vec{r})$. Therefore, PCs operate most effectively in an intermediate frequency (positive energy) range that is higher than the highest potential (dielectric) barrier.

5.2.1.2 Semiconductors

A PBG is often likened to the electronic band gap of a semiconductor. The latter feature corresponds to energy regions and directions within a semiconducting crystal where analogous to photons in a PC, de Broglie electron waves do not propagate because of Bragg reflections. Specifically, the periodic potential corresponding to the periodicity of the positively charged ion cores within a semiconducting crystal produces two standing waves at the band edges of the gap, one with its peak amplitude overlapping the periodic potential maxima and the other with its peak amplitude found at the potential minima. The energy region between these states is the electronic band gap. The lower energy state below the gap is labeled the valence band while the energy state above the gap is the conduction band.

The minimum frequency of a photonic band can be derived from variational theory:

$$\omega_1(\vec{k})^2 = \frac{\min}{H_{\vec{k}}} \frac{\int \langle | (\vec{\nabla} + i\vec{k} \times \vec{H}_{\vec{k}}) | \rangle^2 / \varepsilon}{\int |H_{\vec{k}}|^2} c^2. \tag{5.9}$$

The numerator in Eq. (5.9) is the expectation value of the operator in Eq. (5.5). Since the numerator is inversely proportional to the dielectric constant ε the lowest frequency band is obtained when the curl of the magnetic field; that is, the electric field (Eq. (5.2)), is concentrated in regions of highest ε. This lowest energy state is therefore designated as the "dielectric band" and is the photonic analogue of the valence band. In principle the second band should also be concentrated in regions with a large dielectric constant. However as this band must be orthogonal to the lower energy state, the electric field will as a result be found in the lower dielectric constant regions of the PC. For this reason the higher energy level is labeled the "air band" and is the photonic analogue of the conduction band. As shown in Eq. (5.9) a lower dielectric constant will result in an increase in frequency, resulting in a PBG; that is a frequency range where light propagation through the PC is forbidden.

Compared to a homogeneous medium the density of states (DOS), defined as the number of photon states per frequency interval, is significantly modified by the periodic structure of a PC. Specifically, PBGs correspond to regions of vanishing DOS [28].

5.2.2 1D PCs

Consider a 1D quarter wave stack shown schematically in Figure 5.2. The resultant band gaps for these systems are often referred to as stop bands. The index of refraction,

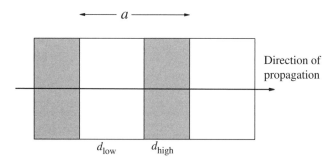

FIGURE 5.2 Schematic of a 1D photonic crystal with relevant scales indicated.

n, of each material is related to its dielectric constant, ε, through the relation: $n = \sqrt{\varepsilon}$. A band gap will open (Fig. 5.3) when two conditions are satisfied. First, the lattice periodicity, a, should be on the order of a multiple m of one-half of the vacuum wavelength of the incident light, λ:

$$m\lambda \sim 2a, \qquad (5.10)$$

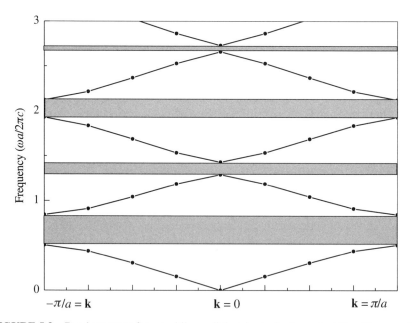

FIGURE 5.3 Band structure for a multilayer dielectric stack of equal width with $n_{\text{low}} = 2.9$ and $n_{\text{high}} = 6.7$ [8]. The gray regions correspond to photonic band gaps. The first band gap is centered at 677 nm when $a = 1$ μm. The band diagram is generated by MIT photonic band gap (MPB) software [26].

where $m = \pm 1, \pm 2, \ldots$. This condition leads to Bragg scattering and with enough periods, complete destructive interference of light within the structure.

Second, the optical thickness of each segment of dielectric material should be equal to one-quarter of the incident wavelength; that is,

$$n_{low} \cdot d_{low} = \frac{\lambda}{4} = n_{high} \cdot d_{high}, \qquad (5.11)$$

where n_{low} and n_{high} are the refractive indices of the two materials with the lowest and highest values, respectively. Here, d_{low} and d_{high} are the widths of the each section of the dielectric materials with n_{low} and n_{high}, respectively, and $d_{low} + d_{high} = a$. This maximizes the reflections within each unit cell.

For a 1D lattice with N bilayers at normal incidence, the mid gap frequency ω_0 can be calculated from

$$\omega_0 = \left(\frac{2\pi c}{a}\right) \cdot \left(\frac{n_{low} + n_{high}}{4 n_{low} \cdot n_{high}}\right), \qquad (5.12)$$

while the width of the band gap, $\Delta \omega$, centered at frequency ω_0, and the stack reflectivity, R, are related to the refractive indices by the relation

$$\frac{\Delta \omega}{\omega_0} = \frac{4}{\pi} \arcsin\left(\frac{n_{high} - n_{low}}{n_{high} + n_{low}}\right) \qquad (5.13)$$

and

$$R = \left(\frac{n_0 - n_s \left(n_{low}/n_{high}\right)^{2N}}{n_0 + n_s \left(n_{low}/n_{high}\right)^{2N}}\right)^2, \qquad (5.14)$$

respectively. n_0 and n_s in Eq. (14) are the indices of refraction of the surrounding environment and the substrate, respectively [29]. The size of the PBG is best characterized by $\frac{\Delta \omega}{\omega_0}$ because unlike $\Delta \omega$ alone this measure of the size of the band gap is invariant to size scaling. The band gap edges are found at $\omega_0 \pm \Delta \omega$.

It should be noted that the width of the band gap depends on the index of refraction difference of the dielectric layers. Hence any nonzero index difference will yield a PBG, albeit one that is small if Δn is small. Furthermore, band gaps will appear for a layered two dielectric arrangement which does not satisfy Eq. (5.11); for example, $d_{low} = d_{high} = 0.5a$. For this reason 1D PCs are usually considered separately from 2D and 3D arrangements.

Near the PBG edges the group velocity $v_g = \frac{\partial \omega}{\partial k}$, defined as the velocity at which the envelope of light propagates within the material, approaches zero; that is, "slow photons" are generated. The group velocity v_g is distinct from the phase velocity of a wave with frequency ω, $v_p = \frac{\omega}{k}$, although the two are related: $v_g = v_p \left(1 - \frac{k}{n}\frac{\partial n}{\partial k}\right)$.

The physical reason for slow light is that at the band edges photons undergo multiple reflections and therefore only propagate slowly through the photonic structure with a mean velocity, v_g. The analog phenomenon at the band edge of a semiconductor corresponds to the case where the de Broglie electron wave is nearly standing (equal probability of moving in either direction).

One can also understand this phenomenon by considering the local photon DOS within a PC. The photon DOS corresponds to the number of photon states per frequency interval. A vanishing DOS within the PBG leads to strong localization of photons and suppressed spontaneous emission. It can be shown that the group velocity, v_g, is inversely proportional to the large local photon DOS found at the band edges

5.2.3 2D and 3D PCs

The situation is more complex for 2D and 3D periodic structures which, unlike 1D PCs, can reflect light at angles different from normal incidence. Here, light reflection can be polarization dependent, resulting in possible band gaps for either transverse electric (TE; no electric field in the direction of propagation), transverse magnetic (TM; no magnetic field in the direction of propagation) modes, or both.

The criteria for the appearance of band gaps in 2D or 3D PCs are more stringent. Although there will be gaps in each dimension (as expected for any 1D structure) they need not overlap or be between the same bands. Several conditions will optimize the chances for a complete (omnidirectional) band gap. First, the dielectric contrast must be large. Second, the best geometries will have periodicities which are the approximately the same in all directions. In such cases, the frequency of the band gap will be expected around $\omega = (\frac{c}{n})k = (\frac{c}{n})\frac{\pi}{a}$, where a is the lattice parameter. In 2D this corresponds to a hexagonal lattice with a circular Brillouin zone, and in 3D to a face-centered cubic (fcc) arrangement which has a spherical Brillouin zone. Lastly, it can be shown that it is more difficult to contain the electric field within the high dielectric regions of a PC when the electric field lines cross a dielectric boundary than when they are parallel. Therefore larger band gaps are favored when the PC consists of thin continuous high dielectric veins which the electric field lines can run surrounded by the lower dielectric material.

In this work only the highly symmetrical hexagonal (triangular) and square 2D geometries will be considered in detail due to their importance for applications. It is common to label the possible directions of light propagation in photonic band structure diagrams by the standard reciprocal space notation listed in Table 5.1. The directions listed correspond to vertices of the respective polyhedral making up their irreducible Brillouin zones. All other points in reciprocal space are related to these positions by rotational symmetry in the plane [8].

For this discussion the periodicity of a 2D PC will be defined as being in the xy-plane. The direction of light propagation is also in the xy plane. The TE (or H) modes then have their $\vec{E}$ vector in the xy plane and their $\vec{H}$ vector perpendicular to the plane in the z-direction while the TM (or E) modes have their $\vec{H}$ vector in the xy-plane and their $\vec{E}$ vector perpendicular to the plane in the z-direction.

TABLE 5.1 Symbols and wave vector directions for square and hexagonal lattices

Square lattice		Hexagonal lattice	
Symbol	Reciprocal space	Symbol	Reciprocal space
Γ	$\vec{k} = 0$	Γ	$\vec{k} = 0$
X	$\vec{k} = \dfrac{\pi}{a}\hat{x}$	M	$\vec{k} = \dfrac{2\pi}{\sqrt{3}a}\hat{y}$
M	$\vec{k} = \dfrac{\pi}{a}\hat{x} + \dfrac{\pi}{a}\hat{y}$	K	$\vec{k} = \dfrac{2\pi}{3a}\hat{x} + \dfrac{2\pi}{\sqrt{3}a}\hat{y}$

When considering which 2D configurations will exhibit band gaps for TE versus TM modes, a useful rule-of-thumb is that TM band gaps are favored for isolated high dielectric regions while TE band gaps are favored for connected lattices.

A nearly isolated hexagonal lattice connected by small triangular dielectric veins is shown in Fig. 5.4a. The resultant photonic band diagram calculated using the MIT photonic band gap (MPB) software is shown in Fig. 5.4b. A PBG is present for both TE and TM modes. The size of band gap will be affected by many parameters including the structural symmetry of the crystal, the dielectric constant ratio, the air-filling ratio, and so on. Meade et al. theoretically explored square, hexagonal, and honey-comb structures composed of air holes drilled into a dielectric or dielectric rods in air as a function of filling fractions (ratio of hole radius, r to in-plane lattice constant, a) [30] and deduced that a hexagonal lattice exhibits the largest band gap, $\left(\dfrac{\Delta\omega}{\omega_0}\right) = 0.186$, or 18.6% for both TE and TM polarization for a dielectric ratio ≥7.2 and $r = 0.48$ a or a filling factor of 48%. It should be noted however, that a hexagonal array of isolated high index columns exhibits a band gap for the TM modes only, in line with the rule-of-thumb mentioned above.

Square lattices can also exhibit photonic bad gaps. An example of a 2D lattice and associated first Brillouin zone are shown in Figures 5.5a and 5.5b. The resultant photonic band diagrams show a band gap for TM modes only as expected (Fig. 5.5c). If however the square lattice were a grid of dielectric veins versus isolated air islands then the TE modes would exhibit a band gap, in agreement with the rule-of-thumb above. A polarization-independent PBG has been found for a square lattice having a filling fraction of ~65% although the width of the band gap is smaller than that of the hexagonal structure even at larger index contrasts [31].

Several strategies have been devised to increase the band-gap width and to ensure that the band gaps for both TE and TM modes overlap. The first is to use materials whose dielectricity is anisotropic [32]. This can be challenging because both the materials' indices of refraction and their anisotropy difference must be large. One promising material that has been proposed is tellurium (Te) which is a uniaxial crystal which in the wavelength region between 3.5 and 35 μm has an index $n_e = 6.2$ along its extraordinary axis and an index $n_0 = 4.8$ along its ordinary axis. It is also important that PCs are fabricated with the different axes aligned across its periodic structure. This could be accomplished by using a strong electric field to pole the extraordinary

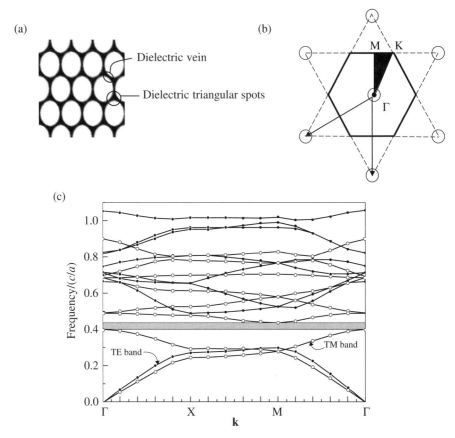

FIGURE 5.4 (a) A 2D hexagonal photonic crystal with a complete band gap for both TE and TM modes. The structure is a column array of air holes ($\varepsilon_{low} = 1$) in a dielectric substrate ($\varepsilon_{high} = 6.7$), with $r/a = 0.45$. A dielectric triangular spot and vein are indicated. (b) The first Brillouin zone. The filled in triangle is the irreducible Brillouin zone. (c) Photonic band diagram for the structure shown in (a). The open circles correspond to TM modes while the closed circles to TE modes. The gray box shows the PBG [8].

axes (largest index of refraction) of colloidal Te spheres making up the PC along the axis of applied field.

The second approach is to lower the symmetry of the dielectric rods. Specifically, the size of the PBG is expected to maximize when the geometric shape of the rods is the same of the lattice [33].

The first 3D PC exhibiting a full band gap was devised by Yablonovitch and coworkers for the microwave region by drilling holes into a dielectric slab forming a fcc set of air holes [34]. The resultant cylindrical air voids served to break the degeneracy expected between the dielectric and air bands when the air voids are spherical. This PC is now referred to as Yablonovite. The conditions for a complete band gap in 3D

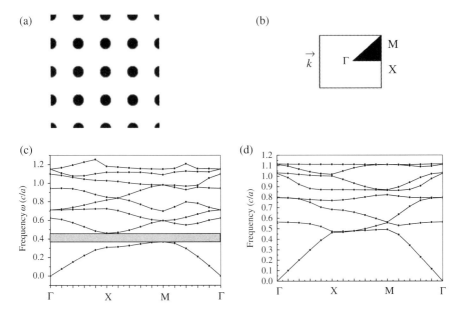

FIGURE 5.5 (a) A 2D rod lattice with $r/a = 0.2$ (r = radius of each rod) and $\varepsilon_{high}/\varepsilon_{low} = 12$. (b) The first Brillouin zone. The filled in triangle is the irreducible Brillouin zone. (c) Photonic band diagram for the TM modes. The gray box indicates the photonic band gap. (d) Photonic band diagram for the TE modes which do not exhibit a photonic band gap [8].

PCs are stringent. To date, Yablonovite and other structures discovered are related to a diamond-like lattice [35]. However, as with 2D PCs, the use of a sufficiently anisotropic dielectric such as Te is expected to open full band gaps for all the usual lattices including fcc, body-center cubic, and simple cubic [36].

5.2.4 Group Velocity Effects

In a homogeneous material, the dispersion, which relates the frequency ω to the wave vector k, can be expressed as $\omega = \frac{ck}{n}$ where c is the speed of light and n is the material index of refraction. Typically n ranges from 1 to 3.5, and has a weak dependence on wavelength except in regions of absorption where anomalous dispersion is important.

An equifrequency surface (EFS) is defined as the intersection of a constant ω plane through the dispersion surface, and displays the locus of allowed wave vectors at a given frequency in k-space. In a 2D or 3D homogenous medium the EFS is circular or spherical, respectively, meaning that light propagates isotropically.

The dispersion however is strongly and nonlinearly modified by the resultant band structure of a PC. Here, the refractive index is more appropriately expressed as $n = {}^c/(\frac{d\omega}{dk})$ versus $n = {}^c/(\frac{\omega}{k})$. One important result is that the derivative, $\frac{d\omega}{dk}$, corresponding to the group velocity, v_g, tends to zero at the band edges as a result of the photon undergoing many multiple Bragg reflections within the PC [37]. Notomi

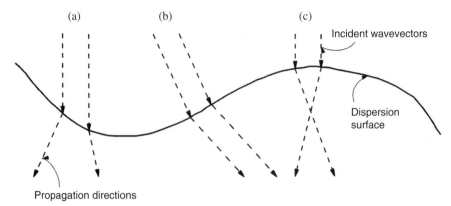

FIGURE 5.6 A schematic illustration of the effect of the curvature of an equifrequency dispersion surface on the resultant propagation of incident light wave vectors. (a) divergent (concave lens) effect; (b) a collimator lens effect, and (c) convergent (convex lens) effect. See Figure 1 of Reference 40.

and co-workers measured the group velocity of waveguiding modes in a PC that decreased by two orders of magnitude relative to propagation speed in air as the frequency approached the band edge [38]. Such "slow photons" are expected to have a significant impact on linear and nonlinear optical effects that depend on the effective interaction times of light with active matter embedded within a PC [39].

Another manifestation of the strongly modified dispersion is that an EFS can become highly distorted. In a multidimensional PC the direction of $\vec{v}_g = \nabla_{\vec{k}} \omega(k)$ and hence the energy flux which is normal to the EFS, can change dramatically with small changes in light frequency and propagation direction. Indeed, $\vec{v}_g$ need not point in the same direction as $\vec{k}$. As shown in Figure 5.6 depending on the curvature of the EFS the PC can act as a collimator (linear dispersion), a divergent lens (negative EFS curvature), or a convergent lens (positive EFS curvature) [40].

Directional emitters, beam splitters including those which are Y-shaped, or have one-to-three and one-to-five structures can be made by self-collimation in 2D square lattice PCs whose output surfaces have been modified appropriately [41]. These devices are compact, exhibit high transmittance and possess symmetrical intensity output distributions.

One refractive property based on the structure of the EFS which PCs can exhibit is superprism behavior; that is, dispersion far beyond that observed by most materials. In one report the output beam propagation direction in a PC was shown to change from $-90°$ to $+90°$ as a result of relatively small ($\pm 12°$) changes in incident angle [42]. Similarly, an angular swing of $10°$ was achieved by changing the incident wavelength from 1290 nm to 1310 nm [43].

At the interface between two homogenous dielectrics Snell's Law, $n_1 \sin \theta_1 = n_2 \sin \theta_2$, which is derived by conserving the tangential component of the wave vector ($k \sin \theta$) across the boundary, states that the angle of refraction will be positive since n is positive. In a PC however, $\vec{v}_g$ can have the opposite sign of the wave

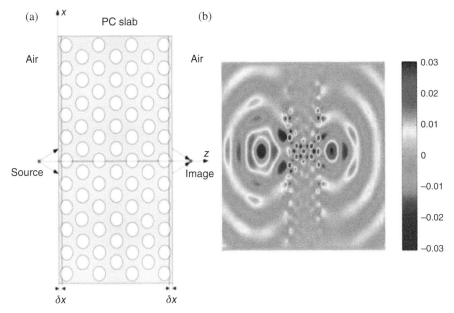

FIGURE 5.7 (a) Schematic diagram of an imaging system formed from a photonic slab utilizing negative refraction. (b) Field distribution for a point source and its image across a photonic slab. *(For a color version of this figure, see the color plate section.)*

vector $\vec{k}$; that is, the sign of the group velocity and therefore energy flux can become inverted. This leads to the phenomenon of negative refraction which when described by Snell's law leads to the conclusion that the effective refractive index of the PC is negative. However, the origin of the negative refractive index here is distinct from that of metamaterials (also called left-handed materials because of the behavior of the Poynting vector in these systems) which in the latter case arises because their dielectric permittivity, ε, and magnetic permeability, μ, are both negative [44].

The implication of negative refraction using a PC is that the flat surface of a crystal with no focal length or principal axis can behave like a lens focusing light originating from a point source with a frequency approaching that of the band edge to form a real inverted image within the PC. Negative refraction will take place twice if two boundaries exist. In this way a positive real image can be generated on the other side of a slab PC [45] (Fig. 5.7).

5.3 FABRICATION METHODS AND MATERIALS

5.3.1 Microfabrication Techniques

The most common strategy for fabricating PCs is to use dielectrics with a sufficiently high index of refraction relative to air ($n = 1$). In this regard there are many potential candidates [46]. However, silicon (Si; index of refraction ~3.45 in the infrared [47];

~4.1 at 546.1 nm [48]) is arguably the material of choice for 2D PCs because conventional semiconductor microfabrication techniques (electron beam lithography, optical lithography, and pattern transfer by etching) can be used [10, 49]. Extension to 3D is also possible by building up structures layer-by-layer via repetitive deposition and etching processes [50]. 3D PCs with a complete band gap have also been made from III–V semiconductors (GaAs or InP) [51].

2D PCS formed from porous silicon (pSi) can be made by electrochemically etching a Si substrate in hydrofluoric acid which has been first been templated with etch pits by standard lithography and alkaline etching [52]. This approach can be extended to 3D PCs by combining pSi formation with focused ion-beam milling [53].

Overall however, semiconductor microfabrication is relatively expensive and can typically only generate 3D structures a few unit cells deep. Furthermore, while semiconductor PCs have the potential to be integrated into existing optoelectronic devices, their successful operation will be limited to wavelengths longer than the materials' electronic band gap where light absorption is minimal. The Moss rule for semiconductors states that the wavelength of the electronic band edge is proportional to n [4]. Thus, PCs made from semiconducting materials with large indices of refraction are expected to operate best in the infrared [54].

5.3.2 Other Physical Techniques

Other physical methodologies devised to fabricate 3D Si-based PCs include bias sputtering which involves simultaneous radio frequency (rf) sputter deposition and sputter etching to build up alternating layers of Si and SiO_2 on a hole-patterned substrate formed by electron beam lithography, followed by dry etching [55].

Glancing angle deposition or GLAD [56] employs conventional physical vapor deposition (PVD) to deposit material onto a substrate which can be orientated at different oblique angles with respect to the PVD source. Structures initially formed at certain locations on the substrate prevent growth from taking place at other positions by shadowing. Better control of the nucleation process can be obtained by providing seed sites on the substrate through lithographic patterning. In addition, the substrate rotation rate can be varied to control the growth and shape of the individual structures. This approach has been used to fabricate novel tetragonal square spiral lattices [57,58] of Si which have PBGs of up to 15% in the visible, near infrared (NIR), and infrared (IR).

5.3.3 Chemical Techniques

5.3.3.1 Colloidal Crystals Several chemical approaches have been successfully used to fabricate 3D PCs. Colloidal suspensions of silica [59] or polystyrene latex spheres [60, 61] are known to self-organize due to electrostatic interactions into fcc crystals which exhibit stop bands (Fig. 5.8). Xia and co-workers were able to produce large-scale assemblies of uncharged polystyrene spheres in this way with a cubic close packed (ccp) structure having mid-band gap wavelengths that depend linearly on the

FIGURE 5.8 SEM images of a crystalline array that were assembled in 12-μm-thick packing cells. (a, b) ~220 nm polystyrene beads [62].

diameter of the spheres, and which as a result, exhibit a wide range of well-defined colors due to diffraction [62].

The resultant crystals however can be sensitive to cracking upon drying due to lateral shrinkage as a result of their attachment to the solid substrate. Crack-free crystals with base areas of the order of 1 cm [2] can be prepared by using high density liquids such as Ga or Hg as substrates [63]. In this way lateral shrinkage can be accommodated since the crystal is not firmly attached to the liquid surface.

3D PCs can also be fabricated using colloidal crystals as template structures [64]. For example, in one report an fcc (opal) arrangement of close-packed silica spheres (diameters between 600 and 1000 nm) self-assembled from solution was sintered to create a silica structure having small necks connecting the spheres [65]. The voids of the resultant template were subsequently filled with Si by chemical vapor deposition of a disilane (Si_2H_6) precursor. This approach provided a sufficiently high dielectric filling ratio to open up a complete band gap [66]. The silica frame was then etched away to form an Si inverse opal PC with a complete band gap near the 1.5 μm telecommunications window.

A similar approach was used to make 3D porous arrays of silica (SiO_2) from colloidal crystals whose voids were filled by a polymerization reaction [67,68]. Other semiconductors such as liquid Te can be infused into the pores of a colloidal crystal at reasonably low temperatures. However, many wide band-gap semiconductors melt at temperatures that are too high to also maintain the structural integrity of a colloidal crystal [69]. Other strategies for filling the template voids include using deposition of colloidal semiconductor nanocrystals from solution, and electrodeposition. While nanocrystals form densely packed solids as solvent is slowly evaporated, the degree of pore filling can be limited. Electrodeposition on the other hand is capable of completely filling the void spaces of a colloidal crystal and has been used successfully to make inverse opals of high index group II–VI CdS ($n = 2.5$ at 600 nm) and CdSe ($n = 2.75$ at 750 nm) [70]. Group II–VI semiconductors are of interest due to their transparency in the infrared to visible regions of the spectrum. It should be noted that group II–VI quantum dots can also self-assemble into 3D superlattices with fcc packing [71]. Lastly, 3D Ge ($n = 4$) PCs have been made by repeatedly hydrolyzing tetramethyoxygermane ($Ge(OCH_3)_4$) deposited into the void space of a colloidal crystal to form GeO_2 which was subsequently reduced to Ge at moderately high temperatures [72].

Metal oxides are particularly attractive for PC applications due to their large dielectric constants [73]. One approach is to infuse the voids with a metal alkoxide sol-gel precursor. Inverse crystals of silica, titania (TiO_2, anatase phase), zirconia (ZrO_2), and other ceramics have been synthesized in this way [74–77]. In a similar approach NiO arrays were made using a metal oxalate precursor [78].

One limitation of using metal alkoxides is that the resultant metal oxide gels can shrink by ~25–30% during the heating step required to remove the colloidal spheres. This issue was found to be mitigated significantly by initially mixing a colloidal suspension of polystyrene spheres with a colloidal dispersion of ultrafine (<100 nm) oxide particles which pack into the voids between the polystyrene spheres upon assembly [68].

Only the rutile phase of TiO_2 has an index of refraction which is sufficiently large to open a complete band gap. Yet, attempts to convert inverse opal replicas made of anatase formed by sol-gel methods to rutile at high temperatures and still preserve the PC architecture were unsuccessful. One solution to this problem is emulsion templating [79]. Here a suspension of oil drops is suspended in an immiscible titania sol composed of titanium oxide oligomers. The volume fraction of oil droplets is increased by sedimentation or centrifugation until an ordered close packed arrangement is achieved. After the sol has been gelled by the addition of a catalyst or by cross-linking the oligomers, the oil droplets can be removed by immersion into a solvent that dissolves the oil. While similar to colloidal crystal templating the oil template is deformable which minimizes stress cracking caused by shrinkage during gelation. The gel can then be dried and calcined to form the rutile phase. The primary challenge here is to produce monodisperse emulsions. Manoharan et al. [79] showed that this problem could be solved by using polymer microspheres with a low glass transition temperature, T_g, instead of oil droplets. At temperatures $>T_g$ the spheres, like an oil emulsion, are deformable and easily dissolved.

Overall, inverse opals, while very promising, are sensitive to disorder arising either from stacking faults or small deviations in sphere size in the colloidal template, or from inhomogeneous void filling during the infiltration step [69].

5.3.3.2 Direct Write Assembly Colloidal gels can serve as inks for direct writing assembly. In this maskless technique, solutions of concentrated polyelectrolytes with the requisite viscosity are delivered through a fine nozzle or syringe onto a substrate which is immersed in a coagulation reservoir. The composition of the reservoir determines the ink's elasticity and coagulation mechanism. The elasticity of the resultant gel should be high enough to retain the shape being produced and still allow the gel to flow and adhere to the substrate and the underlying patterned layers. This approach can generate 3D structures of arbitrary design which are amenable to robotic construction using a three-axis micropositioning stage [80].

5.3.3.3 Polymers Block copolymers (BCPs) are long chains made up of blocks of different polymerized monomers. Depending on the nature of the monomers, the polymer can microphase separately to form 1D, 2D, or 3D structures with periodicities on the order of 10–100 nm [81]. The relative composition and architectures of the BCP domains determine the periodic morphology while the number of repeat units of the microdomains depends on the overall molecular weight. Usually the index contrast between the BCP domains is too small for PC activity. However, in certain instances one of the microphases can either be selectively etched leaving behind air voids which makes the assembly inherently photonic or backfilled by a precursor which can form a high index ceramic [82].

5.3.4 Lithography Techniques

One of the most established methodologies for fabricating 2D planar photonic crystal slabs [83] is electron beam lithography where a focused beam of electrons is scanned across a resist substrate to create a pattern [84]. The resolution of the holes produced (10–20 nm) is limited essentially by electron scattering within and from the resist material. Patterns are then transferred onto a semiconductor by dry (reactive ion) etching. Dry etching can be viewed as having a physical and a chemical basis; the former pertaining to the bombardment of the ions while the latter involves the chemical reactions between the etched semiconductor and the ions. The main limitation of electron beam lithography is that it is a serial process which limits high volume mass production.

Other strategies for fabricating PCs are presented below.

5.3.4.1 Soft Lithography Soft lithography is a nonoptical technique where a patterned elastomer stamp is used to create structures within materials having lateral dimensions ranging from ~30 nm to 500 μm [85]. The stamp is made by cast molding an elastomer precursor onto a master having a relief pattern fabricated by either photo- or electron lithography, or micromachining. The most common elastomer used is poly(dimethylsiloxane) (PDMS) which is chemically and thermally stable, nonhydroscopic, permeable to gas, optically transparent down to ~300 nm, isotropic and homogeneous, and able to pattern nonplanar surfaces. The material however does shrink slightly upon curing and can swell in solvents such as hexane or toluene. Its elastic and thermal properties can also be limiting if creating large area patterns with aspect ratios (element height/element lateral width) outside the range 0.2–2.

Soft lithography can be used to pattern sol-gels made from metal alkoxide precursors to create metal oxide structures which are photonic due to their inherently high refractive indices. This approach has been used to successfully pattern a wide variety of simple and complex metal oxides [86]. A similar approach has been extended to novel materials including a reasonably high index ($n = 1.623$ at 633 nm) photocurable methacryl silica–titania resin [87], a luminescent titania sol-gel doped with europium chelates for light extraction, and polyacrylamide hydrogels [88].

5.3.4.2 Interference Lithography

Mask-free interference lithography (IL) (or holographic lithography) is a means of creating patterns in a photoresist when the optical paths of two or more electromagnetic waves overlap in space [89]. N-dimensional periodic structures ($N \leq 3$) can be generated using ($N + 1$) non-coplanar beams. When there is such a superposition of waves, the intensity distribution, I, of the interference field at a given point in space, $\vec{r}$, can be written as

$$I = \sum_{i=1}^{N} |\vec{E_i}| + \sum_{i \neq j}^{N} \vec{E_i^*} \cdot \vec{E_j} \exp\left[i\{(\vec{k_j} - \vec{k_i}) \cdot \vec{r} + (\phi_j - \phi_i)\}\right], \quad (5.15)$$

where $\vec{E_i}$, $\vec{k_i}$, and ϕ_i are the complex electric field amplitude, wave vector and phase of the ith beam. The intensity in the superposition region for coherent monochromatic waves varies from point to point between maxima which exceed the sum of the individual intensities of the beams, and minima, which may be zero.

The application of IL for PC application was pioneered by Turberfield and coworkers who used dielectric beam splitters to derive four non-coplanar beams from single pulses of an injection-seeded, frequency-tripled Q-switched Nd:YAG laser at 355 nm [90]. The orientation of the beams were arranged to produce an fcc interference pattern within an epoxy-type photoresist. IL can be used to generate PCs with other geometries and is size scalable [91]. Indeed, it has been predicted that all 14 Bravais lattices can be generated by multibeam IL [92]. 3D interference patterns can also be made using a single diffraction element instead of beam splitters [93]. Like colloidal crystals the resultant 2D and 3D structures can be used as templates which can incorporate higher refractive index contrast materials for PC applications [94]. Alternatively, IL can be used to directly fabricate 3D PCs from photosensitive high index chalcogenide glasses [95].

5.3.4.3 Direct Laser Writing

The availability of relatively inexpensive femtosecond lasers has allowed the fabrication of PC templates by direct laser writing (DLW) to be realized [96]. Here the output pulses of a femtosecond laser are focused within a photoresist or liquid resin to initiate a two-photon or higher nonlinear absorption at a specific spatial location. The absorption can either induce optical damage [97] or a photopolymerization reaction [98]. The photoresist is chosen to be transparent to the laser wavelength at the one-photon level but photo-active upon nonlinear absorption. This ensures that the optically induced modification occurs at the focal plane chosen. The structures generated can serve as templates for inverse PCs by infusing the void spaces with higher index materials [99]. Alternatively, 3D PCs can be fabricated by using LDW to pattern high index photosensitive materials such as chalcogenide glasses [100].

IL can be effectively used in combination with DLW by using the former technique to make the PC template and the latter to create defects by either removing lattice elements from the PC by femtosecond ablation or adding lattice elements by two-photo polymerization [101].

5.3.5 Other Types of PCs

5.3.5.1 Metallic PCs Although most PCs are made of dielectric material there are some advantages in constructing periodic structures using metals [102]. Specifically, such PCs are expected to be relatively inexpensive compared to dielectric assemblies, and to weigh less. However unlike ceramics whose dielectric constants are monotonically increasing functions of frequency, metals have dielectric functions, $\varepsilon(\omega)$, which can be expressed as

$$\varepsilon(\omega) = 1 - \left(\frac{\omega_p^2}{\omega^2}\right), \tag{5.16}$$

where ω_p is the plasma frequency of the charge carriers in the metal. The plasma frequency of a metal which is proportional to the electron density in the material typically lies in the ultraviolet region of the spectrum. Equation (5.16) shows that at frequencies $\omega < \omega_p$ the dielectric function $\varepsilon(\omega) < 1$ and the complex index of refraction becomes completely imaginary. Under these conditions the metal is a perfect reflector meaning that electromagnetic fields are not supported within the material. The resultant skin depth, δ, the dimension from the surface over which the incident field intensity drops to $1/e$ of its surface value is defined as

$$\delta = \sqrt{\frac{2}{\omega \mu_0 \sigma}}, \tag{5.17}$$

where σ is the metal conductivity and μ_0 is the permeability constant. It can be seen that δ depends inversely on the square root of the metal conductivity; that is, the better the metal is as a conductor the steeper the attenuation as a function of distance. However, at frequencies $\omega > \omega_p$ the frequency behavior of dielectric function resembles that of a dielectric.

This dispersion relationship means that metallic PCs exhibit behaviors that differ from those made of dielectrics. In general, the first PBG of a metallic PC starts at zero frequency and continues up to a cut-off frequency $\omega \sim c/2a$ where a is the periodicity and c is the speed of light. Unlike dielectric PCs which exhibit their first stop band at $\lambda \sim 2a$, at normal incidence metallic PCs exhibit a transmission gap centered at that wavelength. The position of this window can be rationalized by assuming that the periodic metal and air voids behave like a series of coupled Fabry–Pérot interferometers in the direction of field propagation. At higher frequencies other transmission windows and band gaps are present but with a decreased contrast.

In 2D systems the propagation of light within a metallic PC is very sensitive to its angle of incidence and polarization. In particular the Fabry–Pérot waveguide model works well for modes with their E vectors parallel to the z-axis for a 2D arrangement in the xy-plane (s-polarization) while modes with their E-vectors perpendicular to the z-axis (p-polarization) behave in a manner similar to waves propagating in a PC with a positive and frequency-independent dielectric constant [103]. In addition to having similar polarization characteristics, 3D PCs are also very sensitive to topographic details of the structures. Interestingly however, any periodic structure composed of metallic spheres is expected to exhibit a PBG [104]. Experimentally, 3D metallic PCs have been realized using traditional lithographic techniques and femtosecond LDW followed by metallization [105].

5.3.5.2 Tunable PCs There is an interest in fabricating PCs whose properties can be changed in response to external stimuli because of their many potential uses including optical switching and sensing. This burgeoning field has been recently reviewed in detail [106] and so only selected examples are provided here. Tunable PCs exhibit different diffraction wavelengths or intensities when exposed to a physical or chemical stimulus. In general several strategies can be used to fabricate tunable PCs. One approach is to fabricate the photonic structure using building blocks that are themselves responsive. A second approach is to introduce a responsive material into the void spaces of a PC to create a composite structure. Alternatively, one can use a PC as a template for creating inverse structures of a responsive material first introduced into the void spaces of the PC.

The materials used in or for the PC will determine in large part the most appropriate external stimulus.

Physical Stimuli Temperature change is a common stimulus which can be applied by heating or by laser irradiation. For example, colloidal crystals embedded within a hydrogel will exhibit a reversible structural change which manifests itself as a diffraction wavelength change when the hydrated gel undergoes a volume phase transition to a dehydrated state. Alternatively, PCs made of the hydrogel itself exhibit an intensity change in the diffracted light as a function of temperature [107]. Inorganic materials usually exhibit a reversible phase change around a specific temperature which manifests itself as a change in index of refraction. As one example the thermochromic material VO_2 undergoes an ultrafast phase change from semiconductor to metal near $T = 68°C$ with an accompanying change in dielectric constant. 3D opals of SiO_2 whose pores were filled with VO_2 exhibit an ultrafast reversible blue-shift in Bragg diffraction upon femtosecond irradiation and heating which could prove useful for optical switching [108].

One of the simplest physical stimuli is mechanical pressing or stretching. In one experiment Fudouzi and Sawada imbedded colloidal crystals of polystyrene into a sheet of PDMS to make a photonic rubber sheet (Fig. 5.9). The film color could be blue-shifted reversibly upon horizontal stretching due to a reduction of the lattice spacing in the vertical direction [109].

Both electric and magnetic fields can be used to drive responsive PCs. In the former case, an electric field applied to inverse Si PC impregnated with a low index liquid crystal will induce a reorientation of the anisotropic liquid crystal with respect to the inverse opal backbone thereby opening or closing the full 3D PBG [110]. As an example of the latter, Yin and co-workers synthesized superparamagnetic colloidal particles of Fe_3O_4 coated with SiO_2 that were dispersed in a resin, and then further dispersed in a viscous silicone or mineral oil. The particles within the resultant emulsion microsphere droplets were found to self-assemble in a magnetic field to

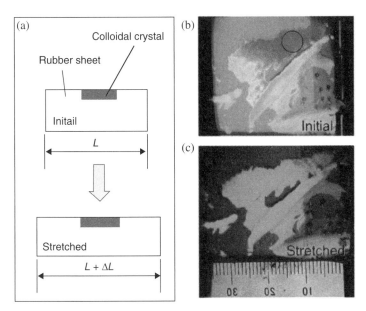

FIGURE 5.9 Changes in the structural color of the colloidal crystal film covering a silicone rubber sheet. (a) Elastic deformation when the rubber is stretched from L to $L + \Delta L$. (b) Photographic image of the initial sheet (L). (c) Photographic image of the stretched sheet ($L + \Delta L$). Reprinted with permission from Fudouzi and Sawada [109]. Copyright 2006 American Chemical Society. *(For a color version of this figure, see the color plate section.)*

form 1D chains which could be fixed in position within the droplet by polymerizing the resin using ultraviolet irradiation [111]. The chains diffract light, exhibiting a field-induced color which can be switched off by rotating the microspheres with an external magnetic field and thereby change the orientation of the chains with respect to the direction of the incident light.

Chemical Stimuli One of the simplest chemical approaches to tune the properties of a PC is to infiltrate the porous structure with solvents and vapors that lead to a change in the index of refraction of the assembly. The response of the PC can be precise and reproducible for different compounds although it can be challenging to distinguish between solvents which have similar indices of refraction [112].

PCs made of hydrogels and functionalized with a crown ether to selectively bind Pb^{2+}, Ba^{2+} and K^+ onto the gel network were found to swell due to an increased osmotic pressure as a result of a Donnan potential arising from counter ions interacting with the cations. The degree of swelling and corresponding red-shift in the diffraction peak wavelength was found to be a function of the number of covalently attached charged groups [113]. PC response can also be modulated based essentially on the same mechanism. Lee and Braun used mixtures of 2-hydroxy-ethyl methacrylate (HEMA) and acrylic acid (AA) to create inverse opal hydrogel PCs that exhibited diffraction wavelength red-shifts with increasing pH due to hydrogel swelling [114].

Ozin, Manners, and co-workers have developed a PC made of an iron-based metallopolymer whose dimensions can be reversibly changed electrochemically. When an oxidative potential is applied electrons are extracted from the Fe atoms in the polymer backbone which drives anions in the solvent into the pore structure to neutralize the buildup of positive charge. This swells the assembly leading to a red-shifted diffraction peak. Conversely, a reducing potential provides electrons for the iron atoms which leads to anions being expelled from the PC. The oxidation state of the system can be continuously tuned leading to a continuous range of voltage-dependent colors [115].

5.4 APPLICATIONS

Although the challenge of fabricating 3D PCs has stimulated the large number of innovative physical and chemical strategies described above, many of the most important applications can be realized using 2D PCs which can be patterned using well-established semiconductor nanofabrication techniques [10]. Most applications rely on the introduction of defects into the periodic structure of PC to support optical modes. A point defect within a 2D PC constitutes a cavity, while a line defect functions as a waveguide through which light travels, with the surrounding PC serving as a cladding (Fig. 5.10).

Although in principle there is no light confinement in the third (vertical) direction this problem can be overcome by using a PC slab which is a high index semiconductor membrane such as Si with a width of the order of λ/n, and which has been patterned using standard semiconductor microfabrication techniques. Guided modes within the 2D plane below the light line of a low index cladding material such as air, arises by total internal reflection (TIR). The main losses are due to two main factors:

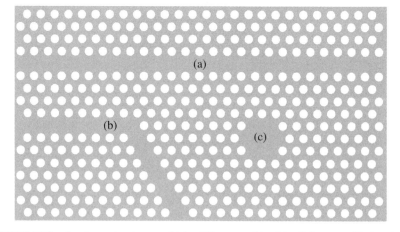

FIGURE 5.10 A schematic picture of (a) a PC waveguide, (b) a PC waveguide bend, and (c) a cavity in a 2D PC [135].

out-of-plane leakage since the holes are not infinitely deep, and diffraction because the holes do not contribute to waveguiding [84].

A plethora of PC devices and applications have been explored. Only a select number are briefly described here.

A cavity formed by fabricating a PC with a point defect has the dual benefit of being small and having a high quality factor (Q). Q is defined as $\omega/\Delta\omega$ where ω is the frequency of the light and $\Delta\omega$ is the line width of the supported mode. Alternatively, Q can be expressed as $Q = \omega\tau$ where τ is the photon lifetime within the cavity. Since the group velocity scales as L/Q where L is the length of the cavity, light–matter interactions in a PC are enhanced though slow light effects when photons are localized within smaller high Q volumes (of the order of their wavelength, λ) for longer times. In one experiment optical pulses were delayed in a PC nanocavity by 1.45 ns which corresponds to light propagation at $\sim 2 \times 10^{-5}c$, where c is the speed of light in vacuum [116]. Multiple point defects ($N > 100$) can be incorporated onto one 2D PC slab to form a coupled-resonator optical waveguide (CROW) [117, 118] (Fig. 5.11). Such low loss configurations maintain a high Q-value ($\sim 10^6$) and exhibit ultraslow light propagation with a group velocity below $0.01c$. Ultimately high Q PC cavities offer the potential of realizing smaller devices for photonic systems integration. In contrast, traditional cavities which rely on multiple reflections are extremely difficult to construct with dimensions of the order of λ. Furthermore, the larger number of reflections which takes place during the photon lifetime within a small-volume high Q cavity requires extraordinarily high mirror reflectivities.

It has been shown that the Q of a cavity formed from a single point defect in a hexagonal PC is not particularly high ($\leq 10^3$) since the supported dipole modes suffer substantial radiative losses. The situation can be dramatically improved however by structurally modifying the PC to support hexapolar modes. These modes have a sixfold symmetrical nodal pattern with alternating phases, leading to destructive interference in the vertical direction. Such cavities are characterized by Q-factors $\geq 10^6$ and a mode volume of $\sim 1.18(\lambda/n)^3$. A similar approach has been used to support high Q quadrupole modes in a 2D square lattice PC [119]. Interestingly, high Q nanocavities within a PC can also be made by creating a line defect where some of the air holes have been shifted out of registry by only a few nanometers [120] or by modifying the radii of the nearest neighbor holes in an air-clad hexagonal PC slab [121].

Another approach is to achieve a high Q nanocavity by fabricating a double-heterostructured planar PC [122, 123]. Essentially a short segment of one PC waveguide is sandwiched between photonic crystals with a different spacing. This approach allows the electric field within the center portion to be controlled. In this regard light localization arises as much from a mode-gap effect than a band gap effect in that only photons of a specific energy can be supported by the center portion and are excluded from the outer regions of the device.

5.4.1 Fundamental Effects

PCs provide opportunities to test and utilize quantum electrodynamic (QED) effects. When an optically active medium such as a semiconductor quantum dot is placed

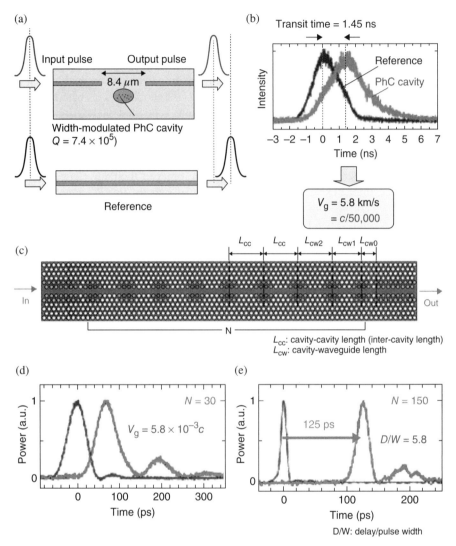

FIGURE 5.11 Slow light in a cavity. (a), (b) Slow light propagation in a single ultrahigh Q nanocavity; (c) schematic of coupled nanocavities; (d), (e) slow light propagation in coupled nanocavities [118]. *(For a color version of this figure, see the color plate section.)*

inside a cavity its spontaneous emission rate can be modified as it depends on the optical mode density at the emission wavelength [124]. In the weak coupling limit, the Rabi frequency which scales as $1/\sqrt{V}$ is less than the photon loss rate which scales as $1/Q$. Here, the spontaneous emission rate can be enhanced by an amount corresponding to the Purcell factor which scales as Q/V [125] when the frequency of a supported mode frequency in a PC nanocavity is resonant with the emission frequency. Conversely, PC nanocavities can be designed to suppress spontaneous emission.

Such systems have been shown to be effective on-demand single photon emitters for quantum computing and communication applications because the spontaneous emission output coupling efficiency into a cavity mode can approach 100% [126].

Interesting single photon nonlinear effects are observed when the Rabi frequency exceeds the photon loss rate. Selecting the energy of a single photon captured within the cavity appropriately can decrease the probability of admitting a second photon; a photon antibunching phenomenon called photon blockade. At the same time, at other energies of the captured photon the probability of admitting a second photon can be increased; a photon bunching phenomenon which has been termed photon-induced tunneling [127].

5.4.2 Lasers

Lasers based on PC nanocavities which incorporate a gain medium such as semiconductor quantum wells, and operate at the 1.5 μm telecommunications wavelength have been realized [128]. These devices can be optically pumped or electrically driven. The latter devices exhibit ultralow threshold operating currents since that parameter, which is proportional to the product of the active volume and photon lifetime, scales as V/Q. An electrically pumped CROW having quantum well active regions can also be used to generate stable single-mode lasing [129].

There are other laser designs which incorporate PC elements as part of their architecture. One example which does not involve a resonator is the photonic band edge laser [130]. This device involves optically pumping a 1D PC close to its band gap edge where the group velocity approaches zero. In the presence of an active medium embedded into one of the dielectric regions the effective increase in optical path length due to multiple reflections leads to an enhancement of laser gain. Traditional lasers have also been built that use 2D PCs as one of the end-mirrors [131]. Single-mode operation can be achieved by inserting an additional 2D PC mirror within the cavity.

5.4.3 Sensors

Nanocavities are excellent environments for sensors [132]. A high Q enhances the analyte detection sensitivity while a small V is conducive for single particle detection. Sensing is achieved by modifying the reflectance or transmittance of a PC due to the interaction between light and the analyte. Specifically, the introduction of analyte into a nanocavity of a PC laser changes the index of refraction of the ambient surroundings, causing a change in the mode resonance wavelength. Spectral shifts of 1 nm have been observed for a 0.0056 change in n [133] and index changes can be observed using sample volumes on the order of femtoliters. This approach can be used in large arrays to analyze several reagents simultaneously.

5.4.4 Add/Drop Filters

Add/drop filters are key elements of wavelength-division multiplexed (DWDM) systems. They can be constructed from a PC slab which combines point and line defects

[134]. In this device the cavity defect which serves as an optical resonator for a particular frequency is placed in the vicinity of the waveguide. The in-plane Q-value of the cavity increases with distance from the waveguide while the out-of-plane Q-value is determined by index contrast between the defect and the air cladding. This latter parameter can be controlled by changing the thickness of the dielectric slab. Photons having the resonant frequency propagating down the waveguide are trapped by the defect and emitted into free space. Such devices can be used to drop (add) photons of specific energies to (from) optical communications systems. The concept can be extended to several point defects near the waveguide which have been structurally modified for different optical frequencies. The output power from the cavity (cavities) can be controlled by adjusting the values of their in-plane and out-of-plane Q-values. The challenge of these devices is that one requires Angstrom control (of the order of the lattice constant of many semiconductors) over the fabrication of the cavities to achieve dense demultiplexing capabilities [135].

5.4.5 Directional Couplers

Directional couplers are devices designed to overcome the obstacles associated with fabricating nanocavities having the proper dimensions for dense demultiplexing applications. Here a slab PC is fabricated with two waveguides. In the original demonstration of this concept [136] a long-line waveguide was run the length of a slab PC. A second short coupling line runs parallel to the long line one hole-row distance away, and is connected to an output arm 9 hole-rows away through a 60° bend. When light is introduced into the long arm, transmission at the frequency of the guided mode drops while the intensity of the output from the second waveguide increases. There is little radiative loss at the bend due to the PBG effect. A comparable coupling efficiency can be achieved using a significantly shorter coupling line in the PC device than that required for a silica-based planar lightwave-circuit directional coupler.

Contra-directional PC waveguide couplers have also been fabricated using multi-mode waveguides of different dimensions [137]. The filter bandwidth is determined by the degree of coupling between the waveguides and the group velocities of the guided modes. As PC waveguides can support very slow modes the effective bandwidth of the photons dropped from one waveguide into the other can be very narrow even when the coupling is large.

5.4.6 PC Fibers

PCs with line defects exhibit remarkably effective waveguiding capabilities [138] and are capable of transmitting light through angles as large as 90° turns with low loss. One can imagine introducing a cavity defect into a 2D PC and guiding the light through that defect in the third dimension. A PC fiber is formed if the depth of the defect in the third direction is long [139].

There are two variants of such "holey" fibers. The first is a fiber with a solid high-index core surrounded by a hexagonal array of air holes. The light is constrained within the core in part due to TIR and partly through the PBG effect. The second is a hollow or low index core surrounded by a hexagonal array of air holes. Here TIR is not possible and therefore light localization is entirely a band gap phenomenon. The first type of fiber has a number of intriguing properties including single-mode behavior regardless of wavelength which arises because the core acts as a "sieve" which filters out higher modes which have their transverse intensity distributions further off axis. They also exhibit ultralow dispersion over hundreds of nanometer, and by appropriately engineering the microstructure of the core, can be highly birefringent. The strong PBG effect in hollow core PC fibers means that only certain wavelengths will be transmitted. Conversely, launching 800 nm fs pulse down a low dispersion, nonlinearity solid core PC fiber can generate ultrabright supercontinuum pulses having the bandwidth of sunlight [139].

The main losses which in both cases are due to absorption and scattering phenomena, are less important for the hollow core fibers since there light interacts less with the cladding.

5.5 CONCLUSIONS

Moore's law states that the number of transistors that can be placed on an integrated circuit is doubling approximately every 2 years [140]. Furthermore integrated circuit performance has been increasing at a rate of more than two orders of magnitude per decade as transistor gate widths have decreased [141]. The question exists if Moore's law will hold indefinitely; that is, is transistor scaling approaching a limit? One estimate suggests that another 30 years of progress in Si nanoelectronics is still possible before reaching fundamental limits imposed by quantum mechanics, thermodynamics, and electromagnetics [142]. One particular challenge is the large interconnect capacitance which must be charged or discharged during a binary transition involving relatively high voltage changes.

As photonic devices come to forefront it is envisaged that PC-based waveguides will be able to provide the optical interconnects between optical elements which can influence light using light through nonlinear effects [143]. One can then imagine devices limited only by the speed of light. While predictions are always fraught with uncertainty there seems little doubt that even before this is achieved PCs are poised to be a transformative component in the future evolution of communication and computing systems.

REFERENCES

[1] http://www.wave-scattering.com/pbgheadlines.html.
[2] K. Ohtaka, "Energy band of photons and low-energy photon diffraction," *Phys. Rev. B* **19**, 5057–5067 (1979).
[3] E. Yablonvitch, "Inhibited spontaneous emission in solid-state physics and electronics," *Phys. Rev. Lett.* **58**, 2059–2062.
[4] S. John, "Strong localization of photons in certain disordered dielectric superlattices," *Phys. Rev. Lett.* **58**, 2486–2489.

[5] E. Yablonovitch, "Photonic band-gap crystals," *J. Phys. Condens. Matter* **5**, 2443–2460 (1993).

[6] J. D. Joannopoulos, P. R. Villeneuve, and S. Fan, "Photonic crystals: putting a new twist on light," *Nature* **386**, 143–149 (1997).

[7] T. F. Krauss and R. M. De La Rue, "Photonic crystals in the optical regime-past, present and future," *Prog. Quant. Electron.* **23**, 51–96 (1999)

[8] R. H. Lipson and C. Lu, "Photonic crystals: a unique partnership between light and matter," *Eur. J. Phys.* **30**, S33–S48 (2009).

[9] K. Sakoda and J. W. Haus, "Science and engineering of photonic crystals," *Prog. Optics* **54**, 271–317 (2009).

[10] M. Notomi, "Manipulating light with strongly modulated photonic crystals," *Rep. Prog. Phys.* **73**, 09501 (57 pp.) (2010).

[11] J. D. Joannopoulos. S. G. Johnson, J. N. Winn, and R.D. Meade, *Photonic Crystals Molding the Flow of Light*, 2nd ed. (Princeton University Press, Princeton and Oxford, 2008).

[12] S. G. Johnson and J. D. Joannopoulos, *Photonic Crystals: The Road from Theory to Practice* (Springer, New York, 2002).

[13] K. Sakoda, *Optical Properties of Photonic Crystals* (Springer, Berlin, Heidelberg, New York, 2005).

[14] L. Rayleigh, "On the maintenance of vibrations by forces of double frequency, and on the propagation of waves through a medium endowed with a periodic structure," *Phil. Mag.* **24**, 145–159 (1887).

[15] P. Vukusic and J. R. Sambles, "Photonic structures in biology," *Nature* **424**, 852–855 (2003).

[16] S. Kinoshita and S. Yoshioka, "Structural colors in nature: the role of regularity and irregularity in the structure," *Chem. Phys. Chem.* **6**, 1442–1459 (2005).

[17] A. van Blaaderen, "Opals in a new light," *Science* **282**, 887–888 (1998).

[18] H. Fudouzi, "Tunable structural color in organisms and photonic materials for design of bioinspired materials," *Sci. Technol. Adv. Mat.* **12**, 064704 (7 pp.) (2011).

[19] B. J. Glover and H. M. Whitney, "Structural colour and iridescence in plants: the poorly studied relations of pigment colour," *Ann. Bot.* **105**, 505–511 (2010).

[20] Z. Vértesy, Z. Bálint, K. Kertész, J. P. Vigneron, V. Lousee, and L. P. Biró, "Wing scale microstructures and nanostructures in butterflies-natural photonic crystals," *J. Microsc.* **224**, 108–110 (2006).

[21] J. Zi, X. Yu, Y. Li, X. Hu, C. Xu, X. Wang, X. Liu, and R. Fu, "Coloration strategies in peacock feathers," *Proc. Nat. Acad. Sci. USA.* **100**, 12576–12578 (2003).

[22] J. W. Galusha, L. R. Richey, J. S, Gardner, J. N. Cha, and M. H. Bartl, "Discovery of a diamond-based photonic crystal structure in beetle scales," *Phys. Rev. E* **77**, 050904(R) (2008).

[23] J. W. Galusha, L. R. Richey, M. R. Jorgensen, J. S. Gardner, and M. H. Bartl, "Study of natural photonic crystals in beetle scales and their conversion into inorganic structures via a sol-gel bio-templating route," *J. Mater. Chem.* **20**, 1277–1284 (2010).

[24] E. Yablonovitch, "Photonic crystals: what's in a name?" *Opt. Photon. News* **18**, 12–13 (2007).

[25] S. G. Johnson and J. D. Joannopoulos, "Block-iterative frequency-domain methods for Maxwell's equations in a planewave basis," *Opt. Exp.* **8**, 173–190 (2000).

[26] S. G. Johnson and J. D. Joannopoulos, "The MIT Photonic-Bands Package Home Page," http://ab-initio.mit.edu/mpb/mpb-1.4.2.tar.gz.

[27] S. John, "Localization of light," *Phys. Today* **44**, 32–40 (1991); D. G. Angelakis, P. L. Knight, and E. Paspalakis, "Photonic crystals and inhibition of spontaneous emission: an introduction," *Contemp. Phys.* **45**, 303–318 (2004).

[28] K. Busch, S. F. Mingaleev, M. Schillinger, D. Hermann, and L. Tkeshelashvili, "Solid state theory meets photonics: the curious optical properties of photonic crystals," *Lect. Notes Phys.* **658**, 1–22 (2005).

[29] L. D. Bonifacio, B. V. Lotsch, D. P. Puzzo, F. Scotognella, and G. A. Ozin, "Stacking the nanochemistry deck: structural and compositional diversity in one-dimensional photonic crystals," *Adv. Mater.* **21**, 1641–1646 (2009).

[30] R. D. Meade, K. D. Brommer, A. M. Rappe, and J. D. Joannopoulos, "Existence of a photonic band gap in two dimensions," *Appl. Phys. Lett.* **61**, 495–497 (1992).

[31] P. R. Villeneuve and M. Piché, "Photonic band gaps in two-dimensional square and hexagonal lattices," *Phys. Rev. B* **46**, 4969–4972 (1992).

[32] Z.-Y. Li, B.-Y. Gu, and G.-Z. Yang, "Large absolute band gaps in 2D anisotropic photonic crystals," *Phys. Rev. Lett.* **81**, 2574–2577 (1998).

[33] R. Wang, X.-H. Wang, B.-Y. Gy, and G.-Z. Yang, "Effects of shapes and orientations of scatterers and lattice symmetries on the photonic band gap in two-dimensional photonic crystals," *J. Appl. Phys.* **90**, 4307–4313 (2001).

[34] E. Yablonovitch, T. J. Gmitter, and K. M. Leung, "Photonic band structure: the face-centered-cubic case employing nonspherical atoms," *Phys. Rev. Lett.* **67**, 2295–2298 (1991).

[35] C. T. Chan, S. Datta, K. M. Ho, and C. M. Souloulis, "A7 structure: a family of photonic crystals," *Phys. Rev. B* **50**, 1988–1991 (1994).

[36] Z.-Y. Li, J. Wang, and B.-Y. Gu, "Creation of partial band gaps in anisotropic photonic-band-gap structures," *Phys. Rev. B* **58**, 3721–3729 (1998).

[37] J. P. Dowling, M. Scalora, M. J. Bloemer, and C. M. Bowden, "The photonic band edge laser: a new approach to gain enhancement," *J. Appl. Phys.* **75**, 1896–1899 (1994).

[38] M. Notomi, K. Yamada, A. Shinya, J. Takahashi, C. Takahashi, and I. Yokohama, "Extremely large group-velocity dispersion of line-defect waveguides in photonic crystal slabs," *Phys. Rev. Lett.* **86**, 253902-1–253902-4 (2001).

[39] J. I. L. Chen, G. von Freymann, S. Y. Choi, V. Kitaev, and G. A. Ozin, "Slow photons in the fast lane of chemistry," *J. Mater. Chem.* **18**, 369–373 (2008).

[40] H. Kosaka, T. Kawashima, A. Tomita, M. Notomi, T. Tamamura, T. Sato, and S. Kawakami, "Self-collimating phenomena in photonic crystals," *Appl. Phys. Lett.* **74**, 1212–1214 (1999).

[41] W. Y. Liang, J. W. Dong, and H. Z. Wang. "Directional emitter and beam splitter based on self-collimation effect," *Opt. Exp.* **3**, 1234–1239 (2007).

[42] H. Kosaka, T. Kawashima, A. Tomita, M. Notomi, T. Tamamura, T. Sato, and S. Kawakami, "Superprism phenomena in photonic crystals," *Phys. Rev. B* **58**, R10096–R10099 (1998).

[43] L. Wu, M. Mazilu, T. Karle, and T. F. Krauss, "Superprism phenomena in planar photonic crystals," *IEEE J. Quantum Electron.* **38**, 915–918 (2002).

[44] V. M. Shalaev, "Optical negative-index metamaterials," *Nat. Photonics* **1**, 41–48 (2007); Y. Liu and X. Zhang, "Metamaterials: a new frontier of science and technology," *Chem. Soc. Rev.* **40**, 2494–2507 (2011).

[45] M. Notomi, "Theory of light propagation in strongly modulated photonic crystals: refraction like behavior in the vicinity of the photonic band gap," *Phys. Rev. B* **62**, 10696–10705 (2000).

[46] C. López, "Materials aspects of photonic crystals," *Adv. Mater.* **15**, 1679–1704 (2003).

[47] C. D. Salzberg and J. J. Villa, "Infrared refractive indexes of silicon germanium and modified selenium glass," *J. Opt. Soc. Am.* **47**, 244–246 (1957).

[48] E. A. Taft, "The optical constants of silicon and dry oxygen oxides of silicon at 5461 Å," *J. Electrochem. Soc.* **125**, 968–971 (1978).

[49] A. Birner, R. A. Wherpohn, U. M. Gösele, and K. Busch, "Silicon-based photonic crystals," *Adv. Mat.* **13**, 377–388 (2001).

[50] S. Y. Lin, J. G. Fleming, D. L. Hetherington, B. K. Smith, R. Biswas, K. M. Ho, M. M. Sigalas, W. Zubrzycki. S. R. Kurtz, and J. Bur, "A three-dimensional photonic crystal operating at infrared wavelengths," *Nature* **394**, 251–253 (1998).

[51] S. Noda, K. Tomoda, N. Yamamoto, and A. Chutinan, "Full three-dimensional photonic bandgap crystals at near-infrared wavelengths," *Science* **289**, 604–606 (2000).

[52] U. Grüning, V. Lehmann, S. Ottow, and K. Busch, "Macroporous silicon with a complete two-dimensional band gap centered at 5 µm," *Appl. Phys. Lett.* **68**, 747–749 (1996).

[53] U. Grüning, V. Lehmann, and C. M. Engelhardt, "Two-dimensional infrared photonic gap structure based on porous silicon," *Appl. Phys. Lett.* **66**, 3254–3256 (1995); A. Chelnokov, K. Wang. S. Rowson, P. Garoche, and J.-M. Lourtioz, "Near-infrared Yablonovite-like photonic crystals by focused-ion-beam etching of macroporous silicon," *Appl. Phys. Lett.* **77**, 2943–2945 (2000).

[54] P. Hervé and L. K. J. Vandamme, "General relation between refractive index and energy gap in semiconductors," *Infrared Phys. Technol.* **35**, 609–613 (1994).

[55] S. Kawakami, T. Kawashima, and T. Sato, "Mechanisms of shape formation of three-dimensional periodic structures by bias sputtering," *Appl. Phys. Lett.* **74**. 463–465 (1999).

[56] M. J. Brett and M. M. Hawkeye, "New materials at a glance," *Science* **319**, 1192–1193 (2008).

[57] O. Toader and S. John, "Proposed square spiral microfabrication architecture for large three-dimensional photonic band gap crystals," *Science* **292**, 1133–1135 (2001).

[58] S. R. Kennedy, M. J. Brett, O. Toader, and S. John, "Fabrication of tetragonal square spiral photonic crystals," *Nano Lett.* **2**, 59–62 (2002).

[59] W. L. Vos, R. Sprik, A. van Blaaderen, A. Imhof, A. Lagendijk, and G. H. Wegdam, "Strong effects of photonic band structures on the diffraction of colloidal crystals," *Phys. Rev. B* **53**, 16231–16235 (1996).

[60] İ. İ. Tarhan and H. H. Watson, "Photonic band structure of fcc colloidal crystals," *Phys. Rev. Lett.* **76**, 315–318 (1996).

REFERENCES   163

[61] Y. Xia, B. Gates, and S. H. Park, "Fabrication of three-dimensional photonic crystals for use in the spectral region from ultraviolet to near-infrared," *J. Lightwave Technol.* **17**, 1956–1962 (1999).

[62] Y. Xia, B. Gates, Y. Yin, and Y. Lu, "Monodispersed colloidal spheres: old materials with new applications," *Adv. Mat.* **12**, 693–713 (2000).

[63] B. Griesebock, M. Egen, and R. Zentel, "Large photonic films by crystallization on fluid substrates," *Chem. Mater.* **14**, 4023–4025 (2002).

[64] A. Stein and R. C. Schroden, "Colloidal crystal templating of three-dimensionally ordered macroporous solids: materials for photonics and beyond," *Curr. Opin. Solid State Mater. Sci.* **5**, 553–564 (2001).

[65] A. Blanco, E. Chomski, S. Grabtchak, M. Ibisate, S. John, S. W. Lennard, C. Lopez, F. Meseguer, H. Miguez, J. P. Mondia, et al., "Large-scale synthesis of a silicon photonic crystal with a complete three dimensional band gap near 1.5 micrometres," *Nature* **405**, 437–440 (2000).

[66] K. Busch and S. John, "Photonic band gap formation in certain self-organizing systems," *Phys. Rev. E.* **58**, 3896–3908 (1998).

[67] O. D. Velev, T. A. Jede, R. F. Lobo, and A. M. Lenhoff, "Porous silica via colloidal crystallization," *Nature* **389**, 447–448 (1997); O. D. Velev, T. A. Jede, R. F. Lobo, and A. M. Lenhoff, "Microstructured porous silica obtained via colloidal crystal templates," *Chem. Mater.* **10**, 3597–3602 (1998).

[68] G. Subramanium, V. N. Manoharan, J. D. Thorne, and D. J. Pine, "Ordered macroporous materials by colloidal assembly: a possible route to photonic bandgap materials," *Adv. Mat.* **11**, 1261–1265 (1999).

[69] D. J. Norris and Y. A. Vlasov, "Chemical approaches to three-dimensional semiconductor photonic crystals," *Adv. Mat.* **13**, 371–376 (2001).

[70] P. V. Braun and P. Wiltzius, "Electrochemically grown photonic crystals," *Nature* **402**, 603–604 (1999).

[71] C. B. Murray, C. R. Kagan, and M. G. Bawendi, "Self-organization of CdSe nanocrystallites into three-dimensional quantum dot superlattices," *Science* **270**, 1335–1338 (1995).

[72] H. Míguez, F. Meseguer, C. López, M. Holgado, G. Andreasen, A. Mifsud, and F. Fornés, "Germanium fcc structure from a colloidal crystal template," *Langmuir* **16**, 4405–4408 (2000).

[73] J. Robertson, "High dielectric constant oxides," *Eur. Phys. J. Appl. Phys.* **28**, 265–291 (2004).

[74] B. T. Holland, C. F. Blanford, and A. Stein, "Synthesis of macroporous minerals with highly ordered three-dimensional arrays of spheriodal voids," *Science* **281**, 538–540 (1998).

[75] B. T. Holland, C. F. Blanford, T. Do, and A. Stein, "Synthesis of highly ordered, three-dimensional, macroporous structures of amorphous or crystalline inorganic oxides, phosphates, and hybrid composites," *Chem. Mater.* **11**, 795–805 (1999).

[76] B. Gates, Y. Yin, and Y. Xia, "Fabrication and characterization of porous membranes with highly ordered three-dimensional periodic structures," *Chem. Mater.* **11**, 2827–2836 (1999).

[77] R. C. Schroden, M. Al-Daous, C. F. Blanford, and A. Stein, "Optical properties of inverse opal photonic crystals," *Chem. Mater.* **14**, 3305–3315 (2002).

[78] H. Yan, C. F. Blanford, B. T. Holand, M. Parent, W. H. Smyrl, and A. Stein, "A chemical synthesis of periodic macroporous NiO and metallic Ni," *Adv. Mat.* **11**, 1003–1006 (1999).

[79] V. N. Manoharam, A. Imhof, J. D. Thorne, and D. J. Pine, "Photonic crystals from emulsion templates," *Adv. Mater.* **13**, 447–450 (2001).

[80] G. M. Gratson, M. Xu, and J. A. Lewis, "Direct writing of three-dimensional webs," *Nature* **428**, 386 (2004); R. B. Rao, K. L. Krafcik, A. M. Morales, and J. S. Lewis, "Microfabricated deposition nozzles for direct-write assembly of three-dimensional periodic structures," *Adv. Mater.* **17**, 289–293 (2005).

[81] A. C. Edrington, A. M. Urbas, P. DeRege, C. X. Chen, T. M. Swager, N. Hadjichristidis, M. Xenidou, L. J. Fetters, J. D. Joannopoulos, Y. Fink et al., "Polymer-based photonic crystals," *Adv. Mater.* **13**, 421–425 (2001).

[82] Y. Fink, A. M. Urbas, M. G. Bawendi, J. D. Joannopoulos, and E. L. Thomas, "Block copolymers as photonic bandgap materials," *J. Lightwave Technol.* **17**, 1963–1969 (1999).

[83] E. Chow, S. Y. Lin, S. G. Johnston, P. R. Villeneuve, J. D. Joannopoulos, J. R. Wendt, G. A. Vawter, W. Zubrzycki, H. Hou, and A. Alleman, "Three-dimensional control of light in a two-dimensional photonic crystal slab," *Nature* **407**, 983–986 (2000).

[84] T. F. Krauss, "Planar photonic crystal waveguide devices for integrated optics," *Phys. Stat. Sol. (a)* **197**, 688–702 (2003).

[85] Y. Xia and G. M. Whitesides, "Soft lithography," *Annu. Rev. Mater. Sci.* **28**, 153–184 (1998).

[86] P. Yang, T. Deng, D. Zhao, P. Feng, D. Pine, B. F. Chmelka, G. M. Whitesides, and G. D. Stucky, "Hierarchically ordered oxides," *Science* 2244–2246 (1998); S. Seraji, Y. Wu, N. E. Jewell-Larson, M. J. Forbess, S. J. Limmer, T. P. Chou, and G. Cao, "Patterned microstructure of sol-gel derived complex oxides using soft lithography," *Adv. Mater.* **12**, 1421–1424 (2000); M. J. Hampton, S. S. Williams, Z. Zhou, J. Nunes, D.-H. Ko, J. L. Templeton, E. T. Samulski, and J. M. DeSimone, "The patterning of sub-500 nm inorganic oxide structures," *Adv. Mater.* **20**, 2667–2673 (2008).

[87] W.-S. Kim, K. B. Yoon, and B.-S. Bae, "Nanopatterning of photonic crystals with a photocurable silica-titania organic-inorganic hybrid material by a UV-based nanoimprint technique," *J. Mater. Chem.* **15**, 4535–4539 (2005).

[88] F. Di Benedetto, A. Biasco, D. Pisignano, and R. Cingolani, "Patterning polyacrylamide hydrogels by soft lithography," *Nanotechnology* **16**, S165–S170 (2005).

[89] C. Lu and R. H. Lipson, "Principles and applications of interference lithography," *Laser Photonics Rev.* **4**, 568–580 (2010).

[90] M. Campbell, D. N. Sharp, M. T. Harrison, R. G. Denning, and A. J. Turberfield, "Fabrication of photonic crystals for the visible spectrum by holographic lithography," *Nature* **404**, 53–56 (2000); A, J. Turberfield, "Photonic crystals made by holographic lithography," *MRS Bull.* **26**(8), 632–636 (2001).

[91] C. K. Ullal, M. Maldovan, E. L. Thomas, G. Chen, Y.-J. Han, and S. Yang, "Photonic crystals through holographic lithography: simple cubic, diamond-like and gyroid-like structures," *Appl. Phys. Lett.* **84**, 5434–5436 (2004).

[92] L. Z. Cai, X. L. Wang, and Y. R. Wang, "All fourteen Bravais lattices can be formed by interference of four noncoplanar beams," *Opt. Lett.* **27**, 900–902 (2002).

[93] I. Divliansky, T. S. Mayer, K. S. Holliday, and V. H. Crespi, "Fabrication of three-dimensional polymer photonic crystal structures using single diffraction element interference lithography," *Appl. Phys. Lett.* **82**, 1667–1669 (2003).

[94] I. B. Divliansky, A. Shishido. I.-C. Khoo, T. S. Mayer, D. Pena, S. Nishimura, C. D. Keating, and T. E. Mallouk, "Fabrication of two-dimensional photonic crystals using interference lithography and electrodeposition of CdSe," *Appl. Phys. Lett.* **79**, 3392–3394 (2001); Y. Xu, X. Zhu, Y. Dan, J. H. Moon, V. W. Chen, A. T. Johnson, J. W. Perry, and S. Yang, "Electrodeposition of three-dimensional titania photonic crystals from holographically patterned microporous polymer templates," *Chem. Mater.* **20**, 1816–1823 (2008); M. Miyake, Y.-C. Chen, P. V. Braun, and P. Wiltzius, "Fabrication of three-dimensional photonic crystals using multibeam interference lithography and electrodeposition," *Adv. Mater.* **21**, 3012–3015 (2009); S.-G. Park, M. Miyake, S.-M. Yang, and P. V. Braun, "Cu_2O inverse woodpile photonic crystals by prism holographic lithography and electrodeposition," *Adv. Mater.* **23**, 2749–2752 (2011).

[95] A. Feigel, Z. Kotler, B. Sfez, A. Arsh, M. Klebanov, and V. Lyubin, "Chalcogenide glass-based three-dimensional photonic crystals," *Appl. Phys. Lett.* **77**, 3221–3223 (2000); A. Feigel, M. Veinger, B. Sfez, A. Arsh, M. Klebanov, and V. Lyubin, "Three-dimensional simple cubic woodpile photonic crystals made from chalcogenide glasses," *Appl. Phys. Lett.* **83**, 4480–4482 (2003).

[96] S. Kawata, H.-B. Sun, T. Tanaka, and K. Takada, "Finer features for functional microdevices," *Nature* **412**, 697–698 (2001); M. Deubel, G. von Freymann, M. Wegener, S. Pereira, K. Busch, and C. M. Soukoulis, "Direct laser writing of three-dimensional photonic-crystal templates for telecommunications," *Nature Mater.* **3**, 444–447 (2004); S. Juodkazis, V. Mizeikis, and H. Misawa, "Three-dimensional microfabrication of materials by femtosecond lasers for photonics application," *J. Appl. Phys.* **106**, 051101 (14 pp.) (2009).

[97] G. Zhou, M. J. Ventura, M. R. Vanner, and M. Gu, "Fabrication and characterization of face-centered-cubic void dots photonic crystals in a solid polymer material," *Appl. Phys. Lett.* **86**, 011108 (3 pp.) (2005).

[98] H.-B. Sun, S. Matsuo, and H. Misawa, "Three-dimensional photonic crystal structures achieved with two-photon-absorption photopolymerization of resin," *Appl. Phys. Lett.* **74**, 786–788 (1999).

[99] F. Heinroth, S. Münzer, A. Feldhoff, S. Passinger, W. Chang, C. Reinhardt, B. Chichkov, and P. Behrens, "Three-dimensional titania pore structures produced using a femtosecond laser pulse technique and a dip coating procedure," *J. Mater. Sci.* **44**, 6490–6497 (2009).

[100] S. Wong, M. Deubel, F. Pérez-Willard, S. John, G. A. Ozin, M. Wegener, and G von Freymann, "Direct laser writing of three-dimensional photonic crystals with a complete photonic bandgap in chalcogenide glasses," *Adv. Mater.* **18**, 265–269 (2006).

[101] H.-B. Sun, A. Nakamura, K. Kaneko, and S. Shoji, "Direct laser writing defects in holographic lithography-created photonic lattices," *Opt. Lett.* **30**, 881–883 (2005).

[102] J.-M. Lourtioz and A. de Lustrac, "Metallic photonic crystals," *C. R. Physique* **3**, 79–88 (2002).

[103] M. M. Sigalas, C. T. Chan, K. M. Ho, and C. M. Soukoulis, "Metallic photonic band-gap materials," *Phys. Rev. B* **52**, 11744–11751 (1995).

[104] W. Y. Zhang, X. Y. Lei, Z. L. Wang, D. G. Zheng, W. Y. Tam, C. T. Chan, and P. Sheng, "Robust photonic band gap from tunable scatterers," *Phys. Rev. Lett.* **84**, 2853–2856 (2000).

[105] N. Vasilantonakis, K. Terzaki, I. Sakellari, V. Purlys, D. Gray, C. M. Soukoulis, M. Vamvakaki, M. Kafesaki, and M. Farsari, "Three-dimensional metallic photonic crystals with optical bandgaps," *Adv. Mater.* **24**, 1101–1105 (2012).

[106] J. Ge and Y. Yin, "Responsive photonic crystals," *Angew. Chem. Int. Ed.* **50**, 1492–1522 (2011) and references within.

[107] J. M. Weissman, H. B. Sunkara, A. S. Tse, and S. A. Asher, "Thermally switchable periodicities and diffraction from mesoscopically ordered materials," *Science* **274**, 959–960 (1996).

[108] A. B. Pevtsov, D. A. Kurdyukov, V. G. Golubev, A. V. Akimov, A. A. Meluchev, A. V. Sel'kin, A. A. Kaptyanskii, D. R. Yakolev, and M. Bayer, "Ultrafast stop band kinetics in a three-dimensional opal-VO$_2$ photonic crystal controlled by a photoinduced semiconductor-metal phase transition," *Phys. Rev. B* **75**, 1531010 (4 pp.) (2007).

[109] H. Fudouzi and T. Sawada, "Photonic rubber sheets with tunable color by elastic deformation," *Langmuir* **22**, 1365–1368 (2006).

[110] K. Busch and S. John, "Liquid-crystal photonic-band-gap materials: the tunable electromagnetic vacuum," *Phys. Rev. Lett.* **83**, 967–970 (1999).

[111] J. Ge, H. Lee, J. Kim, Z. Lu, H. Kim, J. Goebl, S. Kwon, and Y. Yin, "Magnetochromatic microspheres: rotating photonic crystals," *J. Am. Chem. Soc.* **131**, 15687–15694 (2009).

[112] C. F. Blanford, R. C. Schroden, M. Al-Daous, and A. Stein, "Tuning solvent-dependent color changes of three-dimensionally ordered macroporous (3DOM) materials through compositional and geometric modifications," *Adv. Mater.* **13**, 26–29 (2001).

[113] J. H. Holtz and S. A. Asher, "Polymerized colloidal crystal hydrogel films as intelligent chemical sensing materials," *Nature* **389**, 829–832 (1997).

[114] Y.-J. Lee and P. V. Braun, "Tunable inverse opal hydrogel pH sensors," *Adv. Mater.* **15**, 563–566 (2003).

[115] A. C. Arsenault, D. P. Puzzo, I. Manners, and G. A. Ozin, "Photonic-crystal full-color displays," *Nat. Photonics* **1**, 468–472 (2007).

[116] T. Tanabe, M. Notomi, E. Kuramochi, A. Shinya, and H. Taniyama, "Trapping and delaying photons for one nanosecond in an ultrasmall high-Q photonic-crystal nanocavity," *Nat. Photonics* **1**, 49–52 (2007).

[117] A. Yariv, Y. Xu, R. K. Lee, and A. Scherer, "Coupled-resonator optical waveguide: a proposal and analysis," *Opt. Lett.* **24**, 711–713 (1999); M. Notomi, E. Kuramochi, and T. Tanabe, "Large-scale arrays of ultrahigh-Q coupled nanocavities," *Nat. Photonics* **2**, 741–747 (2008).

[118] M. Notomi, "Manipulating light by photonic crystals," *NTT Tech. Rev.* **7**(9), 1–10 (2009).

[119] H.-Y. Ryu, S.-H. Kim, H.-G. Park, J.-K. Hwang, Y.-H. Lee, and J.-S. Kim, "Square-lattice photonic band-gap single-cell laser operating in the lowest-order whispering gallery mode," *Appl. Phys. Lett.* **80**, 3883–3885 (2002).

[120] E. Kuramochi, M. Notomi, S. Mitsugi, A. Shinya, T. Tanabe, and T. Watanabe, "Ultrahigh-Q-photonic crystal nanocavities realized by the local width modulation of a line defect," *Appl. Phys. Lett.* **88**, 041112-1–041112-3 (2006).

REFERENCES 167

[121] M. Notomi, A. Shinya, S. Mitsugi, E. Kuramochi, and H.-Y. Ryu, "Waveguides, resonators and their coupled elements in photonic crystal slabs," *Opt. Exp.* **12**, 1551–1561 (2004).

[122] B.-K. Song, S. Noda, T. Asano, and Y. Akahane, "Ultra-high-Q photonic double-heterostructure nanocavity," *Nat. Mater.* **4**, 207–210 (2005).

[123] Y. Akahane, T. Asano, B.-S. Song, and S. Noda, "High-Q photonic nanocavity in a two-dimensional photonic crystal," *Nature* **425**, 944–947 (2003).

[124] S. Noda, M. Fujita, and T. Asano, "Spontaneous-emission control by photonic crystals and nanocavities," *Nat. Photonics* **1**, 449–458 (2007).

[125] D. Englund, D. Fattal, E. Waks, G. Solomon, B. Zhang, T. Nakaoka, Y. Arakawa, Y. Yamamoto, and J. Vučković, "Controlling the spontaneous emission rate of single quantum dots in a two-dimensional photonic crystal," *Phys. Rev. Lett.* **95**, 013904-1–013904-4 (2005).

[126] W.-H. Chang, W.-Y. Chen, H.-S. Chang, T.-P. Hsieh, J.-I. Chyi, and T.-M. Hsu, "Efficient single-photon sources based on low density quantum dots in photonic-crystal nanocavities," *Phys. Rev. Lett.* **96**, 117401-1–117401-4 (2006)

[127] A. Faraon, I. Fushman, D. Englund, N. Stolz, P. Petroff, and J. Vučković, "Coherent generation of non-classical light on a chip via photon-induced tunnelling and blockade," *Nat. Phys.* **4**, 859–863 (2008).

[128] O. Painter, R. K. Lee, A. Scherer, A. Yariv, J. D. O'Brien, P. D. Dapkus, and I. Kim, "Two-dimensional photonic band-gap defect mode laser," *Science* **284**, 1819–1821 (1999); H.-G. Park, S.-H. Kim, S.-H. Kwon, Y.-G. Ju, J.-K. Yang, J.-H. Baek, S.-B. Kim, and Y.-H. Lee, "Electrically driven single-cell photonic crystal laser," *Science* **305**, 1444–1447 (2004).

[129] T. D. Happ, M. Kamp, A. Forchel, J.-L. Gentner, and L. Goldstein, "Two-dimensional photonic crystal couple-defect laser diode," *Appl. Phys. Lett.* **82**, 4–6 (2002).

[130] J. P. Dowling, M. Scalora, M. J. Bloemer, and C. M. Bowden, "The photonic band edge laser: a new approach to gain enhancement," *J. Appl. Phys.* **75**, 1896–1899 (1994).

[131] J. D. O'Brien, O. Painter, R. Lee, C. C. Cheng, A. Yariv, and A. Scherer, "Lasers incorporating 2D photonic bandgap mirrors"" *Electron. Lett.* **32**, 2243–2244 (1996); T. D. Happ, A. Markand, M. Kamp, A. Forchel, and S. Anand, "Single-mode operation of coupled-cavity lasers based on two-dimensional photonic crystals," *Appl. Phys. Lett.* **79**, 4091–4093 (2001).

[132] R. V. Nair and R. Vijaya, "Photonic crystal sensors: an overview," *Prog. Quantum Electron.* **34**, 89–134 (2010).

[133] M. Lončar, A. Scherer, and Y. Qiu, "Photonic crystal laser sources for chemical detection," *Appl. Phys. Lett.* **82**, 4648–4650 (2003).

[134] S. Noda, A. Chutian, and M. Imada, "Trapping and emission of photons by a single defect in a photonic bandgap structure," *Nature*, **407**, 608–610 (2000).

[135] L. Thylén, M. Qui, and S. Anand, "Photonic crystals-a step towards integrated circuits for photonics," *Chem. Phys. Chem.* **5**, 1268–1283 (2004).

[136] M. Tokushima and H. Yamada, "Photonic crystal line defect waveguide directional coupler," *Electron. Lett.* **37**, 1454–1455 (2001).

[137] M. Qiu, M. Mulot, S. Anand, B. Jaskorzynska, M. Kamp, and A. Forchel, "Photonic crystal optical filter based on contra-directional waveguide coupling," *App. Phys.* **83**, 5121–5123 (2003).

[138] S. G. Johnston, P. R. Villeneuve, S. Fan, and J. D. Joannopoulos, "Linear waveguides in photonic crystal slabs," *Phys. Rev. B* **62**, 8212–8222 (2000).

[139] P. Russell, "Photonic crystal fibers," *Science* **299**, 358–362 (2003); J. C. Knight, "Photonic crystal fibers," *Nature* **424**, 847–851 (2003), and references within.

[140] G. E. Moore, "Cramming more components onto integrated circuits," *Electronics* **38**, 114–117 (1965).

[141] L. C. Kimerling, "Photons to the rescue: microelectronics becomes microphotonics," *Electrochem. Soc. Interface* **9**, 28–31 (2000).

[142] M. Lundstrom, "Moore's law forever?" *Science*, **299**, 210–211 (2003); J. D. Meindl, Q. Chan, and J. A. Davis, "Limits on silicon nanoelectronics for terascale integration," *Science* **293**, 2044–2049 (2001).

[143] M. Soljačić and J. D. Joannopoulos, "Enhancement of nonlinear effects using photonic crystals," *Nat. Mater.* **3**, 211–219 (2004).

6

QUANTUM DOTS

STANLEY TSAO AND MANIJEH RAZEGHI

Center for Quantum Devices, Department of Electrical Engineering and Computer Science, Northwestern University, Evanston, IL, USA

6.1 INTRODUCTION

The strong interest in low dimensional semiconductor structures originates from their exciting electronic properties, which can have an important impact on the performance of electronic and photonic devices. The quantum dots (QDs), also known as quantum boxes, are nanometer scale islands in which electrons and holes are confined in three-dimensional (3D) potential boxes. They are expected to show a zero-dimensional (0D), δ-function density of states and are able to quantize an electron's free motion by trapping it in a quasi-0D potential confinement. As a result of the strong confinement imposed in all three spatial dimensions, quantum dots are similar to atoms. They are frequently referred to as "artificial atoms." Due to this confinement, novel physical properties will emerge, which can lead to new semiconductor devices as well as drastically improved device performance.

As the particles are confined in all three dimensions, there is no dispersion curve and the density of states is just dependent on the number of confined levels. For one single dot, only two (spin-degenerate) states exit at each energy level and the plot of the density of states versus energy will be a series of δ-functions. Figure 6.1 shows the change of the density states from a bulk system to the low dimensional systems of quantum wells (QWL), quantum wires (QWR), and QDs. The calculation of these density of states can be found in an introductory solid state text [1].

In QDs, the width of the electron energy distribution is zero in an ideal case. This means that electrons in those structures are distributed in certain discrete energy

---

Photonics: Scientific Foundations, Technology and Applications, Volume II, First Edition.
Edited by David L. Andrews.
© 2015 John Wiley & Sons, Inc. Published 2015 by John Wiley & Sons, Inc.

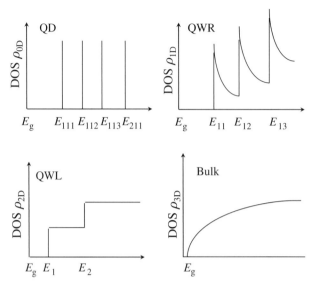

FIGURE 6.1 Density of state of zero-dimensional (upper left), one-dimensional (upper right), two-dimensional (lower left), and bulk (lower right) systems.

levels, and the energy distribution width is fundamentally independent of temperature. In real semiconductor structures, due to many interaction processes such as electron–electron and electron–phonon scattering (which can also be reduced by QDs due to the lack of phonons to satisfy the energy conservation, which is the so-called phonon bottleneck [2]), a certain width in the electron energy distribution exists. However it is expected to be much smaller compared to bulk and QW systems.

The condition for novel and interesting electronic properties to occur in a QD-based device is that the lateral size of the QD should be smaller than the coherence length and the elastic scattering length of the carriers. Additional quantum-size effects require the structural features to be reduced to the range of the de Broglie wavelength. The advantages in operation depend not only on the absolute size of the nanostructures in the active region, but also on the uniformity of size and shape. A large distribution of sizes would "smear" the density of states of QDs thus making it more like that of bulk material. Therefore, the repeatable fabrication of these nanometer 3D quantum structures requires methods with atomic scale accuracy, which presents a major challenge for current nanostructure material fabrication technologies.

The fabrication techniques of quantum dots can be categorized into "top-down" methods using lithography and etching and "bottom-up" methods utilizing self-assembly. Examples of top-down methods usually include electron beam lithography, dry etching, and, sometimes, patterned substrate regrowth. QDs can be etched from QW structures via low energy electron beam lithography. Another method of creating quantum dots is realized by applying voltage to nanoelectrodes. The spatially modulated electric field created by the voltage localizes electrons in a small area. Quantum dots can also be created through the selective growth of a narrow gap semiconductor material on a patterned wide gap substrate. One problem with such "top-down"

INTRODUCTION

methods is the low optical efficiency of the resultant dots: the high surface-to-volume ratios of these nanostructures and associated high surface recombination rates, plus the damage introduced during the fabrication, prevents the successful formation of high quality QD devices.

A past breakthrough in QD fabrication techniques was self-assembly based on the Stranski–Krastanow growth mode [3]. In this method, the lattice constants of the substrate and the crystallized material differ greatly, so only the first deposited monolayer crystallizes in the form of epitaxial strained layers where the lattice constant is equal to that of the substrate. When the critical thickness of the epitaxial layer is exceeded, the significant strain that occurs in the epitaxial layer leads to the breakdown of this ordered structure and to the spontaneous creation of randomly distributed islets (e.g., quantum dots) of regular shapes and similar sizes. The small sizes of the self-assembled quantum dots, the homogeneity of their shapes and sizes in a macroscopic scale, the perfect crystal structure, and the fairly convenient growth process, without the necessity to precisely deposit electrodes or etching, are among this method's advantages.

QDs are expected to lead to novel semiconductor devices and improve existing devices' performance. One successful example is the QD laser. The main advantages of QD lasers over conventional QW lasers are lower threshold current density, high gain, weak temperature dependence (high characteristic temperature T_0) and low chirp [4,5]. Another application of QDs is in quantum dot infrared detectors (QDIPs), and it is this area that will be the main focus of this chapter.

6.1.1 Infrared Detection Basics

Before moving into the discussion of quantum dot detectors it is necessary to give a brief introduction to some infrared detection concepts. More detailed treatments can be found elsewhere [6,7].

QWIPs and QDIPs are intersubband (ISB) devices, which are monopolar devices involving only electrons or holes. Most ISB detectors, particularly QDIPs, are made n-type (to take advantage of the higher electron mobility). Basic QWIPs and QDIPs have the structures shown in Figure 6.2. The structures are basically the same except

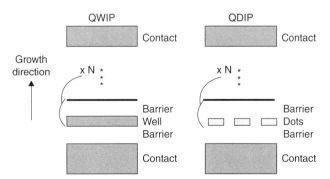

FIGURE 6.2 Schematic device structure of a QWIP (right) and a QDIP (left). From Reference 8.

172 QUANTUM DOTS

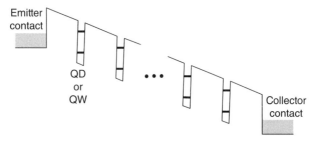

FIGURE 6.3 Band structure of a QWIP or QDIP under bias.

with the QWs swapped for QDs. A schematic of the band structure of a QWIP or QDIP is shown in Figure 6.3. In interpreting Figure 6.3 it is important to remember that the dimensionality of the electron confining potential is not illustrated since the QD actually has 3D confinement in such a band diagram. This is why we can say the picture applies to either a QWIP or QDIP. With these two pictures in mind let us now consider the basic processes in our detectors.

6.1.1.1 Photocurrent We will start by looking at where the signal comes from in a QWIP or QDIP. Photon detection occurs when incident infrared light excites an electron in the ground state out of the QW or QD and into the continuum. There are two possible paths for the photoexcited electron to escape out of the confining potential, which are depicted in Figure 6.4.

In the first path, the light directly excites the electron from the ground state to the continuum. In the second path, the light first excites the electron from the ground state to an excited state. Then, from the excited state the photoexcited carrier can thermally escape to or tunnel to the continuum. While under an applied bias, once an electron has been excited to the continuum it will be swept toward the contacts and contribute to the photocurrent.

Also depicted in Figure 6.4 are the relaxation or recapture paths, which occur when a photocurrent electron does not make it out of the well or dot or does not make it to a contact. While the photoexcitation process just described is caused by electrons,

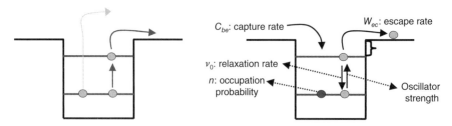

FIGURE 6.4 (Left) Schematic of the photoexcitation mechanisms in an intersubband detector. (Right) Schematic of the factors that affect the photoexcitation process.

INTRODUCTION

these relaxation processes are caused by phonons. The time it takes for a carrier traveling in the continuum to be recaptured into a well or dot is the carrier lifetime. The ratio of the carrier lifetime to the time it takes for an electron to travel across the entire device from contact to contact is defined as the gain g. Gains greater than 1 means that a photogenerated electron can travel through the device more than once creating a greater signal per photoexcitation event which increases the responsivity.

$$g = \frac{\tau_{\text{lifetime}}}{\tau_{\text{transit}}} = \frac{\text{carrier lifetime}}{\text{carrier transit time}}. \quad (6.1)$$

6.1.1.2 Dark Current The dark current is a nonsignal current flow in the detector or current that flows without light. In imaging applications low dark current is desired because the amount of current that can be read at one time is fixed, and it is preferable for as much of that fixed amount of current to be photocurrent not dark current. In an ISB detector there are three primary dark current mechanisms, as illustrated in Figure 6.5. They are

A. Thermionic emission—electrons in the ground state are thermally excited to the continuum
B. Thermally assisted tunneling—electrons are thermally excited to an excited state and then tunnel out of the excited state into the continuum
C. Sequential tunneling—electrons tunnel directly between well or dot ground state and eventually to the contact

Mechanisms A and B are the dominant processes for most of the devices and operating conditions relevant to our discussion. Mechanisms A and B are most strongly effected by the energy level structure of the dot or well, the operating temperature, and the applied bias.

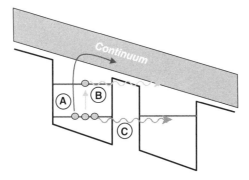

FIGURE 6.5 Dark current mechanisms in QWIPs and QDIPs. (a) Thermionic emission; (b) thermally assisted tunneling; (c) sequential tunneling. *(For a color version of this figure, see the color plate section.)*

6.1.1.3 Noise For typical devices and operating conditions in QWIPs and QDIPs, generation–recombination (GR) noise is the dominant noise source. The GR noise is determined by the following equation:

$$i_{GR}^2 = 4egAf I_D, \quad (6.2)$$

where i_{GR} is the GR noise, e is the electron charge, g is the gain, Δf is the noise bandwidth, and I_D is the dark current. GR noise is a result of the statistical fluctuation in the generation and recombination rates of electrons between different energy states in the material.

6.1.1.4 Detector Metrics A few of the infrared detector metrics more relevant to this chapter are briefly described here. More detailed information on IR detector characterization can be found elsewhere [9].

Responsivity R is a signal strength metrics defined as the ratio of photocurrent to incident optical power and can be written as

$$R = g \frac{e(1 - \exp(-\alpha L))}{\hbar \omega} \left\{ \frac{v_0 e^{-E_{\text{eff}}/kT}}{v_{eg} + v_0 e^{-E_{\text{eff}}/kT}} \right\}, \quad (6.3)$$

where e is the magnitude of the charge, α is the absorption coefficient, L the width of the active part of the device, ω is the photon frequency, g is the gain (given in Eq. 6.1), and v_0 is an attempt frequency for carrier thermal/tunnel escape out of the excited states with effective activation energy E_{eff}. One can see that good performance requires a high absorption coefficient. The responsivity (and photocurrent) is often characterized as a function of wavelength as well, giving the detector's spectrum.

Related to the responsivity is the quantum efficiency (QE), which is essentially the responsivity but in quantized units, namely, electrons per photon. There are two types of quantum efficiency relevant here, external QE (also sometimes called conversion efficiency) and internal QE. The external QE η_{ext} can be defined very simply as

$$\eta_{\text{ext}} = \frac{i_p}{e\Phi}, \quad (6.4)$$

where i_p is the photocurrent and Φ is the incident photon flux. The internal QE is simply the external QE divided by the gain, and so the internal QE is an indicator of the effectiveness of the combined effectiveness of photon absorption and subsequent electron escape in the QW or QD.

The overall performance of a detector is characterized using a signal-to-noise metric called specific detectivity D^* defined as

$$D^* = \frac{R\sqrt{A\Delta f}}{I_n}, \quad (6.5)$$

where A is the area, I_n is the noise current, R is the responsivity, and Δf is the noise bandwidth. D^* is intended to account for variations in detector operation parameters (size and noise bandwidth) to allow for easy comparison of the sensitivity of different devices and device technologies.

The final metric to be discussed here is noise equivalent difference temperature (NEDT). NEDT is often used as a measure of FPA performance to compare different FPAs. It gives the minimum resolvable temperature difference when viewing a scene. It can be defined in terms of D^* by the expression

$$\text{NEDT} = \frac{1}{C_d(\Delta\lambda)} \frac{\sqrt{A\Delta f}}{D^*} = \frac{1}{\int_{\Delta\lambda} d\lambda \frac{dR}{d\lambda}} \frac{\sqrt{A\Delta f}}{D^*}, \quad (6.6)$$

where a thermal variation in an object of ΔT gives a change in the blackbody emittance of $C_d(\Delta\lambda)\Delta T$ over a spectral range $\Delta\lambda$, and $dR/d\lambda$ is the emittance. A more practical definition in terms of measuring the NEDT of an actual FPA uses the following two relations:

$$\text{SNR} = \frac{\text{Signal}_{T_2} - \text{Signal}_{T_1}}{\text{Noise}}, \quad (6.7)$$

$$\text{NEDT} = \frac{\Delta T}{\text{SNR}}, \quad (6.8)$$

where SNR is the signal-to-noise ratio, Signal_{T_1} and Signal_{T_2} are the FPA signal levels corresponding to the FPA viewing a blackbody source at the two different temperatures T_1 and T_2, Noise is the temporal noise of the FPA signal, and ΔT is the difference in temperature between T_1 and T_2.

6.2 QUANTUM DOTS FOR INFRARED DETECTION

Infrared photodetectors have been extensively investigated during the past decades for use as the building blocks of focal plane array (FPA) imagers that are useful in military, medical, and civilian applications of infrared detection [10]. For earth-based systems, it is important to have devices that detect in the mid-wavelength infrared (MWIR, 3–5 μm) and long-wavelength infrared (LWIR, 8–12 μm) spectral regions where the atmosphere is mostly transparent to the infrared radiation.

Most MWIR and LWIR photodetector FPAs are based either on HgCdTe (MCT) [11] or quantum well infrared photodetectors (QWIPs) [3, 10]. Although bulk materials such as MCT dominate the detector market, they suffer from major challenges for large two-dimensional arrays that stimulated the search for new technologies. MCT FPAs suffer from difficult material growth, device instability, high array nonuniformity, and very high cost. QWIPs have also been used successfully for commercial FPAs. They utilize ISB transitions for detection and can take advantage of mature III–V compound growth and fabrication technology. However, one major limitation of QWIPs is that, due to the transition selection rules, the most widely used n-type QWIPs are not sensitive to normally incident light and typically have a narrow response range in the infrared. P-type QWIPs are able to detect normal incidence light due to band mixing, however, their low detectivity limits their practical use. Also QWIPs require cryogenic cooling to get rid of their high intrinsic dark current [12].

QDIPs extend the 1D confinement in QWIPs to 3D confinement. QDIPs utilize ISB absorption between bound states in the conduction or valence band in QDs. With high uniformity and high density quantum dot layers, QDIPs are actually predicted to outperform QWIPs [13, 14] due to their (i) intrinsic sensitivity to normal incidence light, (ii) longer photoexcited electron lifetime due to the phonon bottleneck, and (iii) lower dark and noise currents [15]. These benefits will ultimately allow for higher operating temperatures that will reduce the cost and complexity of detector and imaging systems by reducing the cooling requirements normally associated with cryogenically cooled detector systems.

This potential of QD-based detectors has spurred a great deal of research activity in the area [16–21]. QDIPs can be used in FPA-based infrared imaging systems, which have been widely investigated for middle wavelength infrared (3–5 µm) and LWIR (8–12 µm) applications [22–25]. So far, most QDIPs reported have showed inferior or at best comparable performance to that of QWIPs with similar parameters, though high temperature demonstrations with moderate performance are becoming more common. The major challenge facing QDIPs is the QD fabrication. To achieve its potential advantages, QDIPs need to have uniform and high density QD layers. New device designs for QDIPs are also required to further improve the technology's performance as a competitive infrared photodetector platform.

The Center for Quantum Devices (CQD) at Northwestern University has many years of experience in QD-based detector and FPA development. Some of this work is presented later in this chapter. After a brief introduction to some of the experimental methods, GaAs-based QD devices and FPAs are discussed followed by InP-based devices and FPAs. Within each section the self-assembled quantum dot growth by metalorganic chemical vapor deposition (MOCVD) is discussed along with device design and results, and where applicable FPA imaging results are also presented.

6.2.1 Benefits of Quantum Dots for Intersubband Detectors

In the mid-1990s, Ryzhii published the first detailed analysis of the benefits of using quantum dots for infrared detection. [15] Phillips and Martyniuk et al. [26,27] further developed models of QDIP performance and showed that QDIPs have the potential for excellent IR detector performance. The interest in QDIPs comes primarily from their predicted ability to achieve high performance at high operating temperatures (near room temperature). High temperature operation is an important but difficult technological hurdle for current photon detectors, which typically require some level of cryogenic cooling. The high operating temperature capability of QDIPs comes from two quantum dot-related effects: low dark current and high photoelectric gain. QDIPs also have two other important technological benefits which we will also discuss here.

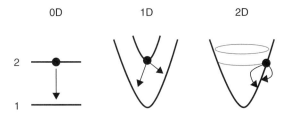

FIGURE 6.6 Illustration of the differences in phonon-mediated electron relaxation in 0D, 1D, and 2D structures.

6.2.1.1 High Gain and the Phonon Bottleneck For the operation of ISB detectors, carrier relaxation is an important factor in detector performance, and the dominant relaxation paths in ISB devices are phonon-mediated ones. Early theoretical studies of low dimensional structures predicted a phenomenon called the *phonon bottleneck*, where the delta function–like density of states significantly slows down carrier transitions between energy levels in a QD system. This is illustrated for carrier relaxation in Figure 6.6. For the 1D and 2D cases there is a continuum of states in at least one dimension, whereas in the 3D system there is only the fully quantized, delta function–like ground state. As a result, phonon-mediated processes are less likely to occur since the phonon energy must exactly match the energy spacing between the QD levels. In addition, in a typical QD system, the energy level spacing for QDs with only one or two levels ends up being larger than the typical LO phonon energy so there is little opportunity for phonon scattering. Since the phonon-mediated relaxation paths are blocked the excited state carrier lifetimes should lengthen from the 1 to 10 ps measured in QW systems to the nanosecond range predicted for QD systems.

In a detector context, this means that photoexcited carriers will have a longer lifetime both in excited states and in the continuum. So a longer lifetime gives higher gain, which then yields higher responsivity, since responsivity is proportional to gain.

6.2.1.2 Low Dark Current The other primary advantage for high operating temperature is the expectation that QDIPs will have low dark current. The dominant dark current mechanism in ISB devices is typically thermionic emission, when an electron is thermally excited out of the well. Other possible mechanisms are tunneling-assisted thermionic emission and dot-to-dot tunneling. With 3D confinement and the resultant well-defined quantized state, the thermionic emission path and the photoexcitation path become one and the same, competing with each other. Photoexcitation is typically faster than thermionic emission, causing a reduction in dark current. Also, since there are only one or two quantized states in the QD, any thermal excitation must take the electron out of the QD in only one step, making the activation energy higher for QD systems. In contrast, in materials with bands, like quantum well detectors, there is a distribution of available states at each energy level thus giving more available initial and final states for thermal excitation. Also, the QWs can accommodate a higher carrier density. Martyniuk et al. [8] calculated and compared the ideal dark current values for HgCdTe and QDIP detectors and found that in the MWIR and

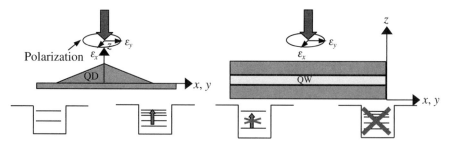

FIGURE 6.7 Schematic of the polarization sensitivity of QD and QW absorption.

beyond at operating temperatures above 200 K QDIP detectors should have lower dark current than HgCdTe. This advantage is more than one order of magnitude at room temperature and in the LWIR regime. This clearly shows the potential for QDIP performance to outpace HgCdTe.

6.2.1.3 Normal Incidence Absorption In comparison to QWIPs, QDIPs have the advantageous capability to absorb normally incident light. Most detector applications utilize a normal incidence light configuration where the incoming light signal is traveling perpendicular to the growth plane of the detector elements and is therefore randomly polarized in the growth plane. This is illustrated in Figure 6.7. The strength of the absorption of the incident light is related to the oscillator strength, which can be written as,

$$f_{12} \propto |\langle F_1|\vec{\varepsilon}\cdot\vec{p}|F_2\rangle|^2 \propto |\langle F_1|[\varepsilon_x(\partial/\partial x) + \varepsilon_y(\partial/\partial y) + \varepsilon_z(\partial/\partial z)]|F_2\rangle|^2, \quad (6.9)$$

where F_1 and F_2 are the electron envelope functions, ε is the polarization vector for the incident infrared light, and the p is the momentum operator. For normally incident light, ε_x and ε_y are nonzero and ε_z is zero. In the QW case, F_1 and F_2 are only functions of z (the confined dimension). So the for oscillator strength, in the x and y directions the partial derivative goes to zero and in the z direction ε_z is zero, therefore f_{12} is zero. In the QD case the, F_1 and F_2 are functions of x, y, and z, therefore even though ε_z is zero, the first two terms are nonzero and contribute to the oscillator strength for normal incidence.

In QWIPs the normal incidence limitation is dealt with by placing a grating on the detector surface that redirects the light, changing the polarization orientation. In QDIPs with normal incidence absorption this extra, nontrivial processing step could be removed, simplifying the device processing. Also the polarization sensitivity limits the ultimate quantum efficiency of a QWIP to 50%, whereas an ideal QDIP would not have this limit on the quantum efficiency.

6.2.1.4 Versatility This last advantage of QDIPs is not a performance advantage but a design or technology advantage. Compared with bulk and QW detectors, QDIPs have more numerous adjustable design parameters and thus greater potential versatility. For example, to tune the wavelength, the main adjustment parameter for bulk systems is material composition. In a QW system the parameters are material composition and well thickness, with strain being a secondary parameter. Finally, in a QD system material composition, strain, and QD size and shape can all be used to tune the wavelength. In QD systems there are more available "knobs to turn" to achieve the desired outcome. However, this can be a double-edged sword if the parameters are difficult to control—which is, in fact, the case for QDs as discussed in the section on QD growth.

6.2.2 The Potential of QDIPs

The above advantages if fully realized make QDIPs a very promising technology for infrared imaging applications. This can be seen in the calculations that take these advantages into account by Martyniuk et al. [26] in a review of QD-based detector technology where the authors showed that the ultimate predicted QDIP detectivity would be comparable to or better than HgCdTe detectors, especially at an operating temperature of 300 K and for LWIR detection where the D^* for QDIP was predicted to be more than 10× higher than HgCdTe. However, they also showed that the demonstrated QDIPs in the literature are all still achieving D^* values at least an order of magnitude lower than this ideal, predicted value.

6.3 QUANTUM DOT GROWTH

The biggest hurdle in the study of low dimensional semiconductor systems has always been the physical realization of such systems. Since the relevant features sizes for these systems are at the nanometer level, even slight imperfections in the material can mask or eliminate any quantum-size effects. For example, in early experimental investigations of QW systems, low quality well material and rough well–barrier interfaces prevented the observation of quantum-size effects. These problems were overcome with the advent of MBE and MOCVD epitaxy technologies. After much progress was made in QWs, attention turned toward structures with lower dimensionality, but QDs faced similar fabrication challenges.

For detector applications, QD systems should have the following qualities:

1. QDs of appropriate size
2. Uniform QDs in high density arrays
3. Defect-free QDs

Creating quantum dots with these qualities proved difficult by methods such as direct patterning using electron beam lithography. A key innovation in QD formation was the rediscovery and application of the Stranski–Krastanow (SK) growth mode, which is a 3D crystal growth mode that can occur in the epitaxy of III–V materials. Of the available methods for creating quantum dots, the use of the SK growth mode comes closest to meeting the aforementioned three main requirements for QDs in

FIGURE 6.8 Illustration of the Stranski–Krastanow growth mode.

detectors. This section looks more closely at the SK growth mode, which is also referred to as quantum dot fabrication by self-assembly.

6.3.1 The Formation of Quantum Dots in the SK Growth Mode

The SK growth mode occurs for lattice-mismatched materials where the QD material is grown on a substrate or matrix layer with smaller lattice constant. The first few deposited monolayers of the QD material grow in a flat, layer-by-layer fashion. This flat layer is called the wetting layer. Since the QD material is lattice mismatched, the wetting layer is pseudomorphically strained and the strain builds up with increasing material deposition. Beyond a certain critical thickness the wetting layer spontaneously reorganizes and continued deposition of material results in the growth of 3D dot features on top of the thin wetting layer. This process is illustrated in Figure 6.8. SK growth typically occurs when the lattice mismatch is around 3–10% (though these limits are not well defined). The most studied SK QD growth in III–V materials is In(Ga)As on GaAs. Other systems that have been studied include InAs on InP, Sb-based III–Vs, and III–Nitrides.

This growth process can be understood by considering the interplay between the surface, interface, and strain energies in this situation. Initially, the sum of the epilayer surface energy and the interface energy is lower than the surface energy of the matrix, therefore it is favorable to have layer-by-layer growth of the QD material. After several monolayers of deposition, the buildup of strain energy makes it no longer favorable to have flat, layer-by-layer growth. With this built-up energy the material needs to relieve the strain and does so by forming 3D dot structures.

The key factor here for optoelectronic devices is that the relaxation process in SK growth can be controlled to produce defect-free (also called coherent) QDs. This coherent relaxation was first noted in the mid-1980s [3]. Since the dot material is experiencing a compressive strain, when the coherent relaxation takes place the lattice constant expands toward the dot edges, as illustrated in Figure 6.9. This coherent relaxation means that the QDs grown by SK mode do not have the defects that reduce the 3D confinement in other QD fabrication methods.

6.3.2 Properties of SK Grown Dots and Their Effect on QDIP Performance

The most important QD parameters are size, shape, density, and uniformity. We will discuss the significance of each parameter for detector performance and the typically achieved growth result from SK growth.

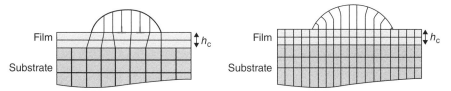

FIGURE 6.9 (Left) Schematic illustrating an SK dot that has relaxed via defect formation and (Right) schematic of a dot that has coherent relaxed without defects. *(For a color version of this figure, see the color plate section.)*

6.3.2.1 Quantum Dot Size For QDs to be useful they must be of an appropriate size. The absolute minimum size for a dot is the smallest size for which there will still be one electron energy level in the dot. For a spherical dot this size is [28]

$$D_{\min} = \frac{\pi \hbar}{\sqrt{(2m_e^* \Delta E_c)}}. \tag{6.10}$$

In practice the dot must actually be larger than this minimum size in order to provide strong carrier localization at nonzero temperature. This is easily achievable in SK growth. Instead the challenge is in not surpassing the maximum useful dot size. The maximum size has been surpassed when the QDs have more than one or two energy levels within the dot and no longer show strong 3D confinement effects. When there are more than two energy levels in a dot, these additional levels can break the phonon bottleneck and dark current reduction. Also if the dot is too large the energy levels will become more widely spread further reducing the phonon bottleneck. QDIP device models showing good performance have been based on dot sizes around 15 nm [26,27]. In the literature QDs grown by SK self-assembly vary from about 10 to 50nm for coherent dots.

Size control of the quantum dots is also important to control the position of the excited state energy level of the QD relative to the continuum. If the excited state is too deep in the dot a high bias may be required to extract a photoexcited electron from a QD. This is the case for the calculated dot configurations shown in Figure 6.10.

6.3.2.2 Quantum Dot Shape Along with the size, the shape of the QD is important in determining the energy level structure. "Ideal" QDs are spherical or at least symmetrically shaped, but often this is not the case for SK grown dots, which tend to have pyramidal or lens-like shapes that are flattened in the growth plane (i.e., wider than they are tall). If large enough, this asymmetry can result in a polarization-dependent absorption because the confinement will become more QWIP like. The shape of the QD is not as easily controlled as QD size.

6.3.2.3 Quantum Dot Density The QD density in SK growth is determined by the QD nucleation process and subsequent ripening. During SK growth, the adatom mobility on the growth surface strongly determines how close together the QDs will nucleate. For example, most growth experiments have shown that decreasing the

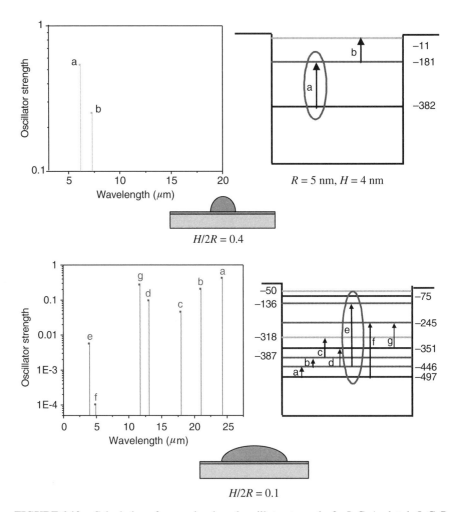

FIGURE 6.10 Calculation of energy levels and oscillator strengths for InGaAs dots in InGaP barriers. The dots have different aspect ratios and fixed height. *(For a color version of this figure, see the color plate section.)*

growth temperature can increase the quantum dot density. Since the adatom mobility is reduced at lower temperature, a given adatom cannot travel as far to find a favorable nucleation center so on average the nucleation centers will be closer together. After the nucleation finishes, additional deposition of material will only grow the existing dots and not nucleate new ones. After nucleation and dot growth and without further deposition of dot material or a cap layer, the dot array will either tend to stabilize or undergo ripening, depending on the growth conditions. Under stable conditions the dots will reach an equilibrium size and spacing. Under conditions that favor ripening the dots will migrate and merge with one another reducing the density and creating dots that are either too large for 3D confinement effects or relax incoherently via defects.

For detector applications a high density of quantum dots above 5×10^{10} dots cm^{-2} is desired. For SK growth of In(Ga)As dots on GaAs substrate, densities of up to 10^{11} dots cm^{-2} have been demonstrated though densities in the low or mid-10^{10} range are more typical.

The QD density affects both the absorption and transport characteristics of a QD detector. Phillips [27] formulated the absorption coefficient α in a QDIP as follows:

$$\alpha(E) = A \frac{n_1}{D} \frac{\sigma_{QD}}{\sigma_{ens}} \exp\left(-\frac{(E-E_G)^2}{\sigma_{ens}^2}\right) \text{[cm}^{-1}\text{]}, \tag{6.11}$$

where E is the energy, A is the maximum theoretical absorption coefficient, n_1 is the areal density of electrons in the QD ground state, D is the QD density, E_G is the ground to excited state transition energy (for a two-level QD system). σ_{QD} and σ_{ens} are the standard deviations in the Gaussian lineshape for ISB absorption in a single quantum dot and for the distribution in energies for the QD ensemble, respectively. The n_1/D term deals with the reduction in absorption due to lack of electrons in the ground state and the σ_{QD}/σ_{ens} term deals with the reduction in absorption due to dot-to-dot nonuniformity, which will be discussed later. We have stated that a high density of QDs is desired for good performance, however, in Eq. (6.11) the density is in the denominator. This is because density cannot be considered independently of the electron occupation in the dots, which comes into the equation via the n_1 term.

The density also relates to the interdot spacing, which effects the amount of lateral interaction between dots in the same layer. Ryzhii's QDIP model [15, 29] predicts that the best QDIP performance occurs for QD arrays that have dots that have a small coupling giving rise to a miniband. This assures that the charge distribution and potential in the plane will be a uniform one. For low dot densities, the QDs behave as localized changes in the potential with the spaces in between acting like punctures where current can flow easily.

6.3.2.4 Quantum Dot Uniformity

In a QDIP device the uniformity of the quantum dots has a very strong correlation with device performance. The nonuniformity of the dots can occur in any of the material parameters such as strain, composition, shape, or size, but size is usually considered the most dominant and is also the most easily quantifiable via methods like AFM or TEM. One source of size nonuniformity is the presence of incoherent, relaxed dots in an array. These are usually of much lower density than the coherent dots but are significant due to their very large size and potential to create defects in subsequent layers. These defect dots can be almost completely avoided under good growth conditions. The main source of nonuniformity is the Gaussian distribution of the size of the coherent dots. There are two possible "dimensions" of nonuniformity to consider, we first discuss the layer-to-layer nonuniformity, then we look at the nonuniformity within a single layer.

In a real QDIP, in order to get a high enough volume density of dots it is necessary to stack QD layers on top of one another in the same way QWIPs consist of stacks of QW. The layer grown on top of the QDs to cover them is called the cap layer. Fortunately

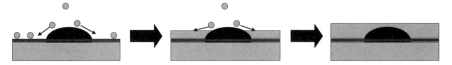

FIGURE 6.11 Schematic illustrating the mechanism for flat cap layer growth.

for device applications, the strain profile of the SK QDs results in preferential growth such that the cap layer returns to a flat surface once the QDs have been completely covered, which facilitates the stacking of more dot layers on top. This flattening process is illustrated in Figure 6.11.

For the stacking of dots the layer-to-layer spacing is important because below a certain thickness the strain field of the buried dot layer will cause the QDs in subsequent layers to vertically align with the dots below. For general detector applications this is not desired because aligned dots tend to increase in size with stack number, have lower spatial coverage due to nonrandom positioning, and may increase the probability of dot-to-dot tunneling increasing the dark current. For thick enough spacings and if there are neither excess strain nor defects, each dot layer grows more or less independent of one another with each layer having similar in-layer dot uniformity.

In the literature, the standard deviation in size for SK grown QDs is around 10%. Optical, electrical, and structural measurements on QD arrays suggest that the variation is a Gaussian distribution. One possible cause for this nonuniformity is asynchronous nucleation of islands due to local variations in the thickness of the QD layer [30]. This amount of nonuniformity can signifcantly affect the detector performance. Recall from Eq. (6.11) that the relevant measure of nonuniformity is the ratio of the linewidth of a single dot to that of the entire array σ_{QD}/σ_{ens}. Increased nonuniformity in the dot array will broaden the spectral response resulting in a lower peak responsivity (although the integrated responsivity will be similar to a more uniform array). Values of 0.01 for σ_{QD}/σ_{ens} are typical for current fabrication technology, and according to Phillips' modeling results [27], cause an order of magnitude decrease in detectivity from the ideal case of $\sigma_{QD}/\sigma_{ens} = 1$.

6.4 DEVICE FABRICATION AND MEASUREMENT PROCEDURES

To test the detector performance, $400 \times 400 \, \mu m^2$ mesas were fabricated using electron cyclotron resonance reactive ion etching to etch through the top contact and active region down to the bottom contact layer. For GaAs-based devices AuGe/Ni/Au bottom and top metal contacts were made via lift-off technique and alloyed at 375°C for 3 min. For InP-based devices Ti/Pt/Au bottom and top metal contacts were made via lift-off technique and alloyed at 400°C for 2 min. The sample was then mounted on a copper heat sink or leadless chip carrier and attached to the cold finger of a cryostat cooled with liquid nitrogen and capable of varying the sample temperature between 77 K and room temperature. The spectral response of the QDIP was tested on a Fourier transform infrared (FTIR) spectrometer. The peak responsivity was determined using a calibrated blackbody source at 800°C modulated at 400 Hz. A 2–12 μm optical band pass filter plus a ZnSe or Ge cryostat window suppressed near-infrared radiation from the blackbody. The noise current was measured with a fast Fourier transform spectrum analyzer and a low noise current amplifier. The dark current was extracted using a semiconductor parameter analyzer. For both the noise current and dark current measurements the device was covered with a cold shield held at the detector operating temperature. The specific detectivity D^* was calculated from $D^* = R_p (A \cdot \Delta f)^{1/2}/i_n$, where R_p is the peak responsivity, A is the detector area, Δf is the bandwidth and i_n is the noise current. The noise gain g was extracted from the noise (i_n) and dark current (I_d) using the expression $g = i_n^2/4eI_d$. The quantum efficiency (η), which can be obtained from the relation $\eta = R_p h\nu/qg$ where $h\nu$ is the incoming photon energy, q is the charge of the carrier, and g is the photoconductive gain. As a good approximation, the noise gain can be used instead for the photoconductive gain [31].

The FPA was fabricated as follows. The pixel definition and metallization of the FPA were essentially the same as the test detector array fabrication, utilizing conventional UV photolithography, electron cyclotron resonance reactive ion etching, and metallization via electron beam metal evaporation. After array fabrication, the readout integrated circuit (ROIC) needs to be hybridized to the FPA. The first step in this process was the creation of indium bumps on both the FPA and the ROIC dies. A thick photoresist layer with undercut profile suitable for lift-off processing of a thick indium layer was applied to the dies using a multilayer resist and chlorobenzene soak method. After the photoresist patterning, indium bumps were deposited on each die via thermal evaporation. Then lift-off was performed by soaking in acetone. After a thorough sample cleaning process, the dies were flip chip bonded and underfilled with epoxy. Then, the FPA substrate was thinned using mechanical lapping and polishing. Finally, the hybridized die was mounted in and wire bonded to a leadless ceramic chip carrier. A more detailed description of the FPA fabrication can be found in References 22 and 17.

The FPAs were tested on a CamIRa infrared FPA evaluation system made by SE-IR Corp. The FPA hybrid was mounted on the cold finger in liquid nitrogen cryostat. The cryostat window was a 3–12 μm broadband Ge filter. A Janos ASIO MWIR lens with $f/2.3$ was used. In the following tests, the background temperature was 300 K. Varying operating temperatures, biases, and frame rates were used. For the GaAs work a Litton ROIC was used, whereas in our InP FPA work a Indigo 9705 ROIC was used. The FPA response was characterized using an extended area blackbody from CI systems. The current injection efficiency was estimated from the single detector measurements as described in Reference 22. The temporal noise of the FPA was measured by taking the standard deviation of the FPA signal. A common metric for FPA performance is the NEDT. In order to measure the NEDT, the FPA signal is measured for two different temperature targets provided by the extended area blackbody (25°C and 35°C). The differential signal for the two temperatures divided by the temporal noise gives the SNR, and the target temperature difference divided by the SNR gives the NEDT.

6.5 GALLIUM ARSENIDE–BASED QUANTUM DOT DETECTORS

6.5.1 InGaAs/InGaP QDIP

The earliest demonstration of a QDIP at CQD was reported in 1998 using InGaAs quantum dots in InGaP barriers lattice matched to GaAs substrate [32]. The InGaAs/InGaP quantum dots were grown on semi-insulating (100) GaAs substrates by LP-MOCVD. A 5000 Å thick Si doped n+ GaAs bottom contact layer ($n = 1 \times 10^{18}$ cm^{-3}) was first grown on the GaAs substrate, then 1000 Å of undoped lattice matched InGaP was deposited. The active region was composed of 10 stacks of InGaAs quantum dots separated by 350 Å of InGaP barrier. The InGaAs quantum dots were formed on the InGaP surface by flowing the sources for several seconds and then interrupting growth for 60 s. The growth rate was 0.8 ML s^{-1} and the V/III ratio was 300. The dots were doped with silicon by supplying SiH$_4$. After the deposition of the 10 stacks of dots, a 1500 Å undoped InGaP layer and a GaAs top contact layer ($n = 8 \times 10^{17}$ cm^{-3}) were grown. The entire structure was grown at 480°C. Based on these growth conditions, a single layer of InGaAs dots on InGaP was grown in order to examine the formation of the dots. A planar image of atomic force microscopy (AFM) of the dots is shown in Figure 6.12. The average size of the dots was measured at 16 nm in radius, and the shape of the dots was spherical rather than pyramidal. The areal density of the dots was estimated at about 3×10^{10} cm^{-2}.

A Fourier transform photoluminescence spectrometer was used to measure the luminescence from the dots at various temperatures. Samples were excited by a 488 nm Ar+ laser with an excitation power of 2 W cm^{-2}, and the signal was collected

FIGURE 6.12 The planar AFM scan images of closely packed InGaAs quantum dots on an InGaP matrix (1 µm × 1 µm) showing spherical dots.

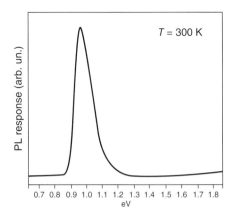
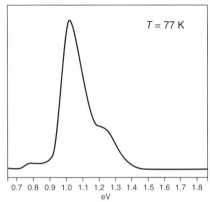

FIGURE 6.13 The photoluminescence spectrum of the 10 stack InGaAs quantum dot photodetector structure at 77 and 300 K.

by a cooled Ge detector. At low temperature, a well-resolved luminescence peak from the wetting layer was observed at 1.23 eV (1.0 μm) in addition to the central peak at 1.024 eV (1.21 μm) which results from the $e1$–$hh1$ electron and hole ground state of the InGaAs QDs. The peak from the wetting later decreased with increasing temperature and it disappeared above 150 K. Strong room temperature luminescence was observed at 0.96 eV (1.29 μm) from $e1$–$hh1$ ground state. The full-width at half-maximum of the peak was 76 meV and was constant throughout the temperature range from 77 to 300 K. The photoluminescence spectra are shown in Figure 6.13.

Figure 6.14 shows the detector responsivity of a sample at various temperatures. The spectra show the ISB transition at 5.5 μm at 77 K with the peak responsivity of 0.067 A W^{-1} at −2 V bias. The cutoff wavelength was 6.5 μm at 77 K. The photoconductive signal was observed up to 130 K. Beyond this temperature, the spectral response degraded due to the rapid increase of noise with temperature. The spectral FWHM of 48 meV ($\delta\lambda/\lambda = 20\%$) was independent of temperature up to 110 K. The large broadening of the spectrum was attributed to the dispersion of the electron confinement energies of dots due to the size fluctuation.

The measured noise current, $i_n = 1.49 \times 10^{10}$ A Hz$^{-1/2}$, resulted in a peak detectivity of 4.74×10^7 cm Hz$^{1/2}$ W^{-1} at 5.5 μm at 77 K with −2 V bias. The measured responsivity increases linearly with bias at low bias voltage and saturates at around −4.5 V as shown in Figure 6.15. This may be due to the saturation velocity of the generated carriers in large electric field.

6.5.2 First QDIP FPA

Several years after the initial demonstration of a QDIP [33] and the demonstration at CQD by Kim [32], the investigation of QDIP devices increased and performances were reaching levels where imaging demonstrations would be possible. The first QDIP imaging came from a raster scan system in 2002 [34]. By leveraging its prior

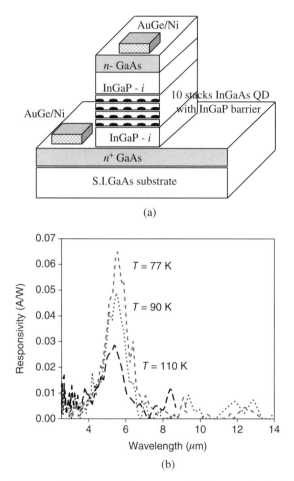

FIGURE 6.14 (a) Schematic diagram of the quantum dot photoconductive detector structure; (b) responsivity of the quantum dot photoconductive detector at various temperatures.

development of QWIP FPAs [35], CQD produced the first QDIP-based FPA in 2004 [22]. The 256 × 256 MWIR FPA was based on an InGaAs/InGaP QDIP structure reported by Jiang [17] and described below.

A LP-MOCVD reactor was used to grow the InGaAs quantum dots on semi-insulating (100) GaAs substrate. The device structure was as follows. The first grown layer was a 0.5 μm bottom GaAs contact layer doped with SiH_4 to $n = 1 \times 10^{18}$ cm^{-3} followed by a 0.1 μm lattice-matched InGaP thick barrier. Next grown was the three active regions consisting of 10 barrier layers of undoped lattice-matched InGaP confining 10 GaInAs quantum dot layers. The nominal thickness of the barriers was 350 Å. The InGaAs quantum dots were formed on top of the InGaP matrix (barrier) by self-assembly based on the Stranski–Krastanow epitaxial growth mode. The growth time for InGaAs quantum dots was 5 s and the ripening time was 30 s with AsH_3 flow.

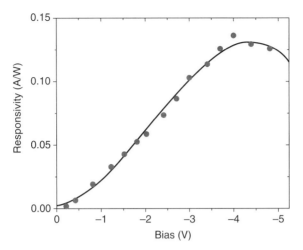

FIGURE 6.15 The bias voltage-dependent responsivity of the InGaAs/InGaP quantum dot photoconductive detector at 77 K.

The growth rate and V/III ratio of InGaAs quantum dots were 0.68 ML s^{-1} and 480, respectively. The dot density, as given by AFM was 2.7×10^{10} cm^{-2}. The InGaAs dots had a disc-like shape 20 nm in diameter and 4 nm in height. The InGaAs quantum dots were doped n-type with dilute SiH$_4$ (200 ppm) with flow rate of 50 sccm. Last grown was a 0.15 μm lattice-matched InGaP thick barrier and a 0.5 μm top contact layer of GaAs doped to $n = 1 \times 10^{18}$ cm^{-3}. The whole structure was grown at 480°C except the active region, which was grown at 440°C. The major improvements over our previous InGaAs/InGaP QDIP structure [32] were that the active region was grown at lower growth temperature and the doping level of quantum dot layer was optimized to give the maximum photoresponse [36]. Also, the InGaP barrier was grown with a slower growth rate (1.3 Å s^{-1}) to improve the morphology and reduce the occurrence of defects.

The dark current (I_d) of the QDIP mesa was measured as a function of bias (V_b) at different temperatures (see Fig. 6.16). Also shown in Figure 6.16 is the 300 K background photocurrent with a 150° field of view (FOV). The background limited performance (BLIP) temperature was measured to be 140 K for the -2.6 V $< V_b <$ 3.7 V range. For 0 V $< V_b <$ 2 V, the dark current did not show any increase from $T = 30$ K to $T = 140$ K. At $V_b = -1$ V, dark current increased less than two orders of magnitude from $T = 30$ to 120 K. Positive threshold voltages seemed to exist. For $T = 30, 50, 77, 95, 120,$ and 140 K, the threshold voltages were 4.8, 4.2, 3.2, 2.5, 2.1, and 1.7 V, respectively. Above threshold voltages, dark current increased exponentially. This behavior is similar to the dark current shown in bound-to-bound QWIPs and may be due to the complex tunneling process caused by the high field domain formation [13]. Overall, very low dark current was observed for this QDIP. At a bias of 0.1 V and $T = 120$ K, dark current of 0.98 pA was observed, which corresponded to a current density of 6×10^{-10} A cm^{-2}. A very asymmetric I–V

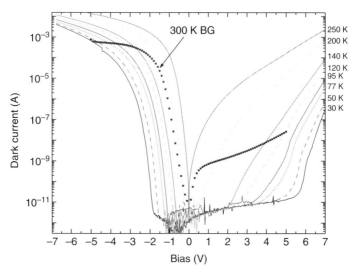

FIGURE 6.16 Dark current measured as a function of bias for InGaAs/InGaP QDIP at different temperatures. *(For a color version of this figure, see the color plate section.)*

relationship was also observed in this QDIP, especially at lower temperature ranges. At $T = 77$ K and bias of 2 V, the I_d was 5.1 pA, while I_d was five orders of magnitude higher at bias of -2 V (0.43 µA). The properties observed in our QDIP were very different from the dark current characteristics of QDIPs reported by other groups [37]. According to Ryzhii's QDIP dark current model, the dark current increases rapidly with temperature due to its exponential dependence on temperature [15, 36]. Except for the existence of a threshold voltage, which was predicted in Ryzhii's model, the dark current behavior of our QDIP was not in agreement with Ryzhii's model. At temperatures below 140 K, the major contributor to the dark current seems to be from tunneling current instead of thermionic emission current. In addition, the strong asymmetry of the dark current for our QDIP also does not agree with Ryzhii's model, and it could be attributed to several factors. First, the asymmetry of the device heterostructure—different bottom and top thick InGaP barrier thickness, contributed part of the asymmetry of dark current. Second, the dopant diffusion in the quantum dot layer during the material growth created a small built-in electric field. Third, the metal contact for this QDIP might not be ideally ohmic. This factor was eliminated after we verified the I–V curve of metal contact, which showed very good ohmic behavior.

The QDIP photoresponse was measured and showed a peak at 4.7 µm with a cutoff at 5.2 µm. The spectral width ($\Delta\lambda/\lambda_{peak}$) was 14%, which originated from the bound-to-bound ISB absorption (see Fig. 6.17). The shape, peak and cutoff of this QDIP showed negligible change with varying temperature (from $T = 30$ K to $T = 160$ K) and bias (from $V_b = 1$ V to $V_b = 3$ V). At a bias of zero volts and $T = 95$ K, a nonzero peak responsivity of 24 mA W^{-1} was observed, which further proved that a built-in electric field exists in our QDIP structure (see Fig. 6.18). Interestingly, our QDIP showed higher peak responsivity at $T = 95$ K than that of at $T = 77$ K in almost

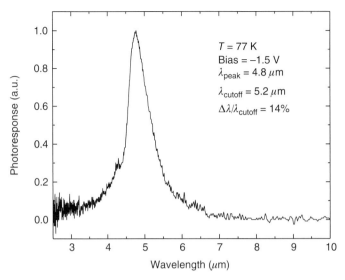

FIGURE 6.17 Relative spectral response of InGaAs/InGaP QDIP structure.

the whole bias range (-5 V $< V_b <$ $+5$ V). At $T = 95$ K and bias of -5 V, very high peak responsivity of 3.1 A W^{-1} was observed for our QDIP. Due to the low dark current, photoresponse with a peak responsivity of 6.7 mA W^{-1} was observed at a temperature of $T = 200$ K with $V_b = -0.7$ V.

After measuring the noise current, the detectivity of our QDIP as a function of bias was measured at both $T = 77$ K and $T = 95$ K and is shown in Figure 6.19. The highest detectivities of our QDIP were 3.67×10^{10} cm Hz$^{1/2}$ W^{-1} at bias -1.6 V and

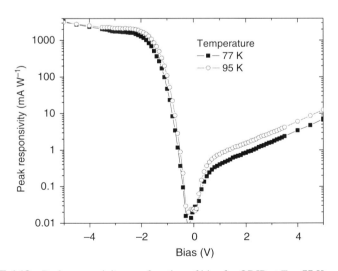

FIGURE 6.18 Peak responsivity as a function of bias for QDIP at $T = 77$ K and $T = 95$ K.

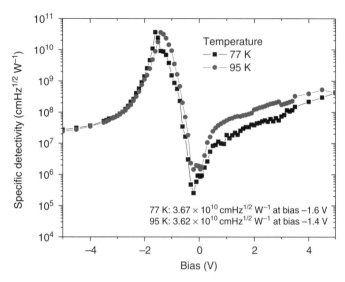

FIGURE 6.19 Peak detectivity as a function of bias for QDIP at $T = 77$ K and $T = 95$ K.

3.62×10^{10} cm Hz$^{1/2}$ W^{-1} at bias -1.4 V for $T = 77$ K and $T = 95$ K, respectively. Figure 6.19 showed that our QDIP performance did not degrade from temperature 77 K to 95 K. At $T = 95$ K and bias -1.6 V, g was found to be 823 and η was found to be 2.0×10^{-4}. Very high gain value can be attributed to the very long lifetime of photocarriers inside quantum dots [38]. The very low quantum efficiency for our QDIP was due to the smaller fill factor of quantum dots across a single epitaxial layer. In addition, the electrons are better confined in dots than in quantum wells.

After the single element QDIP detector measurement described above, a 256×256 detector array was fabricated from the same wafer. Our first InGaAs/InGaP QDIP FPA gave acceptable imaging with 90% of the pixels operational. The measured NEΔT was 509 mK at an operating temperature of 77 K and bias of -1.6 V. The higher NEΔT shown in the actual FPA might be due to nonuniformity of the array pixels [13]. Since only 2 µm tall indium bumps were used for the QDIP array pixels, some pixels did not have a good connection with the readout units during the flip chip bonding (with bonding pressure 12 kg). This resulted in about 10% nonoperational pixels. In addition, since only mechanical polishing was used for the FPA substrate thinning, some pits and defects on the surface might create scattering centers that reduce the photoresponse of QDIP pixels. Similar to QWIPs, our QDIPs are also high impedance devices. They yield a very high current injection efficiency, based on the QDIP test mesa, the current injection efficiency was calculated to be 90% at a bias of -1.6 V, temperature of 77 K, and frame rate of 35 Hz.

Due to the high temperature performance of our QDIPs, thermal imaging can be achieved at temperatures as high as 120 K. Figure 6.20 shows a thermal image of a hot soldering iron head, which was taken at $T = 120$ K without bad pixel replacement or two-point correction.

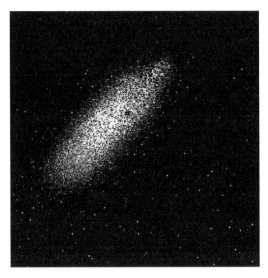

FIGURE 6.20 Thermal image of a hot soldering iron head taken by QDIP FPA camera operating at $T = 120$ K.

It should be noted that our first GaInAs/GaInP QDIP FPA has not been fully optimized. First, the Litton ROIC with direct-injection readout was designed for MWIR InSb photodiode FPAs so that this ROIC is definitely not the best choice for our QDIP FPA. Second, no anti-reflection coating was used on the backside of FPA substrate (GaAs) and this results in about a 30% reflection loss. Third, no passivation layer was used for the QDIP detector pixels.

6.5.3 Two Temperature Barrier Growth for Morphology Improvement

One of the challenges faced while developing the InGaP/InGaAs/GaAs QDIP system was maintaining a smooth surface morphology over several stacks of QD layers. When the QDs were covered by InGaP grown at the same temperature as the QDs, the morphology of the surface after the deposition of the InGaP material was generally poor (see Fig. 6.21b), and thus the roughness of the layers kept increasing after each period of the multistack device structure. Although a detector grown in such conditions had a good performance, as previously demonstrated [39], successive QD layers of the structure are not grown in the same conditions and the final roughness of the device would make it difficult to fabricate a resonant cavity if one desired to increase the quantum efficiency of such a device. To improve the morphology, one can cover the QDs with a 6 nm InGaP layer at 440°C, and then increase the growth temperature up to 480°C to provide better growth conditions for the remainder of the barrier. In this way, the roughness of the surface after the barrier growth can be reduced from 6.0 nm (when the entire barrier was grown at low temperature) to 1.9 nm (see Figs. 6.21b and 6.21c).

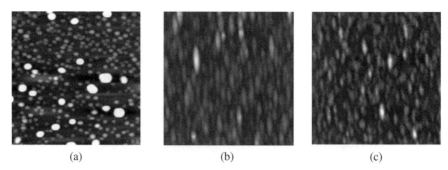

FIGURE 6.21 (a) 1×1 μm^2 AFM of a single layer of uncapped InGaAs QDs on a InGaP matrix, showing a dot density of about 3×10^{10} cm^2; 5×5 μm^2 AFM images of the surface of Figure 6.21a when the InGaAs QDs were capped (b) with a 35 nm InGaP layer grown at 440°C and (c) with a 6-nm-thick InGaP layer grown at 440°C and then 29 nm of InGaP layer grown at 480°C. The root-mean-square (RMS) roughness of the surface of (b) and (c) are 6.0 nm and 1.9 nm respectively.

Photoluminescence (PL) measurements can provide useful information on the interband transitions in the QDs, although it does not reveal any information about the ISB transitions, which are directly related to the QDIP operation. However, the PL data can be used to check the optical quality of the buried QDs. Room temperature measurements were performed on the capped samples described in Figures 6.21b and 6.21c. The peak luminescence wavelength was the same for both samples, which means that the two-step barrier growth technique does not have a dramatic effect on the size or In composition of the QDs, which could result from an enhancement of In–Ga intermixing. However, when the InGaP barrier was grown at two different temperatures, the intensity of the PL spectrum was two times higher than for the single, low temperature barrier sample, confirming that the growth of the upper part of the barrier at a higher temperature improved the material quality (see Fig. 6.22).

To test the effect of this growth technique on the device performance, two QDIP devices were grown and based on the two dot and barrier growth conditions described in Figures 6.21b and 6.21c. A 0.5-μm-thick GaAs:Si ($n = 5 \times 10^{17}$ cm^{-3}) bottom contact layer was initially deposited at 480°C on a semi-insulating GaAs (100) substrate, followed by a 100-nm-thick In$_{0.49}$Ga$_{0.51}$P layer lattice matched to the GaAs bottom contact layer. The active region consisted of 10 In$_{0.68}$Ga$_{0.32}$As QD layers separated by 35-nm-thick In$_{0.49}$Ga$_{0.51}$P barriers. Finally, an InGaP layer followed by a top GaAs:Si contact layer identical to the bottom ones were deposited at 480°C to complete the structure. The InGaP barriers of device A were grown at the same temperature as the QDs themselves (440°C), which is the so-called "single-step barrier growth." In device B, 6 nm of InGaP material were deposited at 440°C, followed by 29 nm of InGaP grown at 480°C, which is "two-step barrier growth." The morphology of the samples was investigated with a Dimension 3100 atomic-force microscope operating in ambient conditions and tapping mode. Figure 6.23 shows the schematics of the device structures and AFM images of the surfaces of the first period of a QD layer capped with the InGaP barrier and top surface of the devices.

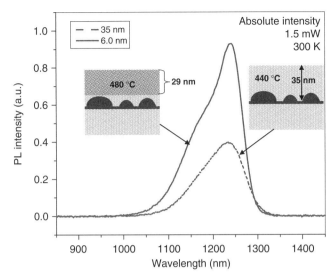

FIGURE 6.22 Photoluminescence spectra of the two samples described in Figures 6.21b and 6.21c, whose upper InGaP barrier was grown (a) at 400°C (dashed line) and (b) at two different temperatures (440°C and 480°C, continuous line).

The spectral response and peak responsivity measured at 120 K are shown in Figure 6.24 for the two devices. Both devices had a response peak around 4.7 μm which was consistent with the fact that the QDs were always directly covered by an InGaP cap layer deposited at low temperature. The spectral width ($\Delta\lambda/\lambda_{peak}$) was 7% and 19% for devices A and B, respectively. This increase was most probably related to the higher temperatures used for the second part of the InGaP barrier of device B. The peak responsivities at 120 K are also shown in Figure 6.24. At low bias, the signal varies by several orders of magnitude, which was consistent with a bound-to-bound transition and with the extraction of the photoexcited carriers out of the QDs by voltage-assisted tunneling. Device B exhibited a much higher responsivity than device A at positive bias. Under such experimental conditions, the photoexcited electrons escape from the QDs into the upper barriers that have a better quality in devices B as a consequence of the higher growth temperature.

The noise of the dark current was measured at 120 K and is shown in Figure 6.25. A floor around 4×10^{-14} A is observed in all the curves and was due to the instrumental limitations of the experimental setup. At negative bias the dark current and noise of devices B was significantly lower than those of device A, at both 77 K (not shown here) and 120 K. The noise of device A takes off from the floor around −0.5 V, whereas the one of devices B only starts to increase below −1.5 V.

The noise and detectivity are shown in Figure 6.25 for devices A and B at 77 K and 120 K. It is worth noting that, in this kind of device, the maximum peak detectivity value always corresponds to the onset of the noise floor because it will be at that point that the ratio of the peak responsivity over the noise was maximized. The highest peak detectivity of sample B at 77 K was 3.35×10^{12} cm Hz$^{1/2}$ W^{-1} with responsivity of

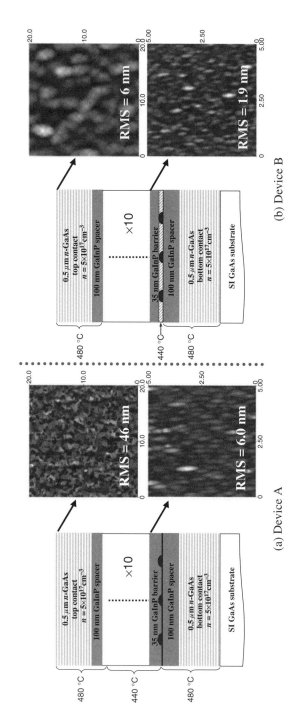

FIGURE 6.23 Schematic diagram of QDIP structures with an active region grown by (a) single-step barrier growth and (b) two-step barrier growth. AFM images of the surfaces of the first period of a QD layer capped with the InGaP barrier and top surface of the devices are shown.

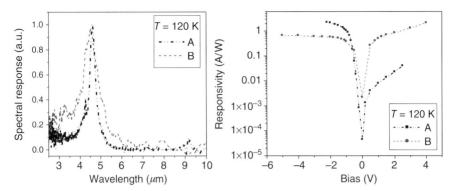

FIGURE 6.24 (Left) Normalized spectral response of QDIP devices A and B at 120 K. (Right) peak responsivity of the QDIPs as a function of bias voltage at 120 K.

3.4 A W^{-1} at −1.9 V, which was three times higher than 1.13×10^{12} cm Hz$^{1/2}$ W^{-1} with responsivity of 1.15 A W^{-1} at −0.9 V of sample A. The improvements can be seen more clearly in higher temperature tests because we can leave the instrument-limited regime. The highest peak detectivity at 120 K is still 4.5×10^{11} cm Hz$^{1/2}$ W^{-1} with responsivity of 0.53 A W^{-1} at −1.5 V of sample B, which was more than one order higher than 5.75×10^{10} cm Hz$^{1/2}$ W^{-1}, 186 mA W^{-1} at −0.6 V of sample A. The high detectivity was attributed to the lower noise in high bias regions brought by the double barrier structure.

The background-limited performance (BLIP) temperature of the QDIP was found to be 220 K under a FOV of 45° by comparing temperature dependent dark current measurements [17] with the photocurrent generated by a 300 K background as shown in Figure 6.26.

The quantum efficiency η was determined to be 1.0% at 77 K. The quantum efficiency of QDIPs is known to be low, but it is generally compensated by a very high photoconductive gain. Such low quantum efficiency is generally attributed to the small

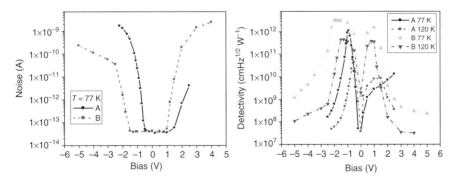

FIGURE 6.25 (Left) Noise of the dark current as a function of bias voltage at 120 K. At low bias, the signal is limited by the experimental setup around 4×10^{-14} A. (Right) Specific peak detectivity of the QDIPs as a function of bias voltage at 77 and 120 K.

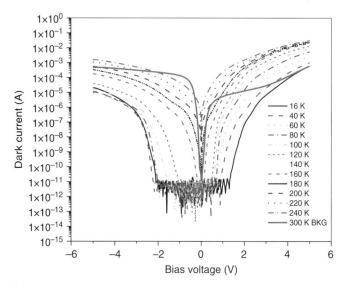

FIGURE 6.26 Dark current at various temperatures and 300 K background photocurrent. *(For a color version of this figure, see the color plate section.)*

fill factor and to the low oscillator strength of the principal ISB transition involved in the optical process. However, it could potentially be considerably improved by incorporating the QDIP structure into a resonant cavity.

QDIPs are supposed to exhibit very good performance at high temperature, as predicted theoretically. Although the detectivity of the present QDIP is still 4.5×10^{11} cm Hz$^{1/2}$ W^{-1} at 120 K, as the temperature goes up further, the detectivity decreases rapidly due to the increase of the dark current (and consequently of the noise) and to the decrease of the photocurrent (responsivity). The reason why the photocurrent decreases with increasing temperature was due to the relatively large volume and flat lens-like shape of the quantum dots obtained by self-assembly which have many electronic energy levels in the conduction band and therefore whose spacing can be very small. If the energy spacing was not different from the LO phonon energy in the quantum dots, the relaxation rate to lower energy states can be dramatically increased by multiphonon emission as the temperature goes up. This relaxation rate competes against the escape rate of the photoexcited electrons that generate the photocurrent. Above a certain temperature, the fast relaxation process starts to dominate the escape process and leads to the dramatic decrease of the photocurrent.

In conclusion, we found that growing the barrier in two different steps improved both the morphology and performance of our QDIP devices.

6.6 INDIUM PHOSPHIDE-BASED QUANTUM DOT DETECTORS

MWIR and LWIR QDIPs based on the InAs/GaAs system have been reported by several groups. These QDIP structures were usually grown by self-assembled method via molecular beam epitaxy (MBE) technique [40–43]. Compared to the GaAs system, only a limited amount of work has been done on QDIP grown on InP substrates, and no device detectivity results have been reported [44, 45]. The study of the InAs/InP quantum dots will expand our understanding of the basic mechanisms of dot formation, enrich the existing quantum dot systems such as InAs/GaAs and Si/Ge [46], and enable devices with new applications or better performances. The lattice mismatch between the quantum dot material and the substrate, which is a critical parameter to determine the formation of the quantum dots, is only roughly half (3.2%) that of the InAs/GaAs system (7%). Different matrix materials such as InP, InGaAs, and AlGaAs can be used to modify the dot formation and engineer the bandgap [47]. In the case of QWIPs, high performance devices have been demonstrated with the InGaAs/InP system grown on InP substrates using MOCVD [48, 49]. The high mobilities and low effective masses of the InGaAs/InP system give rise to high responsivity and long wavelength devices. In addition, when compared to MBE epitaxy, MOCVD epitaxy has advantages such as relative simplicity, easy adaptability to industrial fabrication, and lower cost [50].

6.6.1 InAs/InP QDIP

An Emcore LP-MOCVD reactor was used to grow the InAs QDs on semi-insulating (100) InP substrate. First, a 0.5 μm undoped InP buffer was grown at 590°C followed by a 0.5 μm bottom InP contact layer doped with dilute SiH_4 to $n = 1 \times 10^{18}$ cm^{-3}. Next the active region was grown at 500°C. The active region consisted of 10 periods of the following structure: 400 Å InP barrier, 10 Å GaAs, InAs QDs, and 30 Å $Al_{0.48}In_{0.52}As$ current blocking layer (CBL). The InAs QDs were formed on top of the strained GaAs/InP matrix (barrier) by a self-assembly method which is based on the Stranski–Krastanow epitaxial growth mode. For InAs QDs, the nominal growth rate was 0.42 ML s^{-1} and the growth time was 12 s. After the QD layer was deposited, 60 s of ripening time was given with dilute AsH_3 flowing. The InAs QDs were doped with dilute SiH_4 (200 ppm) with a flow rate of 35 sccm. Finally, a 0.2 μm n-type InP ($n = 1 \times 10^{18}$ cm^{-3}) top contact layer was grown at 590°C.

Single-layer and multistack QDs had been characterized and optimized before the QDIP device growth. It was found that the InAs QD density and uniformity were improved by inserting a thin (10 Å) strained GaAs layer between the InP barrier and InAs QDs. A 10 Å lattice-matched $In_{0.53}Ga_{0.47}As$ layer has also been used and compared to the results with a GaAs layer. Inserting the GaAs layer produced both a higher dot density and better uniformity. It has been shown that a thin GaAs layer can improve the uniformity and the photoluminescence intensity from InAs QDs grown on an InGaAs/InP matrix [51]. The thin GaAs layer prevented indium migration from the InGaAs layer to the InAs QDs. AFM was used for the structural characterizations (see Fig. 6.27). At the optimum growth conditions, the dot density was about 4×10^{10} cm^{-2}. The InAs QDs had a lens form with a typical base diameter of 50 nm and a height of 5 nm.

The dark current (I_d) of a QDIP mesa was measured as a function of bias (V_b) at different temperatures, as shown in Figure 6.28. Also shown in Figure 6.28 is the

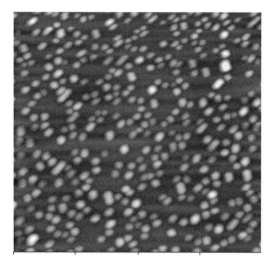

FIGURE 6.27 Atomic force microscopy image (1 μm × 1 μm) of a single layer of InAs quantum dots on GaAs/InP matrix. The scan shows a dot density of about 4×10^{10} cm^{-2}.

300 K background photocurrent with a 45° FOV. The BLIP was obtained at 100 K for the -1.9 V $< V_b <$ 3.7 V range. A very low dark current was observed for this QDIP due to the Al$_{0.48}$In$_{0.52}$As CBL. At $T = 77$ K, a dark current below the pA range was observed between -0.8 V and 1.8 V. An asymmetric I–V relationship was also observed in this QDIP, especially at lower temperatures. The asymmetry of the dark

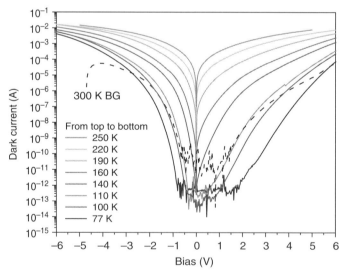

FIGURE 6.28 Dark current measured as a function of bias for InAs/GaAs/AlInAs/InP QDIP at different temperatures. Also shown is the 300 K background photocurrent with a 150° field of view (dashed line). *(For a color version of this figure, see the color plate section.)*

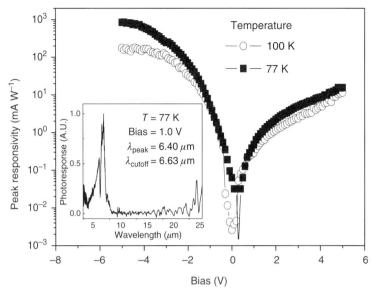

FIGURE 6.29 Peak responsivity as a function of bias for the QDIP at 77 K and 100 K. The inset shows the relative spectral response measured at bias of 1 V and temperature of 77 K.

current for our QDIP can be attributed to several factors. First, it could be due to the asymmetry of the device structure as every GaAs/InAs/AlInAs/InP period of the active region is asymmetric. Secondly, the InAs QD has a lens shape which was not symmetric in the growth direction. Thirdly, dopant diffusion into the quantum dot layer during the material growth might create a small built-in electric field.

The spectral response of the QDIP peaked at 6.4 µm with a cutoff at 6.6 µm (see the inset in Figure 6.29). The spectral width ($\Delta\lambda/\lambda_{peak}$) was 12%, which indicated bound-to-bound ISB absorption. The shape, peak, and cutoff of this QDIP showed negligible change with varying temperature (from $T = 77$ K to $T = 160$ K) and bias (from $V_b = 1$ V to $V_b = 3$ V). The absolute magnitude of the blackbody responsivity (R_{bb}) was determined by measuring the photocurrent (I_p) with a calibrated blackbody source at 800°C. The test mesa was illuminated from the top of the mesa with normally incident infrared radiation. The peak responsivity (R_p) results at both 77 K and 100 K are shown in Figure 6.29. An asymmetry was also observed for the peak responsivity, both at 77 K and 100 K. The asymmetry of the responsivity was caused by the asymmetry of the potential in the QD itself. At $T = 77$ K and bias of −5 V, a peak responsivity of 1.0 A W^{-1} was observed for our QDIP.

The noise current (i_n) was measured at both $T = 77$ K and $T = 100$ K. As shown in Figure 6.30a, very low noise was observed and was almost constant for biases −1.0 V < V_b < 2.5 V at 77 K and −0.8 V < V_b < 2.3 V at 100 K. Beyond these ranges, the noise was dominated by generation–recombination noise of the dark current [31]. Based on the measured noise current and dark current (i_d), we extracted the gain g of the device. The gain of the device depended strongly on the bias and

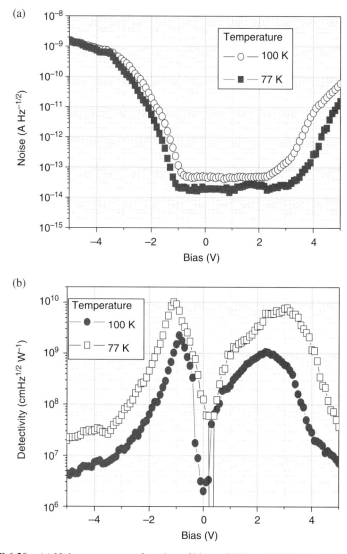

FIGURE 6.30 (a) Noise current as a function of bias at 77 K and 100 K; (b) peak detectivity as a function of bias for at 77 K and 100 K.

increases from 5 to 5000 when the bias changes from −1.2 to −3.5 V. The high value of the gain in a given voltage range was a unique feature of this QDIP technology. The gain in our QDIP devices was much higher than the gain measured in typical quantum well infrared detectors. The detectivities of our QDIP as a function of bias at both $T = 77$ K and $T = 100$ K are shown in Figure 6.30b. The highest detectivities of our QDIP were 1.0×10^{10} cm Hz$^{1/2}$ W^{-1} at a bias of −1.1 V and 2.3×10^9 cm Hz$^{1/2}$ W^{-1}

at a bias of −0.9 V for $T = 77$ K and $T = 100$ K, respectively. The internal quantum efficiency (η) of our QDIP was less than 0.1%. The very low quantum efficiency was (i) due to the low oscillator strength for s-polarized light and (ii) less-than-unity fill factor of the quantum dot layer. A low oscillator strength must be expected for s-polarized light in a flat (50 nm × 5 nm) lens-like QD at this wavelength. The quantum efficiency can be improved by increasing dot density, optimizing the dot shape, size and uniformity [27], and also by using a resonant cavity [52].

6.6.2 Detection Wavelength Tuning Using Quantum Dot Engineering

The device just discussed showed the potential of InP-based QDIP devices, however, the detection wavelength of that device was centered at 6.4 μm, outside the MWIR (3–5 μm) or LWIR (8–12) μm window where there is low absorption in the atmosphere. To shift the detection wavelength into the MWIR window, quantum dot engineering was used to tune the peak detection wavelength.

In order to tune the QDIP detection wavelength to a shorter wavelength, smaller dot sizes were needed. The dot size directly affects the detection wavelength by determining the energy levels in the conduction band (for n-type) of QD, where the photoexcited ISB transitions between those levels correspond to the detection of infrared light. The detector's performance is also dependent on the dot density and uniformity. In other words, in order to have a QDIP working at the preferred wavelength with a desirable performance, QDs with high density, high uniformity, and a specific size need to be formed. The formation of self-assembled QDs is determined by the growth parameters such as growth temperature, growth rate, V/III ratio, and ripening time. The matrix material on which the QDs are formed also affects the dot formation. We have found that a thin (10 Å) strained GaAs layer inserted between the InP barrier and InAs QDs improves the uniformity and photoluminescence (PL) intensity of the InAs QDs compared with those grown on InGaAs/InP or directly on the InP matrix [19]. Such a GaAs layer was used for all the growths that discussed below.

First, the influence of growth temperature was studied. Growths of single layer InAs QDs were done on the GaAs/InP matrix with a growth temperature of 520°C., 500°C., 480°C., 460°C., and 440°C. All the other growth conditions were kept the same. AFM was used for the structural characterizations, including the dot size, shape, density, and uniformity. At 520°C, the QDs have a low dot density with a large dot size. With the decrease of the growth temperature, the dot density increases while the dot sizes decreases. Beyond a certain point, the dot density stopped increasing. However, the dot size still changed. For example, the QDs at 500°C and 440°C have similar average dot diameter, which was 50 nm, however, the height of the dots grown at 440°C was smaller, and this can affect device performance, as shown later.

The growth rate was also optimized by varying the flow rate of TMIn during the QD growth while keeping the other conditions the same. 50, 100, and 150 sccm flow rates of TMIn were used. The dot density was checked with AFM. Although 150 sccm

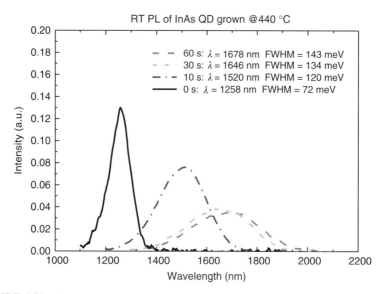

FIGURE 6.31 Room temperature PL from a single layer of InAs QDs grown and capped with InP at 440°C with different ripening times. *(For a color version of this figure, see the color plate section.)*

gave slightly higher dot density, PL showed that 100 sccm has better quality. Thus 100 sccm was used in the following optimization steps.

Next, the V/III ratio was optimized by varying the flow rate of dilute arsine. Flow rates of 25, 50, 75, 100, and 200 sccm were used. 100 sccm was used for TMIn, as determined before. 50 sccm of dilute arsine gave the highest dot density and was chosen for the next growth.

The ripening time is the wait time after the deposition of the dot material but before capping is carried out. It was used to allow the dot to fully form, or to "ripen". 60 s, 30 s, 10 s, and 0 s ripening times were used to study the effect on the dot formation. For this optimization, the above single layer QD growth cannot be used because during the cooling down after growth, the QD on the surface may still change. As a result, AFM is not a usable characterization technique anymore. Instead, the QDs were capped and then studied using PL. From the room temperature PL of these samples, as shown in Figure 6.31, it was seen that the ripening time had a dramatic effect on the QDs. With decreased ripening time, there was a blue-shift of the peak wavelength (indicating smaller dots), reduced FWHM (indicating improved uniformity) and higher intensity (indicating higher dot density.)

While PL provides useful information on the effect of ripening time on the interband transitions in the QDs, it does not reveal information about the ISB transitions, which are directly related to QDIP operation. Instead, the measurement of absorption was done by an FTIR spectrometer. QDIPs with peak detection wavelengths from 4 to 6.4 μm were obtained by choosing the growth temperature and ripening time.

At 500°C, the minimum wavelength that could be reached was around 6 μm, even with 0 s ripening time. At 440°C, 4 μm and 5 μm QDIPs were obtained with 10 and 60 s ripening times. Therefore to get MWIR QDIPs, at least the QD and its adjacent layers had to be grown at 440°C.

Finally, with all the growth parameters of QDs optimized, a single layer InAs QDs on top of the GaAs/InP was obtained with a dot density of about $3\text{–}5 \times 10^{10}$ cm^2. The InAs QDs had a lens shape with a typical base diameter of 50 nm and a height of 5 nm for the QDs grown at 500°C, and a height of 4 nm for the QDs grown at 440°C.

A QDIP device structure will consist of multiple layers of QDs. In order to guarantee the uniformity of the QDs in the different layers as well as improve the quality of the barrier material, a two-step growth of the InP barrier after each QD layer was used. First the InAs QDs were covered with 10 nm of InP grown at the same temperature as the QDs, then the temperature was ramped up to a higher temperature to grow the rest of the barrier. AFM shows that the InP barrier surface was much improved, with clear atomic steps, while PL showed no change of peak wavelength or FWHM, which indicates that this process did not affect the QDs.

We compared three devices (here denoted A, B, and C) with different QD growth conditions and different structures. The typical device structures were as follows. First, a 0.5-μm-thick undoped InP buffer layer was grown at 590°C followed by a 0.5 μm bottom InP contact layer doped with dilute SiH$_4$ to $n = 1 \times 10^{18}$ cm^{-3}, then the active region with 10 QD layers separated by InP barriers, and finally a 0.2-μm-thick top Ga$_{0.47}$In$_{0.53}$As contact layer doped with dilute SiH$_4$ to $n = 5 \times 10^{17}$ cm^{-3}. The difference of these three devices was only in the active region, as shown in Figure 6.32. Device A was based on our previous 6.4 μm QDIP and acts as a reference. Its whole active region was grown at 500°C. The detailed structure of device A was described in Reference 19. The active regions of the other two devices (B and C) were grown at two different temperatures. The InAs QDs of devices B and C were grown at 440°C. The nominal growth rate was 0.84 monolayer per second

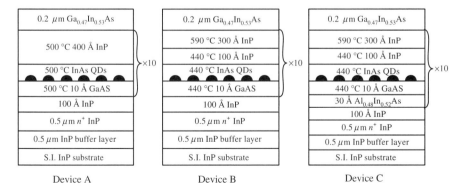

FIGURE 6.32 Schematic diagram of the three QDIP structures with a different active region.

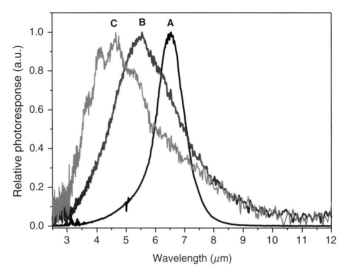

FIGURE 6.33 Normalized spectral response of QDIP devices A, B, and C.

(ML s^{-1}) and the growth time was 6 s. The InAs QDs of device B were grown on InP barriers which were grown at 590°C. On the other hand, the QDs of device C were grown on 3 nm $Al_{0.48}In_{0.52}As$/InP barriers which were grown at 590°C. As mentioned above, before the deposition of the InAs layers in devices B and C, thin 1 nm GaAs layers were inserted on the InP or $Al_{0.48}In_{0.52}As$ surfaces in order to produce more uniform QDs as in device A [51]. The growth-interruption time (ripening time) after the QD growth was 30 s under dilute AsH_3 flow. After the 30-s interruption, the first 10 nm of the InP capping layers were grown at 440°C, and the remaining 30 nm InP barriers were then deposited at 590°C.

The spectral response of the QDIPs was tested by FTIR with the source light was normally incident. The bias was applied to the top contact and the bottom contact was always grounded. As shown in Figure 6.33, the photoresponse of device A peaked at 6.5 μm (190 meV) at 77 K with a bias of 0.4 V and the half width ($\Delta\lambda/\lambda_{peak}$) was only 18%. The narrow photoresponse was a strong indication of a bound-to-bound transition between the QD energy levels. On the other hand, the photoresponses from devices B and C had peaks centered at 5.5 μm (224 meV) and 4.7 μm (266 meV), respectively. The half widths of devices B and C were 46% and 57%, respectively. These broad responses were very different from that of device A. Multiple peaks appeared on the high energy side above 0.23 eV for device C.

In order to explain this change of the optical transition scheme from device A to device C, we used the effective mass embedding method to calculate the QD energy levels [53]. The main photo-transition of device A was identified to be one from a bound state to a higher bound state (bound-to-bound), and therefore it was expected to give a fairly narrow response. When the height of the QDs decreases, as from the QDs of device A to those of device B, the spatial confinement becomes stronger and pushes the ground state upward toward the band edge of the barrier. Our calculation showed that the energy level configuration had changed. As a result, the most probable photo-transition changed from a bound-to-bound transition to a bound-to-continuum transition. In device B, the main transition was from the ground state to the continuum state of the barrier. For device C, the $Al_{0.48}In_{0.52}As$ layer below the QDs increases the electron confining potential by 300 meV in our calculation [54], thus pushing slightly the ground state downward, when compared to device B. In the spectral response of device C, in Figure 6.33, multiple peaks were seen and were believed to be due to transitions originating from the ground state of the QDs to the minibands formed by the thin $Al_{0.48}In_{0.52}As$ layers and InP barriers. Note that all the transitions were broadened by the inhomogeneous size distribution of the self-assembled QDs. In summary, we were able to shift the peak detection wavelengths of QDIPs from 6.5 to 4.7 μm by changing the QD growth conditions and the confinement potential generated by the material surrounding the QD layers.

The blackbody responsivities (R_{bb}) of devices A, B, and C were calculated by measuring the photocurrent (I_p) with a calibrated blackbody source at 800°C. At $T = 77$ K and a bias of 1.5 V, peak responsivities (R_p) of 77 and 65 mA/W were obtained in device B and device C, respectively (see Fig. 6.34). We believe that the reason for the lower responsivity of devices B and C compared to that of device A was that the optical transition was changing from a sharp bound-to-bound transition with a relatively high oscillator strength, to a much more diffuse bound-to-continuum transition.

Once the noise currents (i_n) were measured at $T = 77$ K, the peak detectivities (D^*) were calculated. The peak detectivities of devices B and C, which are MWIR QDIPs, were 1.0×10^9 cm $Hz^{1/2}$ W^{-1} at 77 K and bias of 0.2 V, while it was 1.0×10^{10} cm $Hz^{1/2}$ W^{-1} at the same temperature but with a bias of 0.4 V for device A. The lower detectivity of devices B and C were mainly due to their lower peak responsivity compared to that of device A.

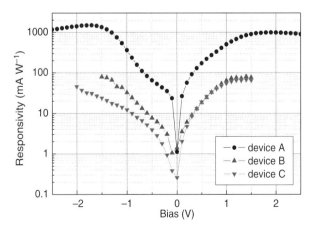

FIGURE 6.34 Peak responsivities as a function of bias at 77 K.

The performances of our QDIPs were still inferior to what we might have expected. One reason was the low quantum efficiency. For example, the 6.4 µm QDIP with a detectivity of 1.0×10^{10} cm Hz$^{1/2}$ W^{-1} at $T = 77$ K only had a quantum efficiency of 0.1%. The very low quantum efficiency may first be attributed to the nature of the ISB transitions when compared with the interband ones. Secondly, compared with QWIPs, QDIPs have a less than unity fill factor. In theory, the internal quantum efficiency can be increased by increasing the number of QD layers. However, in reality, it was limited by the growth of a thick highly strained quantum dot structure. Also, our calculations show that the oscillator strength of the ISB transition was small due to the "flat" geometry of our quantum dots. The quantum efficiency can be improved by increasing the dot density, optimizing the dot shape, size, and uniformity [27], and by growing a thicker active region without degrading the quality. All this requires further optimization of the growth conditions. On the other hand, it might be possible to improve the quantum efficiency by using an external enhancement structure such as a grating or a resonant cavity [52]. The effects of the cavity include wavelength selectivity and a large enhancement of the resonant optical field. The increased optical field allows the detector active region to remain thin and therefore fast, while simultaneously increasing the quantum efficiency at the resonant wavelengths.

The effects of the growth parameters including the growth temperature, growth rate, V/III ratio, and ripening time were thoroughly studied. As a result, InAs/InP QDs with high density and uniformity and preferred size were obtained, which led to an MWIR InAs/InP QDIP. The change of the peak wavelength and shape of spectral response were explained.

6.6.3 High Operating Temperature Quantum Dot Detector and Focal Plane Array

Achieving higher operating temperatures will reduce the cost and complexity of detector and imaging systems by reducing the cooling requirements normally associated with cryogenically cooled detector systems. In addition, in the specific application of high temperature imaging, a low dark current is important to avoid saturation of the readout circuitry. Many of the QDIPs reported so far in the literature have been working at temperatures in the range 77–200 K [17–20]. We developed a high performance, room temperature operating mid-wavelength infrared (MWIR) photodetector based on InAs QDs grown on $Al_{0.48}In_{0.52}As$ barriers and capped by $Ga_{0.47}In_{0.53}As$ quantum wells (QWs) grown on top of an InP substrate. A similar detector structure was also applied to a 320 × 256 FPA that operated at temperatures up to 200 K.

The quantum dot and device growth was carried out by low pressure MOCVD in an Emcore Discovery reactor. The material system investigated here was self-assembled InAs QDs grown directly on an $Al_{0.48}In_{0.52}As$ matrix. The quantum dot self-assembly was based on the Stranski–Krastanow growth mode. The InAs QDs were grown at a nominal growth rate of 0.64 monolayers per second for 3.6 s. The quantum dots were grown at the same temperature, 590°C, as the matrix and cap layer. For the quantum dot growth 5% dilute arsine was used as the group V source. Based on AFM measurements, the quantum dots were around 50 nm in diameter and 4 nm in height.

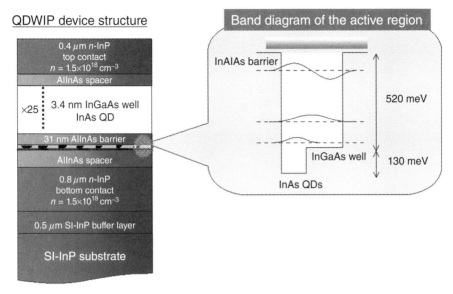

FIGURE 6.35 Schematic of the detector device structure.

The dot density was around 4×10^9 cm^{-3}. For the actual device structure, the dots were capped with a 3.5-nm-thick Ga$_{0.47}$In$_{0.53}$As layer without any interruption time between the end of the dot growth and the start of the cap layer growth.

After finding suitable quantum dot growth conditions, we grew a photoconductive device structure, shown in Figure 6.35, based on the QD/QW structure. A 1-μm-thick n-doped InP layer was grown for the bottom contact. Then the active region was grown, consisting of 25 stacks of InAs QD/GaInAs QW layers with 29 nm AlInAs barrier layers. The 3.5-nm GaInAs QW layer on capping each QD layer was doped to a level of $n = 1 \times 10^{18}$ cm^{-3}. After the final AlInAs barrier layer was grown, a 0.5-μm-thick n-doped InP layer was grown for the top contact.

We measured the spectral response of the detector at several temperatures and applied biases, in the normal incidence configuration, without any optical coupling structures. In this device structure, both the InAs QD layers and GaInAs QW layers were involved in the infrared absorption process. The coupling of QDs and QWs has been used in other QDIP device structures, such as dot-in-a-well (DWELL) [18, 55], where the ISB transition occurred between the hybrid states of the quantum dot and the quantum well. The device here differed from the typical DWELL design in that the quantum well was only on the top of the quantum dots, as opposed to the dot being surrounded on both sides by the quantum well. Figure 6.36 shows the photocurrent spectra of the device at various operating temperatures. The spectrum was clearly measurable at 280 K. For applied bias of smaller than 2 V, there were two peaks in the detection spectrum, around 3.2 and 4.0 μm. Both peaks were visible in the two high temperature spectra in Figure 6.36. The two peaks were present at lower temperatures as well, but in the spectra in Figure 6.36 the more intense longer

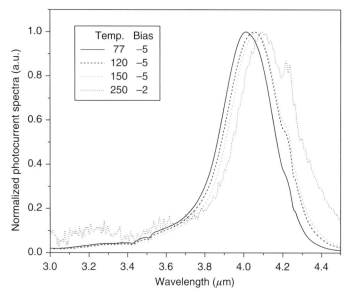

FIGURE 6.36 Normalized temperature dependent photocurrent spectrum. At higher temperatures the increased dark current and noise required operation at lower bias.

wavelength peak drowned out the shorter wavelength peak. Also, measurements showed that the intensity of the peak near 3.2 μm did not increase significantly as the temperature increases. The peak around 3.2 μm came from a bound-to-continuum transition where the electrons are photoexcited from the ground state to a continuum state. That was the reason why the increase of the temperature does not improve the photoresponse around 3.2 μm. On the other hand, the photoresponse around 4.0 μm increased significantly with the temperature because it came from a bound-to-bound transition in the InAs QD/GaInAs QW hybrid states and thus the temperature can help the photoexcited electrons escape to the continuum. Both the temperature dependence and the bias sensitivity of the peak near 4.0 μm indicated that the transition of the photoexcited electrons took place between bound states of the QD/QW hybrid. Looking more closely at the 4.0 μm peak, it showed a red-shift and broadening with increased operating temperature. This behavior was similar to that observed in QWIP structures as documented in Reference 56. A model of the QD/QW hybrid system is required to gain an understanding of where this effect comes from and whether or not the causes are similar to those in QWIP structures.

The peak responsivity (R_p) was measured as a function of bias and temperature and is shown in Figure 6.37. The responsivity increased with temperature from 77 to 200 K and started decreasing above 200 K. The highest measured peak responsivity was 778 mA W^{-1} at 150 K and 5 V. The peak responsivity at 250 K was 54 mA W^{-1} at 2 V. In QDIPs or QWIPs, the photocurrent can increase or decrease with the temperature depending on the balance of the competition between phonon-induced relaxation of the photoexcited carrier and temperature-aided escape of the photoexcited carrier

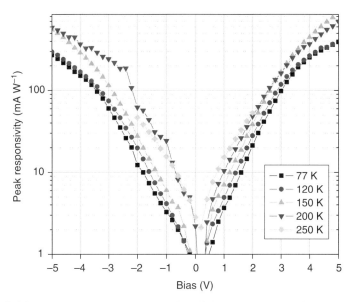

FIGURE 6.37 Peak responsivity as a function of bias at several operating temperatures.

to the continuum state. Above a certain temperature, the adverse thermal increase of the relaxation of the photoexcited electrons back to the lower state dominates any improvement in escape created by the temperature [53]. In the device, the data showed that turnover took place around 200 K, above which the responsivity started decreasing with increasing temperature. The responsivity increased with operating temperature for temperatures up to 200 K. Above 200 K the responsivity decreased with temperature.

The dark current density of this device was measured as functions of bias and temperature. This data is shown in Figure 6.38. A low dark current density was obtained in this device. At 200 K and 5 V, the dark current density was measured to be 128 mA cm^{-2}. High dark current can limit the capability for high temperature operation in photoconductors. Therefore, it is crucial to achieve a low dark current with a reasonable photocurrent at high temperature. In QDIPs, low dark currents have been engineered by introducing CBLs [57]. But CBLs will also decrease the photocurrent because the dark current and photocurrent follow the same transport path. In our device, the QD layers decrease the dark current without significantly compromising the photocurrent.

The highest detectivity at each measured temperature is shown in Figure 6.39. The detectivity decreased with increasing temperature due to the constantly increasing noise. The 77 and 120 K measurements showed similar D^* because the noise level was below the system limit. A maximum D^* of 2.3×10^{11} cm Hz$^{1/2}$ W^{-1} was measured at 120 K. The detectivity at 250 K was 1.7×10^8 cm Hz$^{1/2}$ W^{-1}.

A very high quantum efficiency of 48% was obtained in this device for normal incidence at 150 K and 5 V. This was a significant improvement over our previous

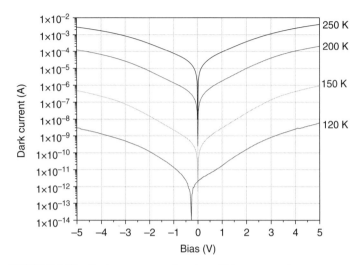

FIGURE 6.38 Dark current as a function of bias at several temperatures.

result of 35% for our earlier design of a similar device [58]. The primary change for this device was improved optimization of the quantum dot growth conditions.

6.6.4 High Operating Temperature FPA

An FPA was fabricated based on an earlier version of the QDWIP device structure discussed above. The format of the FPA was 320 × 256 with a 30 μm pitch and 25 × 25 μm² detector size. The ROIC used was an Indigo 9705.

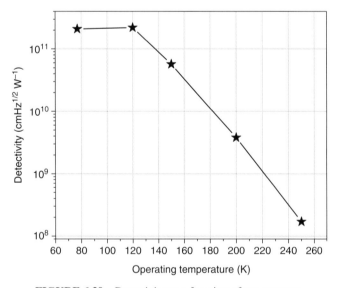

FIGURE 6.39 Detectivity as a function of temperature.

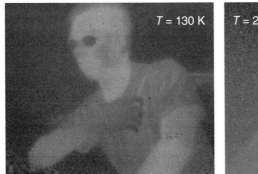 

FIGURE 6.40 Focal plane array imaging at an operating temperature of 130 and 200 K.

The FPA was tested using a CamIRa FPA testing system from SE-IR Corp. The imaging system cryostat was equipped with a Ge window with 94% transmission and a MWIR, f/2 ASIO series lens from Janos with 90% transmission. All imaging and measurements were taken with a 300 K background.

Images obtained from the FPA are shown in Figure 6.40. During testing, imaging was achieved at operating temperatures from 77 to 200 K. The imaging tests were carried out at a fixed frame rate of 32.64 Hz and depending on the operating temperature with the bias varying from -1 to around -3 V and integration times from 0.34 up to 30.41 ms. Two-point nonuniformity correction was used for the imaging. Low contrast imaging of human targets was possible up to around 150 K, and a soldering iron could be imaged up to 200 K. Above 200 K the dark current of the detector became too high for any imaging. While the single detector measurements showed that the responsivity was still appreciable at temperatures up to room temperature, the dark current set the upper limit on the operating temperature because at temperatures above 200 K the dark current saturated the ROIC for any integration times long enough to collect any appreciable photocurrent signal. The capacity of the ROIC used was 18 million electrons at its lowest gain setting. In order to operate the FPA at even higher temperatures the detector dark current should be further reduced or an ROIC with a larger electron capacity could be used.

The FPA photocurrent was measured using an extended area blackbody source from CI Systems. The photocurrent of the FPA was measured with the extended area blackbody set at 35°C. The responsivity and conversion efficiency could then be calculated by dividing the measured photocurrent by the appropriate optical input from the system radiometry. The mean responsivity and mean conversion efficiency of the FPA were 34 mA W^{-1} and 1.1%, respectively. The temporal noise of the FPA was measured by taking the standard deviation of the FPA signal. At 120 K the average NEDT was 344 mK and the percentage of connected pixels was greater than 99%. The histogram of the NEDT is shown in Figure 6.41. The number of pixels with NEDT within two standard deviations of the mean NEDT was 91%. It should be noted that the NEDT measurements were taken without two-point nonuniformity correction.

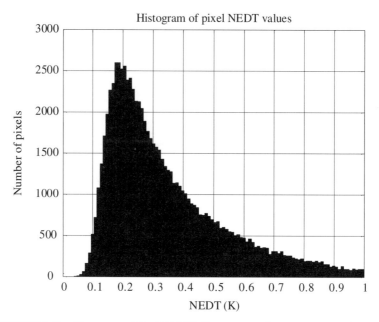

FIGURE 6.41 Histogram of the FPA NEDT at an operating temperature of 120 K.

The injection efficiency was 99% at 120 K. At 200 K the injection efficiency was still 98% due to the high differential resistance even at high temperature.

The imaging quality in terms of contrast and uniformity for this FPA is not yet comparable to that of high quality QWIPs. From the single device measurement, the optimum bias condition for this detector was at (or beyond) the limit of the test setup's biasing capability, −5 V. However, since the ROIC cannot apply more than −3 V nominally, the detector had to be operated under conditions where the photocurrent was at such a low level that it was barely detectable by the camera system. The same low photocurrent limitation also meant that operating the FPA at 77 K actually gave lower performance compared to 120 K because the detector's responsivity decreases with temperature below 180 K. This low responsivity was why the NEDT was 344 mK even though the corresponding single detector D^* was greater than 10^{10} cm $Hz^{1/2}$ W^{-1} under similar operating conditions. The detector properties were no longer dominant and any noise and nonuniformity in the ROIC and the camera system itself will be more significant at these low signal levels. Of course, any nonuniformities in the detector material and processing will also factor in. As a result, one important step for the application of this detector structure to FPAs will be optimizing the device to run at a lower bias without sacrificing its high SNR and quantum efficiency. The new single device results reported here still require a high operating bias, and thus the device is not suitable for our particular imaging application.

We have demonstrated a high performance InAs quantum dot mid-infrared photodetector grown on InP substrate, which operates up to room temperature. The peak detection wavelength was observed at 4.0 μm. The peak responsivity and the specific detectivity at 150 K were 778 mA W^{-1} and 6×10^{10} cm Hz$^{1/2}$ W^{-1}, respectively. Low dark current density and a high quantum efficiency of 48% were obtained in this device. High temperature FPA was also demonstrated. Imaging was achieved at temperatures up to 200 K, with human body imaging up to 150 K. The very low dark current of the device enabled the high temperature imaging capability.

6.7 COLLOIDAL QUANTUM DOTS

For completeness let us briefly mention another way of arriving at quantum confinement on a nanoscale. This is the so-called colloidal semiconductor nanocrystal CNC and quantum dot CQD fabrication method and technology. The subject area is vast and of great importance. It will be impossible to do justice to this subject and the many authors and prestigious papers published in a brief overview of this type. The reader is advised to do his own research as well if he is convinced that this is an area of interest to him(her).

Colloidal semiconductor nanocrystals are synthesized from precursor compounds dissolved in solutions, this is very much like traditional chemical processes [59]. The synthesis of colloidal quantum dots is based on a three-component system composed of precursors, organic surfactants, and solvents. When heating a reaction medium to a sufficiently high temperature, the precursors chemically transform into monomers. Once the monomers reach a high enough supersaturation level, the nanocrystal growth starts with a nucleation process. The temperature during the growth process is one of the critical factors in determining optimal conditions for the nanocrystal growth. It must be high enough to allow for rearrangement and annealing of atoms during the synthesis process while being low enough to promote crystal growth. Another critical factor that has to be stringently controlled during nanocrystal growth is the monomer concentration. The growth process of nanocrystals can occur in two different regimes, "focusing" and "defocusing." At high monomer concentrations, the critical size (the size where nanocrystals neither grow nor shrink) is relatively small, resulting in growth of nearly all particles. In this regime, smaller particles grow faster than large ones (since larger crystals need more atoms to grow than small crystals) resulting in "focusing" of the size distribution to yield nearly monodisperse particles. The size focusing in the growth process is optimal when the monomer concentration is kept such that the average nanocrystal size present is always slightly larger than the critical size. There are colloidal methods to produce many different types of semiconductors. Typical quantum dots are made of binary alloys such as cadmium selenide, cadmium sulfide, indium arsenide, and indium phosphide. Dots may also be made from ternary alloys such as cadmium selenide sulfide [CQD *on line dictionary*]. These quantum dots can contain as few as 100–100,000 atoms within the quantum dot volume, with a diameter of 10–50 atoms. This corresponds to about 2–10 nm, and at 10 nm in diameter, nearly 3 million quantum dots could be lined up end-to-end and fit within the width of a human thumb. Colloidal QDs are the subject of intensive research. They have a wide range of interesting and useful properties which are discussed in numerous reviews and publications [59–64]. To distinguish them (CQD) from the self-assembled QD just described above, here are the noteworthy features. With CQD one can have much better size and shape control [59, 60]. The CQD can be doped with spin and charge. Magnetic impurities such as Mn can be built in. The CQD can by designed to exhibit well-defined excitonic features in the visible and can reach an amazing 99% PL efficiency [63, 64]. Another exciting and very useful property is associated with the sensitivity of the PL to charging of the QD [63, 64]. An extra charge can completely quench the PL and this property can have a range of possible technical applications [63, 64]. The CQD can be made to self-assemble and close pack in very regular lattice or superstructures [60–62]. The excitons can be efficiently transferred to other QDs or conducting channels by the Foerster mechanism or vice versa from an LED to the CQD emitter layer [65]. The PL blinking caused by long time illumination has been identified as being due to trapped charges appear to have been overcome by using multishell QD or implant designs. Given the structural density and energy tuneability advantages, high absorption cross-sections and efficient PL, one can now ask the question: can CQD also provide us with high performance photon detectors? The answer seems to be yes [61] in the short, visible, and near IR but not yet in the longer wavelength regime. The reason is that colloidal QD suffer from one serious drawback: in order to process them from solution and use self-assembly techniques, the surface needs to be passivated by an organic layer. The effect is to introduce an insulating barrier, and a concomitant lowering of the QD to QD banding or tunnel energy. This unfortunately means that even though everything else seems "to be right" in these materials, the (doped) ground state and even the (undoped) photo-excited state has a weak overlap to its neighbors (<30 meV), and the excited carrier will most likely return to the ground states. If it does get away and conduct, it will do so with low mobility [60–64], and will be subject to Anderson localization by disorder. Having said that, it is noteworthy that very recently, the Talapin group [60, 61], Sargent et al. [62] and importantly Carey et al. [66] are overcoming this problem, and learning how to shorten or even get rid of the organic passivation making wide bands. There is now serious hope [[66] to also reach the wide bandwidths required not only for efficient LWIR photodetection with suitably tuned and selected arrays, but importantly also for designer conductive and photoconductive systems, including multicolor photovoltaic cells [67–69].

6.8 CONCLUSION

Quantum dots are extremely interesting nanostructures to study both from a basic physics standpoint and a device technology standpoint. In infrared detector applications, in particular, quantum dots show great potential to enable the next generation of devices with high quantum efficiency and high operating temperature.

REFERENCES

[1] M. Razeghi, *Fundamentals of Solid State Engineering*, (Springer, New York, 2010).
[2] K. Mukai and M. Sugawara, *Semiconductor and Semimetals*, (Academic Press, San Francisco, 1999).
[3] L. Goldstein, F. Glas, J. Y. Marzin, M. N. Charasse, and G. L. Roux, *Appl. Phys. Lett.* **47**(10), 1099–1101 (1985).
[4] Y. Arakawa and H. Sakaki, *Appl. Phys. Lett.* **40** (1982).
[5] M. Asada, Y. Miyamoto, and Y. Suematsu, *IEEE J. Quantum Electron.* **22**, 1887 (1986).
[6] M. Razeghi, *Technology of Quantum Devices* (Springer, New York, 2009).
[7] A. Rogalski, *Infrared Detectors*, 2nd ed. (CRC Press, Boca Raton, 2010).
[8] P. Martyniuk and A. Rogalski, *Prog. Quantum Electron.* **32**(3–4), 89–120 (2008).
[9] E. L. Dereniak and G. D. Boreman, *Infrared Detectors and Systems*, Wiley Series in Pure and Applied Optics (Wiley-Interscience, New York, 1996).
[10] A. Rogalski, *Prog. Quantum Electron.* **27**(2–3), 59–210 (2003).
[11] A. Piotrowski, P. Madejczyk, W. Gawron, K. Klos, J. Pawluczyk, M. Grudzien, J. Piotrowski, and A. Rogalski, *Proc. SPIE.* **5732** (2005).
[12] K. K. Choi, *The Physics of Quantum Well Infrared Photodetectors* (World Scientific, New Jersey, 1997).
[13] B. F. Levine, *J. Appl. Phys.* **74**(8), R1–R81 (1993).
[14] S. D. Gunapala, S. V. Bandara, J. K. Liu, C. J. Hill, S. B. Rafol, J. M. Mumolo, J. T. Trinh, M. Z. Tidrow, and P. D. LeVan, *Semicond. Sci. Technol.* **20**, 473–480 (2005).
[15] V. Ryzhii, *Semicond. Sci. Technol.* **11**, 759–765 (1996).
[16] S. Chakrabarti, A. D. Stiff-Roberts, and P. Bhattacharya, *IEEE Photonics Technol. Lett.* **16** (2004).
[17] J. Jiang, S. Tsao, T. O'Sullivan, M. Razeghi, and G. J. Brown, *Infrared Phys. Technol.* **45**(2), 143–151 (2004).
[18] E.-T. Kim, A. Madhukar, Z. Ye, and J.C. Campbell, *Appl. Phys. Lett.* **84**(17), 3277–3279 (2004).
[19] W. Zhang, H. Lim, M. Taguchi, S. Tsao, B. Movaghar, and M. Razeghi, *Appl. Phys. Lett.* **86**, 191103 (2005).
[20] X. Lu, J. Vaillancourt, and M. J. Meisner, *Appl. Phys. Lett.* **91**, 051115 (2007).
[21] J. Shao, T. E. Vandervelde, A. Barve, W.-Y. Jang, A. Stintz, and S. Krishna. *Enhanced Normal Incidence Photocurrent in Quantum Dot Infrared Photodetectors* (AVS, 2011).
[22] J. Jiang, K. Mi, S. Tsao, W. Zhang, H. Lim, T. O'Sullivan, T. Sills, M. Razeghi, G. J. Brown, and M. Z. Tidrow, *Appl. Phys. Lett.* **84**(13), 2232–2234 (2004).
[23] S. Krishna, D. Forman, S. Annamalai, P. Dowd, P. Varangis, T. Tumolillo, A. Gray, J. Zilko, K. Sun, M. Liu et al., *Appl. Phys. Lett.* **86** (2005).
[24] S.-F. Tang, C.-D. Chiang, P.-K. Weng, Y.-T. Gau, J.-J. Luo, S.-T. Yang, C.-C. Shih, S.-Y. Lin, and S.-C. Lee, *IEEE Photonics Technol. Lett.* **18**(8) (2006).
[25] E. Varley, M. Lenz, S. J. Lee, J. S. Brown, D. A. Ramirez, A. Stintz, S. Krishna, A. Reisinger, and M. Sundaram *Appl. Phys. Lett.***91**, 081120 (2007).
[26] P. Martyniuk, S. Krishna, and A. Rogalski, *J. Appl. Phys.* **104**(3), 034314 (2008).
[27] J. Phillips, *J. Appl. Phys.* **91**(7), 4590–4594 (2002).

[28] D. Bimberg, M. Grundmann, and N. N. Ledenstov, *Quantum Dot Heterostructures* (John Wiley & Sons Ltd, Chichester, 1999).

[29] V. Ryzhii, K. Irina, R. Maxim, I. P. Victor, V. M. Vladimir, and W. Magnus, *Proc. SPIE* **4288**(1), 396–403 (2001).

[30] R. Notzel, *Semicond. Sci. Technol.* **11**, 1365–1379 (1996).

[31] Z. Ye, J. C. Campbell, Z. Chen, E.-T. Kim, and A. Madhukar, *Appl. Phys. Lett.* **83**(6), 1234–1236 (2003).

[32] S. Kim, H. Mohseni, M. Erdtmann, E. Michel, C. Jelen, and M. Razeghi, *Appl. Phys. Lett.* **73**(7), 963–965 (1998).

[33] D. Pan, Y. P. Zeng, M. Y. Kong, J. Wu, Y. Q. Zhu, C. H. Zhang, J. M. Li, and C. Y. Wang, *Electron. Lett.* **32**(18), 1726 (1996).

[34] A. D. Stiff-Roberts, S. Chakrabarti, S. Pradhan, B. Kochman, and P. Bhattacharya, *Appl. Phys. Lett.* **80**(18), 3265–3267 (2002).

[35] J. Jiang, K. Mi, R. McClintock, M. Razeghi, G. J. Brown, and C. Jelen, *IEEE Photonics Technol. Lett.* **15**(9), 1273–1275 (2003).

[36] V. Ryzhii, V. Pipa, I. Khmyrova, V. Mitin, and M. Willander, *Jpn. J. Appl. Phys. Part 2: Letters*, **39**(12B), L1283–L1285 (2000).

[37] S. J. Xu, S. J. Chua, T. Mei, X. C. Wang, X. H. Zhang, G. Karunasiri, W. J. Fan, C. H. Wang, J. Jiang, S. Wang et al., *Appl. Phys. Lett.* **73**(21), 3153–3155 (1998).

[38] H. Jiang and J. Singh, *Physica E* **2**(1–4), 720–724 (1998).

[39] J. Szafraniec, S. Tsao, W. Zhang, H. Lim, M. Taguchi, A. A. Quivy, B. Movaghar, and M. Razeghi, *Appl. Phys. Lett.* **88**(12), 121102–121103 (2006).

[40] S. Raghavan, P. Rotella, A. Stintz, B. Fuchs, S. Krishna, C. Morath, D. A. Cardimona, and S. W. Kennerly, *Appl. Phys. Lett.* **81**(8), 1369–1371 (2002).

[41] L. Jiang, S. S. Li, N.-T. Yeh, J.-I. Chyi, C. E. Ross, and K. S. Jones, *Appl. Phys. Lett.* **82**(12), 1986–1988 (2003).

[42] D. Pan, E. Towe, and S. Kennerly, *Appl. Phys. Lett.* **73**(14), 1937–1939 (1998).

[43] Z. Chen, E.-T. Kim, and A. Madhukar, *Appl. Phys. Lett.* **80**(14), 2490–2492 (2002).

[44] E. Finkman, S. Maimon, V. Immer, G. Bahir, S. E. Schacham, F. Fossard, F. H. Julien, J. Brault, and M. Gendry, *Phys. Rev. B* **63** (2001).

[45] H. Hwang, K. Park, S. Yoon, E. Yoon, H. Cheong, and Y. Kim, *Proc. SPIE* **4999** (2003).

[46] Y. W. Mo, D. E. Savage, B. S. Swartzentruber, and M. G. Lagally, *Phys. Rev. Lett.* **65**(8), 1020 (1990).

[47] V. M. Ustinov, E. R. Weber, S. Ruvimov, Z. Liliental-Weber, A. E. Zhukov, A. Y. Egorov, A. R. Kovsh, A. F. Tsatsul'nikov, and P. S. Kop'ev, *Appl. Phys. Lett.* **72**(3), 362–364 (1998).

[48] M. Erdtmann, A. W. Matlis, C. L. Jelen, M. Razeghi, and G. J. Brown, *Proc. SPIE* **3948** (2000).

[49] J. Jiang, K. Mi, R. McClintock, M. Razeghi, C. Jelen, and G. Brown, *IEEE Photonics Technol. Lett.* **15** (2003).

[50] M. Razeghi, *The MOCVD Challenge* (Institute of Physics Publishing, Bristol and Philadelphia, 1995), Vol. 2.

[51] Y. M. Qiu and D. Uhl, *J. Cryst. Growth* **257** (2003).

[52] M. S. Ünlü and S. Strite, *J. Appl. Phys.* **78** (1995).

[53] H. Lim, W. Zhang, S. Tsao, T. Sills, J. Szafraniec, K. Mi, B. Movaghar, and M. Razeghi, *Phys. Rev. B* **72** (2005).
[54] M. Allovon and M. Quillec, *IEE Proc. Optoelectron.* **139**, 148–152 (1992).
[55] S. Krishna, *J. Phys. D: Appl. Phys.* **38**, 2142–2150 (2005).
[56] X. L. Huang, Y. G. Shin, E.-K. Suh, H. J. Lee, Y. G. Hwang, and Q. Huang, *J. Appl. Phys.* **82**(9), 4394–4399 (1997).
[57] S. Y. Wang, S. D. Lin, H. W. Wu, and C. P. Lee, *Appl. Phys. Lett.* **78** (2000).
[58] H. Lim, S. Tsao, W. Zhang, and M. Razeghi, *Appl. Phys. Lett.* **90**(13), 131112–131113 (2007).
[59] C. B. Murray, C. R. Kagan, and M. G. Bawendi, *Ann. Rev. Mat. Res.* **30**, 545–610 (2000).
[60] D. V. Talapin, J. S. Lee, M. V. Kovalenko, and E. V. Shevchenko, *Chem. Rev.* **110**, 389 (2010).
[61] D. V. Talapin, *MRS Bull.* **37**, 63 (2012).
[62] G. Konstantatos, E. H. Sargent, *Nat. Nanotechnol.* **5** (2010).
[63] C. Wang, M. Shim, and P. Guyot-Sionnest, *APL* **80**, 4, 28 (2002).
[64] P. P. Jha and P. Guyot-Sionnest, *J. Phys. Chem. C* **111**, 15440 (2009).
[65] M. Achermann, M. Petruska, D. Koleske, M. Crawford, and V. I. Klimov, *Nanoletters* **6**, 1396 (2006).
[66] G. H. Carey, K. W. Chou, B. Yan, A. R. Kirmani, A. Amassian, and E. H. Sargent, *MRS Comm.* **3**, 83–90 (2013).
[67] M. V. Jarosz, V. J. Porter, B. R. Fisher, M. A. Kastner, and M. G. Bawendi, *Phys. Rev. B* **70**, 195327 (2004).
[68] A. F. G. Monte et al. *Microelectr. J.* **39**, 667 (2003).
[69] S. Chanyawadee, R. T. Harley, M. Henini, D. V. Talapin, and P. G. Lagoudakis, *Phys. Rev. Lett.* **102**, 077402 (2009).

7

MAGNETIC CONTROL OF SPIN IN MOLECULAR PHOTONICS

EITAN EHRENFREUND[1] AND Z. VALY VARDENY[2]

[1]*Department of Physics and Solid State Institute, Technion-Israel Institute of Technology, Haifa, Israel*
[2]*Department of Physics and Astronomy, University of Utah, Salt Lake City, UT, USA*

We review recent advances in magnetic field spin manipulation of the electroluminescence from organic light emitting diodes based on small molecules and π-conjugated polymers. The review contains a summary of our recent results and understanding of various spin-mixing mechanisms responsible for the magneto-electroluminescence in these devices. The main mechanisms that we discuss here include the hyperfine interaction in pairs of positive and negative spin $1/2$ charge polarons, the collision between triplet excitons and spin $1/2$ charge polarons, and the difference in the g-factor of positive and negative polarons.

7.1 INTRODUCTION

Since their inception in 1990 [1] organic light emitting diodes (OLEDs) have become efficient devices that are routinely used at the present time in solid state flat panel displays, such as mobile phones and television screens. Manipulating the device electroluminescence (EL) intensity or color by small magnetic fields may enrich their usage in various applications. Due to the inherently small spin orbit coupling in organics, the spin relaxation rate of the excited species borne out by the injected charges (e.g., trapped polarons, polaron pairs, triplet excitons) is relatively small, allowing small magnetic fields to manipulate their spin configuration before spin

Photonics: Scientific Foundations, Technology and Applications, Volume II, First Edition.
Edited by David L. Andrews.
© 2015 John Wiley & Sons, Inc. Published 2015 by John Wiley & Sons, Inc.

coherence is lost. Indeed, it has been shown that by applying a ~10 mT magnetic field on an aluminum tris(8-hydroxyquinoline) [Alq_3]-based OLED, its EL intensity could be changed by as much as ~10% at room temperature [2, 3]. For the understanding of such a magneto-EL (MEL) effect there must exist a spin selective mechanism that adjusts the precursor excitation to the singlet exciton (such as polaron pair, PP), or their nonradiative–radiative recombination rate upon the application of an external magnetic field B. An important spin-mixing mechanism of the PP species is the hyperfine interaction (HFI) between the electron (hole) spin ($S = \frac{1}{2}$) and the proton nuclear spin ($I = \frac{1}{2}$) in organic molecules [4–9]. The electron–proton HFI strength a_H in typical organic molecules is of the order $a_H/2\mu_B$ ~3 mT (where μ_B is the Bohr magneton) [10], which enables considerable spin-mixing variations for $B < $ ~10 mT. Another important mechanism involves triplet excitons (TE) (that are produced by PP in the triplet configuration), where spin-dependent quenching of TE by polarons [11] or polaron–TE collisions [9] renders the conductance and/or EL magnetic field sensitive. A third spin-mixing mechanism that we consider here is the so called "Δg mechanism" [12–15] in which the difference between the g-value of positive and negative polarons causes PP singlet–triplet mixing with B; as B increases the spin configuration mixing rate increases, leading to MEL effect.

In this chapter we review recent advances in understanding the phenomenon of MEL and magneto-conductance (MC) in OLEDs based on small molecules and π-conjugated polymers. The review contains a brief summary of existing data and models, and more in-depth description of our most recent results including the spin-mixing mechanisms mentioned above that are mainly responsible for the magnetic field effects (MFE).

7.2 A SURVEY OF THE MAGNETO-ELECTROLUMINESCENCE IN OLEDS

The basic structure of an OLED is shown in Figure 7.1, where an organic layer is sandwiched between hole-injecting (anode) and electron-injecting (cathode) electrodes. The anode is usually an indium tin oxide (having large work function), whereas the cathode is usually Al or Ca (having smaller work function). The oppositely charged injected carriers (or polarons in organic semiconductors) meet inside the organic layer, form polaron pairs that are precursors to excitons, which subsequently form singlet or triplet excitons (SE or TE, respectively) that may recombine radiatively as EL radiation. We emphasize that the recombination process involves several intermediate slower steps. These steps include the formation of long-lived PP species that eventually fuse into SE, which may recombine radiatively. The SE recombination is also influenced by a myriad of nonradiative processes. Among them are SE-polaron and SE–TE quenching. Another possible avenue for the injected polarons is the formation of TE via the intermediate PP in the triplet spin configuration. In polymer-based OLED the TE cannot recombine radiatively (to a singlet ground state); hence, TE formation reduces the device EL yield. At high TE density, however, the EL intensity may increase due to delayed EL [16] caused by the process known as triplet–triplet annihilation (TTA) [17], in which two TEs are annihilated by forming SE that may

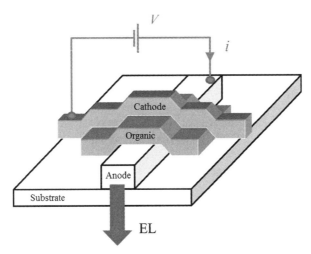

FIGURE 7.1 Schematics of an OLED structure. The organic layer is sandwiched between the anode and cathode. Voltage is applied as shown and the EL emerges from the device through the transparent anode (usually made of indium tin oxide alloy, ITO) and substrate (made of glass). A hole transport layer is usually put in between the anode and the organic layer for better hole injection into the organic HOMO levels. The cathode is made from a metal with suitable work function to allow electron injection into the organic LUMO levels; Ca or Al are often used depending on the material of the organic layer. The organic layer is made of a film of small organic molecules, such as Alq_3 (see Fig. 7.2), or π-conjugated polymers, such as poly phenylene vinylene (PPV) or one of its derivatives (see Fig. 7.5).

emit light. These intermediate recombination stages have dramatic effects on the OLED efficiency. When the only fast direct electron–hole (each having spin $S = \frac{1}{2}$) recombination is allowed, then the maximum light output efficiency is limited to 25% since electron–hole pairs can recombine radiatively only from their singlet spin state. However, when long-lived PP are formed then spin mixing between the singlet and triplet configurations takes place via, for example, the HFI or exchange interaction [15], and the EL efficiency may dramatically increase. In addition, the TTA process that produces delayed EL also increases the overall device efficiency.

When an OLED is placed in a magnetic field, the rate at which SE are formed either from PP or via the TTA process changes because the singlet content of each quantum state changes. Other mechanisms, such as TE-polaron collision or TE quenching by polarons, are also magnetic field dependent. Since TE and free polarons may act as nonradiative recombination centers, then the change in their respective density with B may also change the EL output. These latter mechanisms also may modulate the current density in the device that controls the EL intensity; and this also may change the EL output. The MEL is defined as

$$\mathrm{MEL}(B) = \frac{\mathrm{EL}(B) - \mathrm{EL}(0)}{\mathrm{EL}(0)}, \quad (7.1)$$

where B is the magnetic field strength.

7.2.1 Early Organic-MEL Studies: Small Molecule (Alq$_3$) and π-Conjugated Polymer (Polyfluorene)

One of the most studied organic molecules that show efficient EL and sizable MEL in an OLED is Alq$_3$, where the centered Al atom is bonded to three quinoline units, as shown in Figure 7.2. Kalinowski et al. [2] first reported MEL studies of OLED devices based on Alq$_3$ emitting layer, semitransparent ITO anode, and Ca cathode. A hole transporting layer made of a 75–25% mixture of N,N'-diphenyl-N,N'-bis(3-methylphenyl)-(1,1'-bi-phenyl)-4,4 diamine (TPD) and the polymer bisphenol-A-polycarbonate (PC) was spin coated between the Alq$_3$ layer and the anode. The room temperature MEL response versus B is shown in Figure 7.3 for two driving voltages. As can be seen, MEL(B) steeply increases with increasing |B| (for |B| < 0.1 T), and subsequently tends to saturate at larger fields. This positive magnetic field effect (MEL(B) > 0) was explained by the authors in terms of magnetic field modulation of the ratio between electron–hole pairs with singlet and triplet characters, due to the hyperfine coupling of the carriers forming pairs in a bimolecular recombination process of the injected holes and electrons in the Alq$_3$. The EL emission is controlled by the field modulated population of singlet excitons, reflecting directly their concentration. The authors also concluded that the contribution of other magnetic field sensitive processes such as singlet exciton fission into two triplets, triplet–triplet or triplet–charge carrier interactions can be excluded on the basis of singlet–triplet energy interrelations and the MEL(B) response. Singlet exciton fission requires that the SE energy is comparable to twice the TE energy, which is not the case in Alq$_3$ [2].

Subsequently, there appeared several reports on MEL studies in OLED based on π-conjugated polymers. Figure 7.4 shows the MEL(B) response at room temperature reported by Mermer et al. [3] in a PEDOT/polyfluorene/Ca device measured at a constant voltage, or at a constant current. Also shown is the magneto-conductance (MC(B)) response, which is seen to be quite similar to that of MEL(B). Similar to the MEL(B) response of the the Alq$_3$-based device described in the preceding paragraph,

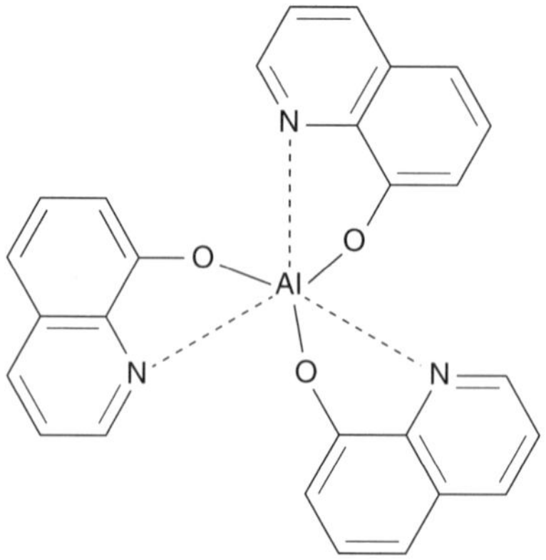

FIGURE 7.2 The chemical structure of Alq$_3$.

A SURVEY OF THE MAGNETO-ELECTROLUMINESCENCE IN OLEDS 225

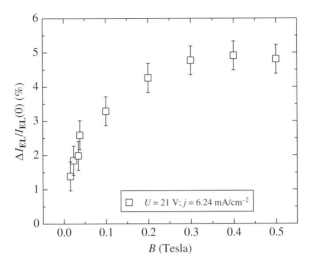

FIGURE 7.3 MEL of an OLED device made of ITO/75% TPD: 25% PC(60 nm)/Alq$_3$ (60 nm)/Ca/Ag at bias $U = 21$ V and current density $j = 6.24$ mA cm^{-2}. Data taken from Reference 2.

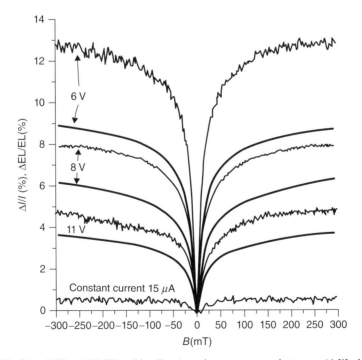

FIGURE 7.4 MEL (ΔEL/EL, thin lines) and magneto-conductance ($\Delta I/I$, bold) of ITO/PEDOT/polyfluorene(100 nm)/Ca device at constant voltage. MEL at constant current of 15 µA is also shown. From Reference 3.

MEL(B) here first increases with B (for $|B| < 50$ mT), followed by saturation at larger fields. The half width at half maximum (HWHM) of the narrow feature is ~5mT, very close to the proton hyperfine field in this polymer.

7.2.2 Isotope Effect in MEL: the Case of Poly(Dioctyloxy) Phenyl Vinylene (DOOPPV)

Various models have been suggested for explaining the MEL response in OLED, where the active layers are π-conjugated organic semiconductors. Most of these models are based on the HFI between the injected spin $\frac{1}{2}$ carriers and nuclear spins in the organic layer [4, 18–22]. The most common model considers the HFI mixing of spin sublevels of PP species, where PP spin singlet/triplet level mixing becomes less effective as B increases [18].

Experimental verification for the importance of the HFI in determining the MEL(B) response has been obtained [5,6] by replacing skeletal protons (nuclear spin $I = \frac{1}{2}$) with deuterons (D, $I = 1$ and with smaller hyperfine constant, $a_D < a_H$); or the naturally abundant spinless nucleus ($I = 0$) ^{12}C by ^{13}C ($I = \frac{1}{2}$, $a_{13C} > a_{12C} = 0$) in the π-conjugated polymer interlayer. In Figures 7.5a and 7.5b the fully protonated and deuterated DOOPPV polymer structures used in these studies are shown. All ^{1}H atoms near the backbone carbon atoms on the polymer chains were replaced by ^{2}H (D) atoms. The isotope exchange was concluded from extensive nuclear magnetic resonance (NMR) and Raman scattering spectra, where the two main Raman-active vibrational modes at ~1300 cm^{-1} (C–C stretching) and 1500 cm^{-1} (C=C stretching) are red shifted by about 3% upon deuteration. This is in agreement with the expected isotope shift due to the "square root of the mass ratio rule" [27], namely $[m(CD)/m(CH)]^{1/2} \approx 1.037$. In the ^{13}C-enriched DOOPPV about 30% of the skeletal ^{12}C carbon atoms were replaced by ^{13}C atoms. For the MEL investigations, OLED devices based on the three isotope-enriched polymers as active layers with various film thicknesses were fabricated. The OLEDs were transferred to an optical cryostat that was placed in between the two poles of an electromagnet producing magnetic field, B, up to 300 mT. The devices were driven at constant bias voltage, V, using a Keithley 236 apparatus; and EL was measured while sweeping B. Electroluminescence in the device results from recombination of PP in the spin singlet configuration, PP$_S$. The PP in the two spin configurations, namely PP$_S$ and PP$_T$, exchange spins via intersystem crossing (ISC), and their steady state populations are also determined by their spin exchange interaction, as well as by the individual recombination and dissociation rates [23].

Figures 7.6a and 7.6b show the MEL response of two OLED devices based on H- and D-DOOPPV having the same thickness d_f, measured at the same bias V [33]. Figures 7.7a and 7.7b show the MEL responses of ^{13}C-rich DOOPPV have similar thickness, driven at $V = 3$ V. The MEL(B) response is narrow in D-DOOPPV device and broader for ^{13}C-DOOPPV device; in fact the field, $B_{1/2}$, at half the MEL maximum for H-DOOPPV is about twice as large as that of the D-DOOPPV device and ~30% smaller than that of the ^{13}C-DOOPPV device. It was also found that $B_{1/2}$ increases with V (Fig. 7.6a inset for H- and D-DOOPPV) [25]; in fact, $B_{1/2}$ increases approximately linearly with the device electric field, $E = (V - V_{bi})/d_f$, where V_{bi} is

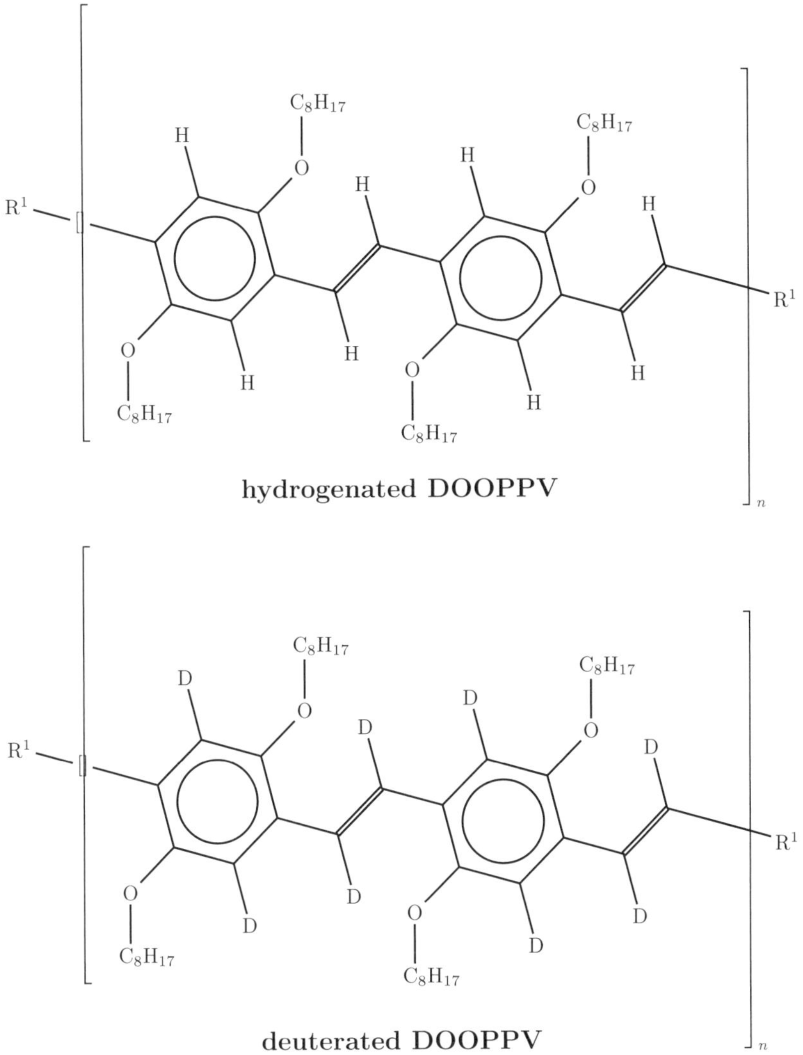

FIGURE 7.5 D- and H-polymer structures. The basic repeat unit of the synthesized H-DOOPPV (top) and D-DOOPPV (bottom) polymers. H- and D- stand for protonated (^{1}H) and deuterated hydrogen (^{2}H), respectively; R^1 is an end-capped molecule used to terminate the polymerization process.

the built-in potential in the device; which is related to the onset bias where EL and MEL are observed [21]. Importantly, in all cases it was found that $B_{1/2}(H) > B_{1/2}(D)$ for devices having the same value of the electric field, E (Fig. 7.6a inset).

MEL in OLEDs may be considered as an example of a much broader research field that deals with MFE in Physics [24], Chemistry, and Biology [15, 25]. For MFE that results from pairs of radical ions it was empirically realized [26] that $B_{1/2}$ scales

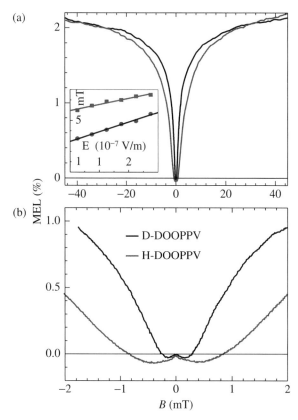

FIGURE 7.6 Room temperature MEL response of H- and D-DOOPPV (red and blue lines, respectively) measured at bias voltage $V = 2.5$ V, plotted on large (**a**) and small (**b**) magnetic field scales, where the respective regular and ultra-small-field MEL responses are separated. **Inset to a**: The field, $B_{1/2}$, at half the MEL maximum for the two polymers as a function of the applied bias voltage, V, given in terms of the internal electric field in the polymer layer; the lines are linear fits to guide the eye. *(For a color version of this figure, see the color plate section.)*

with the HFI constant, a. In fact, a semiempirical law was advanced [26], which for opposite charged radicals with same HFI coefficient a reads

$$B_{1/2} \approx 2a[I(I+1)]^{1/2}, \tag{7.2}$$

where I is the nuclear spin quantum number. In the analysis that led to Eq. (7.2) the nuclear spin is treated classically [26] and is therefore a crude approximation for $I = \frac{1}{2}$ or 1. Assuming that the HFI constant a scales with the nuclear magnetic moment [10], g_N, one obtains from Eq. (7.2): $B_{1/2}(H)/B_{1/2}(D) \approx 0.6 g_H/g_D \approx 3.9$ (note that $g_H/g_D = 6.51$), compared with $B_{1/2}(H)/B_{1/2}(D) \approx 6/3 = 2$ obtained experimentally at small voltages (Fig. 7.3a, inset). Given this considerable experimental reduction in

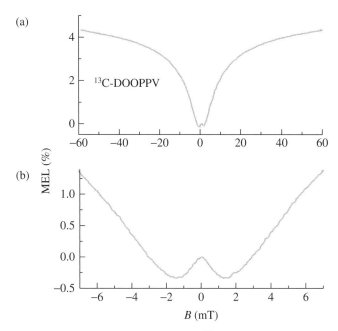

FIGURE 7.7 Room temperature MEL response of ^{13}C-DOOPPV measured at bias voltage $V = 3$ V, plotted on large (**a**) and small (**b**) magnetic field scales, where the respective regular and ultra-small-field MEL responses are separated. Note: $B_{1/2} \approx 10$ mT, and $B_m \approx 1.4$ mT are larger than those obtained for H-DOOPPV (Fig. 7.6).

$B_{1/2}$ upon deuteration and enhancement upon ^{13}C enrichment, it was concluded [6] that the MEL response in DOOPPV OLED *is mainly due to the HFI*.

Surprisingly, Figures 7.6b and 7.7b also show that the MEL has yet another component at low B (dubbed "ultra-small-field MEL", or USMEL), which has an opposite sign to that of the positive MEL at higher fields. Similar USMFE components have been observed before in some biochemical reactions [27] and anthracene crystals [28]. It is argued that the USMEL component is also due to the HFI, since its width is isotope dependent [5]. One can see that the dip in the USMEL(B) response is isotope dependent: it occurs at $B_{min} \sim 0.2$ mT in D-DOOPPV, at $B_{min} = 0.7$ mT in H-DOOPPV and at $B_{min} = 1.4$ mT in ^{13}C-DOOPPV. The positive, high field MEL(B) component was interpreted as due to less effective ISC between PP spin sublevels as B increases [18, 29]; therefore, it was concluded that the negative USMEL(B) component is due to ISC *rate increase* between PP spin sublevels at small B.

7.2.3 Isotope Effect in MEL: The Case of Alq$_3$

As mentioned Section 7.2.1, Alq$_3$ is a common active molecular layer used in OLED due to its efficient EL emission and high electron mobility [30–32]. It is thus not surprising that MFE in Alq$_3$-based OLED devices have been extensively studied in the last few years [2, 3, 11, 29, 33–35]. As a result, several basic models were originally proposed to explain the obtained MFE(B) response. Basically, all models agree that the underlying mechanism for the MFE is the magnetic field dependence of spin sublevel mixing; but there is no agreement as to the excitation species where the spin-mixing occurs. The competing models include: (i) spin mixing in oppositely charged PP, and in pairs of same-charge polarons (or bipolarons, BP) by the HFI [2, 4, 29, 36]; (ii) spin mixing within TTA process [33]; and (iii) spin mixing during the process of triplet exciton quenching by spin $^1\!/_2$ charge polarons [11]. Importantly, the HFI proposed models should differ substantially from the other models in their response to isotope exchange in the Alq$_3$ molecule active layer, where all hydrogen atoms (nuclear spin $I_H = {}^1\!/_2$, nuclear g-factor $g_H = 5.586$) are exchanged by deuterium atoms ($I_D = 1$, $g_D/g_H = 0.154$). This should happen since the HFI constant, a_{HF}, scales with the nuclear g-factor [10], whereas the other proposed interactions are isotope insensitive. Consequently, the isotope exchange effect on the MFE(B) response of Alq$_3$-based OLED was recently studied [34, 37]. Because of the lack of any isotope effect on the measured MFE(B) response it was concluded that the MFE in Alq$_3$-based OLED is not dominated by PP or BP species and MEL(B) responses. These measurements and conclusions are surprising, because similar MFE measurements in devices based on DOOPPV (see previous section) did show a substantial isotope effect [5, 6]. It was thus important to investigate in more detail the influence of the isotope exchange on the MFE in Alq$_3$-based devices in order to identify the underlying spin exchange mechanisms.

The MEL(B) response of fully protonated H$_{18}$Alq$_3$ and fully deuterated D$_{18}$Alq$_3$ OLED are shown with various field resolutions in Figures 7.8a and 7.8c; a clear isotope-dependent response is seen. First, as seen in Figure 7.8b and the inset of Figure 7.8a, the width, ΔB, of the MEL(B) response in the H$_{18}$Alq$_3$ device is $\sim$40% larger than that in the D$_{18}$Alq$_3$ device. This observation is at variance with the earlier study in which much less dependence on the isotope exchange was reported [34, 37]. Second, as seen in Figure 7.8c, the MEL(B) response shows another feature at low fields ($B \sim$ <2 mT): as $|B|$ increases from $B = 0$, MEL(B) is negative, reaches a minimum value at $|B| = B_m$, then increases through a zero crossing, and monotonically increases thereafter. Clearly, B_m is isotope dependent: $B_m = 0.2$ mT and 0.4 mT for D$_{18}$Alq$_3$ and H$_{18}$Alq$_3$, respectively. Similar features, namely the USMFE, were observed previously in DOOPPV-based OLEDs, where the isotope dependence has been shown to originate from the HFI in PP species [5, 6] (see preceding section). It has therefore been concluded that the HFI in PP species plays an important role also in the low-field MEL response in Alq$_3$ devices. Note, however, that the experimentally measured ratios $B_m(H)/B_m(D) \approx 2$ and $\Delta B(H)/\Delta B(D) \approx 1.4$ for the Alq$_3$ OLED are $\sim$30–40% smaller than those measured in devices based on D- and H-isotopes of DOOPPV [6]. This observation indicates that, in addition to the HFI, other interactions that are isotope insensitive need be taken into account for explaining the full MEL response in Alq$_3$. A more detailed discussion of the isotope effect in the low field MEL response is presented in Section 7.5.1.

At higher fields ($|B| \sim$ 50–250 mT), the MEL response in Alq$_3$-based devices does not level off, but instead continues to increase, in contrast to what is expected

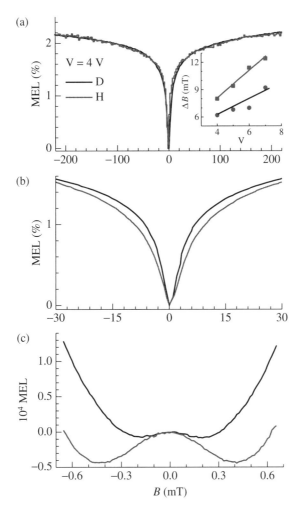

FIGURE 7.8 (a)–(c) MEL(B) response of OLEDs based on $H_{18}Alq_3$ (red line) and $D_{18}Alq_3$ (blue line) measured at room temperature and bias $V = 4$ V, plotted at three *different B scales*. The $D_{18}Alq_3$ response was normalized to that of $H_{18}Alq_3$ at $B \sim 250$ mT. Inset in (a): The full width, ΔB at MEL = 0.8% plotted versus V for $H_{18}Alq_3$ (red) and $D_{18}Alq_3$ (blue). *(For a color version of this figure, see the color plate section.)*

for MFE response governed by the HFI [4, 6]. This characteristic behavior indicates that a different mechanism is dominant for the high field response of both MEL and MC here. Alq_3 is known to have phosphorescence emission from triplet excitons and delayed fluorescence (DF) caused by TTA [38]. Therefore, it is likely that TE are involved in the MFE response at intermediate high fields. In order to examine this hypothesis the Alq_3-based OLED devices were exposed to oxygen atmosphere, which is known to quench TE [39]. Figure 7.9 shows the MEL(B) response of

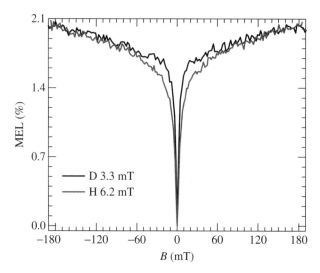

FIGURE 7.9 MEL(B) response of OLEDs based on $H_{18}Alq_3$ and $D_{18}Alq_3$ saturate-exposed to oxygen, measured at $V = 4$ V and room temperature. The full width, ΔB, at MEL = 1% is 12.4 mT (6.6 mT) for the $H_{18}Alq_3$ ($D_{18}Alq_3$) device. The response of $D_{18}Alq_3$ was normalized to that of $H_{18}Alq_3$ at $B = 200$ mT. *(For a color version of this figure, see the color plate section.)*

oxygen-exposed OLED devices of both Alq_3 isotopes. The MEL response is similar to that shown in Figure 7.8, but with much clearer difference between the responses of the two isotopes: the MEL width, as defined in Figure 7.9, of $H_{18}Alq_3$ OLED is now two times larger than that of $D_{18}Alq_3$ OLED. The MEL results for OLED exposed to oxygen indicate that the intermediate high field MEL response seen in unexposed devices comes from a process involving TE, which is insensitive to isotope exchange. When this component is quenched by exposure to oxygen, then the HFI-dominated component prevails, and consequently the isotope-dependent response becomes clearer.

7.3 ORGANIC MEL AT SMALL MAGNETIC FIELDS; COMPASS EFFECT

Recently, both MC and MEL responses of OLED based on many organic semiconductors were found to contain a non-monotonic magnetic field response at small fields $B_m \sim 0.2$ mT [5, 6, 40] (see also Section 7.2.2). Similar non-monotonic field responses were also found in the singlet yield of biochemical reactions [25, 41, 42]. The proximity in the value of B_m to that of the Earth magnetic field, B_E (~0.05 mT), points toward the possibility that B_E might influence the conductance and electroluminescence of OLEDs, as well as the outcome of biochemical reactions in living creatures. Indeed, it was shown [43] that B_E influences the low field response of both MC and MEL in OLED, yielding a way to determine the magnitude of the local B_E.

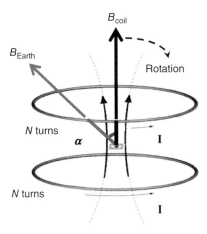

FIGURE 7.10 The experimental setup for the direction-sensitive MC and MEL measurements.

Subsequently, a method was introduced for determining the direction of B_E using a small, fixed magnetic field, B_0, by measuring MC and MEL in the presence of B_E. This "compass behavior" was demonstrated [43] at room temperature in OLEDs based on various isotopes of DOOPPV (chemical structure shown in Fig. 7.5).

The 0.5×0.5 mm^2 OLEDs used for this demonstration were encapsulated and placed at the center of a homemade Helmholtz coil-pair (see Fig. 7.10). The magnetic coils with $N = 50$ turns each and radii of 6 mm were wound around a cylindrical wood frame. The induced magnetic field at the device position as a function of current source was calibrated using a Lakeshore Gauss probe with 2 µT field resolution. The coil and power supply arrangement could produce magnetic field, $\mathbf{B}_{coil}$, with 5 µT field resolution. For the directionally sensitive studies, the Helmholtz coil-pair with the device was placed on a non-magnetic rotation axis of a high precision motor enabling 360° tilt angle variation between $\mathbf{B}_{coil}$ and $\mathbf{B}_E$. The magnetic field from the motor was shielded using mu-metal shield foils. The direction of the Earth magnetic field was determined with an uncertainty of 2° using two magnetic compasses. One compass was used to determine the projection of Earth field direction in the horizontal plane while the other compass was used to identify the inclination angle of the field. A Keithley 238 apparatus was used as a constant bias voltage source for the OLED devices. The device current density was controlled at ~1 µA/mm^2 in all measurement for optimizing the MC and MEL magnitudes. EL intensity was detected by a photovoltaic Si diode. The MC and MEL responses were measured at constant applied voltage while sweeping the field B. For the angle-dependent MC and MEL the applied voltage and $\mathbf{B}_{coil}$ were kept constant while the conductivity and EL were recorded as a function of the tilt angle, α.

Figures 7.11a and 7.12a show, respectively, the MEL(B) and MC(B) responses of the (H-, D-, and C13)-DOOPPV OLEDs up to applied field $B_{coil} = 0.5$ mT with a fixed angle α. In these experiments, the apparatus was shielded using mu-metal

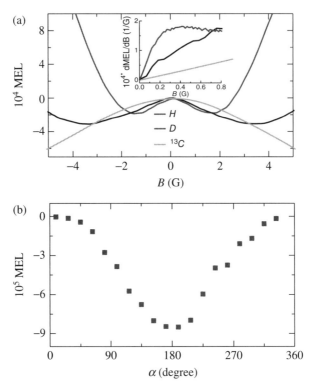

FIGURE 7.11 MEL(B) response of OLEDs based on D-, H-, and C13-DOOPPV (red, blue, green, respectively). (a) Up to 0.5 mT when B_E is shielded; it shows a maximum at $B_{coil} = 0$ and minima at $B_{coil} = \pm B_m$. Inset: The first derivative $|d(MEL)/dB|$ of the response in (a) up to 0.08 mT; $|d(MEL)/dB|$ is largest for the D-DOOPPV-based OLED. (b) MEL(α) response for H-DOOPPV at $B_{coil} = 0.05$ mT. The line through the data points is a fit using the data in Figure 7.11a and a procedure described in the text. *(For a color version of this figure, see the color plate section.)*

shield foils so that the influence of the Earth magnetic field ($B_E \approx 0.05$ mT) was excluded; that is, the total magnetic field B on the OLED was $B = B_{coil}$. In all cases a local maximum of MEL(B) and MC(B) response at $B = 0$ and two local minima at small fields $\pm B_m$ was obtained. As is clearly seen, B_m is smaller for weaker HFI: $B_m = 0.14, 0.33$, and 1.5 mT for D-, H-, and C13-DOOPPV, respectively, in agreement with previous reports [5, 6, 40]. As an indicator for the sensitivity of the MEL(B) response to small changes in B, the inset of Figure 7.11a shows the first derivative (absolute value) of MEL versus B ($dMEL(B)/dB$) close to zero field. The magnetic field sensitivity, as measured by the MEL slope is the largest for D-DOOPPV (having the weakest HFI) and the smallest for ^{13}C-DOOPPV (strongest HFI).

The "compass action" in MC and MEL responses is revealed by using the OLED as a sensitive magnetic sensor in conjunction with a static fixed $\mathbf{B}_{coil} = \mathbf{B}_0$, that is applied perpendicular to the OLED plane (Fig. 7.10). The device apparatus (or

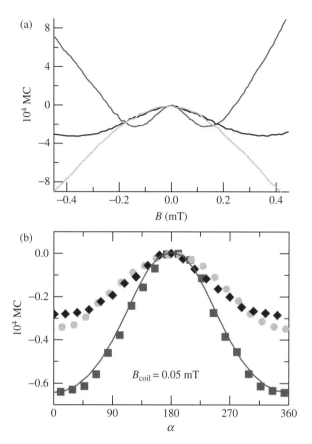

FIGURE 7.12 MC(B) response of OLEDs based on D-, H-, and C13-DOOPPV (red, blue, green, respectively). (a) Up to 0.5 mT, when the Earth's magnetic field, B_E, is shielded, showing maximum at $B_{coil} = 0$ and minima at $B_{coil} = \pm B_m$. The inset shows the chemical backbone of D-DOOPPV. (b) MC(α) response for the three isotopes using $B_{coil} = 0.05$ mT. The line through the data points is a fit using the data in Figure 7.12a and the procedure described in the text. *(For a color version of this figure, see the color plate section.)*

"sensor"), including the OLED and the coil, is tilted relative to the direction of the vector $\mathbf{B_E}$ (note that $\mathbf{B_E}$ is inclined and is, in general, not horizontal), such that the angle α between the vectors $\mathbf{B_{coil}}$ and $\mathbf{B_E}$ is varied between 0° and 360°. In this case the sensor experiences the total magnetic field $\mathbf{B} = \mathbf{B_E} + \mathbf{B_{coil}}$. Therefore, the magnitude B is a function of α, which is given by the relation,

$$B = \sqrt{B_E^2 + B_{coil}^2 + 2B_E B_{coil} \cos \alpha}. \quad (7.3)$$

Consequently, the device MEL (as well as MC) changes with α. Figures 7.11b and 7.12b show, respectively, the measured MEL(α) and MC(α) (symbols) for all three isotope-rich DOOPPV-based OLED at $B_0 = 0.05$ mT. In order to relate the α-dependent response (i.e., either MEL(a) or MC(α) shown for fixed $\mathbf{B}_{coil}$ in Figure 7.11b or Figure 7.12b) to the field-dependent response (i.e., either MEL(B) or MC(B) shown for fixed α in Figs. 7.11a or 7.12a), for each angle α the corresponding total field from Eq. (7.3) was calculated, and then the value of MEL or MC that matches this field was extracted using Figure 7.11a or 7.12a. The normalized response thus obtained is shown as a solid line for D-DOOPPV in Figure 7.12b, showing that by changing α the total magnetic field experienced by the OLED does indeed change. It is thus concluded that the OLED can be used as a compass sensor for detecting the $\mathbf{B}_E$ direction.

Comparing the angular response of the three isotope-rich DOOPPV (Fig. 7.2b) OLEDs, it is seen that the largest angular variation is obtained for the D-DOOPPV-based OLED, having the smallest HFI coupling constant. This is also in agreement with the highest slope dMEL(B)/dB shown in the inset to Figure 7.11a. Thus, the compass effect is more noticeable for the lowest a_{HFI} in the DOOPPV system.

Although the sensitivity of the OLED as a compass ($\sim 10^{-4}$) is too small for contemporary technological applications, it nevertheless shows that measurable effects of the Earth magnetic field can be directly observed in conductivity, electroluminescence, and possibly also in biochemical reaction rates. Bacteria, honey bees, sharks, birds, and many other living creatures are known to use the geomagnetic field as a source of compass information used for navigation. There are two dominating and competing hypotheses for the primary process underlying the organism magnetic compass: one mechanism that involves the ferromagnetic substance magnetite [44, 45], and the other mechanism that utilizes magnetically sensitive chemical reaction [46]. Recently, the first avian magnetic compass, which is based on a magneto-sensitive radical-pair reaction in birds has been proposed [42]. However, the proposed compass operates at very low temperatures that are not feasible for living organisms. The experiments discussed in this section show that by applying a tiny external magnetic field of the order of 0.05 mT, both MC and MEL change by a measurable amount, when tilted relative to the Earth magnetic field direction. In the spirit of these experiments, the constant magnetic field, B_0, needed for the "compass effect" in living creatures can in fact be produced, for example by the magnetite (Fe_3O_4) that has been found in some birds [44]. Electrical signals transmitted by neurons in living creatures may thus be sensitive to the magnetite angle with respect to the Earth's magnetic field.

7.4 MAGNETIC FIELD EFFECT ON EXCITED STATE SPECTROSCOPIES IN ORGANIC SEMICONDUCTOR FILMS

The organic MFE is not limited to current-carrying devices; it has been studied for a long time in biochemical reactions and more recently it has been applied to spectrally resolved photoinduced absorption (PA) and photoluminescence (PL) (dubbed here MPA and MPL, respectively) in organic semiconductor free standing *films*, not connected to external electrical circuits [47]. There are two advantages of this "spectroscopic-sensitive" MFE technique over the previously studied "transport-related" MC and MEL in devices. First, the MPA and MPL are determined by the photoexcitation spin *density* (such as PP or TE) rather than carrier mobility that was hypothesized to dominate the MC and MEL responses [4, 18, 48, 49]. By directly comparing the MPA and MPL responses in films to those of MC and MEL in OLED made of the same active layers, the role of the spin density of the photo- or current-generated excitation can be determined. Second, being a spectroscopic technique the MPA can be used as a new tool to discern various long-lived photo-excitations in organic semiconductor films. A schematic diagram of the philosophy underlying the MPA technique is presented in Figure 7.13a. For obtaining photoinduced absorption the film is photo-excited by a continuous wave laser beam with above-gap photon

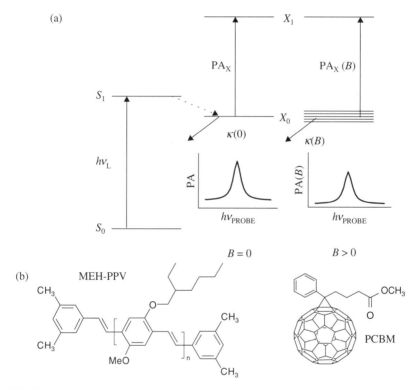

FIGURE 7.13 Schematic illustration of the magnetic field dependent pump-probe PA processes. (a) The pump beam with photon energy $h\nu_L = 2.54$ eV excites MEH-PPV to the SE level ($S_0 \rightarrow S_1$). SE relaxes via ISC to a TE or charge separate to form PP ($S_1 \rightarrow X_0$). The steady state density of the X species is controlled by the decay coefficient κ. The incandescent probe beam monitors the photoinduced absorption ($X_0 \rightarrow X_1$, PA$_X$), which is proportional to the X_0 steady state density. In a magnetic field $B > 0$, X_0 splits according to its spin multiplicity and κ of each spin sublevel becomes field dependent, resulting in a B-dependent density and B-dependent PA$_X$ (MPA$_X$). (b) The backbone structures of MEH-PPV [2-methoxy-5-(2'-ethylhexyloxy)-poly(p-phenylene vinylene) [PPV]] polymer and PCBM [phenyl-C_{61}-butyric acid methyl ester] molecule.

energy that generates steady state singlet excitons (SE; $S_0 \to S_1$). The SE may either radiatively recombine ($S_1 \to S_0$); convert into long-lived TE via ISC; or separate into pairs of positive and negative charges forming transfer complexes or polaron pairs, some of which may form long-lived PP ($S_1 \to X_0$, where X stands for species such as PP, TE, and pairs of TEs). The X species has an excited state transition $X_0 \to X_1$ (PA_X), which is activated by a weak probe beam. PA is defined as the negative fractional change in transmission, T: $PA(E) = (-\Delta T/T) = N_{SS}\beta(E)$, where N_{SS} is the species *steady state density* at the X_0 level and $\beta(E)$ is the photo-excitation optical cross-section, where E is the probe photon energy. Therefore, in a magnetic field, B, $PA_X(B)$ is determined by $N_{SS}(B)$; which, in turn, is controlled by the X species decay rate coefficient, $\kappa(B)$. For $B \neq 0$, the X_0 level splits according to the relevant spin multiplicity, L ($L = 3, 4$, and 9 for $S = 1$ TE, PP composed of two $S = \frac{1}{2}$ polarons, and a pair of TEs, respectively). Consequently, through specific spin mixing processes, the spin content of each sublevel, its decay rate κ, and thus N_{SS} and consequently PA become B dependent, yielding the $MPA_X(B) \equiv [PA_X(B) - PA_X(0)]/PA_X(0)$. Moreover, the optical cross section is spin sublevel dependent (spin quantum numbers are conserved in optical transitions in organics), and this may also affect $PA_X(B)$. In contrast, since it originates from SE, MPL(B) cannot originate from the SE $S = 0$ level (which is B independent); but instead is caused indirectly, through SE collision with TE.

7.4.1 MPA of PP and TE in the MEH-PPV System

The organic polymer "platform" used for the study of MPA and MPL responses is a derivative of poly(phenylene vinylene), namely, MEH-PPV (see backbone structure in Fig 7.13b), which is a well-known π-conjugated polymer [40]. MEH-PPV was used in three different forms, namely: pristine film; film exposed to prolonged UV illumination; and electron donor in MEH-PPV/PCBM blend having weight ratio 1:1. It is advantageous to use the polymer platform of MEH-PPV films since the MPA of both PP and TE can be experimentally studied using the same polymer film. When films of pristine MEH-PPV are kept in the dark for a long time they show fairly strong PL emission, and their PA spectrum consists of long-lived triplet excitons, PA_T (Fig. 7.14a); but do not show long-lived photogenerated polarons. However, if the films are exposed to prolonged UV illumination at low temperature, then a metastable state is formed in which the PA spectrum also contains substantial long-lived photogenerated polarons, having two characteristic PA bands (PA_P) that are formed at the expense of both PL and PA_T [50]. The process is reversible when the films are subjected to elevated temperatures in the dark.

The PA spectra of pristine MEH-PPV film at $B = 0$ and 100 mT are shown in Figure 7.14a. The PA spectrum consists of a broad band centered at ~1.37 eV (marked T) that is assigned to TE transition (PA_T) [50]; no other PA bands were obtained down to 0.2 eV. The $B = 100$ mT spectrum is identical in shape to that of $B = 0$, except that is slightly weaker. The difference, ΔPA, spectrum is similar to PA_T demonstrating that it relates to the TE density. As seen in Figure 7.14b the magnetic field response, $MPA_T(B) \equiv \Delta PA/PA_T$, varies strongly with the laser excitation intensity I_L, and

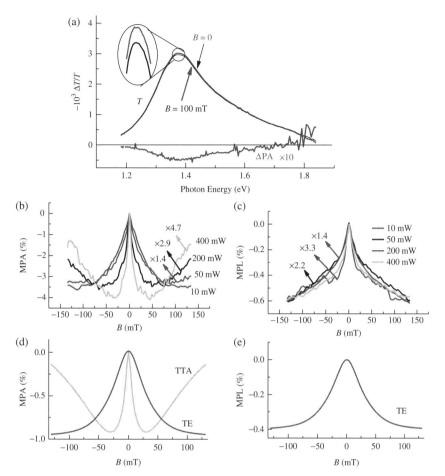

FIGURE 7.14 Excited state spectra and magnetic field effects in pristine MEH-PPV films. (a) The triplet PA band, PA_T at $B = 0$ and 100 mT (black and red lines, respectively), generated by $h\nu_L = 2.54$ eV at $I_L = 0.2$ W, and their difference $\Delta PA_T = [PA_T(100 \text{ mT}) - PA_T(0)]$ (blue line). The region near the peak is magnified (circle). (b) $MPA_T(B)$ measured at 1.37 eV for various laser excitation intensities (normalized). (c) $MPL(B)$ measured at 2.05 eV for various laser excitation intensities (normalized). (d) Model calculations of $MPA_T(B)$ using the TE (blue line, corresponds to the 10 mW data in (b)) and TTA (green line, corresponds to the 400 mW data in (b)) mechanisms (see text, Section 7.5.3). (e) Model calculation of $MPL(B)$ using the SE quenching model (see text, Section 7.5.2). *(For a color version of this figure, see the color plate section.)*

thus with N_{SS} ($\propto I_L$). N_{SS} is also inversely related to the sublevel TE recombination rate constants κ_α ($\alpha = 1, \ldots, L$), which are B dependent because the spin content of each sublevel varies with B (see Section 7.5.2). At small I_L $MPA_T(B)$ monotonically decreases, but it gradually transforms into a more complex response at large I_L, where two components are resolved: a low-field MPA component that decreases with B, and a high-field component that increases with B. It is thus concluded that $MPA_T(B)$ is dominated by two different spin-mixing mechanisms: one that dominates at low I_L and is therefore a "single-TE" mechanism; and the other that increases at large I_L, and therefore involves TTA mechanism.

The same pristine MEH-PPV film also shows MPL response. The PL spectrum consists of several vibronic replicas, with 0–0 transition at 2.05 eV assigned to the $S_1 \rightarrow S_0$ transition (Fig. 7.13a). Unlike MPA(B), however, Figure 7.14c shows that MPL(B) does not change with I_L; it monotonically decreases with B, similar to the low intensity $MPA_T(B)$ (i.e., low-field component). This MPL response was therefore assigned to SE nonradiative decay that is activated by collisions with TEs of which density $N_{SS}(B)$ also determines $MPA_T(B)$ response at low I_L.

Entirely different characteristic PA and MPA properties were measured in the same MEH-PPV film after prolonged UV irradiation (~150 min using a Xenon lamp at ~80 K), which supports photogenerated polaron species [50]. Figure 7.15a shows the PA spectrum of irradiated MEH-PPV film at $B = 0$ and 100 mT at similar I_L as used for the pristine film. The PA spectrum in this case consists of two broad PA bands: one centered at ~0.4 eV, which is assigned to the lower polaron transition (marked P_1), and the other is asymmetric with a peak at ~1.4 eV (marked $T + P_2$), which is composed of the polaron P_2 transition centered at ~1.55 eV and the remnant of the TE transition [50]. The spectrally resolved difference ΔPA (Fig. 7.15a) shows that MPA in this MEH-PPV form is correlated only with the two polaron PA bands, but not with that of PA_T. This is one of the MPA advantages: *its ability to spectrally resolve the dominant spin-dependent process*. The ΔPA spectrum was assigned to magnetic field dependence of the PP density. Unlike the negative ΔPA_T of the pristine sample (Fig. 7.14a), $\Delta PA_{PP} > 0$ in the irradiated sample, which points toward a different spin mixing mechanism altogether. The positive, monotonically increasing $MPA_{PP}(B)$ (Fig 7.15b), is naturally explained by the PP mechanism, in which the spin mixing is governed by the HFI [6] (see Section 7.5.1).

For comparison, MC(B) and MEL(B) obtained in MEH-PPV OLED are shown in Figure 7.15c. The two device MFE responses are identical to each other; and, in addition, *are very similar to the $MPA_{PP}(B)$ response* shown in Figure 7.15b. This indicates that all three MFE responses share a common origin. Since MPA(B) does not involve carrier transport, it is concluded that MC(B) and MEL(B) in OLED *need not involve transport*. All three responses can be explained equally well by the microscopic PP model presented in Section 7.5 below, that involves magnetic field dependence of the excitation spin sublevel character and their *density*, rather than transport related mechanism.

A salient feature of the low field ($B < 1.2$ mT) $MPA_{PP}(B)$ response is shown in Figure 7.15e. Interestingly, this response (dubbed ultra-small MPA, or USMPA) was measured at 1.1 eV, where the PA spectrum actually shows photo-bleaching (Fig. 7.15a). The 1.1 eV MPA is shown on a larger B-scale in Figure 7.15e (inset); it has, in fact the same response as MPA at 1.37 eV. The USMPA response decreases for $B < 0.6$ mT before increasing again to form the monotonic response at larger field. Similar non-monotonic response was previously observed in both MC(B) and MEL(B) in OLED [6,7] (see Section 7.2.2), and was explained as due to level crossing

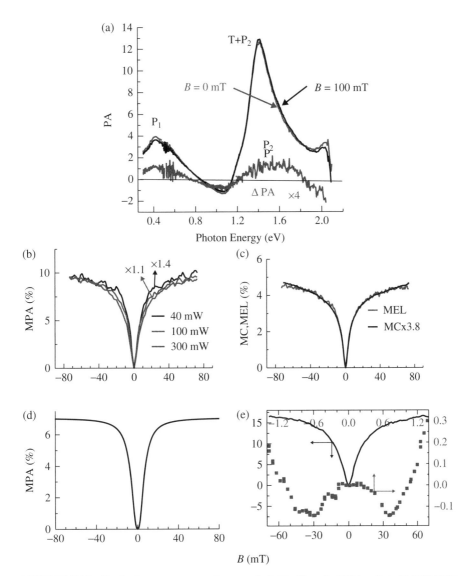

FIGURE 7.15 Excited state spectra and magnetic field effects in UV irradiated MEH-PPV film compared to MFE in OLED device. (a) PA spectrum at $I_L = 0.1$ W for $B = 0$ (black line) and $B = 100$ mT (red line) and their difference $\Delta PA = [PA(100\ mT) - PA(0)]$ (blue line) in MEH-PPV film. (b) MPA(B) measured at 1.4 eV for various laser excitation intensities (normalized). (c) MEL(B) and MC(B) in MEH-PPV OLED device. (d) Model calculations of $MPA_P(B)$ in films using the PP mechanism (see text, Section 7.5.1). (e) MPA(B) at 1.1 eV up to $B = 1.5$ mT (filled squares) and up to $B = 60$ mT (blue line, inset). *(For a color version of this figure, see the color plate section.)*

at $B = 0$ that involves spin sublevels formed by the polaron–proton HFI on the polymer chains. Thus, the same explanation is viable also for the USMPA component here. It is emphasized that the USMPA does not need an electronic device; in addition, it occurs at field values close to the *earth magnetic field* (≈ 0.05 mT). It is thus in principle possible that the USMPA could be used by a variety of living creatures on earth that may take advantage of the earth magnetic field to augment their activity; such as navigation, for example [18] (see also Section 7.3).

Yet a third type of MPA response is formed in films of MEH-PPV/PCBM blend. Upon photoexcitation of the polymer, the SE quickly dissociates into hole-polarons on the MEH-PPV chains and electron–polarons on the PCBM molecules [51], thereby weakening the PL intensity of pristine MEH-PPV, and completely eliminating the TE band from the PA spectrum [52]. The PA spectrum in this case (Fig. 7.16a) consists of positive polaron PA bands on the MEH-PPV chains (P_1 at ~0.4 eV and P_2 at ~1.37 eV), as well as negative polarons on PCBM (C_{61}^-) at ~1.2 eV. Importantly, the positive and negative polarons have different gyro-magnetic g-values [53], with $\Delta g \equiv g(\text{MEH-PPV}) - g(\text{PCBM}) \approx 3 \times 10^{-3}$; this happens since the P^+ and P^- excitations are separated in the blend onto two different environments.

ΔPA in the blend (Fig. 7.16a) is negative, and is assigned to PP transition of both positive and negative polarons. $\text{MPA}_{\text{PP}}(B)$ (Fig. 7.16b) has two components: a low-field component that sharply decreases with B, followed by a high-field component that increases slowly with B, forming an apparent minimum at $B < \sim 10$ mT. For comparison, MC(B) of a photovoltaic device based on the same blend is shown in Figure 7.16c, where again two MC(B) components are visible, except that the MC response is opposite in sign compared to that of MPA. The stunning similarity obtained between $\text{MPA}_{\text{PP}}(B)$ and MC(B) strongly indicates that they share the same underlying mechanism. Owing to the finite Δg for the positive and negative polarons in the blend, both $\text{MPA}_{\text{PP}}(B)$ and MC(B) (Fig. 7.16c) can be accounted for by the PP model that includes the HFI (low-field component) and Δg mechanism (high-field component), as explained in Section 7.5.1. Importantly, similar to the irradiated MEH-PPV films (Fig. 7.15e), a modulated response near $B \sim 0$ is also observed in the blend, as shown in Figure 7.16b (inset). Unlike the USMPA in the irradiated MEH-PPV film, however, in the blend the modulation occurs at much lower fields, $B \sim 0.1$ mT.

7.4.2 Magnetic Field Effect Spectroscopy of C_{60}-Based Films

The MFE has been interpreted as due to spin sensitive processes among pairs of spin bearing excitations [3,4,18,36,54,55]. In many organic semiconductors that contain carbon and hydrogen atoms but lack heavy atoms, the major spin-mixing mechanism has been shown [6, 9] to be the HFI between the protons (nuclear spin $I_H = 1/2$, nuclear g-factor $g_H = 5.585$) and the polaron electronic spin ($S = 1/2$, g-factor ≈ 2.00), with typical HFI coupling constant, $a_H \approx 0.3$ µeV [10]. In contrast, the buckeyball C_{60} molecule is composed of 60 carbon atoms (see Fig. 7.17b inset), of which 98.9% are the naturally abundant ^{12}C isotope having spinless nucleus, and thus zero HFI; and ~1.1% ^{13}C isotope ($I_{C13} = 1/2$, $g_{C13} = 1.405$) with estimated HFI constant

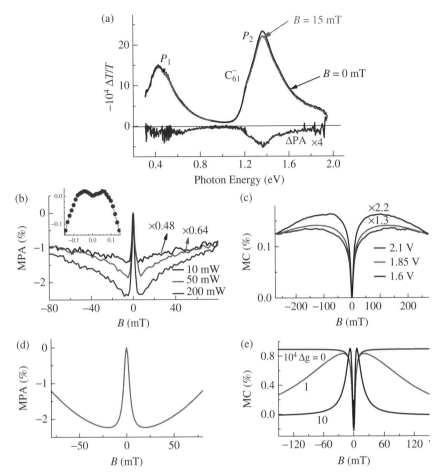

FIGURE 7.16 Excited state spectra and magnetic field effects in MEH-PPV/PCBM film and OLED device. (a) PA spectrum of MEH-PPV film at $I_L = 0.2$ W for $B = 0$ (black line) and $B = 15$ mT (red line) and their difference $\Delta PA = PA(15$ mT$) - PA(0)$ (blue line). (b) MPA(B) measured at 1.37 eV for various laser excitation intensities (normalized). Inset: High resolution data, showing USMPA peaks at $|B| \sim 0.1$ mT. This data was measured under conditions of complete apparatus shielding from the earth magnetic field and any stray field. (c) MC(B) in OLED at various bias voltages, V. (d) and (e) Model calculations of MPA$_{pp}$(B) and MC(B), respectively, using the "Δg + HFI" mechanism (see text, Section 7.5.1). *(For a color version of this figure, see the color plate section.)*

$a_{C13} \approx 0.1$ µeV [10]. Therefore, the HFI constant averaged over the 60 carbon atoms of natural C_{60} molecule should be $a_{C60} \approx 1$ neV (or $a_{C60}/g\mu_B \approx 10$ µT), which is too small to play any significant role in the MFE. Consequently, spin sensitive mechanisms other than the HFI become important for the MFE in fullerene films and devices. These mechanisms may involve [15] radical pairs, or equivalently polaron pairs (PP); or triplet excitons (TE).

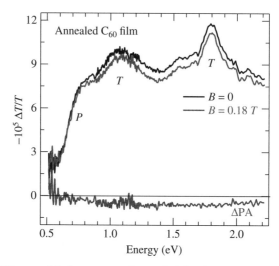

FIGURE 7.17 Photomodulation spectra of annealed C_{60} film at $T = 80$ K and $I_L = 0.2$ W cm^{-2} for $B = 0$ (black line) and $B = 180$ mT (red line). The blue negative line is the difference spectrum $\Delta PA = PA(B = 180$ mT$) - PA(0)$. *(For a color version of this figure, see the color plate section.)*

In Figure 7.17 the photo-modulation (PM) spectrum of an annealed C_{60} film at $B = 0$ (black line) using a laser excitation intensity $I_L = 0.2$ W cm^{-2} at 80 K is shown. The PM spectrum consists of two broad PA bands, E_1 and E_2 that are centered at $E \sim 1.1$ eV and $E \sim 1.8$ eV, respectively, and a low energy shoulder at $E \sim 0.8$ eV. As inferred from PA-detected magnetic resonance, PADMR, studies [56], the E_1 and E_2 bands probably consist of overlapping triplet bands (at ~ 1.1 and ~ 1.8 eV, respectively) and polaron bands (at ~ 1 and ~ 1.9 eV, respectively). The red line in Figure 7.17 shows the PA spectrum at $B = 180$ mT. At these I_L and B values the difference spectrum, $\Delta PA(B_1, B_2, E) \equiv PA(B_2, E) - PA(B_1, E)$, where $B_2 = 180$ mT and $B_1 = 0$, is negative as shown in Figure 7.17 (blue line). However, $\Delta PA(E)$ is very sensitive to B_2, B_1, and I_L, as shown in the next paragraphs.

In Figure 7.18a the magnetic field response, $MPA(B) \equiv \Delta PA/PA$ of the annealed C_{60} film measured at $E = 1.8$ eV is shown for various excitations, I_L. At small I_L, $MPA(B)$ is dominated by a relatively narrow negative component with full width at half maximum (FWHM) ~ 12 mT. At higher fields, however, a second MPA component may be resolved. This MPA component is much broader, and increases monotonically with B up to $B = 0.2$ T, which is the highest field employed here. Also, as I_L increases, the narrow component decreases from $\sim 16\%$ at 0.05 W cm^{-2} to less than 1% at 3 W cm^{-2}, whereas the broader component remains nearly unchanged. It is thus concluded that $MPA(B)$ is dominated by two different spin mixing mechanisms, perhaps related to two different photoexcitation species: one mechanism is responsible for the narrow component that decreases at large I_L, and the other mechanism that is characterized by a much broader $MPA(B)$ response is nearly insensitive to changes in I_L.

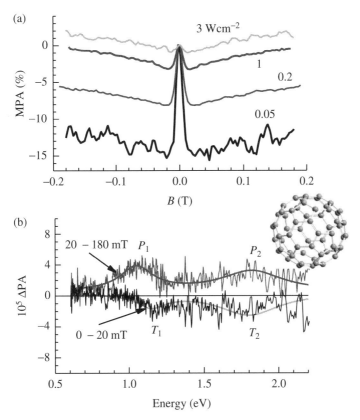

FIGURE 7.18 (a) MPA(B) response of an annealed C_{60} film at various pump excitation intensities, measured at photon energy $E = 1.8$ eV and $T = 80$ K. (b) The spectra $\Delta PA(B_1,B_2,E)$ for $B_1 = 0$, $B_2 = 20$ mT (black line, lower curves), and $B_1 = 20$ mT, $B_2 = 180$ mT (blue line, upper curves) for $I_L = 1.5$ W cm^{-2}. The smooth green and red lines through the data are to guide the eye, and show the TE- and polaron-related MPA bands, respectively. *(For a color version of this figure, see the color plate section.)*

In order to further separate the two MPA components, $\Delta PA(B_1,B_2,E)$ spectra of an annealed C_{60} film at laser excitation intensity $I_L = 1.5$ W cm^{-2} were measured and are shown in Figure 7.18b for two cases. (i) $B_1 = 0$, $B_2 = 20$ mT (negative MPA spectrum), which is sensitive mainly to the narrow MPA component. (ii) $B_1 = 20$ mT, $B_2 = 180$ mT (positive MPA spectrum), which unravels the broad MPA component. The two MPA spectra demonstrate two important differences: (a) The low-energy MPA band in Case (i) (T_1 at ~1.15 eV) is higher in energy than the low-energy MPA band in Case (ii) (P_1 at ~1.05 eV); (b) T_1 is weaker than T_2 in Case (i), whereas P_1 is stronger than P_2 in Case (ii). The two sets of MPA bands are in agreement with the two sets of PA bands obtained in PA versus modulation frequency and PADMR spectra measured by Dick et al [56], who identified the two PA spectra sets as due to polarons (P_1 and P_2) and triplet excitons (T_1 and T_2), respectively. Consequently, based on the agreement

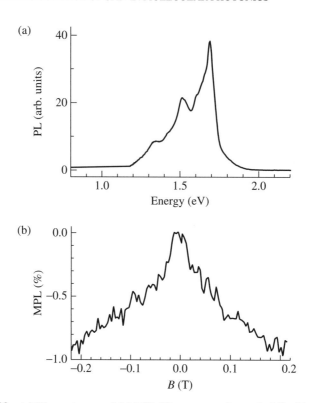

FIGURE 7.19 (a) PL spectrum and (b) MPL(B) response of annealed C_{60} film at $T = 80$ K.

between the previous PADMR and MPA spectra here, the narrow MPA component is assigned as due to TE, whereas the broad MPA component originates from PP species.

The MPL response in C_{60} films was also studied. The PL spectrum at 80 K is shown in Figure 7.19a; it was interpreted as due to radiative transitions of singlet excitons in C_{60}. The PL emission spectrum is composed of a 0–0 line at 1.7 eV, followed by two phonon replica ~0.18 eV apart, which is the energy of the C=C stretching vibration in C_{60}. But, since the singlet excitons in C_{60} are non-emissive, the PL is very weak. Consequently, to increase the system sensitivity the MPL(B) response of the entire PL band was measured using a broad band detector (Fig. 7.19b). Note that the MPL(B) response is also composed of narrow and broad components, typical of the MFE(B) in C_{60}; however, the narrow MPL(B) component is not easily discerned here.

7.5 BASIC QUANTUM MECHANICAL MODELS BASED ON SPIN-MIXING MANIPULATION BY MAGNETIC FIELDS

Because of the inherent very low mobility (<1 cm^2 V^{-1} s^{-1}) of organic semiconductors it is not likely that the MFE described in previous sections are due to Lorentz forces acting on moving charges [55]. The sensitivity of MFE to the magnetic moment of the nuclear isotopes [6], to triplet excitons [11], and to the atomic number in heavy metal containing organic semiconductors [57] point toward the electronic spin as the origin of the various MFE. It is generally accepted that the spin mixing in bound pairs of charge polarons (polaron-pair, PP) becomes less effective as the magnetic field increases, thereby causing a MFE of the measured physical quantity. The PP species consist of negative (P^-) and positive (P^+) polarons; whereas same-charge pairs constitute pairs of P^- or P^+, which may be termed as "loosely bound," doubly charged negative, or positive bipolarons (BP), respectively.

Prior to light emission from singlet and triplet exciton recombination in organic light emitting devices based on π-conjugated polymers, some of the injected free carriers form PP species, which constitute of P^- and P^+, separated by a distance of few nanometer, each having spin $1/2$. The free carriers and PP excitations are in dynamic equilibrium in the device active layer, which is determined by the processes of formation/dissociation, and recombination of PP via intrachain excitons. A PP excitation can be either in spin singlet (PP_S) or in spin triplet state (PP_T), depending on the mutual polarons' spin configuration. The steady state PP density depends on the PP_S and PP_T "effective rate constant," $\kappa_{S,T}$, which is the sum of the formation, dissociation, and recombination rate constants, as well as the triplet–singlet (T–S) mixing via the ISC interaction. If the effective rates κ_S for PP_S and κ_T for PP_T are not identical to each other, then any disturbance of the T–S mixing rate, such as by the application of an external magnetic field, B, would perturb the dynamical steady state equilibrium that results in MEL and MC. In the absence of an external magnetic field, the T–S mixing is caused by the HFI between the polaron spin and several adjacent H nuclei augmented by the exchange interaction between the unpaired spins of the polarons belonging to the PP species. For PP excitation in π-conjugated polymer chains, where the polarons are separated by a distance $R > \sim 1$–2 nm, the HFI with protons is expected to be dominant, and actually determines the T–S mixing rate.

The MFE in organic devices is, in many respects, analogous to the MFE observed in chemical and biochemical reactions, where it was explained using the radical pair (RP) mechanism [25, 58]. In this model the HFI, Zeeman, and exchange interactions are taken into account. It is assumed that the RPs are immobile, and hence the diffusion of the radicals is ignored; but the overall decay rate, κ, of the RP is explicitly taken into account. The steady-state singlet fraction of the RP population (singlet yield, Φ_S) is then calculated from the coherent time evolution of RP wave functions subjected to the above interactions. It is clear that when $\hbar\kappa$ is much larger than the HFI and exchange interactions, then the MFE is negligibly small, since the pairs disappear before any spin exchange between the spin sublevels can occur. In contrast, for a relatively small decay rate, the MFE becomes substantially larger, and for a negligible exchange interaction the singlet yield behaves as $\Phi_S \sim B^2/(B_0^2 + B^2)$ ("upside-down" Lorentzian function of B), with $B_0 \sim a_{HF}/g\,\mu_B$, where a_{HF} is the HFI constant. The half width at half maximum (HWHM) is then $B_{1/2} = B_0$. In organic devices, the PP species play the role of RPs, and the calculated MEL (and MC) response may then be expressed in terms of the singlet Φ_S and triplet Φ_T PP yields in an external magnetic field.

7.5.1 Few Spin Quantum Mechanical Approach

In this section we review simple models that involve few static electronic spins that can naturally account for MEL and other MFE phenomena. Starting with the simple spin Hamiltonian, H_0, appropriate for a pair of polarons, which includes the Zeeman, HFI, and exchange terms, we write:

$$H_0 = H_{\text{Zeeman}} + H_{\text{HF}} + H_{\text{ex}}. \tag{7.4}$$

In Eq. (7.4) H_{HF} is the HFI term,

$$H_{\text{HF}} = \sum_{i=1}^{2} \sum_{j=1}^{N_i} a_{ij} \vec{S}_i \cdot \vec{I}_{ij}, \tag{7.5}$$

where a_{ij} is the isotropic HFI and describing the HFI between polaron spin S_i ($= 1/2$) and N_i neighboring nuclei, each with spin I_{ij}. For protons in organic molecules the HFI constant is of the order of $a(H) \sim 0.3$ μeV (or $a/g\mu_B \sim 3$ mT) [10].

$$H_{\text{Zeeman}} = \mu_B \sum_{i=1}^{2} g_i \vec{S}_i \cdot \vec{B} \tag{7.6}$$

is the electronic Zeeman interaction term; g_i (~ 2) is the respective isotropic g-factor of each of the polarons in the PP; and μ_B is the Bohr magneton.

$$H_{\text{ex}} = J_{\text{ex}} \vec{S}_1 \cdot \vec{S}_2 \tag{7.7}$$

is the isotropic exchange interaction (for simplicity, scalar g-factors, scalar HFI, and scalar exchange interaction were chosen; anisotropic interactions are discussed in Reference 59).

The MEL, as well as the MC, is controlled by the relative singlet and triplet fractions and their spin-dependent decay, either by recombination or dissociation. The decay process is not contained in the spin Hamiltonian, Eq. (7.4), as it is a Hermitian operator that conserves energy. A convenient way to include the spin-dependent decay kinetics is to add to the Hermitian spin Hamiltonian H_0 a non-Hermitian decay (relaxation) term [25, 60]:

$$H_R = -\frac{i\hbar}{2} \sum_\alpha \kappa_\alpha P^\alpha, \tag{7.8}$$

where $\alpha = S, T$, $\kappa_{S,T}$, $P^{S,T}$ are the singlet (triplet) decay rates and projection operators, respectively. The time evolution of the density operator is now expressed in terms of the total Hamiltonian, $H = H_0 + H_R$,

$$\sigma(t) = \exp(-iHt/\hbar)\sigma(0)\exp(iH^\dagger t/\hbar), \tag{7.9}$$

BASIC QUANTUM MECHANICAL MODELS BASED ON SPIN-MIXING MANIPULATION   249

where $H^\dagger$ is the Hermitian conjugate of H and the $t = 0$ density matrix $\sigma(0)$ is determined by the PP generation process. The time-dependent singlet or triplet fraction may now be written as

$$\rho_\alpha(t) = Tr(P^\alpha \sigma(t)) = \frac{4}{M} \sum_{n,m} P^\alpha_{n,m} \sigma_{m,n}(0) \exp[i(E^*_m - E_n)t/\hbar] , \qquad (7.10)$$

where E_m, E_n are the (complex) eigenvalues of H (the asterisk denotes the complex conjugate), the double summation (n, m) is over all M states, $|n\rangle$ and $|m\rangle$, of H, and $\alpha = S, T$. Writing $E_n = \hbar(\omega_n - i\gamma_n)$ Eq. (7.10) becomes

$$\rho_\alpha(t) = \frac{4}{M} \sum_{n,m} P^\alpha_{n,m} \sigma_{m,n}(0) \cos(\omega_{mn}t) \exp(-\gamma_{mn}t) , \qquad (7.11)$$

where $\omega_{nm} = \omega_n - \omega_m$; $\gamma_{nm} = \gamma_n + \gamma_m$. Eq. (7.11) expresses the fact that because of the decay term (Eq. 7.8), the singlet and triplet densities diminish exponentially with time, each with its own rate. It is this unequal reduction in density of triplet and singlet PP that gives rise to either MC or MEL. The measured MEL may be calculated using Eq. (7.11), where the final expression depends on the radiative recombination of the SE and the detailed relaxation route from PP to the SE. For instance, in polymers where the SE–TE gap is relatively large (say, >10% of the SE energy), there is a substantial SE–TE ISC through the spin orbit coupling. As a result, PP^T (PP^S) may transform not only to TE (SE) but also to SE (TE). Let us denote the effective SE (TE) generation rates, from the PP$^\alpha$ ($\alpha = S, T$) configuration, as $\kappa_{\alpha, SE}$ ($\kappa_{\alpha, TE}$). Then, the steady state "SE generation yield" can be defined as $\Phi_{SE} = \Phi_{S,SE} + \Phi_{T,SE}$, where $\Phi_{\alpha,SE}$ is given by

$$\Phi_{\alpha SE} = \int_0^\infty \kappa_{\alpha SE} \rho_\alpha(t) dt = \frac{4}{M} \sum_{n,m} P^\alpha_{n,m} \sigma_{m,n}(0) \frac{\kappa_{\alpha SE} \gamma_{nm}}{\gamma^2_{nm} + \omega^2_{nm}} . \qquad (7.12)$$

Since the EL is proportional to the SE density, the MEL response is thus given by

$$\text{MEL}(B) = \frac{\Phi_{SE}(B) - \Phi_{SE}(0)}{\Phi_{SE}(0)} . \qquad (7.13)$$

The effect of the HFI is shown in Figure 7.20a where the MEL(B) response is calculated for $J_{ex} = 0$; $a_{HF} \sim 0.35$ μeV ($a_{HF}/2\mu_B \sim 3$ mT), $g_1 = g_2 = g = 2$ and relatively slow decay $\kappa_S = 0.003 a_{HF}/\hbar = 2\kappa_T$. The USMFE mentioned in Section 7.2.2 is readily visible at very low fields ($B \ll a/g\mu_B$, see also Fig. 7.15e) followed by a monotonic behavior, MEL$_M(B)$, at larger fields. Such a response is typical to MFE in a variety of π-conjugated polymers and small molecules (see experimental MEL data in Figs. 7.6, 7.8, 7.11, 7.15, and 7.16), that do not contain heavy atoms [5–8]. When the exchange interaction becomes of the order of the HFI or larger, the USMFE feature disappears, giving way to the "level crossing" (LC) effect [15] at the vicinity

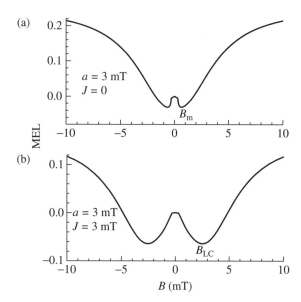

FIGURE 7.20 MEL(B) for PP with HFI and exchange. (a) HFI coupling constant $a/g\mu_B = 3$ mT, and exchange interaction coupling constant $J = 0$, showing the USMFE modulation with minimum at $B_m \sim 0.25 a/g\mu_B$. (b) $a/g\mu_B = 3$ mT, $J = a$, showing the level crossing modulation with minimum at $B_{LC} \sim J/g\mu_B$.

of $B = B_{LC} = J_{ex}/g\mu_B$. At $B = B_{LC}$ singlet and Zeeman split triplet levels cross each other, enhancing thereby the spin mixing due to the HFI, which otherwise is reduced as B increases. At $B \sim B_{LC}$ the MEL(B) response shows a modulation similar to that obtained in Figures 7.7, 7.8c, 7.11a and 7.12a for the USMFE, as shown in Figure 7.20b for $J_{ex} = a_{HF}$. Note that the exchange interaction by itself does not provide singlet/triplet spin mixing; thus the HFI spin-mixing mechanism (or other mechanisms) must be operative. The LC mechanism is less common in organic MEL where the main spin-mixing mechanisms occur within PP since with the relatively large polaron separation in PP ($d_{PP} > 1$–2 nm) the exchange interaction is too small. The LC mechanism may contribute more to the MEL response in charge transfer complexes [61], where d_{PP} is smaller.

Another important spin-mixing mechanism within PP is the "Δg mechanism" [12,13, 15, 40, 62]. When $\Delta g \equiv g_2 - g_1 \neq 0$ (Eq. 7.6) the electronic singlet S and triplet T_0 states mix coherently at a rate $\omega_{S-T_0} = \omega_S - \omega_{T_0}$. The rate of mixing is enhanced by the magnetic field, and for high magnetic fields $\omega_{S-T_0} \cong \Delta g \mu_B B/\hbar$. As long as this rate is smaller than the rate of the singlet/triplet decay, MFE(B) increases with B; but for much higher fields the effectiveness of the $S-T_0$ mixing is reduced [$\omega_{nm} \gg \gamma_{nm}$ in Eq. 7.11] and MFE(B) is diminishing as B is further increased. The effect of the Δg mechanism on MEL is readily obtained using Eqs. (7.4)–(7.13) and is shown for representative values in Figure 7.21 and in Figure 7.16e for MEHPPV/PCBM blend.

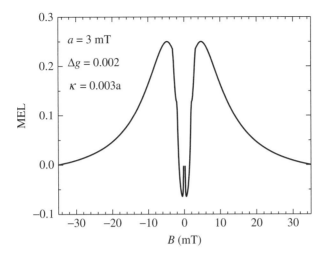

FIGURE 7.21 MEL(B) for PP with HFI and Δg. $a/g\mu_B = 3$ mT, $J = 0$, $\hbar\kappa_S = 2\,\hbar\kappa_T = 0.003a$ and $\Delta g = 0.002$ showing the effect of Δg with maximum at $\sim$5 mT.

The magneto-PA (MPA) can also be calculated using this approach. Referring to Figure 7.13a, in the PA process the X species (polarons or excitons) undergo a specific optical transition; for example, $X_0 \to X_1$. For PA, assuming that the optical cross section R is spin independent, Eq. (7.12) yields

$$\Phi_{PA} = R \sum_\alpha \int \rho_\alpha(t)dt = (2RL/M) \sum_n \sigma_{nn}(0)/\gamma_n \propto N_{SS}, \quad (7.14)$$

where $L = 4$ is the multiplicity of the pair of $S = \frac{1}{2}$ polarons in a PP and N_{SS} is the steady state density of the X species. The MPA response is thus given by an equation similar to Eq. (7.13):

$$\mathrm{MPA}(B) = \frac{\Phi_{PA}(B) - \Phi_{PA}(0)}{\Phi_{PA}(0)}. \quad (7.15)$$

Equations (7.14) and (7.15) show that the magnetic field dependence of the MPA arises from the field dependence of the spin sublevel decay rates: $\gamma_n = \gamma_n(B)$. This is the result of the different singlet/triplet decay rates ($k_S \neq k_T$) and the field dependence of the singlet/triplet content of each state.

Figures 7.22 and 7.15d show the calculated MPA(B) with representative parameters appropriate for UV irradiated MEHPPV.

7.5.2 Triplet Mechanism: Spin-Dependent Recombination

In this section we review triplet mechanisms that affect magneto-photoluminescence (MPL), MPA, and MEL through either spin-dependent decay in a SE–TE ISC process

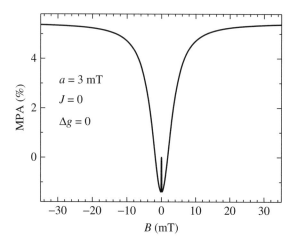

FIGURE 7.22 MPA(B) for PP with HFI. $a/g\mu_B = 3$ mT, $J = 0$, $\hbar\kappa_S = 0.8\,\hbar\kappa_T = 0.003a$ and $\Delta g = 0$ showing the USMFE near $B = 0$.

or delayed emission caused by TTA. The general form of a triplet exciton Hamiltonian may be written as [63]

$$H_T = \vec{S} \cdot \tilde{\tau} \cdot \vec{S}, \quad (7.16)$$

where $S = 1$ is the triplet exciton spin and $\tilde{\tau}$ is a symmetric traceless tensor of rank 2. As a traceless symmetric tensor, it is possible always to find a principal frame of reference (x', y', z') in which $\tilde{\tau}$ is diagonal with only two nonzero independent components, D and E, defined as follows:

$$D = \frac{3}{2}\tau_{z'z'}\,;\ E = \frac{1}{2}(\tau_{x'x'} - \tau_{y'y'})\,. \quad (7.17)$$

D and E are referred to as the "zero field splitting" (ZFS) parameters. In the laboratory frame of reference in which the z-axis makes polar angles (θ, φ) with the principal reference frame Σ', the five independent components of the triplet tensor τ become angle dependent [59] leading to the angular dependence of the observed MFE in organic crystals [17].

The corresponding spin Hamiltonian of a TE in a magnetic field, taking into account spin-dependent decay process, may be written as

$$H = H_T + H_Z + H_R, \quad (7.18)$$

where H_T is given by Eq. (7.16), $H_Z = g\mu_B \vec{S} \cdot \vec{B}$ is the Zeeman term, and the decay term H_R is given by Eq. (7.8) with $\alpha = 0, \pm 1$ for the corresponding 3 TE spin states. Using the ZFS parameters of the TE, the energy levels and wave functions of a TE in a magnetic field applied in a general direction can be calculated, yielding thus the TE steady state density (Eq. 7.14 with $L = 3$). Using Eq. (7.15) and averaging over all directions, the "powder pattern" MPA appropriate for amorphous organic semiconductor films can be obtained.

Using the ZFS parameters obtained in MEH-PPV by the PADMR technique ($D \approx 63$ mT and $E \approx 9$ mT, [64]), the calculated MPA(B) powder pattern is shown in Figure 7.14d (TE) for $\kappa_1 = \kappa_{-1} = 0.25\kappa_0 = 1.3 \times 10^7 \text{s}^{-1}$. Note that this model also explains MPL(B), since it is due to TE–SE interaction, which directly involves the TE density.

7.5.3 Triplet Mechanism: Triplet–Triplet Annihilation

In order to calculate TTA consider a pair of TEs. The spin Hamiltonian of the two TEs in a magnetic field $\vec{B}$ can be written as a simple sum:

$$H_{TT} = \sum_{n=1}^{2} [g_n \mu_B \vec{S}_n \cdot \vec{B} + \vec{S}_n \cdot \tilde{\tau}_n \cdot \vec{S}_n] . \quad (7.19)$$

(In Eq. 7.19 the TE–TE pairing interaction term has been omitted for simplicity). A pair of interacting triplets has total angular momentum of $\vec{S} = \vec{S}_1 + \vec{S}_2$; the configuration space of H_{TT} is of dimension 9 and for an interacting pair it is composed of a singlet ($S = 0$), triplet (T, $S = 1$) and quintet (Q, $S = 2$) [65].

The wave functions and energies in a magnetic field have been calculated explicitly for two identical and parallel TEs [17]. However, for a general case of non-identical and/or non-parallel TEs in a magnetic field of random orientation, the analytic formulae become cumbersome and numerical calculations are more practical. The TTA process is described schematically in the following equation:

$$T + T \underset{\kappa_{-P}}{\overset{\kappa_P}{\rightleftarrows}} (TT) \equiv (QTS) \overset{\kappa_S}{\rightarrow} \text{DF}. \quad (7.20)$$

In Eq. (7.20), two TEs ($T + T$) that were generated from the injected carriers via intermediate PPs, combine together at a rate κ_P, to form TE-pair (TT in Eq. 7.20). The TT pair is composed of quintet (^5Q), triplet (^3T), and singlet (^1S) states as described above. Out of the nine possible states and for $B = 0$, the spin species that are in the singlet state recombine at a rate κ_S to yield delayed fluorescence (DF); the other states either decay away or dissociate back to two separate TEs at a rate κ_{-P} (Eq. 7.20). As the magnetic field becomes finite, the spin character of each level varies causing B-dependent recombination and, consequently, B-dependent delayed EL.

As for the single TE case described by Eq. (7.18), a non-Hermitian relaxation term is added to the Hamiltonian H_{TT} (Eq. 7.20):

$$H_R = -\frac{i\hbar}{2}\kappa_S P^S - \frac{i\hbar}{2}\kappa_T P^T - \frac{i\hbar}{2}\kappa_Q P^Q , \quad (7.21)$$

where κ_α are the respective decay rate constants and P^α ($\alpha = S,T,Q$) are the respective projection operators. Following the procedure outlined in Eqs. 7.9–7.12, the yield of delayed singlet excitons, Φ_{DSE}, is given by

$$\Phi_{DSE} = \int_0^\infty \kappa_S \rho_S(t) dt = \sum_{n,m} P^S_{n,m} \sigma_{m,n}(0) \frac{\kappa_S \gamma_{nm}}{\gamma_{nm}^2 + \omega_{nm}^2}, \quad (7.22)$$

where the double summation, nm, is over the nine TT states. γ_{nm} and ω_{nm} are obtained by solving the TTA spin Hamiltonian: $H = H_{TT} + H_R$ (Eqs. 7.19 and 7.22). The MEL due to the magnetic field dependent DEL is then given by

$$\text{MEL}_{TT}(B) = \frac{\Phi_{DSE}(B) - \Phi_{DSE}(0)}{\Phi_{DSE}(0)}. \quad (7.23)$$

As for PP and single-TE, TTA also affects the triplet state PA. The MPA_{TT} is again given by Eqs. (7.14) and (7.15), where N_{SS} is determined by the individual decay rates of the quintet, triplet, and singlet states that are formed in the TE pair [17].

When pristine MEH-PPV films are subjected to high laser intensity (Fig. 7.14b, $I_L = 400$ mW), the TE density is sufficiently high that the TTA process needs be considered. First, the energy levels and wave functions of a pair of randomly oriented TEs in a magnetic field of a general direction were calculated. Subsequently, the powder pattern $\text{MPA}_{TT}(B)$ was calculated as shown in Figure 7.14d (TTA) for $\kappa_Q = \kappa_T = \kappa_S/30 = 10^6 \text{s}^{-1}$.

7.6 SUMMARY

In this review we have summarized the advance that our group made in the field of MFE in organic films and OLED devices. This field had originally started by measuring MC and MEL in OLED, but more recently other types of MFE have been shown to exist such as MPA, MPL in thin films, and magneto-photocurrent (MPC) in organic photovoltaic (OPV) solar cells [66]. We showed that at the heart of the MFE is spin manipulation by an external magnetic field. Most of the mechanisms that play an important role in MFE(B) response are based on two spins, either spin $\frac{1}{2}$ polaron pairs, spin 1 triplet pairs, or spin 1 paired with spin $\frac{1}{2}$ polaron (TE-polaron collision) leading to triplet–doublet pair. The spin energy levels in such pairs are populated either via carrier injection in OLED or laser excitation in films and OPV devices. Subsequently the spin population evolves in time according to the Hamiltonian that contains various interactions such as HFI, SOC, exchange, and Zeeman. The steady state spin populations are therefore sensitive to the applied field, and their field-dependent density can be monitored by a physical output that is field-dependent such as current density and electroluminescence in OLED, PA, and PL in films and PC in OPV solar cells. This leads to a magnetic field sensor that may be used to detect the magnitude and direction of very small fields (such as the Earth magnetic field).

We have given examples that show the powerful spectroscopy that can be achieved by the MPA method. The MPA was obtained on films of MEHPPV in various forms, including pristine, UV irradiated, and blend of polymer/fullerene used in OPV applications, as well as C_{60} films. Important information about the photoexcitation spin interaction was unraveled. We envision that the MPA method will evolve even further, by measuring its transient behavior in the time domain from picoseconds to tens of milliseconds. From this type of measurements the spin dynamics of photoexcitations will be readily revealed. We thus conclude that the future of MFE in organics is very promising.

ACKNOWLEDGMENTS

We thank L. Wojcik for the isotope-rich compounds, Dr. Tho D. Nguyen and Bhoj R. Gautam for various experiments. This work was supported over the years 2006–2013 by the DOE Grant No. DE-FG02-04ER46109 (ZVV), and from 2011 also by the NSF grant No. DMR-1104495, (ZVV); the Israel Science Foundation Grant No. ISF 472/11 (E.E.), and the US–Israel BSF Grant No. 2010135 (Z.V.V. and E.E.).

REFERENCES

[1] J. H. Burroughes, D. D. C. Bradley, A. R. Brown, R. N. Marks, K. Mackay, R. H. Friend, P. L. Burns, and A. B. Holmes, "Light-emitting diodes based on conjugated polymers," *Nature* **347**, 539–541 (1990).

[2] J. Kalinowski, M. Cocchi, D. Virgili, P. Di Marco, and V. Fattori, "Magnetic field effects on emission and current in Alq_3-based electroluminescent diodes," *Chem. Phys. Lett.* **380**, 710–715 (2003).

[3] O. Mermer, G. Veeraraghavan, T. L. Francis, Y. Sheng, D. T. Nguyen, M. Wohlgenannt, A. Kohler, M. K. Al-Suti, and M. S. Khan, "Large magnetoresistance in nonmagnetic p-conjugated semiconductor thin film devices," *Phys. Rev. B* **72**, 205202 (2005).

[4] P. A. Bobbert, T. D. Nguyen, F. W. A. van-Oost, B. Koopmans, and M. Wohlgenannt, "Bipolaron mechanism for organic magnetoresistance," *Phys. Rev. Lett.* **99**, 216801 (2007).

[5] T. D. Nguyen, B. R. Gautam, E. Ehrenfreund, and Z. V. Vardeny, "Magnetoconductance response in unipolar and bipolar organic diodes at ultrasmall fields," *Phys Rev Lett.* **105**, 166804 (2010).

[6] T. D. Nguyen, G. Hukic-Markosian, F. J. Wang, L. Wojcik, X. G. Li, E. Ehrenfreund, and Z. V. Vardeny, "Isotope effect in spin response of p-conjugated polymer films and devices," *Nat. Mater.* **9**, 345–352 (2010).

[7] T. D. Nguyen, B. R. Gautam, E. Ehrenfreund, and Z. V. Vardeny, "Magneto-conductance of unipolar and bipolar diodes based on p-conjugated polymer," *Synth. Met.* **161**, 604–607 (2011).

[8] T. D. Nguyen, G. Hukic-Markosian, F. Wang, X.-G. Li, E. Ehrenfreund, and Z. V. Vardeny, "The hyperfine interaction role in the spin response of p-conjugated polymer films and spin valve devices," *Synth. Met.* **161**, 598–603 (2011).

[9] T. D. Nguyen, T. P. Basel, Y.-J. Pu, X.-G. Li, E. Ehrenfreund, and Z. V. Vardeny, "Isotope effect in the spin response of aluminum tris(8-hydroxyquinoline based devices," *Phys. Rev. B* **85**, 245437 (2012).

[10] A. Carrington and A. D. McLachlan, *Introduction to Magnetic Resonance* (Harper & Row, New York, 1967).

[11] P. Desai, P. Shakya, T. Kreouzis, W. P. Gillin, N. A. Morley, and M. R. J. Gibbs, "Magnetoresistance and efficiency measurements of Alq_3-based OLEDs," *Phys. Rev. B* **75**, 094423 (2007).

[12] Y. Tanimoto, H. Hayashi, S. Nagakura, H. Sakuragi, and K. Tokumaru, "The external magnetic field effect on the singlet sensitized photolysis of dibenzoyl peroxide," *Chem. Phys. Lett.* **41**, 267–269 (1976).

[13] K. Schulten and I. R. Epstein, "Recombination of radical pairs in high magnetic fields: a path integral—Monte Carlo treatment," *J. Chem. Phys.* **71**, 309–316 (1979).

[14] U. E. Steiner and T. Ulrich, "Magnetic field effects in chemical kinetics and related phenomena," *Chem. Rev.* **89**, 51–147 (1989).

[15] H. Hayashi, *Introduction to Dynamic Spin Chemistry; Magnetic Field Effects on Chemical and Biochemical Reactions* (World Scientific Publishing Co., Singapore, 2004), Vol. **8**.

[16] S. Sinha, C. Rothe, R. Güntner, U. Scherf, and A. P. Monkman, "Electrophosphorescence and delayed electroluminescence from pristine polyfluorene thin-film devices at low temperature," *Phys. Rev. Lett.* **90**, 127402 (2003).

[17] R. E. Merrifield, "Magnetic effects on triplet exciton interactions," *Pure Appl. Chem.* **27**, 481–498 (1971).

[18] J. D. Bergeson, V. N. Prigodin, D. M. Lincoln, and A. J. Epstein, "Inversion of magnetoresistance in organic semiconductors," *Phys. Rev. Lett.* **100**, 067201 (2008).

[19] F. L. Bloom, W. Wagemans, M. Kemerink, and B. Koopmans, "Separating positive and negative magnetoresistance in organic semiconductor devices," *Phys. Rev. Lett.* **99**, 257201 (2007).

[20] B. Hu and Y. Wu, "Tuning magnetoresistance between positive and negative values in organic semiconductors," *Nat. Mater.* **6**, 985–991 (2007).

[21] F. J. Wang, H. Bassler, and Z. V. Vardeny, "Magnetic field effects in p-conjugated polymer/fullerene blends: evidence for multiple components," *Phys. Rev. Lett.* **101**, 236805 (2008).

[22] Y. Wu, Z. Xu, B. Hu, and J. Howe, "Tuning magnetoresistance and magnetic-field-dependent electro-luminescence through mixing a strong-spin-orbital-coupling molecule and a weak-spin-orbital-coupling polymer," *Phys. Rev. B* **75**, 035214 (2007).

[23] C. G. Yang, E. Ehrenfreund, F. Wang, T. Drori, and Z. V. Vardeny, "Spin-dependent kinetics of polaron pairs in organic light-emitting diodes studied by electroluminescence detected magnetic resonance dynamics," *Phys. Rev. B* **78**, 205312 (2008).

[24] R. P. Groff, A. Suna, P. Avakian, and R. E. Merrifield, "Magnetic hyperfine modulation of dye-sensitized delayed fluorescence in organic crystals," *Phys. Rev. B* **9**, 2655–2660 (1974).

[25] C. R. Timmel, U. Till, B. Brocklehurst, K. A. McLauchlan, and P. J. Hore, "Effects of weak magnetic fields on free radical recombination reactions," *Mol. Phys.* **95**, 71–89 (1998).

[26] A. Weller, F. Nolting, and H. Staerk, "A quantitative interpretation of the magnetic field effect on hyperfine-coupling-induced triplet formation from radical ion pairs," *Chem. Phys. Lett.* **96**, 24–27 (1983).

[27] B. Brocklehurst and K. A. McLauchlan, "Free radical mechanism for the effects of environmental electromagnetic fields on biological systems," *Int. J. Radiat. Biol.* **69**, 3–24 (1996).

[28] R. Belaid, T. Barhoumi, L. Hachani, L. Hassine, and H. Bouchriha, "Magnetic field effect on recombination light in anthracene crystal," *Synth. Met.* **131**, 23–30 (2002).

[29] Y. Sheng, T. D. Nguyen, G. Veeraraghavan, O. Mermer, M. Wohlgenannt, S. Qiu, and U. Scherf, "Hyperfine interaction and magnetoresistance in organic semiconductors," *Phys. Rev. B* **74**, 045213 (2006).

[30] C. W. Tang and S. A. VanSlyke, "Organic electroluminescent diodes," *Appl. Phys. Lett.* **51**, 913–915 (1987).

[31] C. W. Tang, S. A. VanSlyke, and C. H. Chen, "Electroluminescence of doped organic thin films," *J. Appl. Phys.* **65**, 3610–3616 (1989).

[32] W. Li, R. A. Jones, S. C. Allen, J. C. Heikenfeld, and A. J. Steckl, "Maximizing Alq$_3$ OLED internal and external efficiencies: charge balanced device structure and color conversion outcoupling lenses," *J. Disp. Technol.* **2**, 143–152 (2006).

[33] C. Gärditz, A. G. Mückl, and M. Cölle, "Influence of an external magnetic field on the singlet and triplet emissions of tris-(8-hydroxyquinoline) aluminumIII(Alq3)," *J. Appl. Phys.* **98**, 104507 (2005).

[34] N. J. Rolfe, M. Heeney, P. B. Wyatt, A. J. Drew, T. Kreouzis, and W. P. Gillin, "Elucidating the role of hyperfine interactions on organic magnetoresistance using deuterated aluminium tris(8-hydroxyquinoline)," *Phys. Rev. B* **80**, 241201R (2009).

[35] J. A. Gómez, F. Nüesch, L. Zuppiroli, and C. F. O. Graeff, "Magnetic field effects on the conductivity of organic bipolar and unipolar devices at room temperature," *Synth. Met.* **160**, 317–319 (2010).

[36] B. Koopmans, W. Wagemans, F. L. Bloom, P. A. Bobbert, M. Kemerink, and M. Wohlgenannt, "Spin in organics: a new route to spintronics," *Phil. Trans. R. Soc. A* **369** 3602–3616 (2011).

[37] N. J. Rolfe, M. Heeney, P. B. Wyatt, A. J. Drew, T. Kreouzis, and W. P. Gillin, "The effect of deuteration on organic magnetoresistance," *Synth. Met.* **161**, 608–611 (2011).

[38] M. Cölle, C. Gärditz, and M. Braun, "The triplet state in tris-(8-hydroxyquinoline) aluminum," *J. Appl. Phys.* **96**, 6133–6141 (2004).

[39] H. D. Burrows, M. Fernandes, J. S. de-Melo, A. P. Monkman, and S. Navaratnam, "Characterization of the triplet state of tris(8-hydroxyquinoline)aluminium(III) in benzene solution," *J. Am. Chem. Soc.* **125**, 15310–15311 (2003).

[40] B. R. Gautam, T. D. Nguyen, E. Ehrenfreund, and Z. V. Vardeny, "Magnetic field effect on excited-state spectroscopies of π-conjugated polymer films," *Phys. Rev. B* **85**, 205207 (2012).

[41] T. Ritz, P. Thalau, J. B. Phillips, R. Wiltschko, and W. Wiltschko, "Resonance effects indicate a radical-pair mechanism for avian magnetic compass," *Nature* **429**, 177–180 (2004).

[42] K. Maeda, K. B. Henbest, F. Cintolesi, I. Kuprov, C. T. Rodgers, P. A. Liddell, D. Gust, C. R. Timmel, and P. J. Hore, "Chemical compass model of avian magnetoreception," *Nature* **453**, 387–390 (2008).

[43] T. D. Nguyen, E. Ehrenfreund, and Z. V. Vardeny, "Organic magneto-resistance at small magnetic fields; compass effect," *Org. Electron.* **14**, 1852–1855 (2013).

[44] J. L. Kirschvink, M. M. Walker, and C. E. Diebel, "Magnetite-based magnetoreception," *Curr. Opin. Neurobiol.* **11**, 462–467 (2011).

[45] G. Fleissner, B. Stahl, P. Thalau, G. Falkenberg, and G. Fleissner, "A novel concept of Fe-mineral-based magnetoreception: histological and physicochemical data from the upper beak of homing pigeons," *Naturwissenschaften* **94**, 631–642 (2007).

[46] C. T. Rodgers and P. J. Hore, "Chemical magnetoreception in birds: the radical pair mechanism," *Proc. Nat. Acad. Sci.* **106**, 353–360 (2009).

[47] T. D. Nguyen, J. Rybicki, Y. Sheng, and M. Wohlgenannt, "Device spectroscopy of magnetic field effects in a polyfluorene organic light-emitting diode," *Phys. Rev. B* **77**, 035210 (2008).

[48] B. Hu, L. Yan, and M. Shao, "Magnetic-field effects in organic semiconducting materials and devices," *Adv. Mater.* **21**, 1500–1516 (2009).

[49] F. Wang, J. Rybicki, K. A. Hutchinson, and M. Wohlgenannt, "Magnetic-field effect in organic photoconductive devices studied by time-of-flight," *Phys. Rev. B* **83**, 241202 (2011).

[50] T. Drori, E. Gershman, C. X. Sheng, Y. Eichen, Z. V. Vardeny, and E. Ehrenfreund, "Illumination-induced metastable polaron-supporting state in poly(p-phenylene vinylene) films," *Phys. Rev. B* **76**, 033203 (2007).

[51] N. S. Sariciftci, L. Smilowitz, A. J. Heeger, and F. Wudl, "Photoinduced electron transfer from a conducting polymer to buckminsterfullerene," *Science* **258**, 1474–1476 (1992).

[52] X. Wei, Z. V. Vardeny, N. S. Sariciftci, and A. J. Heeger, "Absorption-detected magnetic-resonance studies of photoexcitations in conjugated-polymer/C_{60} composites," *Phys. Rev. B* **53**, 2187–2190 (1996).

[53] A. Konkin, U. Ritter, P. Scharff, H. K. Roth, A. Aganov, N. S. Sariciftci, and D. A. M. Egbe, "Photo-induced charge separation process in (PCBM-C_{120}O)/(M3EH-PPV) blend solid film studied by means of X and K-bands ESR at 77 and 120 K," *Synth. Met.* **160**, 485–489 (2010).

[54] T. L. Francis, Ö. Mermer, G. Veeraraghavan, and M. Wohlgenannt, "Large magnetoresistance at room temperature in semiconducting polymer sandwich devices," *New J. Phys.* **6**, 185 (2004).

[55] V. Prigodin, J. Bergeson, D. Lincoln, and A. J. Epstein, "Anomalous room temperature magnetoresistance in organic semiconductors," *Synth. Met.* **156**, 757–761 (2006).

[56] D. Dick, X. Wei, S. Jeglinski, R. E. Benner, Z. V. Vardeny, D. Moses, V. I. Srdanov, and F. Wudl, "Transient spectroscopy of excitons and polarons in C_{60} films from femtoseconds to milliseconds," *Phys. Rev. Lett.* **73**, 2760–2763 (1994).

[57] I. V. Khudyakov, Y. A. Serebrennikov, and N. J. Turro, "Spin-orbit coupling in free-radical reactions: on the way to heavy elements," *Chem. Rev.* **93**, 537–570 (1993).

[58] B. Brocklehurst and K. A. McLauchlan, "Free radical mechanism for the effects of environmental electromagnetic fields on biological systems," *Int. J. Radiat. Biol.* **69**, 3–24 (1996).

[59] E. Ehrenfreund and Z. V. Vardeny, "Effects of magnetic field on conductance and electroluminescence in organic devices," *Israel J. Chem.* **52**, 552–562 (2012).

[60] J. Tang and J. R. Norris, "Theoretical calculations of kinetics of the radical pair PF state in bacterial photosynthesis," *Chem. Phys. Lett.* **92**, 136–140 (1982).

[61] T. M. Clarke and J. R. Durrant, "Charge photogeneration in organic solar cells," *Chem. Rev.* **110**, 6736–6767 (2010).

[62] N. J. Turro, "Influence of nuclear spin on chemical reactions: magnetic isotope and magnetic field effects," *Proc. Natl. Acad. Sci. USA* **80**, 609–621 (1983).

[63] A. Abragam and B. Bleaney, *Electron Paramagnetic Resonance of Transition Ions* (Clarendon Press, Oxford, 1970).

[64] X. Wei, B. C. Hess, Z. V. Vardeny, and F. Wudl: Studies of photoexcited states in polyacetylene and poly(paraphenylenevinylene) by absorption detected magnetic resonance: the case of neutral photoexcitations," *Phys. Rev. Lett.* **68**, 666–669 (1992).

[65] C. Cohen-Tannoudji, B. Diu, and F. Laloe, *Quantum Mechanics* (John Wiley & Sons, New York, 1977), Vol. **2**.

[66] Y. Zhang, T. P. Basel, X. Yang, D. J. Mascaro, F. Liu, and Z. V. Vardeny, "Spin-enhanced organic bulk heterojunction photovoltaic solar cells," *Nature Commun.* **3**, 1043 (2012).

8

THIN-FILM MOLECULAR NANOPHOTONICS

TETSUZO YOSHIMURA

School of Computer Science, Tokyo University of Technology, Tokyo, Japan

8.1 INTRODUCTION

Organic materials have an outstanding feature that they can be natural two-dimensional (2D) systems like graphene [1], one-dimensional (1D) systems like carbon nanotubes [2, 3] and conjugated polymers [4], and zero-dimensional (0D) systems that are individual molecules themselves [5]. Since photonic and electronic properties of materials are greatly affected by dimensionality, the organic materials have the potential to provide a variety of novel devices.

In order to reach the final goal of photonic/electronic materials, however, it is necessary to control electron wavefunction shapes within the materials in addition to the dimensionality. To do this, we have developed the molecular layer deposition (MLD) [6–8], which performs molecular assembling with monomolecular steps to grow tailored organic thin films with nanoscale structures.

Organic materials have another feature such that dynamics of molecular assembling can be controlled by light beams to construct self-organized nanoscale structures. One of the examples is the self-organized lightwave network (SOLNET), which can form self-aligned coupling waveguides between misaligned optical devices three dimensionally [9–12].

In the present chapter, two subjects related to molecular assembling are reviewed. The first is the "static molecular assembling" by MLD. After concept of MLD and the organic multiple quantum dot (MQD), which is one of the nanoscale structures grown

---

Photonics: Scientific Foundations, Technology and Applications, Volume II, First Edition.
Edited by David L. Andrews.
© 2015 John Wiley & Sons, Inc. Published 2015 by John Wiley & Sons, Inc.

by MLD, are described, experimental results are presented to discuss the quantum confinement effect of electrons and the sensitizing effect in the organic MQD. The second is the "dynamic molecular assembling," which achieves SOLNET. Concept of SOLNET and theoretical and experimental demonstrations for self-organization of optical waveguides targeting reflective or luminescent destinations are described. Finally, potential applications of MLD and SOLNET to photonics are proposed, including sensitized solar cells, artificial photosynthesis, three-dimensional (3D) integrated optical interconnects within computers, electro-optic (EO) thin films, cancer therapy, and molecular circuits.

8.2 MOLECULAR ASSEMBLING FOR NANOSCALE TAILORED STRUCTURES

8.2.1 Static Molecular Assembling

In organic thin films, dimensionality is easily controlled, especially, between 0D and 1D as schematically illustrated in Figure 8.1. A molecule is regarded as a very short quantum dot with a length of 0.1-nm order, in which electron wavefunctions are tightly confined. When the molecules are connected by molecules with larger energy gaps, which are regarded as barriers, to form a polymer wire, a 0D material consisting of short quantum dots is fabricated. When the molecules of short quantum dots are connected by double or triple bonds to spread the electron wavefunctions, a 0.5D material consisting of long quantum dots is fabricated. When the molecules are connected by double or triple bonds further, an 1D material of a conjugated polymer, namely, a quantum wire, is fabricated.

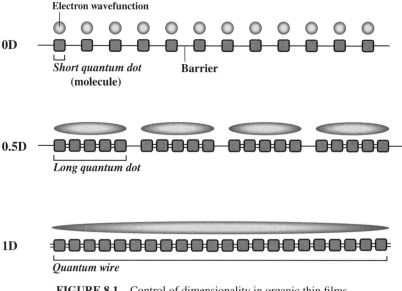

FIGURE 8.1 Control of dimensionality in organic thin films.

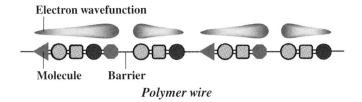

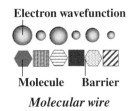

FIGURE 8.2 Control of electron wavefunction shapes in organic thin films.

In order to optimize photonic/electronic properties, electron wavefunction shapes should be controlled within the quantum dots by placing constituent molecules in designated arrangements [13, 14]. For example, as shown in Figure 8.2, non-centrosymmetric wavefunctions can be realized in the quantum dots in a polymer wire by controlling the molecular sequences. The similar wavefunction control may also be achieved in a molecular wire consisting of molecules connected to each other by the electrostatic force. In this case, the individual molecules are regarded as quantum dots and the interfaces between the molecules as barriers.

A process to realize the static molecular assembling is MLD, which enables monomolecular-step growth as described in Section 8.3. The organic MQD described in Section 8.4 is a typical nanoscale structure realized by MLD.

8.2.2 Dynamic Molecular Assembling

As demonstrated in the photopolymer developed by DuPont for holograms [15], diffusion of molecules is induced by light beam exposure to form nanoscale patterns of refractive index. The model is shown in Figure 8.3. The photopolymer contains photoreactive molecules. When it is exposed to write beams, the molecules are combined by the photochemical reactions. Then, the molecules in the surrounding regions diffuse into the exposed region to compensate for the concentration reduction of the photoreactive molecules. This process produces inhomogeneous distribution of the photoreactive molecules, namely, inhomogeneous refractive index distribution according to the pattern of the write beam.

Such a dynamic molecular assembling process enables SOLNET described in Section 8.5.

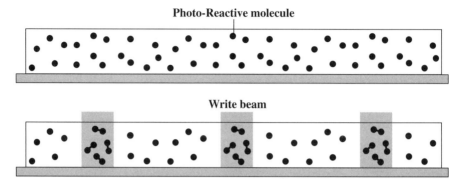

FIGURE 8.3 Control of diffusion of molecules by light beam exposure for nanoscale pattern formation.

8.3 MOLECULAR LAYER DEPOSITION

Tailored materials having designated nanoscale structures grown with atomic/molecular-level control are the final goal of material/device science. So far, a lot of atomic assembling methods have been developed, such as the molecular beam epitaxy (MBE) [16] and the atomic layer deposition (ALD) [17], to build inorganic thin films in a "layer-by-layer" manner.

Meanwhile, MLD is a molecular assembling method to build up organic thin films with designated molecular sequences as shown in Figure 8.4. The concept was inspired by combining the concept of ALD with that of the vacuum deposition polymerization [18, 19], which is a method of polymer film deposition in vacuum. In MLD, the source molecules are designed so that the same molecules cannot be combined while different molecules can be combined by utilizing selective chemical reactions [6–8] or electrostatic force [8, 20, 21].

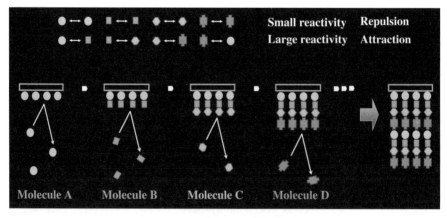

FIGURE 8.4 Concept of MLD. From Reference 20.

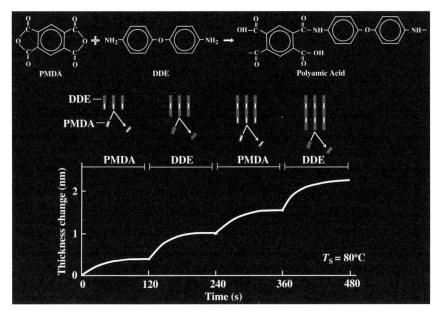

FIGURE 8.5 Experimental demonstration of MLD using PMDA and DDE. From Reference 6. *(For a color version of this figure, see the color plate section.)*

When Molecule A is provided onto a substrate surface, a monomolecular layer of A is formed. Once the surface is covered with A, the deposition of the molecules is automatically terminated by the self-limiting effect similarly to that in ALD [17]. By switching source molecules sequentially as A, B, C, D, ..., thin films with a molecular sequence of A-B-C-D-... are obtained.

Source molecules are usually supplied in the gas phase [6–8]. However, when they are hard to be evaporated, catalysts are required to combine them, or they are difficult to be supplied in the gas phase into objects, the liquid-phase MLD (LP-MLD) is used [8, 20, 21].

Figure 8.5 shows an experimental demonstration of MLD using pyromellitic dianhydride (PMDA) and 4,4′-diaminodiphenyl ether (DDE) as source molecules [6]. The molecules are connected by reactions between the carbonyl-oxy-carbonyl group in PMDA and the amino group in DDE to form polyamic acid. When PMDA is supplied to a DDE surface, the film thickness rapidly increases and saturates. When the source molecule is switched from PMDA to DDE, again, the film thickness rapidly increases and saturates. By repeating the switching operation of the source molecules, step-like film growth is achieved. The change in thickness is 0.4–0.6 nm in each growth step, close to the source molecule sizes. This result indicates that the monomolecular-step growth of polyamic acid is definitely performed by MLD.

It should be noted that MLD can be performed using various kinds of reactive groups such as −CHO, −NCO, −OH, and epoxy ring, as well as the carbonyl-oxy-carbonyl group and the amino group. As described in the early proposal of MLD [22, 23], halogen reactions and photo-assisted processes are also applicable.

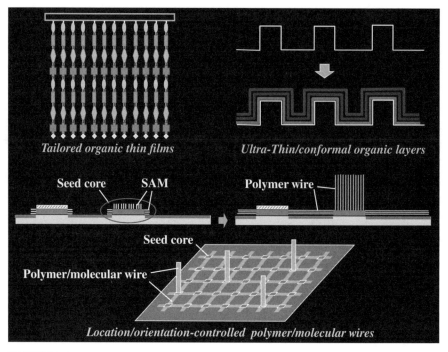

FIGURE 8.6 Expected structures grown by MLD. *(For a color version of this figure, see the color plate section.)*

As summarized in Figure 8.6, MLD can grow tailored organic thin films [7, 8] and ultrathin/conformal organic layers [7, 24] on 3D surfaces such as deep trenches, porous layers, particles. Furthermore, by using seed cores, MLD can achieve 3D growth with location and orientation control [8, 25, 26]. For example, self-assembled monolayers (SAMs) on the top of the seed cores initiate vertical growth, and SAMs on the sidewalls initiate horizontal growth. By distributing the seed cores with designated patterns, polymer/molecular wires will grow with designated configurations, constructing wire networks.

Due to these features, MLD has been applied to various photonic and electronic devices [7, 8], including copper diffusion barriers in large-scale integrated circuits (LSIs) [27], photoresists [28], nanoparticle coatings [29], polyimide thin films for microelectronics [30], conductivity-switching polyimide films [31], photovoltaic devices [8, 20, 32], and optical switches [7, 8, 13, 14]. The use of MLD in cancer therapy has also been proposed [8, 20, 33], in which the human body and cancer cells are respectively the MLD chamber and the substrates.

George et al. combined MLD and ALD to grow a new class of advanced materials; hybrid organic–inorganic polymers [34], which are applied to organic thin-film transistors (TFTs) [35], thin-film encapsulation for organic light emitting diodes (OLEDs) [36], and microelectromechanical systems (MEMSs) and batteries [37].

8.4 ORGANIC MULTIPLE QUANTUM DOTS (MQDs)

One of the tailored nanoscale structures constructed by MLD is the organic MQDs such as polymer MQDs [5, 14, 32] and molecular MQDs [8, 20]. The polymer MQDs are polymer wires containing quantum dots, and the molecular MQDs are molecular wires containing quantum dots.

In this section, growth processes and structures of the organic MQDs are described, and experimental results of absorption spectra, photoluminescence (PL) spectra, surface potential, and photocurrents are presented to discuss the constructed MQD structures and electron confinement/transport in the MQDs.

8.4.1 Polymer MQDs

8.4.1.1 Growth Processes and Structures For polymer MQD fabrication, we used the carrier-gas type MLD [25, 26, 32] schematically shown in Figure 8.7. Source molecules were carried from temperature-controlled molecular cells into the MLD chamber by N_2 carrier gas to give gas blows onto a glass substrate surface. MLD operation was carried out by sequentially opening/closing the valves. Excess molecules were removed with the carrier gas using a rotary pump.

In the present experiment, three source molecules [32] shown in Figure 8.8, terephthalaldehyde (TPA), *p*-phenylenediamine (PPDA), and oxalic dihydrazide (ODH), were used. TPA and PPDA are connected with a double bond, allowing π-electron wavefunctions to be delocalized through the bond. PPDA and ODH are connected with a plurality of single bonds, which block π-electron wavefunctions. Various kinds of polymer MQDs can be constructed using these bond characteristics. Figure 8.9 shows an example of MLD process, in which source molecules are connected in a sequence of -ODH-TPA-PPDA-TPA-ODH-, to construct the polymer MQD of OTPT.

In Figure 8.10, electronic structures of polymer MQDs of OTPTPT, OTPT, OT, and 3QD, as well as a poly-azomethine (AM) quantum wire, are shown. In the

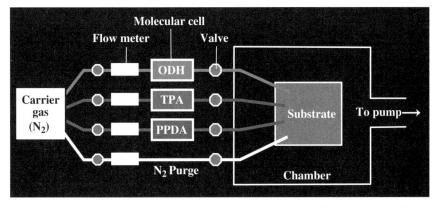

FIGURE 8.7 Schematic illustration of the carrier-gas type MLD. From Reference 32. *(For a color version of this figure, see the color plate section.)*

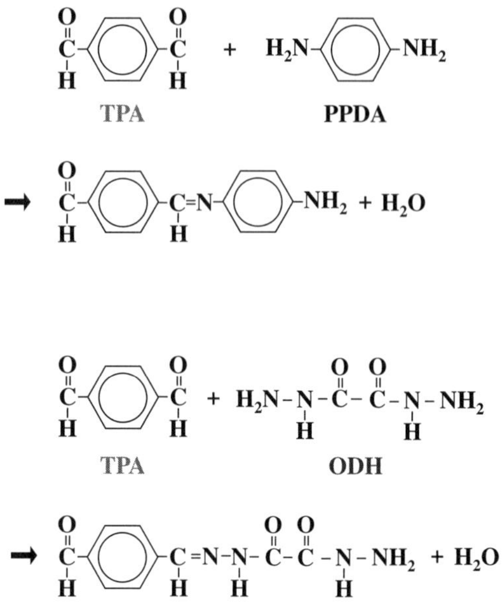

FIGURE 8.8 Source molecules and chemical reactions for growth of polymer MQDs. From Reference 32.

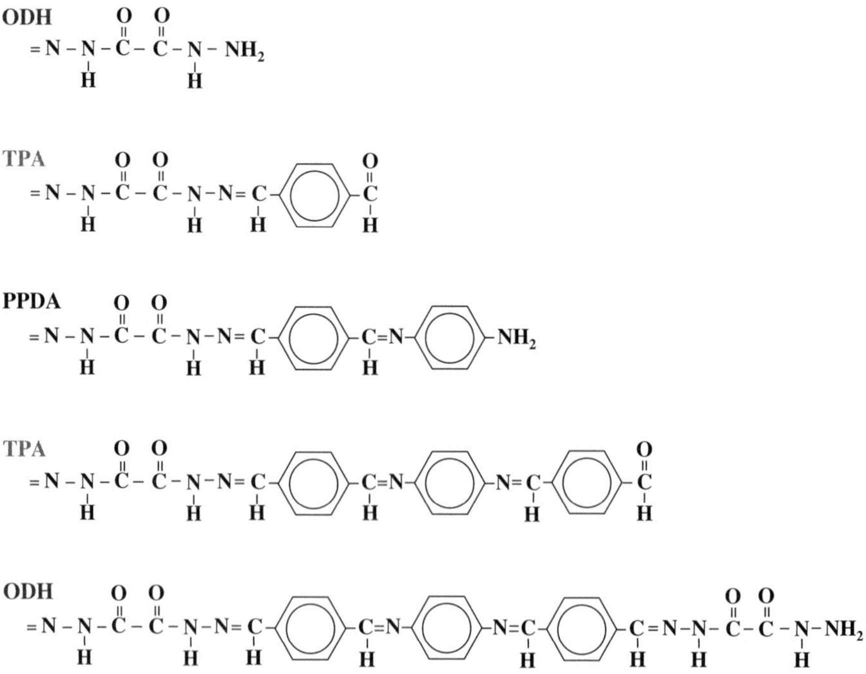

FIGURE 8.9 Example of MLD process for a polymer MQD of OTPT. From Reference 32.

ORGANIC MULTIPLE QUANTUM DOTS (MQDs) 269

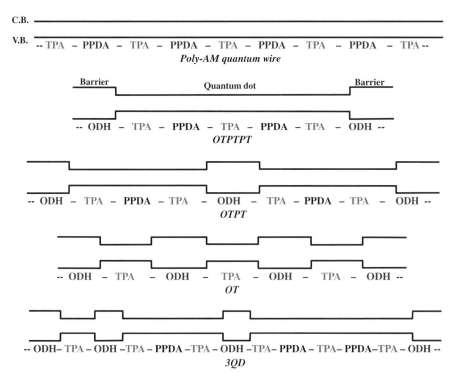

FIGURE 8.10 Schematic energy diagrams of a poly-AM quantum wire and polymer MQDs.

poly-AM quantum wire, TPA and PPDA are alternately connected. The π-electron wavefunctions are delocalized through the polymer wire. In OTPTPT, molecules are connected in a sequence of -ODH-TPA-PPDA-TPA-PPDA-TPA-ODH-. The region between two ODHs is regarded as a quantum dot with a length of ~3.1 nm containing five benzene rings. In OTPT, quantum dots with a length of ~2.0 nm containing three benzene rings are involved. In OT, TPA and ODH are alternately connected, resulting in very short quantum dots with a length of ~0.9 nm containing a single benzene ring.

In the polymer MQD of 3QD, molecules are connected in a sequence of -ODH-TPA-ODH-TPA-PPDA-TPA-ODH-TPA-PPDA-TPA-PPDA-TPA-ODH-. 3QD contains three different-length quantum dots, that is, OT-like [OT], OTPT-like [OTPTP], and OTPTPT-like [OTPTPT], in a single polymer wire.

8.4.1.2 Quantum Confinement Effect in Polymer MQDs

The quantum confinement effect in the polymer MQDs was examined by measuring absorption spectra [32]. As shown in Figure 8.11, the absorption peak shifts toward the higher energy side in the order of poly-AM quantum wire, OTPTPT, OTPT, and OT, namely, with decreasing the quantum dot length. The peak positions are plotted by squares in Figure 8.12 as a function of the quantum dot length. Peak positions calculated by the molecular orbital (MO) method (SCIGRESS developed by Fujitsu Ltd.) for molecular

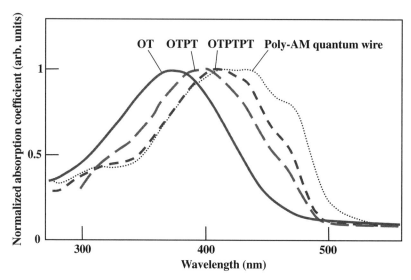

FIGURE 8.11 Absorption spectra of a poly-AM quantum wire and polymer MQDs. From Reference 32.

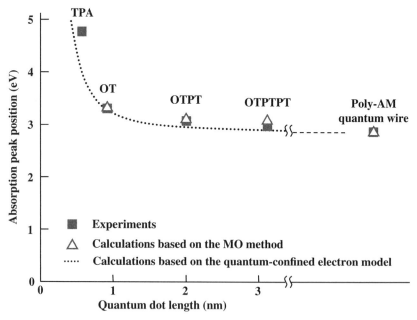

FIGURE 8.12 Dependence of the absorption peak position on the quantum dot length. From Reference 32.

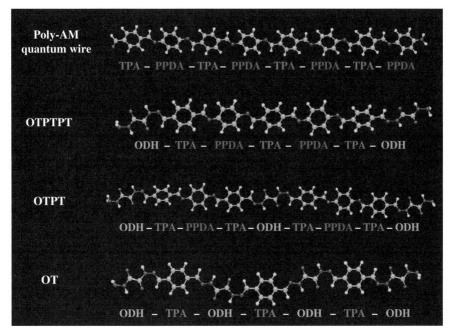

FIGURE 8.13 Molecular structures of a poly-AM quantum wire and polymer MQDs obtained by the MO method. From Reference 32. *(For a color version of this figure, see the color plate section.)*

structures shown in Figure 8.13 are plotted by triangles with taking the experimentally determined peak position of the poly-AM quantum wire as the standard. It is found that the experimental and calculated values fairly agree.

The quantum-dot-length dependence of the absorption peak position, which was derived from the quantum-confined electron model, is also presented in Figure 8.12 by a dotted line [5]. The calculation was carried out using the following expression for the energy of electrons confined in a quantum dot with infinite potential barrier heights:

$$E_n = \hbar^2 \pi^2 n^2 / (2m^* m_0 L^2).$$

Here, $n = 1, 2$, etc., m_0 is the electron mass, m^* is the effective mass in units of m_0, and L is the quantum dot length. The experimental values and calculated values are in good agreement, suggesting that the absorption peak shift shown in Figures 8.11 and 8.12 is attributed to the quantum confinement effect of π-electrons in the quantum dots.

For 3QD [32], as shown in Figure 8.14, a broad absorption band extending from ~300 to ~480 nm appears, which is a superposition of component absorption bands of [OT], [OTPT], and [OTPTPT]. The spectrum of 3QD predicted by adding the spectra of OT, OTPT, and OTPTPT is found to be fairly coincident with the measured one. This result confirms that [OT], [OTPT], and [OTPTPT] are actually formed in a single wire of 3QD by MLD.

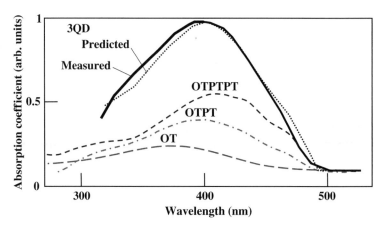

FIGURE 8.14 Absorption spectrum of 3QD. From Reference 32.

In Figure 8.15, the molecular structure and the electron density for 3QD grown on a SAM of amino-alkanethiol are shown. The calculations were carried out by SCIGRESS. Brighter colors indicate higher density. The electron density is found to be high in the quantum dot regions of [OT], [OTPT], and [OTPTPT].

From these results, it is concluded that the molecular arrangements in polymer wires were successfully controlled with designated sequences of three molecules using MLD to fabricate polymer MQDs, which exhibit the quantum confinement effect of π-electrons.

8.4.1.3 Electron Localization in Quantum Dots In order to discuss the degree of π-electron localization in the quantum dots, PL spectra of the polymer MQDs were measured using an excitation light of 365 nm in wavelength [38]. The spectra after subtraction of the PL arising from the glass substrates are shown in Figure 8.16. The PL peak shifts toward the lower energy side with decreasing quantum dot length, indicating that the PL peak and the absorption peak shift toward the opposite directions. This phenomenon is attributed to the Stokes shift as discussed below referring to the small-polaron absorption in WO_3 [39, 40].

As shown in Figure 8.17, when an injected electron is trapped at W_a site in WO_3, the electron distorts the surrounding lattice, forming a small polaron. When the electron hops to the neighboring W_b site by absorbing a photon, the lattice distortion also hops to the W_b site. In the case that the trapped electron is highly localized as in Case II, the lattice distortion concentrates closely around the trapped electron, making the lattice distortion change accompanied with the electron hopping large. In the case that the trapped electron is relatively delocalized as in Case I, the lattice distortion diffuses widely, making the lattice distortion change small.

The small-polaron absorption can be represented by a configuration coordinate diagram shown in Figure 8.17. The lattice distortion change accompanied with the electron hopping corresponds to the horizontal shift of the configuration coordinate curve after the electron hopping. With increasing degree of electron localization the

ORGANIC MULTIPLE QUANTUM DOTS (MQDs) 273

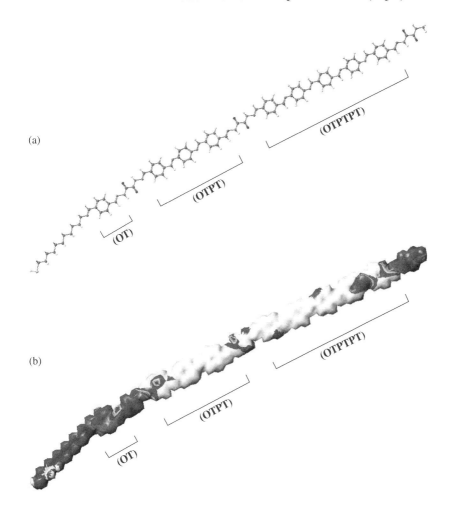

FIGURE 8.15 (a) Molecular structure and (b) electron density for 3QD. From Reference 32. *(For a color version of this figure, see the color plate section.)*

configuration coordinate curve shifts toward the right-hand side, namely, $Q_{bII} - Q_a > Q_{bI} - Q_a$, and accordingly, the photon energy required for the electron hopping becomes large, namely, $E_{0II} > E_{0I}$.

It should be noted that, with increasing the degree of electron localization, the oscillator strength for the small-polaron absorption is expected to decrease since the wavefunction overlap between the initial state "trapped at W_a" and the final state "trapped at W_b" decreases [40].

The degree of electron localization can be represented by the delocalization parameter of polaron wavefunction, ξ, which means a fraction of the wavefunction penetrating to the neighboring W site [39]. ξ dependence of the absorption peak

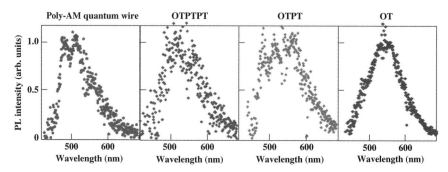

FIGURE 8.16 PL spectra of a poly-AM quantum wire and polymer MQDs.

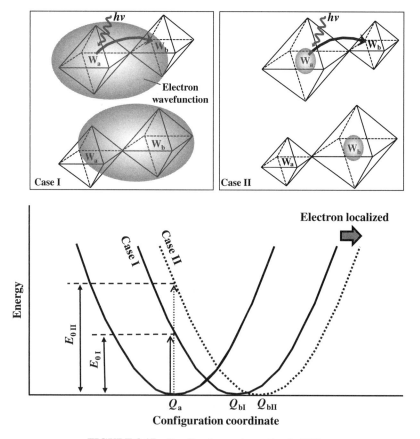

FIGURE 8.17 Small-polaron absorption in WO_3.

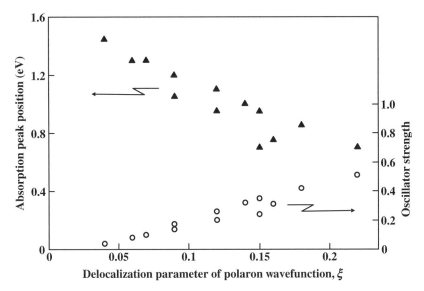

FIGURE 8.18 ξ dependence of the absorption peak position and the oscillator strength in WO_x thin films.

position and the oscillator strength was previously estimated from absorption spectra of 13 kinds of WO_x thin films, which were fabricated under various deposition conditions to vary the atomic arrangements in the films [40]. The results are summarized in Figure 8.18. With decreasing ξ, namely, with increasing the degree of electron localization, the absorption peak moves toward the higher energy side and the oscillator strength decreases as expected, confirming the guideline that the horizontal shift of configuration coordinate curves increases with increasing the degree of electron localization.

According to the above-described argument, a configuration coordinate diagram for light absorption and PL in polymer MQDs can schematically be drawn as Figure 8.19. When electrons are relatively delocalized as in Case I, the horizontal shift of the configuration coordinate curve for the excited state is small, resulting in a small Stokes shift. When electrons are highly localized as in Case II, on the contrary, the horizontal shift is large, resulting in a large Stokes shift.

Thus, it is concluded that the Stokes shift is enhanced by increasing the degree of electron localization. This conclusion leads us to believe that the Stokes shift enhancement caused by decreasing the quantum dot length implies an increase in the degree of electron localization, in other words, electrons become confined strongly in quantum dots with decreasing the quantum dot length as expected.

8.4.2 Molecular MQDs

8.4.2.1 Growth Processes and Structures For molecular MQD fabrication, we used LP-MLD. In the present experiment, p-type and n-type dyes were used as

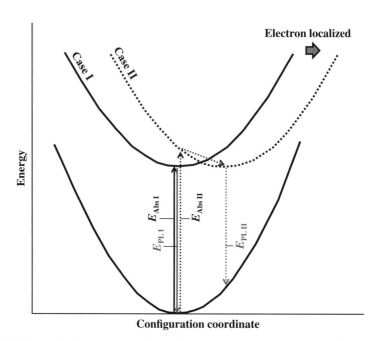

FIGURE 8.19 Configuration coordinate diagram for light absorption and PL in polymer MQDs.

source molecules [8, 20]. The terms "p-type dye" and "n-type dye" were defined by Meier [41]. The former tends to have negative charge by accepting electrons, and the latter tends to have positive charge by donating electrons.

Figure 8.20 shows examples of molecule-switching sequences for LP-MLD. A sequence of p-type dye (p1), n-type dye (n1), and p-type dye (p2) constructs a p1/n1/p2/ structure as shown in (a). When the substrate is an n-type oxide semiconductor like ZnO, due to the attractive force between the positive charge of ionized donors in the n-type semiconductor and the negative charge in the p-type dye molecules, the p-type dye molecules are strongly adsorbed on the substrate. Similarly, the attractive force between the positive charge in the n-type dye molecules and the negative charge in the p-type dye molecules strongly connects the two types of dye molecules. The same type of dye molecules do not connect due to the repulsive force. This interaction scheme satisfies the condition for MLD depicted in Figure 8.4.

MLD can also be performed by switching "molecule groups" containing several kinds of dye molecules as depicted in Figure 8.20b [8]. Molecule Group I containing molecules p1 and p2 and Molecule Group II containing molecules n1 and n2 are prepared. By providing molecules onto a substrate in an order of Molecule Group I and Molecule Group II, stacked structures of p1/n1, p1/n2, p2/n1, and p2/n2 are formed in parallel.

As the p-type dye, rose bengal (RB), eosine (EO), and fluorescein (FL), and as the n-type dye, crystal violet (CV) and brilliant green (BG), were used. The molecules are presented in Figure 8.21.

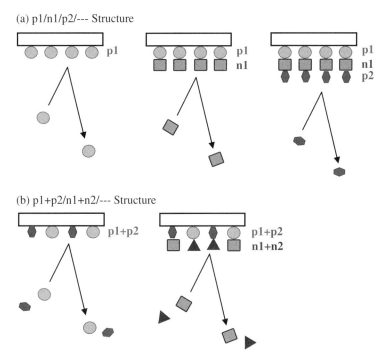

FIGURE 8.20 LP-MLD with molecule-switching sequences of (a) p-type dye (p1), n-type dye (n1), and p-type dye (p2), and (b) Molecule Group I containing molecules p1 and p2, and Molecule Group II containing molecules n1 and n2.

8.4.2.2 Monomolecular-Step Growth by LP-MLD To analyze the monomolecular-step growth of molecular MQDs by LP-MLD, surface potential was measured by an electrostatic voltmeter. The molecular MQDs were grown on ZnO powder layers formed on glass substrates with indium tin oxide (ITO) electrodes.

Surface potential of an undyed ZnO layer was found to be about −200 mV, which is attributed to electric dipole moments generated at the ZnO surface by electrons donated from zinc atoms in interstitial sites to oxygen adsorbed on the surface. When p-type dye molecules are adsorbed on the ZnO layer, as shown in Figure 8.22, the surface potential becomes more negative. This is attributed to the additional electric dipole moments generated by the negative charge in the p-type dye molecules on the ZnO surface [42]. Conversely, when n-type dye molecules are adsorbed on the ZnO layer, the surface potential becomes less negative due to the positive charge in the n-type dye molecules [42].

For the ZnO with a two-dye-molecule MQD of [ZnO/RB/CV], which was fabricated by providing RB and CV sequentially on a ZnO layer, the surface potential becomes less negative compared to that for undyed ZnO. This is caused by the positive charge in the n-type dye molecules on the top of the stacked structure [42]. For the ZnO with a three-dye-molecule MQD of [ZnO/RB/CV/EO], which was fabricated by

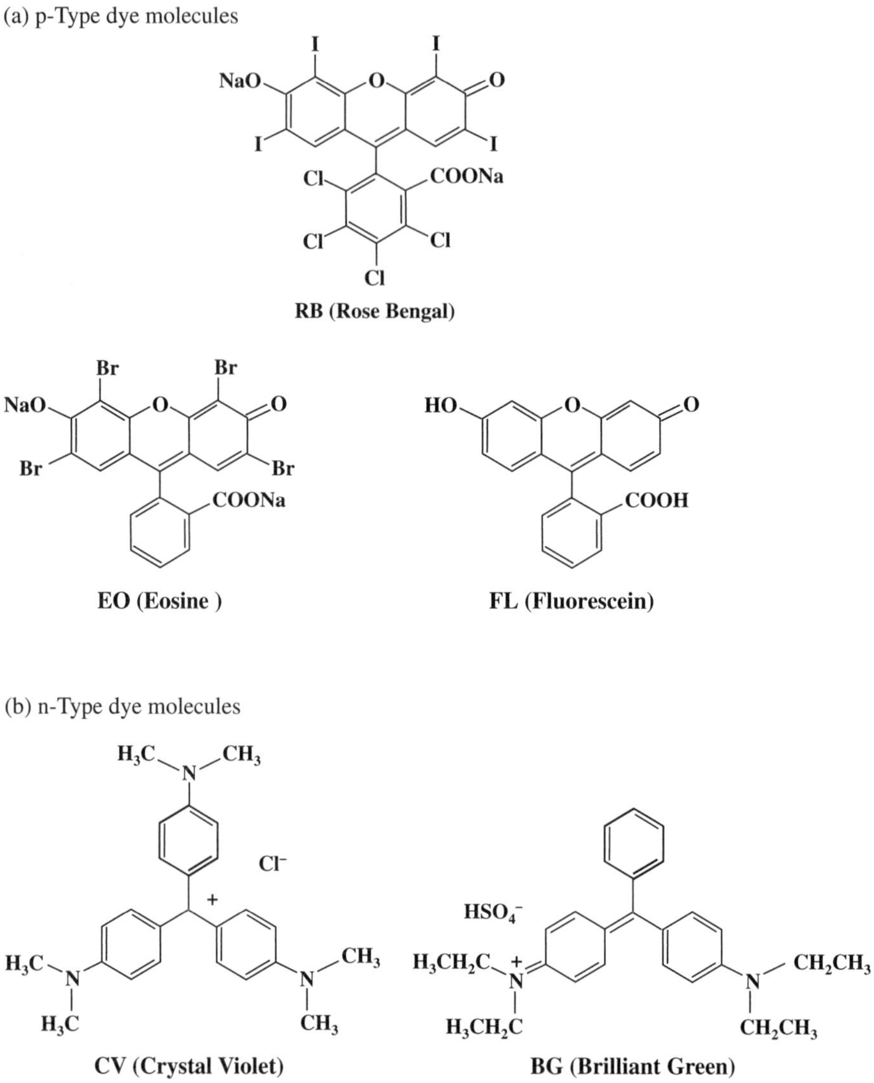

(a) p-Type dye molecules

RB (Rose Bengal)

EO (Eosine)

FL (Fluorescein)

(b) n-Type dye molecules

CV (Crystal Violet)

BG (Brilliant Green)

FIGURE 8.21 (a) p-type dye molecules and (b) n-type dye molecules used in experiments.

providing RB, CV, and EO sequentially, and [ZnO/RB/CV/FL], which was fabricated by providing RB, CV, and FL sequentially, the surface potential becomes more negative compared to that for undyed ZnO. This is caused by the negative charge in the p-type dye molecules on the top.

From these results it is suggested that molecular MQDs of multi-dye-molecule-stacked structures are definitely constructed on ZnO surfaces with monomolecular steps by LP-MLD.

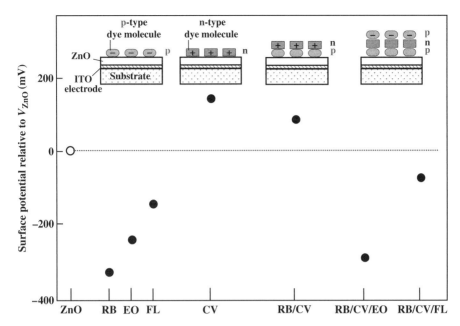

FIGURE 8.22 Surface potential of ZnO layers with dye molecules on the surface.

Figure 8.23 shows PL spectra of the molecular MQDs. The excitation light wavelength was 532 nm. As can be seen in Figure 8.23a, [ZnO/RB/CV] exhibits a wider PL spectrum compared with [ZnO/RB]. This indicates that in [ZnO/RB/CV] the PL spectrum of CV is superposed on the spectrum of RB. In [ZnO/RB/CV/EO], as shown in Figure 8.23b, the PL spectrum of [ZnO/EO] is added on the spectrum of [ZnO/RB/CV].

It should be noted that [ZnO/CV] exhibits very small PL intensity as shown in Figure 8.23a. This is due to the fact that only a small amount of CV molecules can be adsorbed on ZnO surface due to a weak interaction between n-type molecules and n-type ZnO. Meanwhile, in [ZnO/RB/CV], a large amount of CV molecules can be adsorbed by inserting p-type molecules between the n-type molecules and the n-type ZnO to construct an n/p/n structure.

Figure 8.23c shows a PL spectrum of [ZnO/RB+EO/CV+BG], which was fabricated by providing molecules onto a ZnO layer in the order of Molecule Group I containing RB and EO and Molecule Group II containing CV and BG according to the process depicted in Figure 8.20b. Compared to [ZnO/RB/CV], [ZnO/RB+EO/CV+BG] exhibits a wider PL spectrum, which is attributed to superposition of PL spectra of EO and BG on the spectrum of [ZnO/RB/CV].

These results of the PL spectra again suggest that molecular MQDs of multi-dye-molecule-stacked structures are constructed by LP-MLD.

8.4.2.3 Widening of Photocurrent Spectra by Two-Dye-Molecule MQD Photocurrent spectra of ZnO thin films with molecular MQDs consisting of p-type and

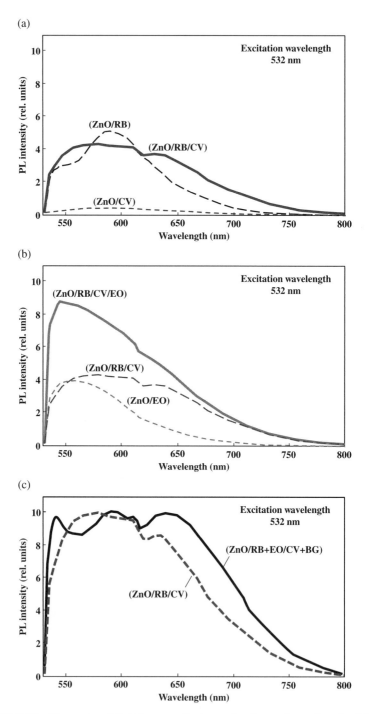

FIGURE 8.23 PL spectra of ZnO layers with dye molecules on the surface for (a) [ZnO/RB/CV], (b) [ZnO/RB/CV/EO], and (c) [ZnO/RB+EO/CV+BG].

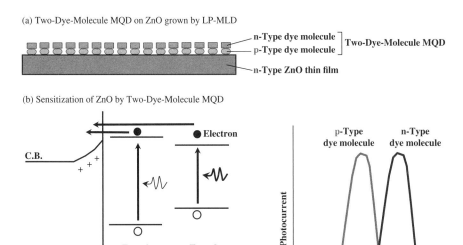

FIGURE 8.24 Models of (a) two-dye-molecule MQD and (b) photocurrent generation in ZnO with a two-dye-molecule MQD of [ZnO/p-type dye/n-type dye]. From Reference 20.

n-type dye molecules were measured. The ZnO thin films were deposited by vacuum evaporation followed by annealing at 400°C for 1 h in air. The film thickness was 600 nm.

Models of a two-dye-molecule MQD and photocurrent generation in ZnO with the two-dye-molecule MQD of [ZnO/p-type dye/n-type dye] are summarized in Figure 8.24. It was reported that the electron injection efficiency into ZnO is much higher for the p-type dye molecule than for the n-type dye molecule. It was also reported that the electron injection efficiency from the n-type dye molecule into ZnO increases by inserting the p-type dye molecule between the n-type dye molecule and the ZnO [43, 44]. Thus, in [ZnO/p-type dye/n-type dye], electrons excited in the p-type dye molecule are directly injected into ZnO, and, at the same time, electrons excited in the n-type dye molecule are injected into ZnO via the p-type dye molecule to give a widened photocurrent spectrum that is a superposition of the spectrum of the p-type dye molecule and the spectrum of the n-type dye molecule.

Figure 8.25a shows absorption spectra of RB and CV in alcohol. Absorption spectrum of CV is located in a longer wavelength region compared with that of RB. Figure 8.25b shows wavelength dependence of photocurrents generated in samples of [ZnO], [ZnO/RB], [ZnO/CV], and [ZnO/RB/CV] [20]. The photocurrents were measured by the "guided light" configuration shown in Figure 8.26 by applying a voltage of 8 V at a slit-type Al electrode with a gap of 60 μm. Laser lights introduced into the ZnO thin film from the edge propagate in the film. In [ZnO], large

282 THIN-FILM MOLECULAR NANOPHOTONICS

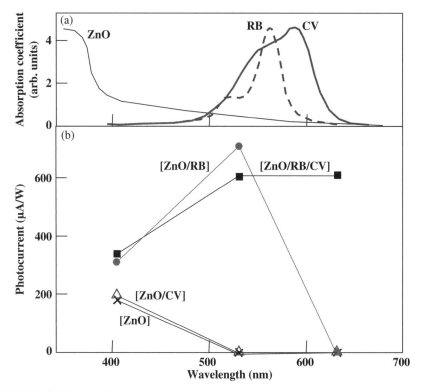

FIGURE 8.25 (a) Absorption spectra of RB and CV in alcohol. (b) Wavelength dependence of photocurrents generated in [ZnO], [ZnO/RB], [ZnO/CV], and [ZnO/RB/CV]. From Reference 20.

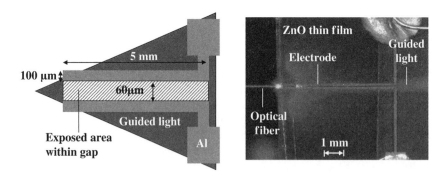

FIGURE 8.26 Setup for photocurrent measurements by the "guided light" configuration. *(For a color version of this figure, see the color plate section.)*

photocurrents are generated at 405 nm, and small photocurrents at 532 nm and 633 nm. In [ZnO/RB], the photocurrent spectrum extends to 532 nm. In [ZnO/RB/CV], the photocurrent spectrum further extends to 633 nm since the absorption band of CV covers 633 nm. Such widening of photocurrent spectra in ZnO with two-dye-molecule MQDs is expected to be useful in sensitization for oxide-semiconductor-based solar cells described in Subsection 8.6.1. By using multi-dye-molecule MQDs with stacked molecule counts of three or more, further widening of the photocurrent spectra is expected.

8.5 SELF-ORGANIZED LIGHTWAVE NETWORK

Automatic construction of optical wiring with designated configurations is one of the goals in the optical science. SOLNET is a candidate to achieve this [9–12]. SOLNET can realize the optical solder of self-aligned optical couplings between misaligned optical devices with different core widths, and can fabricate self-aligned optical waveguides in free spaces three dimensionally. Figure 8.27 shows a concept of a self-organized 3D integrated optical circuit consisting of SOLNET. After a precursor with optical devices distributed three dimensionally is built, optical waveguides are formed between the devices automatically in a self-aligned manner by introducing light beams.

In this section, concept, theoretical analysis, and experimental demonstrations of SOLNET are described, especially emphasizing reflective SOLNET (R-SOLNET), which is the SOLNET targeting reflective or luminescent sites.

8.5.1 Concept

Optical waveguide formation in a photopolymer was first reported by Frisken [45]. An up-tapered waveguide was stretched from a single-mode optical fiber edge by emitting a write beam from the optical fiber to the photopolymer, realizing a mode size conversion up to 50 µm. Such optical waveguides are called "self-written waveguides."

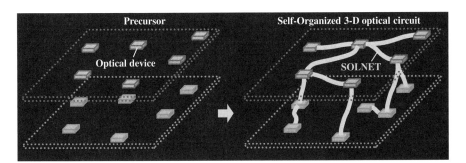

FIGURE 8.27 Concept of a self-organized three-dimensional integrated optical circuit consisting of SOLNET. *(For a color version of this figure, see the color plate section.)*

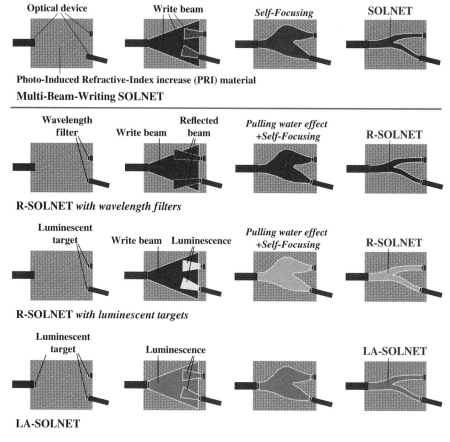

FIGURE 8.28 Concept of SOLNET.

Self-written waveguides were also constructed in photosensitive glass [46], and photorefractive crystals [47], and a wide range of work has been accomplished [48–51].

Meanwhile, SOLNET [9–12] utilizes attractive force induced between a plurality of write beams. The concept of SOLNET is summarized in Figure 8.28. In multi-beam-writing SOLNET, two or more optical devices such as optical fibers, optical waveguides, and laser diodes (LDs) are put in photo-induced refractive index increase (PRI) materials, in which the refractive index increases upon write beam exposure. Write beams are introduced into the PRI material from these optical devices. Since the refractive index increases rapidly in regions, where the write beams overlap, the write beams attract each other to merge into one by the self-focusing. This enables us to construct self-aligned coupling waveguides of SOLNET between the optical devices automatically, even when misalignments and core size mismatching exist between them. SOLNET can also be constructed by write beams from free spaces.

R-SOLNET *with wavelength filters* [11, 12, 52, 53] simplifies the fabrication process of the multi-beam-writing SOLNET. Some of the write beams in the multi-beam-writing SOLNET are replaced with reflected write beams from wavelength filters on edges of the optical devices such as optical fibers, optical waveguides, LDs, photodetectors (PDs), and light modulators. The wavelength filters reflect write beams and transmit signal beams. A write beam from an optical device and the reflected write beams from the wavelength filters on the other optical devices overlap near the wavelength filters, pulling the write beam to the wavelength filter locations more and more. We call this effect the "pulling water" effect. Finally, by self-focusing, self-aligned coupling waveguides are formed between the optical devices.

In R-SOLNET *with luminescent targets* [33, 54, 55], the wavelength filters are replaced with luminescent targets. When the luminescent targets are exposed to a write beam from an optical device, the write beam is absorbed by the targets followed by luminescence from them. The luminescence induces the "pulling water" effect to grow R-SOLNET between the optical devices. Signal beams can propagate through the luminescent targets by choosing appropriate signal beam wavelengths.

In luminescence-assisted SOLNET (LA-SOLNET), luminescent targets are put on edges of all the optical devices to be connected. By introducing excitation lights from outside, for example, from above, the targets emit luminescence to grow SOLNET between the optical devices. The sensitivity spectrum of the PRI material should be adjusted by doping sensitizing agents so that it exhibits low sensitivity to the excitation light and high sensitivity to the luminescence. LA-SOLNET is effective in cases where write beams cannot be introduced from the optical devices.

There are many types of PRI materials, such as photopolymers shown in Figure 8.3, photo-definable materials, photosensitive glass, and photorefractive crystals. Figure 8.29 shows a mechanism of the waveguide construction in a photopolymer [15], in which high refractive index molecules and low refractive index molecules are mixed. The former has higher photoreactivity to write beams than the latter. When the photopolymer is exposed to the write beam, the high refractive index molecules are selectively combined to produce dimmers, oligomers, or polymers, and consequently, the high refractive index molecules diffuse into the exposed region from surrounding regions to raise the refractive index of the exposed region, constructing optical waveguides. Blanket curing or etching provides permanent optical waveguides of SOLNET.

8.5.2 Theoretical Analysis

8.5.2.1 Simulation Procedure Figure 8.30 shows a model of simulation for R-SOLNET *with luminescent targets* [33, 54, 55]. Two input waveguides with a core width of 600 nm are put with a waveguide distance, P. The refractive index of the waveguide core is 1.8 and that of the cladding region is 1.0. Two 600-nm-wide luminescent targets are located with a lateral misalignment, d, from the axis of the input waveguides. Efficiency of luminescence from the targets, η_{Emit}, is assumed 0.36. Between the input waveguides and the luminescent targets, there is a 12.8-μm-long gap filled with a PRI material. The refractive index of the PRI material increases from 1.5 to 1.7 upon write beam exposure, which is expected in the photosensitive sol-gel material developed by Nissan Chemical Industries, Ltd. [56,57]. Wavelengths

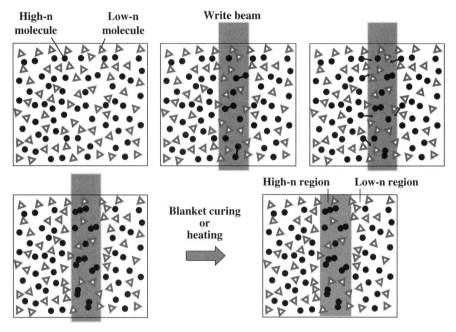

FIGURE 8.29 Mechanism of waveguide construction in a photopolymer.

are 650, 700, and 800 nm, respectively, for the write beam, the luminescence, and the probe beam. The polarization direction is perpendicular to the calculated plane.

For the calculation, the Finite-Difference Time-Domain (FDTD) method was used [12, 55, 58, 59]. Since the refractive index of PRI materials changes with time, the refractive index distribution was rewritten at each time step with duration Δt. Energy density of the write beam exposure on the PRI material over duration Δt is

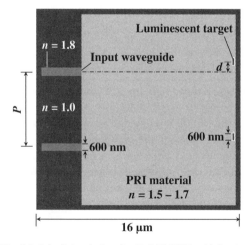

FIGURE 8.30 Model of simulation for R-SOLNET *with luminescent targets*.

$(1/2)\varepsilon E^2 v\Delta t$, where ε is the dielectric constant, v the velocity of the light, and E the electric field of the light. Then, the refractive index change Δn during Δt can be expressed as

$$\Delta n = \gamma(1/2)\varepsilon E^2 \, v\Delta t,$$

where, γ represents the sensitivity of the PRI material.

In real systems, the typical response time of PRI materials ranges from a few seconds to a few minutes. So, we rescaled the time parameter in the simulations as follows:

$$\Delta n = \gamma'(1/2)\varepsilon E^2 \, v\Delta t',$$

where, $\Delta t' = \Delta t \times 10^{12}$.

Mesh sizes for the FDTD calculations are $\Delta x = \Delta y = 20$ nm and $\Delta t = 0.0134$ ps. Amplitude of the write beam electric field at the light source location is 10 V m^{-1} and γ' is assumed to be 1 m^2/J.

The electric field amplitude of the emitted luminescence, E^0_{Emit}, is determined as follows using time average of the write beam electric field $\overline{E_{\text{Write}}}$ at the luminescent target locations and η_{Emit}. $E^0_{\text{Emit}} = \sqrt{2}\eta_{\text{Emit}}\overline{E_{\text{Write}}}$.

8.5.2.2 R-SOLNET with Luminescent Targets
Results of simulation for construction of parallel waveguides with P of 4000 nm are shown in Figure 8.31a for a case of no luminescent targets. The upper and lower figures respectively represent dielectric constant ε, that is, square of refractive index n, and intensity of probe beams E^2. Parallel waveguides of SOLNET are gradually grown from the input waveguides. Accordingly, a probe beam emitted from the upper input waveguide is confined in the SOLNET. When P is reduced to 2000 nm, as can be seen in Figure 8.31b(i), the two optical waveguides attract each other to deviate from the right destinations. When P is further reduced to 1000 nm, the two optical waveguides are merged into one.

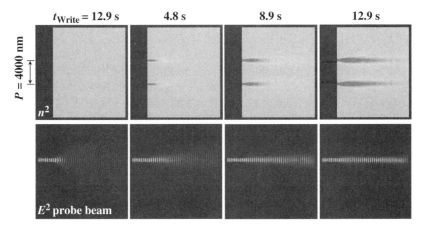

FIGURE 8.31a Simulation for parallel waveguides of SOLNET with no luminescent targets. *(For a color version of this figure, see the color plate section.)*

(i) SOLNET with no luminescent targets

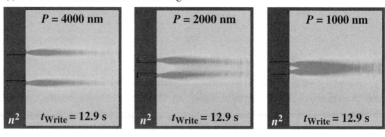

(ii) R-SOLNET *with luminescent targets*

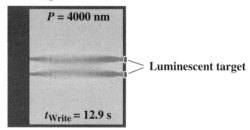

FIGURE 8.31b Simulation for parallel waveguides of (i) SOLNET with no luminescent targets and (ii) R-SOLNET *with luminescent targets*. *(For a color version of this figure, see the color plate section.)*

By using R-SOLNET *with luminescent targets*, the parallel waveguides can be pulled back to the right destinations as shown in Figure 8.31b(ii) for P of 2000 nm. The estimated coupling efficiency to the right destination increases from 30% to 60% by putting the luminescent targets.

Another feature of R-SOLNET *with luminescent targets* is a capability of self-aligned waveguide construction. In Figure 8.31c, results of simulation for R-SOLNET *with luminescent targets* are shown for various lateral misalignments. For d of 0 and 600 nm, parallel waveguides are pulled to the luminescent target locations, and the probe beams are guided toward them. The coupling efficiency and crosstalk were calculated to be ~80% and −25 dB, respectively. For d of 1200 nm, parallel waveguides are pulled to the luminescent target locations, but the probe beam partially leaks. The coupling efficiency and crosstalk were calculated to be ~60% and −23 dB, respectively. For d of 1800 nm, R-SOLNET is not constructed, accordingly, the probe beam goes straight without the self-alignment effect.

8.5.3 Experimental Demonstrations: Targeting Reflective/Luminescent Sites

An experimental demonstration of R-SOLNET was attempted using tris(8-hydroxyquinolinato)aluminum (Alq3) dispersed in polyvinyl alcohol (PVA) as the target. For the photopolymer, Norland Optical Adhesives NOA65 and NOA81 with a mixing ratio of NOA65:NOA81 = 1:2 in wt% was used. As shown in Figure 8.32,

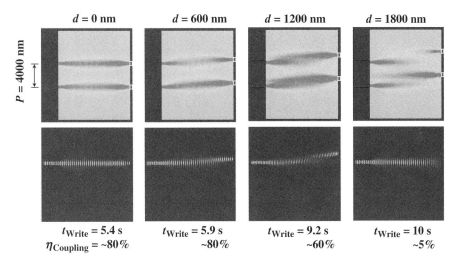

FIGURE 8.31c Simulation for parallel waveguides of R-SOLNET *with luminescent targets* for various lateral misalignments. *(For a color version of this figure, see the color plate section.)*

(a) During write beam exposure

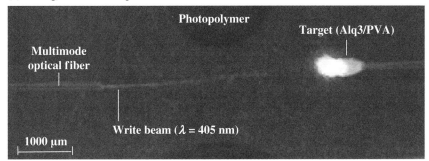

(b) After write beam exposure for 2 min

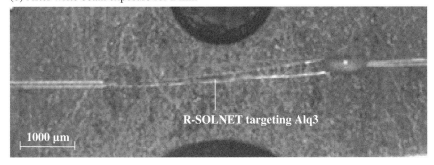

FIGURE 8.32 Experimental demonstration of R-SOLNET targeting Alq3 dispersed in PVA. From Reference 54. *(For a color version of this figure, see the color plate section.)*

(a) Before write beam exposure

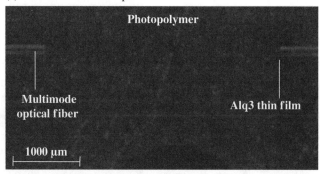

(b) During write beam exposure

FIGURE 8.33 Experimental demonstration of R-SOLNET targeting Alq3 thin film on the edge of the right-hand side optical fiber. (a) Before and (b) during write beam exposure.

the target is placed on an edge of the right-hand-side optical fiber. A write beam of 405 nm in wavelength is introduced from the left-hand-side multimode optical fiber with a core diameter of 50 μm into the photopolymer. Luminescence is generated from the Alq3/PVA target. The write beam is pulled toward the target to construct R-SOLNET [54].

It should be noted that the width of the R-SOLNET gradually varies from the diameter of the optical fiber core to that of the target, suggesting that R-SOLNET provides self-aligned coupling waveguides with a function of mode converters.

Figure 8.33 shows an experimental demonstration using a target of an Alq3 thin film deposited on the edge of the right-hand-side optical fiber by the vacuum evaporation. The 405-nm write beam introduced from the left-hand-side multimode optical fiber is guided targeting the Alq3 thin film to construct R-SOLNET. Since the Alq3 film thickness is less than 1 μm, the film exhibits little absorption for the signal beams of 600–850 nm in wavelength, giving rise to high transmittance.

Figure 8.34 shows an experimental setup and a photograph of R-SOLNET constructed between the left-hand-side multimode optical fiber and an Al micromirror on the edge of the right-hand-side optical fiber [53]. In this case, the micromirror plays

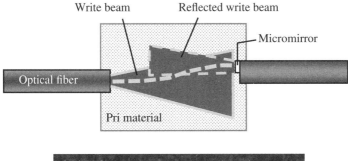

Before R-SOLNET construction

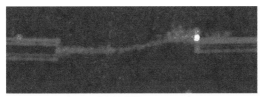

After R-SOLNET construction

FIGURE 8.34 Experimental demonstration of R-SOLNET targeting Al micromirror on the edge of the right-hand side optical fiber. From Reference 53.

a role of the wavelength filter in R-SOLNET *with wavelength filters*. The gap and the lateral misalignment between the optical fiber and the micromirror are ~800 and ~60 μm, respectively. The PRI material is a photopolymer consisting of an epoxy component and an acrylic component with high refractive index and high photochemical activity. When an ultraviolet (UV) write beam of a high pressure mercury lamp was introduced from the left-hand-side optical fiber into the PRI material, R-SOLNET was constructed, targeting the micromirror. It is found in the photograph that a probe beam of 650 nm in wavelength propagates in the S-shaped self-aligned optical waveguide of R-SOLNET with core width of about 50 μm.

A spin-coated photosensitive sol-gel thin film is promising as a PRI material for nanoscale SOLNET since it generates a high refractive index contrast of ~13% at 633 nm in wavelength upon write beam exposure. The sol-gel thin film is made of a silicon oxide-based material incorporating titanium oxide. It enables us to form waveguide cores by selectively exposing it to UV light with a core pattern and postbaking it to promote the hydrolytic condensation of the titanium oxide and raise the refractive index in the exposed area. Figure 8.35a shows examples of optical circuits formed in the sol-gel thin film [56]. Guided beams were found to be confined

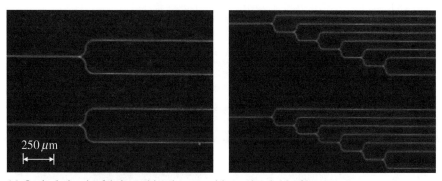

(a) Optical circuits fabricated in photosensitive sol-gel thin films

(b) R-SOLNET targeting a reflective defect in photosensitive sol-gel thin films

FIGURE 8.35 (a) Optical circuits and (b) R-SOLNET targeting a reflective defect formed in the photosensitive sol-gel thin film. From References 56 and 57.

within a cross-section area of 1.2 μm × 0.6 μm. Propagation loss was evaluated to be 1.86 dB cm^{-1} at 633 nm. Figure 8.35b shows R-SOLNET targeting a reflective defect in the photosensitive sol-gel thin film [57]. It was constructed by guiding a 405-nm write beam in the thin film for 30 s under heating at 200°C.

8.6 PROPOSED APPLICATIONS

8.6.1 Sensitized Solar Cells

The dye-sensitized solar cell [60] has received much attention as a next-generation energy conversion system. Energy conversion efficiency of the sensitized solar cell is mainly limited by the following factors:

A. Heat generation due to the excess photon energy in the light absorption process.
B. Limited energy difference between Fermi level of semiconductors and the highest occupied molecular orbital (HOMO) of sensitizing molecules.
C. Resistivity of semiconductors.

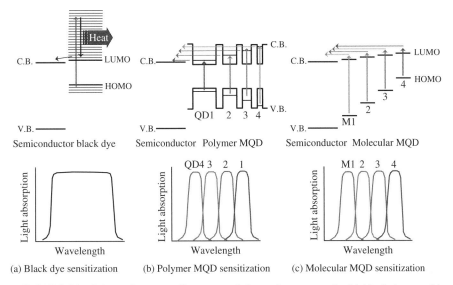

FIGURE 8.36 Schematic energy diagrams and absorption spectra for (a) black dye sensitization, (b) polymer MQD sensitization, and (c) molecular MQD sensitization.

In this subsection, organic MQD sensitization and waveguide-type solar cells are proposed to solve Issues (A)–(C).

8.6.1.1 Organic MQD Sensitization In sensitized solar cells using narrow-energy-gap black dyes, as shown in Figure 8.36a, since the absorption spectra are broad, excess energy of high energy photons is considerably lost as heat in the light absorption process. The organic MQD sensitization such as the polymer MQD sensitization and the molecular MQD sensitization will solve the problem.

In the polymer MQD sensitization shown in Figure 8.36b [8, 32], four different-length quantum dots of QD1, 2, 3, and 4 are contained in a single polymer wire. Since absorption bands of quantum dots are narrow due to their 0D characteristics and shift toward the higher energy side with decreasing dot length, the whole absorption spectrum can be divided into narrow absorption bands of individual quantum dots, suppressing the heat generation arising from the excess photon energy. The result shown in Figure 8.14 presents this situation. Namely, a broad absorption spectrum of 3QD is divided into three absorption bands of [OT], [OTPT], and [OTPTPT].

In order to improve the carrier transport performance in the polymer MQDs, the tunneling effect should be adjusted by choosing appropriate source molecules for the barriers. Furthermore, to expand the wavelength region for sensitization to the longer wavelength side, backbone polymers with narrow band gaps are required.

In the molecular MQD sensitization shown in Figure 8.36c [8, 20], the whole absorption spectrum are similarly divided into absorption bands of individual dye molecules of M1, 2, 3, and 4. Gradual energy steps are preferable to perform smooth electron transport to the semiconductor. The results shown in Figure 8.23, although indirectly, present examples for the situation that a broad absorption spectrum is divided into absorption bands of individual molecules. The result shown in Figure 8.25 presents a proof-of-concept of the molecular MQD sensitization with two dye molecules—RB and CV.

The above-mentioned approach of "absorbing solar beams by superposed narrow absorption bands" will solve Issue (A).

Figure 8.37 shows mechanisms for improved organic MQD sensitization [8,38,61] utilizing the z-scheme process in the photosynthesis of plants. In the molecular MQD sensitization, as shown in Figure 8.37b, an electron excited by a photon with wavelength of λ_1 in M1 is injected into an n-type semiconductor. An electron excited by a photon with λ_2 in M2 is transferred to HOMO of M1. The hole left in HOMO of M2 is compensated by an electron from a p-type semiconductor stacked on M2. Thus, electrons travel cyclically as —[p-type semiconductor] → [M2] → [M1] → [n-type semiconductor] → [p-type semiconductor] →—. This sensitization mechanism suppresses the energy loss arising from the excess photon energy, and at the same time, it increases the energy difference between the Fermi level of the n-type semiconductor and the HOMO of the sensitizing molecules to increase the voltage generated from the solar cell.

In the polymer MQD sensitization, the similar two-step excitation of electrons might be possible by adjusting the band structure as shown in Figure 8.37a. An electron excited by a photon with wavelength of λ_1 in QD1 is injected into the n-type semiconductor. An electron excited by a photon with λ_2 in QD2 is transferred to the valence band of QD1. The hole left in the valence band of QD2 is compensated by an electron from the p-type semiconductor.

The above-mentioned approach of "multi-step excitation of electrons" will solve Issue (B) as well as Issue (A).

In order to make the two-step excitation effective for the sensitization, the rate balance between the excitations at λ_1 and λ_2 is essential. The two rates of excitation at λ_1 and λ_2 should be identical, or at least, the rate at λ_1 should not be smaller than that at λ_2.

8.6.1.2 Waveguide-Type Solar Cells In conventional dye-sensitized solar cells, in order to increase the number of adsorbed dye molecules on the semiconductor surface, porous semiconductors are used. In this case, however, the internal resistivity of the solar cells tends to increase.

In order to reduce the resistivity, we proposed the waveguide-type sensitized solar cell shown in Figure 8.38 [8, 20, 32, 61]. A sensitizing layer of the organic MQD is inserted between n-type and p-type thin-film semiconductors with a flat surface and high crystalline quality. The problem in this structure is that normally incident light beams pass through a very thin sensitizing layer, resulting in very small light absorption. To overcome the drawback, we attempt "guided light" configuration, in which light beams are guided in the semiconductor films and the sensitizing layer to enhance light absorption and photocurrents. In preliminary experiments, photocurrent enhancement by a factor of 4~10 was observed in the "guided light" configuration

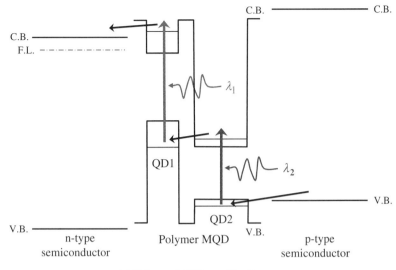

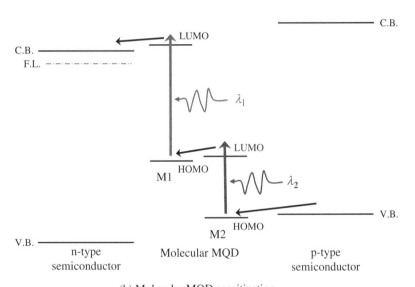

FIGURE 8.37 Mechanisms of the improved organic MQD sensitization utilizing the multi-step excitation of electrons. (a) Polymer MQD sensitization and (b) molecular MQD sensitization.

compared to the "normally incident light" configuration [20]. Thus, the waveguide-type sensitized solar cell will solve Issue (C).

The waveguide-type sensitized solar cell can be embedded in a light beam collecting film to provide a film-based integrated solar cell shown in Figure 8.39 [8, 12, 62].

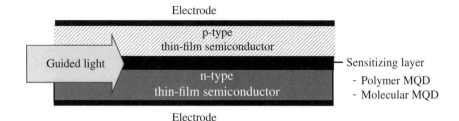

FIGURE 8.38 Concept of the waveguide-type sensitized solar cell.

An array of tapered vertical waveguides and a film with optical waveguides having 45° mirrors are stacked. Incident light beams are introduced into the optical waveguides through the tapered vertical waveguides and the 45° mirrors, and are guided to the waveguide-type sensitized solar cell. This structure enables reduction of semiconductor consumption [8, 12]. By embedding wavelength filters into the light beam collecting film, the solar beam spectrum can be divided, which is effective to reduce the number of divided absorption bands in each solar cell with the organic MQD sensitization shown in Figure 8.36.

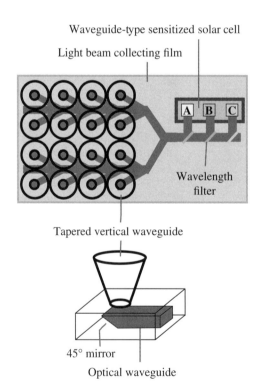

FIGURE 8.39 Concept of the film-based integrated solar cell. *(For a color version of this figure, see the color plate section.)*

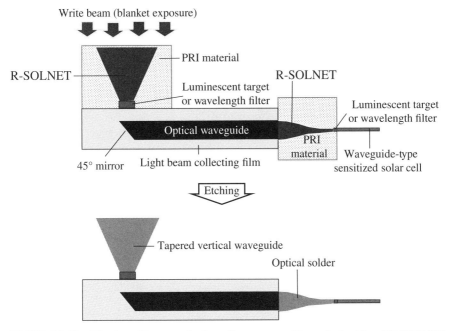

FIGURE 8.40 Film-based integrated solar cells constructed by optical solder of R-SOLNET and tapered vertical waveguides of R-SOLNET self-aligned to vertical mirrors.

A serious concern in the film-based integrated solar cell is the optical coupling between the optical waveguides and the waveguide-type sensitized solar cells. The optical solder of R-SOLNET might provide the solution as illustrated in Figure 8.40 [8, 12]. R-SOLNET might also construct tapered vertical waveguides self-aligned to 45° mirrors by blanket write beam exposure [8, 12].

8.6.1.3 Expansion to Artificial Photosynthesis Artificial photosynthesis is another promising solar energy conversion system. So far, H_2 generation has been achieved with relatively high efficiency, and recently Toyota group succeeded in photosynthesis of formate utilizing semiconductor/complex hybrid photocatalysts [63]. The organic MQD sensitization and the "guided light" configuration might improve the conversion efficiency in the photosynthesis.

When the improved sensitization with multi-step excitation of electrons shown in Figure 8.37 is applied to water splitting, H_2 would be generated from the n-type semiconductor/water interface (left-hand side in Fig. 8.37), where co-catalyst is deposited, and O_2 from the p-type semiconductor/water interface (right-hand side in Fig. 8.37).

In the photosynthesis, furthermore, high power-density light beam exposure is required to achieve high conversion efficiency because the H_2O oxidization is induced by the four-electron process. The waveguide-based light beam collecting film can

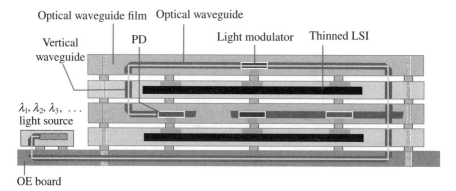

FIGURE 8.41 Concept of the three-dimensional integrated optical interconnects within computers. *(For a color version of this figure, see the color plate section.)*

be used to increase the light beam power density. Meanwhile, it is known that light-harvesting antennas [64] raise density of holes for the H_2O oxidization. The molecular MQD structure with gradual energy steps shown in Figure 8.36c might be applied to the light-harvesting antennas.

8.6.2 Three-Dimensional Integrated Optical Interconnects within Computers

With an increase in clock rates and I/O counts of processors, interconnection bottlenecks appear even in board-level interconnects within computers. Since optical interconnects are promising to overcome these bottlenecks [65, 66], comprehensive accomplishments to implement optics into computers have been done [67–74].

Figure 8.41 shows a concept of the 3D integrated optical interconnects within computers, which we expect to be the final goal of optical interconnects [12, 52, 75–78]. Optical waveguide films, in which thin-film light modulator/PD flakes are embedded, are stacked together with films containing thinned LSIs. The stacked optical waveguide films are connected by vertical waveguides. Light sources are placed on optoelectronic (OE) boards in order to reduce the heat generation inside the 3D structures.

The 3D integrated optical interconnects contribute to size/cost reductions and performance improvements in optical interconnect systems due to the following characteristics:

- Excess module areas for E–O and O–E conversions are not necessary.
- Long electrical lines for connections between LSI pads and E–O/O–E conversion sites are replaced with short vertical electrical lines, resulting in reduction of latency and bit error rates.
- Material consumption for the optical devices is reduced.
- Thin-film optical device flakes are embedded by the photolithographic process [77], avoiding the chip-bonding-based conventional packaging process.

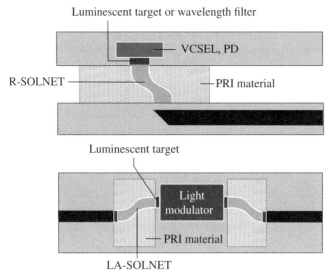

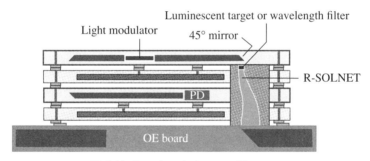

FIGURE 8.42 (a) Optical solder of R-SOLNET and LA-SOLNET for self-aligned optical couplings and (b) self-aligned vertical waveguides of R-SOLNET. *(For a color version of this figure, see the color plate section.)*

However, there are some issues related to the 3D optical wiring. Namely, (1) optical couplings between the thin-film optical device flakes and optical waveguides and (2) vertical waveguides should easily be fabricated. R-SOLNET is expected to provide the solution as depicted in Figure 8.42. For Issue (1), by putting luminescent targets or wavelength filters on active portions of optical devices, optical solder of R-SOLNET for self-aligned optical couplings are constructed. In the case that write beams cannot be introduced, as in the case of light modulators located deep inside the interconnect systems, LA-SOLNET is effective. For Issue (2), R-SOLNET can be applied to the self-aligned vertical waveguide fabrication by depositing luminescent targets or wavelength filters at the 45° mirror locations of optical waveguides.

8.6.3 Electro-Optic (EO) Thin Films

In optical interconnects, high speed/small size optical modulators, optical switches, and tunable wavelength filters are the key components, for which high performance EO materials are required [8]. So far, EO materials such as $LiNbO_3$ (LN) and quantum dots of III–V compound semiconductors [79, 80] have been developed. As the next-generation EO material, organic materials with π-conjugated systems attract interest because they have both "large optical nonlinearity" and "low dielectric constant" characteristics. Lipscomb et al. [81] studied the Pockels effect in a 2-methyl-4-nitroaniline (MNA) single crystal and estimated the EO coefficient to be 67 pm V^{-1}, which is about two times that of LN. The styrylpyridinium cyanine dye (SPCD) thin film was found to exhibit large EO coefficients of 430 pm V^{-1} [82], about 14 times that of LN.

Organic EO materials are classified into poled polymers, molecular crystals, and conjugated polymers. Among these, conjugated polymers having MQDs, namely, polymer MQDs, seem most promising because it enables the wavefunction control between 0D and 1D to enhance the EO effect.

In this subsection, theoretically predicted enhancement of the Pockels effect, which is an EO effect inducing refractive index changes proportional to applied electric fields, in the polymer MQD by controlling wavefunctions is reviewed [8, 13].

Hyperpolarizability β is the measure of the second-order optical nonlinearity of molecules. The EO coefficient for the Pockels effect is proportional to β. In the two-level model, $\beta \alpha r_{gn}^2 \Delta r$. Here, r_{gn} and Δr are respectively the transition dipole moment and the dipole moment difference between the ground state and the excited state. r_{gn} tends to increase with an increase in wavefunction overlap between the ground state and the excited state while Δr tends to increase with an increase in wavefunction separation between the ground state and the excited state.

For the three wavefunction shapes shown in Figure 8.43, from left to right, the wavefunction separation increases while the wavefunction overlap decreases, that is, Δr increases and r_{gn} decreases. Therefore, β is expected to have its maximum in a wavefunction shape of intermediate wavefunction separation with an appropriate dimensionality existing somewhere between 0D and 1D.

Such optimized wavefunction conditions can be obtained by adjusting the quantum dot lengths and donor/acceptor substitution sites in the quantum dots using

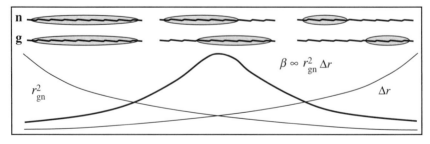

FIGURE 8.43 Guideline to improve the second-order optical nonlinearity. From Reference 13.

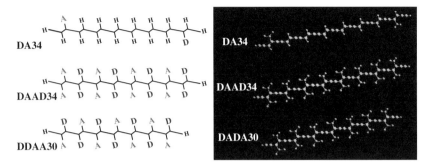

FIGURE 8.44 Models of quantum dots with poly-diacetylene backbones. From Reference 13. *(For a color version of this figure, see the color plate section.)*

the push/pull effects of the donors/acceptors. Three types of models for quantum dots with poly-diacetylene (PDA) backbones are considered as shown in Figure 8.44. $-NH_2$ is the donor (D) and $-NO_2$ the acceptor (A). DA, DAAD, and DDAA represent the types of donor/acceptor substitution, and the numbers following them indicate the number of carbon sites N_C, which corresponds to the quantum dot length. The calculation based on the MO method, for which the details are described in articles published elsewhere [13], revealed that wavefunction separations increase in the order of DA34, DAAD34, and DDAA30.

Figure 8.45a shows Δr, r_{gn}, and ρ_β for these models. Here, ρ_β is β per 1 nm in quantum dot lengths. Δr increases and r_{gn} decreases in the order of DA34, DAAD34, and DDAA30, which makes ρ_β of DAAD34 in between maximum. This parallels the tendency shown in Figure 8.43. It is found from Figure 8.45b that the quantum dot length affects ρ_β and that adjusting the length improves the second-order optical nonlinearity. In DA and DAAD, ρ_β reaches the maximum near $N_C = 20$, corresponding to a quantum dot length of ~ 2 nm. Assuming PDA wire density of 1.3×10^{14} 1/cm^2, the expected maximum EO coefficient of DAAD is about 3000 pm V^{-1}, which is 100 times larger than the EO coefficient r_{33} of LN.

MO calculations revealed that energy gap of DAAD is smaller than that of PDA with no donor/acceptor substitution. Using the phenomena, it is possible to insert many DAAD quantum dots into a PDA backbone to construct a DAAD-type polymer MQD shown in Figure 8.46 [14].

8.6.4 Cancer Therapy

Applications of LP-MLD and SOLNET to the cancer therapy were proposed. One example is a drug delivery system, in which drugs are synthesized *in situ* at cancer sites from nontoxic component molecules by LP-MLD without attacking normal cells [8, 20, 33, 61].

Another example is the SOLNET-assisted laser surgery [8, 20, 33, 61]. The concept is shown in Figure 8.47. First, luminescent molecules are attached to cancer cells by LP-MLD. After setting an optical fiber and a PRI material into a region surrounding

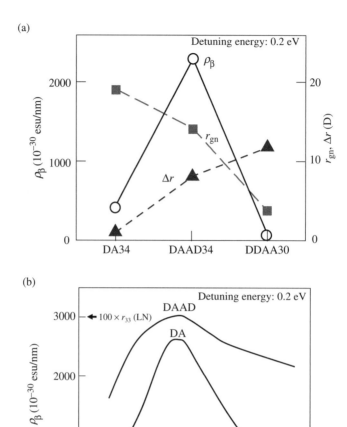

FIGURE 8.45 (a) Δr, r_{gn}, and ρ_β of quantum dots. (b) Quantum dot length dependence of Δr, r_{gn}, and ρ_β. From Reference 13.

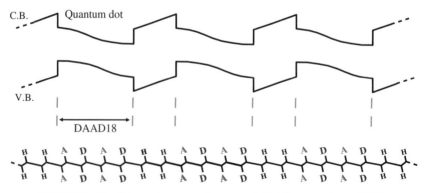

FIGURE 8.46 Schematic energy diagram of DAAD-type polymer MQD.

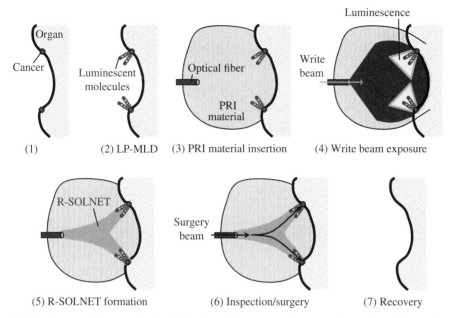

FIGURE 8.47 Concept of SOLNET-assisted laser surgery. From Reference 33. *(For a color version of this figure, see the color plate section.)*

the cancer cells, a write beam is introduced from the optical fiber into the PRI material. Then, R-SOLNET is constructed targeting the cancer sites to connect the optical fiber and the cancer cells. By introducing surgery laser beams into the R-SOLNET via the optical fiber, cancer cells are destroyed selectively.

8.6.5 Molecular Circuits

The polymer/molecular wires grown by MLD, which are schematically illustrated in Figure 8.6, will provide future molecular circuits due to the following reasons [8]:

- High carrier-mobility characteristics of π-conjugated systems.
- Small wire diameter, which is smaller than that of carbon nanotubes.
- Molecular sequence controllability in the wire with monomolecular scale, namely, precise wavefunction shape controllability, which is not available in carbon nanotubes and graphene.

The molecular circuits may consist of molecular LEDs, PDs, photovoltaic devices, and transistors shown in Figure 8.48. Tentative examples of conjugated polymer wires with pn junctions are shown in Figure 8.48. For the donor/acceptor substitution type pn junction, donors and acceptors are substituted in the n-type region and the p-type region. For the backbone type pn junction, an n-type backbone and a p-type

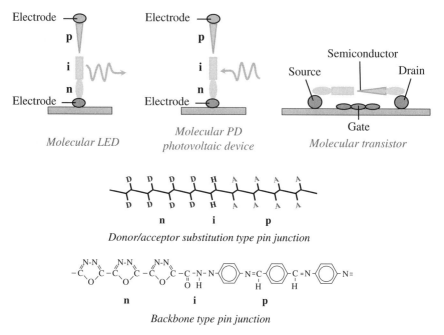

FIGURE 8.48 Concept of molecular circuits and tentative examples of pn junctions in conjugated polymer wires.

backbone are connected. In the example shown in the figure, poly-AM and poly-oxadiazole (OXD) are the backbones to make a junction. Although p/n-type characteristics of poly-AM and poly-OXD are uncertain, possibly, if poly-AM and poly-OXD are formed within a single polymer wire with molecular arrangements like ABABABABADCDCDCDCD using the reactions shown in Figures 8.8 and 8.49 with Molecule A, OA; Molecule B, ODH; Molecule C, TPA; and Molecule D, PPDA; some kind of "pn-like" junctions might be formed in the poly-AM/poly-OXD interface region.

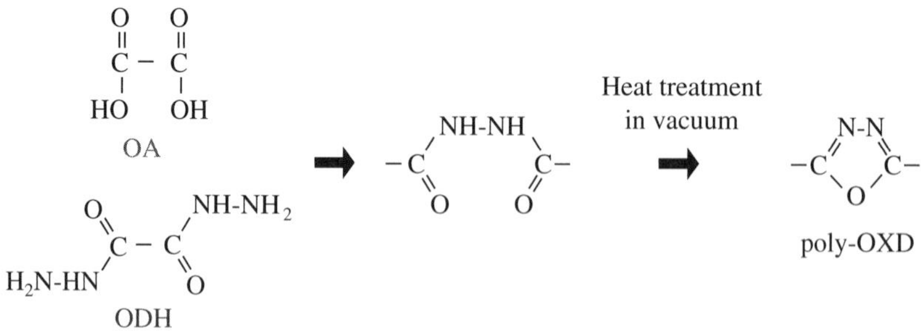

FIGURE 8.49 Reaction between OA and ODH.

8.7 SUMMARY

Tailored organic thin films with designated nanoscale structures can be constructed by the "static molecular assembling" using MLD. One of the nanoscale structures grown by MLD is the polymer MQD and the molecular MQD. The MQDs enable us to control the electron wavefunction shapes and dimensionality between 0D and 1D, which optimizes photonic and electronic properties of thin films and generates novel phenomena. Nanoscale structures can also be constructed by the "dynamic molecular assembling." One of the examples is SOLNET, which enables optical solder with self-aligned optical coupling capability and self-aligned optical waveguides constructed three dimensionally in free spaces filled with PRI materials. MLD and SOLNET are expected to be applied to sensitized solar cells, artificial photosynthesis, optical interconnects within computers, EO thin films, cancer therapy, and molecular circuits. The author hopes that their application fields will expand widely in future.

REFERENCES

[1] K. S. Novoselov, A. K. Geim, S. V. Morozov, D. Jiang, Y. Zhang, S. V. Dubonos, I. V. Grigorieva, and A. A. Firsov, "Electric field effect in atomically thin carbon films," *Science* **306**, 666–669 (2004).

[2] S. Iijima, "Helical microtubules of graphitic carbon," *Nature* **354**, 56–58 (1991).

[3] Z. Yao, H. W. Ch. Postma, L. Balents, and C. Dekker, "Carbon nanotube intermolecular junctions," *Nature* **402**, 273–276 (1999).

[4] H. Shirakawa, E. J. Louis, A. G. MacDiarmid, C. K. Chiang, and A. J. Heeger, "Synthesis of electrically conducting organic polymers: halogen derivatives of polyacetylene, $(CH)_x$," *J. Chem. Soc. Chem. Comm.*, 578–580 (1977).

[5] T. Yoshimura, S. Tatsuura, W. Sotoyama, A. Matsuura, and T. Hayano, "Quantum wire and dot formation by chemical vapor deposition and molecular layer deposition of one-dimensional conjugated polymer," *Appl. Phys. Lett.* **60**, 268–270 (1992).

[6] T. Yoshimura, S. Tatsuura, and W. Sotoyama, "Polymer films formed with monolayer growth steps by molecular layer deposition," *Appl. Phys. Lett.* **59**, 482–484 (1991).

[7] T. Yoshimura, E. Yano, S. Tatsuura, and W. Sotoyama, "Organic functional optical thin film, fabrication and use thereof," US Patent 5,444,811 (1995).

[8] T. Yoshimura, *Thin-Film Organic Photonics: Molecular Layer Deposition and Applications*, (CRC/Taylor & Francis, Boca Raton, FL, 2011).

[9] T. Yoshimura, J. Roman, Y. Takahashi, W. V. Wang, M. Inao, T. Ishitsuka, K. Tsukamoto, K. Motoyoshi, and W. Sotoyama, "Self-organizing waveguide coupling method 'SOLNET' and its application to film optical circuit substrates," Proc. 50th Electronic Components & Technology Conference (ECTC), IEEE, Las Vegas, pp. 962–969 (2000).

[10] T. Yoshimura, J. Roman, Y. Takahashi, W. V. Wang, M. Inao, T. Ishituka, K. Tsukamoto, K. Motoyoshi, and W. Sotoyama, "Self-organizing lightwave network (SOLNET) and Its application to film optical circuit substrates," *IEEE Trans. Comp., Packag. Technol.* **24**, 500–509 (2001).

[11] T. Yoshimura, W. Sotoyama, K. Motoyoshi, T. Ishitsuka, K. Tsukamoto, S. Tatsuura, H. Soda, and T. Yamamoto, "Method of producing optical waveguide system, optical device and optical coupler employing the same, optical network and optical circuit board," US Patent 6,081,632 (2000).

[12] T. Yoshimura, *Optical Electronics: Self-organized Integration and Applications* (Pan Stanford Publishing Pte. Ltd., Singapore, 2012).

[13] T. Yoshimura, "Enhancing second-order nonlinear optical properties by controlling the wave function in one-dimensional conjugated molecules," *Phys. Rev.* **B40**, 6292–6298 (1989).

[14] T. Yoshimura, "Design of organic nonlinear optical materials for electro-optic and all-optical devices by computer simulation," *FUJITSU Sc. Tech. J.* **27**, 115–131 (1991).

[15] R. K. Kostuk, "Dynamic hologram recording characteristics in DuPont photopolymers," *Appl. Opt.* **38**, 1357–1363 (1999).

[16] L. Esaki and R. Tsu, "Superlattice and negative differential conductivity in semiconductors", *IBM J. Res. Dev.* **14**, 61–65 (1970).

[17] T. Suntola, "Atomic layer epitaxy," *Mater. Sci. Rep.* **4**(7), (Elsevier Science Publishers, 1989).

[18] Y. Takahashi, M. Iijima, K. Iinagawa, and A. Itoh, "Synthesis of aromatic polyimide film by vacuum deposition polymerization," *J. Vac. Sci. Technol.* **A 5**, 2253–2256 (1987).

[19] A. Kubono, N. Okui, K. Tanaka, S. Umemoto, and T. Sakai, "Highly oriented polyamide thin films prepared by vapor deposition polymerization," *Thin Solid Films* **199**, 385–393 (1991).

[20] T. Yoshimura, H. Watanabe, and C. Yoshino, "Liquid-phase molecular layer deposition (LP-MLD): potential applications to multi-dye sensitization and cancer therapy," *J. Electrochem. Soc.* **158**, 51–55 (2011).

[21] T. Yoshimura, "Liquid phase deposition," *Japanese Patent, Tokukai Hei* 3-60487 (1991) [in Japanese].

[22] T. Yoshimura and Y. Kubota "Fabrication method of organic films," *Japanese Patent, Tokukai Hei* 1-166528 (1989) [in Japanese].

[23] T. Yoshimura, Y. Kubota, and T. Suzuki, "Fabrication method of organic films," Japanese Patent, Tokukai Hei 1-180531 (1989) [in Japanese].

[24] G. N. Parsons, S. M. George, and M. Knez, "Progress and future directions for atomic layer deposition and ALD-based chemistry," *MRS Bull.* **36**, 865–871 (2011).

[25] T. Yoshimura, S. Ito, T. Nakayama, and K. Matsumoto, "Orientation-controlled molecule-by-molecule polymer wire growth by the carrier-gas-type organic chemical vapor deposition and the molecular layer deposition," *Appl. Phys. Lett.* **91**, 033103-1-3 (2007).

[26] T. Yoshimura and Y. Kudo, "Monomolecular-step polymer wire growth from seed core molecules by the carrier-gas type molecular layer deposition (MLD)," *Appl. Phys. Express* **2**, 015502-1-3 (2009).

[27] P. W. Loscutoff, S. B. Clendenning, and S. F. Bent, "Fabrication of organic thin films for copper diffusion barrier layers using molecular layer deposition," *Mater. Res. Soc. Symp. Proc.* **1249**, F02–F03 (2010).

[28] H. Zhou and S. F. Bent, "Molecular layer deposition of functional thin films for advanced lithographic patterning," *ACS Appl. Mater. Interfaces* **3**, 505–511 (February 2011).

[29] X. H. Liang and A. W. Weimer, "Photoactivity passivation of TiO_2 nanoparticles using molecular layer deposition (MLD) polymer films," *J. Nanopart. Res.* **12**, 135–142 (2009).

[30] M. Putkonen, J. Harjuoja, T. Sajavaara, and L. Niinistö, "Atomic layer deposition of polyimide thin films," *J. Mater. Chem.* **17**, 664–669 (2007).

[31] S. Yoshida, T. Ono, and M. Esashi, "Deposition of conductivity-switching polyimide film by molecular layer deposition and electrical modification using scanning probe microscope," *Micro Nano Lett.* **5**, 321–323 (2010).

[32] T. Yoshimura, R. Ebihara, and A. Oshima, "Polymer wires with quantum dots grown by molecular layer deposition of three source molecules for sensitized photovoltaics," *J. Vac. Sci. Technol. A* **29**, 051510-1-6 (2011).

[33] T. Yoshimura, C. Yoshino, K. Sasaki, T. Sato, and M. Seki, "Cancer therapy utilizing molecular layer deposition (MLD) and self-organized lightwave network (SOLNET) - proposal and theoretical prediction-," *IEEE J. Sel. Top. Quantum Electron.* **18**, 1192–1199 (2012).

[34] S. M. George, B. Yoon, and A. A. Dameron, "Surface chemistry for molecular layer deposition of organic and hybrid organic-inorganic polymers," *Acc. Chem. Res.* **42**, 498–508 (2009).

[35] B. H. Lee, K. H. Lee, S. Im, and M. M. Sung, "Molecular layer deposition of ZrO_2-based organic–inorganic nanohybrid thin films for organic thin film transistors," *Thin Solid Films* **517**, 4056 (2009).

[36] J. S. Park, H. Chae, H. Chung, and S. I. Lee, "Thin film encapsulation for flexible AM-OLED: a review," *Semicond. Sci. Technol.* **26**, 034001 (2011).

[37] Y. C. Lee, "Atomic layer deposition/molecular layer deposition for packaging and interconnect of N/MEMS," *Proc. SPIE* **7928**, 792802 (2011).

[38] T. Yoshimura and S. Ishii, "Effect of quantum dot length on the degree of electron localization in polymer wires grown by molecular layer deposition," *J. Vac. Sci. Technol. A* **31**, 031501 (2013).

[39] B. W. Faughnan, R. S. Crandall, and P. M. Heyman, "Electrochromism in WO_3 amorphous films," *RCA Rev.* **36**, 177–197 (1975).

[40] T. Yoshimura, "Oscillator strength of small-polaron absorption in WO_x ($x \leq 3$) electrochromic thin films," *J. Appl. Phys.* **57**, 911–919 (1985).

[41] H. J. Meier, "Sensitization of electrical effects in solids," *Phys. Chem.* **69**, 719–729 (1965).

[42] T. Yoshimura, K. Kiyota, H. Ueda, and M. Tanaka, "Contact potential difference of ZnO layer adsorbing p-type dye and n-type dye," *Jpn. J. Appl. Phys.* **18**, 2315–2316 (1979).

[43] K. Kiyota, T. Yoshimura, and M. Tanaka, "Electrophotographic behavior of ZnO sensitized by two dyes," *Photogr. Sci. Eng.* **25**, 76–79 (1981).

[44] T. Yoshimura, K. Kiyota, H. Ueda, and M. Tanaka, "Mechanism of spectral sensitization of ZnO coadsorbing p-type and n-type dyes," *Jpn. J. Appl. Phys.* **20**, 1671–1674 (1981).

[45] S. J. Frisken, "Light-induced optical waveguide uptapers," *Opt. Lett.* **18**, 1035–1037 (1993).

[46] T. M. Monro, C. M. de Sterke, and L. Poladian, "Selfwriting a waveguide in glass using photosensitivity," *Opt. Commun.* **119**, 523–526 (1995).

[47] E. Fazio, M. Alonzo, F. Devaux, A. Toncelli, N. Argiolas, M. Bazzan, C. Sada, and M. Chauvet, "Luminescence-induced photorefractive spatial solitons," *Appl. Phys. Lett.* **96**, 091107-1-3 (2010).

[48] A. S. Kewitsch and A. Yariv, "Self-focusing and self-trapping of optical beams upon photopolymerization," *Opt. Lett.* **21**, 24–26 (1996).

[49] M. Kagami, T. Yamashita, and H. Ito, "Light-induced self-written three-dimensional optical waveguide," *Appl. Phys. Lett.* **79**, 1079–1081 (2001).

[50] S. Shoji and S. Kawata, "Self-written waveguides in photopolymerizable resins," *Opt. Lett.* **27**, 185–187 (2002).

[51] N. Hirose and O. Ibaragi, "Optical component coupling using self-written waveguides," *Proc. SPIE* **5355**, 206–214 (2004).

[52] T. Yoshimura, T. Inoguchi, T. Yamamoto, S. Moriya, Y. Teramoto, Y. Arai, T. Namiki, and K. Asama, "Self-organized lightwave network based on waveguide films for three-dimensional optical wiring within boxes," *J. Lightwave Technol.* **22**, 2091–2100 (2004).

[53] T. Yoshimura and H. Kaburagi, "Self-organization of optical waveguides between misaligned devices induced by write-beam reflection," *Appl. Phys. Express* **1**, 062007-1-3 (2008).

[54] M. Seki and T. Yoshimura, "Reflective self-organizing lightwave network (R-SOLNET) using a phosphor," *Opt. Eng.* **51**, 074601-1–5 (2012).

[55] T. Yoshimura and M. Seki, "Simulation of self-organized parallel waveguides targeting nanoscale luminescent objects," *J. Opt. Soc. Am. B* **30**, 1643 (2013).

[56] S. Ono, T. Yoshimura, T. Sato, and J. Oshima, "Fabrication and evaluation of nano-scale optical circuits using sol-gel materials with photo-induced refractive index variation characteristics, " *J. Lightwave Technol.* **27**, 1229–1235 (2009).

[57] S. Ono, T. Yoshimura, T. Sato, and J. Oshima, "Fabrication of self-organized optical waveguides in photo-induced refractive index variation sol-gel materials with large index contrast," *J. Lightwave Technol.* **27**, 5308–5313 (2009).

[58] H. Kaburagi and T. Yoshimura, "Simulation of micro/nano-scale self-organized lightwave network (SOLNET) using the finite difference time domain method," *Opt. Commun.* **281**, 4019–4022 (2008).

[59] T. Yoshimura, K. Wakabayashi, and S. Ono, "Analysis of reflective self-organized lightwave network (R-SOLNET) for Z-connections in three-dimensional optical circuits by the finite difference time domain method," *IEEE J. Sel. Top. Quantum Electron.* **17**, 566–570 (2011).

[60] B. O'Regan and M. Gratzel, "A low-cost, high-efficiency solar cell based on dye-sensitized colloidal TiO_2 films," *Nature* **353**, 737–740 (1991).

[61] T. Yoshimura, "Molecular layer deposition and applications to solar cells and cancer therapy," *Japanese Patent, Tokukai* 2012-45351 (2012) [in Japanese].

[62] R. Shioya and T. Yoshimura, "Design of solar beam collectors consisting of multi-layer optical waveguide films for integrated solar energy conversion systems" *J. Renew. Sust. Energy* **1**, 033106-1-15 (2009).

[63] S. Sato, T. Arai, T. Morikawa, K. Uemura, T. M. Suzuki, H. Tanaka, and T. Kajino, "Selective CO_2 conversion to formate conjugated with H_2O oxidation utilizing semiconductor/complex hybrid photocatalysts," *J. Am. Chem. Soc.* **133**, 15240–15243 (2011).

[64] K. Boeneman, D. E. Prasuhn, J. S. Melinger, M. Ancona, M. H. Stewart, K. Susumu, A. Huston, I. L. Medintz, J. B. Blanco-Canosa, and P. E. Dawson, "Quantum dots as a FRET donor and nanoscaffold for multivalent DNA photonic wires," *Proc. SPIE* **7909**, 79090R (2011).

[65] D. A. B. Miller, "How Large a System Can We Build Without Optics?" Workshop Notes, 8th Annual Workshop on Interconnections Within High Speed Digital Systems, Lecture 1.2, 1997.

[66] Y. S. Liu, "Lighting the way in computer design," *IEEE Circuit. Device.* January, 23–31 (1998).
[67] M. P. Christensen, P. Milojkovic, M. J. McFadden, and M. W. Haney, "Multiscale optical design for global chip-to-chip optical interconnections and misalignment tolerant packaging," *IEEE J. Select. Top. Quantum Electron.* **9**, 548–556 (2003).
[68] N. M. Jokerst, M. A. Brooke, S. Cho, S. Ilkinson, M. Vrazel, S. Fike, J. Tabler, Y. J. Joo, S. Seo, D. S. Wils, et al., "The heterogeneous integration of optical interconnections into integrated microsystems," *IEEE J. Select. Top. Quantum Electron.* **9**, 350–360 (2003).
[69] T. Mikawa, M. Kinoshita, K. Hiruma, T. Ishitsuka, M. Okabe, S. Hiramatsu, H. Furuyama, T. Matsui, K. Kumai, O. Ibaragi et al., "Implementation of active interposer for high-speed and low-cost chip level optical interconnects," *IEEE J. Select. Top. Quantum Electron.* **9**, 452–459 (2003).
[70] C. Choi, L. Lin, Y. Liu, J. Choi, L. Wang, D. Haas, J. Magera, and R. T. Chen, "Flexible optical waveguide film fabrications and optoelectronic devices integration for fully embedded board-level optical interconnects," *J. Lightwave Technol.* **22**, 2168–2176 (2004).
[71] B. S. Rho, S. Kang, H. S. Cho, H.-H. Park, S.-W. Ha, and B.-H. Rhee, "PCB-compatible optical interconnection using 45°-ended connection rods and via-holed waveguides," *J. Lightwave Technol.* **22**, 2128–2134 (2004).
[72] R. M. Kubacki, "Micro-optic enhancement and fabrication through variable in-plane index of refraction (VIPIR) engineered silicon nanocomposite technology," *Proc. SPIE* **5347**, 233–246 (2004).
[73] A. Glebov, M. G. Lee, D. Kudzuma, J. Roman, M. Peters, L. Huang, and S. Zhou, "Integrated waveguide microoptic elements for 3D routing in board-level optical interconnects," *Proc. SPIE* **6126**, 61260L-1-10 (2006).
[74] F. E. Doany, B. G. Lee, S. Assefa, W. M. J. Green, M. Yang, C. L. Schow, C. V. Jahnes, S. Zhang, J. Singer, V. I. Kopp et al., "Multichannel high-bandwidth coupling of ultradense silicon photonic waveguide array to standard-pitch fiber array," *J. Lightwave Technol.* **29**, 475–482 (2011).
[75] T. Yoshimura, Y. Takahashi, M. Inao, M. Lee, W. Chou, S. Beilin, W. V. Wang, J. Roman, and T. Massingill, "Systems based on opto-electronic substrates with electrical and optical interconnections and methods for making," US Patent 6,343,171B1 (2002).
[76] T. Yoshimura, J. Roman, Y. Takahashi, M. Lee, B. Chou, S. I. Beilin, W. V. Wang, and M. Inao, "Proposal of Optoelectronic Substrate with Film/Z-Connection Based on OE-film," *Proceedings of the 3rd IEMT/IMC Symposium*, Tokyo, 1999, pp. 140–145.
[77] T. Yoshimura, M. Ojima, Y. Arai, and K. Asama, "Three-dimensional self-organized micro optoelectronic systems for board-level reconfigurable optical interconnects – performance modeling and simulation," *IEEE J. Select. Top. Quantum Electron.* **9**, 492–511 (2003).
[78] T. Yoshimura, M. Miyazaki, Y. Miyamoto, N. Shimoda, A. Hori, and K. Asama, "Three-dimensional optical circuits consisting of waveguide films and optical Z-connections," *J. Lightwave Technol.* **24**, 4345–4352 (2006).
[79] O. Qasaimeh, K. Kamath, P. Bhattacharya, and J. Phikkips, "Linear and quadratic electro-optic coefficients of self-organized $In_{0.4}Ga_{0.6}As$/GaAs quantum dots," *Appl. Phys. Lett.* **72**, 1275–1277 (1998).

[80] T. Yoshimura and T. Futatsugi., "Non-linear optical device using quantum dots," US Patent 6,294,794B1 (2001).
[81] G. F. Lipscomb, A. F. Garito, and R. S. Narang, "An exceptionally large electro-optic effect in the organic solid MNA," *J. Chem. Phys.* **75**, 1509–1516 (1981).
[82] T. Yoshimura, "Characterization of the EO effect in styrylpyridinium cyanine dye thin-film crystals by an ac modulation method," *J. Appl. Phys.* **62**, 2028–2032 (1987).

9

LIGHT-HARVESTING MATERIALS FOR ORGANIC ELECTRONICS

DAMIEN JOLY,[1] JUAN LUIS DELGADO,[1] CARMEN ATIENZA,[2] AND NAZARIO MARTÍN[1,2]

[1]*IMDEA-Nanoscience, Campus de Cantoblanco, Madrid, Spain*
[2]*Departamento de Química Orgánica, Facultad de Química, Universidad Complutense, Madrid, Spain*

9.1 INTRODUCTION

Energy is nowadays the most important problem faced by our society. The "fire age" in which our civilization has been based from the very beginning is approaching its end. Human beings have been burning a wide variety of materials since earlier times and with the advent of carbon-based fossil fuels in the last two centuries, its combustion has currently become a major problem due to the huge amounts of carbon dioxide emissions produced. Because of the resulting pollution, global warming, and degradation of the planet, a new era based on noncontaminating renewable energies is currently a priority. In this regard, the Sun considered as a giant nuclear fusion reactor represents the most powerful source of energy available in our Solar system and, therefore, its use for providing energy to our planet is among the most important challenges nowadays in science. As is well known, the Sun generates its energy by nuclear fusion of hydrogen nuclei into helium, in a process in which the Sun fuses 620 million metric tons of hydrogen each second!

The huge amount of energy produced by the Sun has been harvested, although in a very small amount, by plants and bacteria for millions of years. In this natural light-harvesting process known as photosynthesis the sunlight is efficiently transformed into chemical energy to be used in the next photosynthetic step, in a process which

---

Photonics: Scientific Foundations, Technology and Applications, Volume II, First Edition.
Edited by David L. Andrews.
© 2015 John Wiley & Sons, Inc. Published 2015 by John Wiley & Sons, Inc.

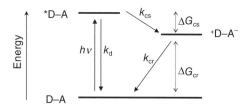

FIGURE 9.1 Energy diagram illustrating the possible pathways available in a photoinduced charge separation process between a donor (D) and an acceptor (A).

is based on an "energy cascade" architecture, where dyes acting as light-absorbing antennae are surrounding the reaction centers and where very efficient, short-range energy and electron transfer processes occurs between well-arranged, light-harvesting organic donor and acceptor pigments [1].

Along the last decades, the strong demand for energy has inspired researchers to design new molecules able to mimic nature, thus harvesting and storing solar energy. A great variety of synthetic light-harvesting compounds have been reported so far [2], involving covalent systems such as linear molecules [3], dendrimers [4], polymers [5], supramolecularly self-assembled compounds [6] as well as nanotubes [7].

At a molecular level, the absorption of light by an organic chromophore promotes an electron from the highest occupied molecular orbital (HOMO) to the lowest unoccupied molecular orbital (LUMO) resulting in the formation of the lowest singlet excited state (S_1) (Fig. 9.1). Even if the light absorption produces a singlet excited state of higher energy (S_n) relaxation to the lowest energy excited state is usually faster than any other kind of process. From that first singlet excited state, excited chromophore C* may relax back to the ground state (S_0) either nonradiatively (vibrational relaxation and internal conversion), or radiatively (fluorescence). It might also undergo an intersystem crossing process leading to the triplet state (T_1) which, eventually, can relax by radiative (phosphorescence) or nonradiative pathways. In the case of photoinduced electron transfer (PET) between two chromophores (or redox centers), the first step involves local excitation of either the Donor (D) or Acceptor (A) chromophore to generate the locally excited state (*D–A) (Fig. 9.1). Exergonic charge separation (k_{CS}) competes with decay modes (k_d) leading to the ground state, to generate the charge-separated (CS) state ($^+$D–A$^-$). In order to obtain efficient and exploitable CS states ΔG_{CS} should be as small as possible and the quantum yield of the charge separation process must be as high as possible, the lifetime of the CS state (τ_{cr}) must be significantly long, ideally in the range of the microsecond to potentially carry out further photochemical work.

Light-harvesting materials are of interest for application in the emergent fields of organic electronics and photonics. In this chapter, we will focus our attention on those systems which precise harvesting light for performing their function. In particular, artificial photosynthetic systems and organic photovoltaic (OPV) solar cells, as well as the study of molecular wires in which electron transfer processes occur upon light irradiation. Considering the exhaustive bibliography existing on these topics, we will mainly discuss some of the most relevant examples reported in recent literature concerning the different topics.

9.2 PHOTOINDUCED ELECTRON TRANSFER (PET) IN ARTIFICIAL PHOTOSYNTHETIC SYSTEMS

Artificial photosynthesis is defined as a chemical process that replicates the natural process of photosynthesis, a process that converts sunlight into chemical energy. Artificial photosynthetic systems are molecules constituted by a light-harvesting moiety covalently or supramolecularly connected to a donor and an acceptor unit. In some cases, the light-harvesting unit is able to behave as an energy or electron donor/acceptor unit, thus simplifying the molecular architecture. In this regard, many chromophores such as porphyrins, phthalocyanines (Pcs), tetrathiafulvalenes (TTF), cyanines, and boron-dipyrromethenes (BODIPYs), have been investigated. Organic synthesis allows fine tuning of their light absorption abilities, redox properties and chemical and photochemical stability, in order to maximize the energy transfer capability. On the other hand, a variety of electron acceptors have also been studied over the years, although fullerenes and their derivatives remain by far the most common systems, due to their singular chemical and redox properties, spherical geometry, and low reorganization energy [8].

In the following, we will discuss about light-harvesting systems based on fullerene derivatives, focusing our attention on fundamental aspects of PET events on fullerene-based donor–acceptor photo- and electroactive systems. In this regard, a variety of representative dyads and triads have been gathered in this chapter and discussed according to the nature of the donor unit.

9.2.1 Covalent and Supramolecular Porphyrin–Fullerene Systems

Porphyrins display a number of characteristics such as light absorption, redox properties, and stability which render them as very attractive building blocks for the design and development of new materials able to capture sunlight. These properties can be finely tuned by using organic synthesis in order to improve their features as components in artificial photosynthetic systems and photovoltaic devices. In this regard, a wide variety of dyads, triads, and tetrads involving porphyrins have been reported in the literature [9]. In the following, we will discuss some representative examples from the recent papers reported along the last years.

De Miguel et al. [10] have recently prepared a series of porphyrin–fullerene dyads (**1**), connected through different triazole-containing linking bridges by means of click chemistry reactions (Fig. 9.2). Light irradiation promotes one electron transfer from the zinc porphyrin (ZnP) to the [60]fullerene electron acceptor. The lifetime of the formed charge-separated radical ion pairs is in the range of 415–850 ns in THF and benzonitrile, which may be explained by the changes on the substitution pattern of the triazole linker.

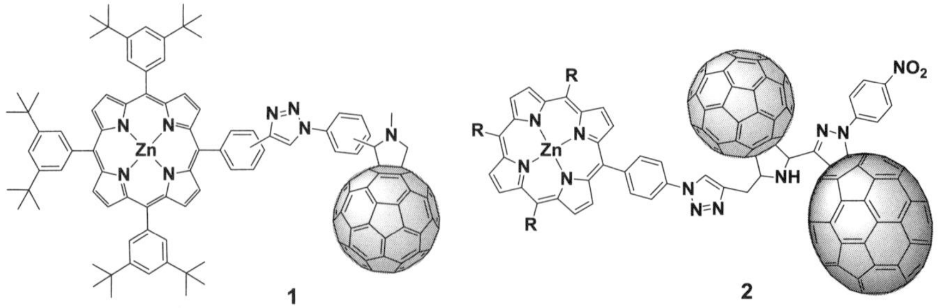

FIGURE 9.2 ZnP–C$_{60}$ dyads and triads with a triazole ring as spacer.

Our research group has recently described the first D$_1$–A$_1$–A$_2$ triad bearing two distinct fullerene units and a ZnP as electron donor [11] (Fig. 9.2, **2**). The different substitution pattern of each fullerene generates a right gradient of redox centers driving the charge eventually from the ZnP to the C$_{70}$. Due to this redox gradient and the distance between the photoactive units, the lifetime of the CS state of **2**, 100 ns in tetrahydrofuran (THF), is two orders of magnitude higher than that of its analog dyad (ZnP–C$_{60}$). Porphyrins have also been used to prepare larger photoactive materials. Thus, Imahori et al. [12] have described the preparation of tetrad **3**, where the correct position of the different photoactive units generates a redox gradient from the ferrocene unit (Fc) to the [60]fullerene (Fig. 9.3). Following the initial excitation of the ZnP, energy transfer occurs from the singlet excited state of the zinc porphyrin to the free base porphyrin. Then, electron transfer from H$_2$P to the fullerene followed by a sequential electron transfer process render Fc$^{•+}$–ZnP–H$_2$P–C$_{60}$$^{•-}$ as the final charge-separated state, showing an outstanding lifetime of 0.38 s in benzonitrile, which is similar to that observed in natural photosynthesis from bacterial photosynthetic reaction centers.

Megiatto et al. [13] reported the synthesis of the first [2] catenate with porphyrin and [60]fullerene subunits (**4**) (Fig. 9.4) showing that supramolecular techniques together with covalent chemistry can efficiently lead to 3D nanostructures containing a variety of electron donor and acceptor moieties. Thus, upon excitation of the

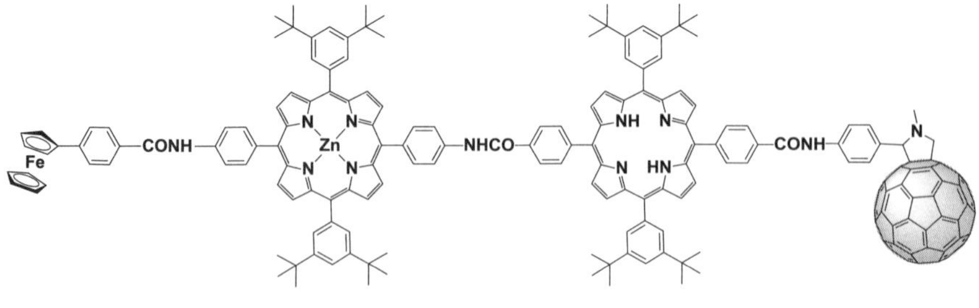

FIGURE 9.3 Fc–ZnP–H$_2$P–C$_{60}$ tetrad (**3**).

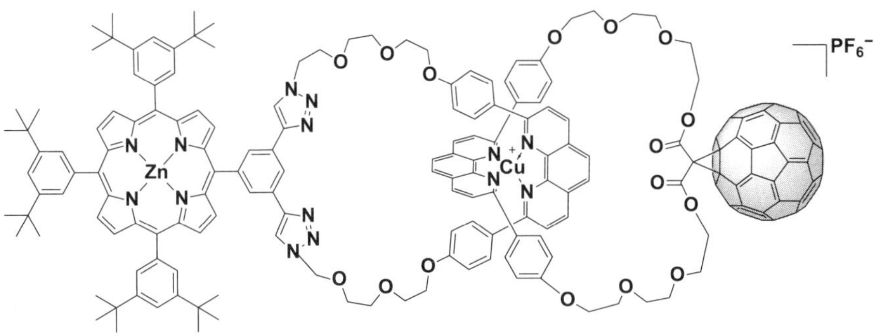

FIGURE 9.4 ZnP–C_{60} dyad with a catenane spacer (**4**).

porphyrin moiety, the long distance ZnP$^{•+}$–[Cu(phen)$_2$]$^+$–$C_{60}$$^{•−}$ charge-separated ion radical pair state was formed through a sequence of energy and electron transfer processes and showing an outstanding lifetime of 1.5 µs in THF.

9.2.2 Covalent and Supramolecular TTF–Fullerene Systems

The first synthesis of TTF [14] and the discovery of its conducting properties either doped with chlorine [15] or associated with a strong acceptor such as tetracyano-*p*-quinodimethane (TCNQ) in charge-transfer complexes [16] showed the great potential of these donor molecules. TTF derivatives display outstanding electronic properties and chemical versatility, thus offering a wide variety of structural modifications [17]. The insertion of a *p*-quinoid spacer between the 1,3-dithiole rings of the TTF molecule produces the proaromatic π-extended TTF (exTTF) molecule, which is readily obtained from 9,10-anthraquinone, and in which the lateral benzene rings result in a highly distorted molecule out of planarity in the ground state. Furthermore, the cyclic voltammetry of exTTF shows only one oxidation wave involving a two-electron process affording the dication species, instead of two separate one-electron oxidation waves to form the radical-cation and dication states found for pristine TTF. Moreover, the oxidation process leads to major changes in both geometry and electronic properties as the formed exTTF dication has a central planar and aromatic anthracene unit and two orthogonal aromatic dithiolium cations. In this regard TTF and exTTF derivatives are very interesting electron donors for the preparation of artificial photosynthetic systems [18].

Martin et al. [19] have synthesized a methanofullerene dyad (**5**) from an exTTF bearing *p*-tosylhydrazone and [60]fullerene under basic conditions (Fig. 9.5). Upon excitation, the singlet excited state of the fullerene is instantaneously formed and gives rise to the charge-separated radical ion pair TTF$^{•+}$–$C_{60}$$^{•−}$. The lifetime of the charge-separated state, which ranges from 95 to 220 ns, depends on the solvent polarity, on the substituent at the periphery of the dithiole rings (R = H and R = SMe) and the excited state of the fullerene derivative ([6-6]-closed vs. [6-5]-open isomers).

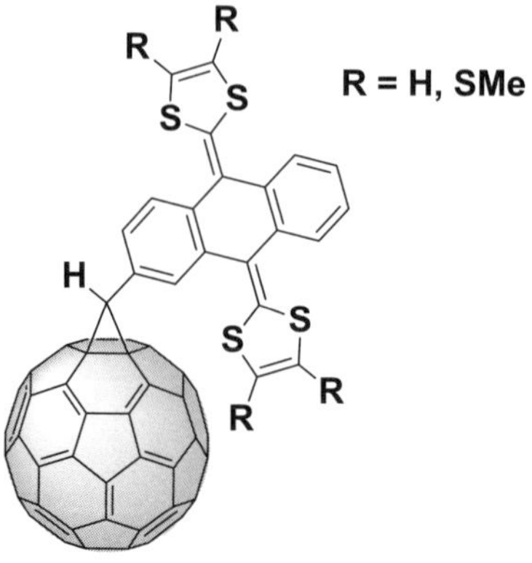

FIGURE 9.5 ExTTF–C$_{60}$ dyad (**5**).

This example illustrates the advantage of exTTF over pristine TTF. The gain of aromaticity and planarity of the donor moiety stabilizes the radical ion pair and gives rise to a longer excited state lifetime. These properties were further exploited by Sanchez et al. [20] in a triad involving two exTTF units and a [60]fullerene. The two exTTFs were linked by using a Wittig reaction and a further 1,3-dipolar cycloaddition between C$_{60}$, the suitably functionalized exTTF dimer derivative, and sarcosine (*N*-methylglycine), led to the formation of this synthetically demanding triad (**6**) (Fig. 9.6). Cyclic voltammetry showed that both TTFs have similar oxidation potential values as shown by a broad four-electron wave centered at 0.42 V. This might be the reason for the equilibrium between the two different charge-separated excited states (exTTF$_1$)–(exTTF$_2$)$^{\bullet+}$–C$_{60}{}^{\bullet-}$ and (exTTF$_1$)$^{\bullet+}$–(exTTF$_2$)–C$_{60}{}^{\bullet-}$ that show two different lifetimes of 664 ns and a remarkable 111 μs, respectively, in dimethylformamide (DMF).

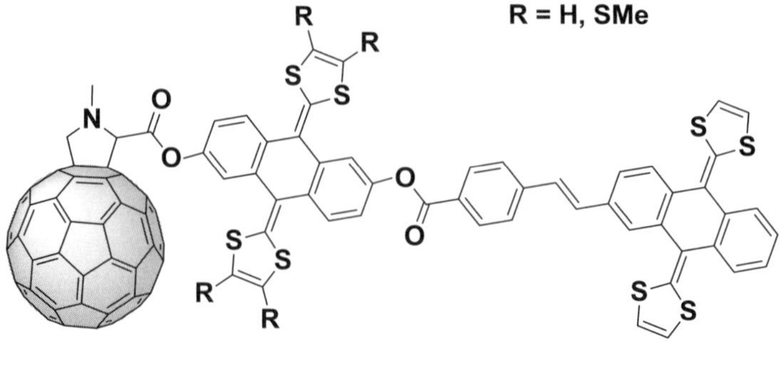

FIGURE 9.6 ExTTF$_1$–exTTF$_2$–C$_{60}$ triad (**6**).

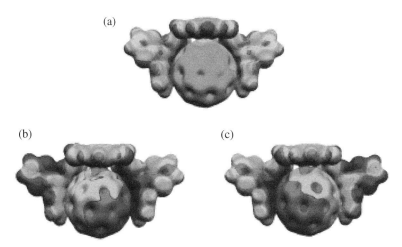

FIGURE 9.7 Electrostatic potential (B3LYP/6-31G**) calculated for 7–C_{60} in (a) the ground electronic state and (b) the charge-separated HOMO→LUMO+4, and (c) HOMO−1→LUMO+4 excited states. Blue: positive potential, red: negative potential (color scale for $\delta^+ \to \delta^-$: blue→green→yellow→orange→red). Reprinted from Reference 21 with permission from Wiley-VCH. (*For a color version of this figure, see the color plate section.*)

Gayathri et al. [21] have taken advantage of the curved geometry of exTTF for the molecular recognition of the convex exterior surface of C_{60} by the concave surface of π-exTTF derivatives to synthesize a variety of supramolecular donor–acceptor structures. Two different molecular tweezers (**7, 8**) based on exTTF bearing either an isophthalic or terephthalic diester spacer were designed to be used as host for C_{60} (Fig. 9.7). The variation in UV–vis absorption spectrum fits well to a 1:1 binding isotherm affording a binding constant of $(2.98 \pm 0.12) \times 10^3$ M^{-1}. In particular, the evolution of a new broad transition at 488 nm in benzonitrile, while the absorption maximum at 438 nm decreases, and an isosbestic point around 450 nm, indicated strong electronic coupling between the tweezers and the fullerene. Transient absorption spectroscopy studies revealed an excited state-centered on exTTF which rapidly evolves to a charge-separated state tweezer exTTF$^{\bullet+}$–$C_{60}{}^{\bullet-}$ with a short lifetime of 12.7 ps.

9.2.3 Covalent and Supramolecular Phthalocyanine–Fullerene Systems

Phthalocyanines (Pcs) which have a two-dimensional 18-π-electron aromatic system possess unique physicochemical properties which make them valuable building blocks to be used as materials for molecular photovoltaics. Materials for molecular photovoltaics Pcs are chemically and thermally stable compounds which present an absorption in the red/near-infrared (NIR) region of the solar spectrum with remarkably high extinction coefficients (200,000 M^{-1} cm^{-1}).

The simplest Pc–C_{60} dyad (**9**) was prepared and photochemically studied by Guldi et al. [22]. This system is the product of a 1,3-dipolar cycloaddition between the formyl-ZnPc, *N*-methylglycine, and [60]fullerene (Fig. 9.8). The authors reported

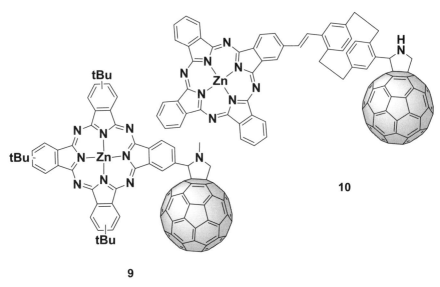

FIGURE 9.8 Zinc phthalocyanine C_{60} dyads **9** and **10**.

electron transfer interactions between the two parts of the dyad upon light irradiation, and the lifetime for the formed charge-separated state was estimated to be 3 ns in the polar solvent benzonitrile.

A great variety of covalently connected bridges between Pc and C_{60} have been used to study its influence on the formation of the charge-separated state including pseudocovalent systems such as 2,2-paracyclophane (Fig. 9.8). Kahnt et al. [23] have described the Pc derivative **10** where the paracyclophane bridge does not prevent the electronic communication between the electroactive units (Pc and C_{60}). On the contrary, light irradiation resulted in the efficient generation of the charge-separated state whose lifetime value was even longer than the phenylenevinylene-bridged analog due to its lower charge recombination rate. Moreover, the charge-separated state lifetime resulted to be strongly dependent on the polarity of the medium, ranging from 458 ns in toluene to 330 ps in the more polar benzonitrile.

A considerably more complex structure has been developed by El-Khouly et al. [24] involving an axially substituted silicon Pc with two different acceptors, naphthalenediimide (NDI) and [60]fullerene (Fig. 9.9, **11**). The resulting photo- and electroactive pentad gives rise to a charge-separated state showing a lifetime of 1000 ps in benzonitrile. Indeed, the addition of [60]fullerene as secondary acceptor to NDI increases the lifetime while favoring electron shift from the NDI to the C_{60}.

Torres et al. [25] have synthesized a supramolecular bridge through Watson–Crick hydrogen bonding with a ZnPc and a [60]fullerene, (Fig. 9.10, **12**). Despite self-aggregation of the Pc-cytidine derivative in organic medium, the final complex can be formed upon addition of the guanosine–C_{60}, showing a good binding affinity (2.6 ± 0.2) × 10^6 M^{-1}. The lifetime of the photoinduced charge-separated state resulted to be 3.0 ns in toluene, which is similar to the covalently linked Pc–C_{60} dyad, thus

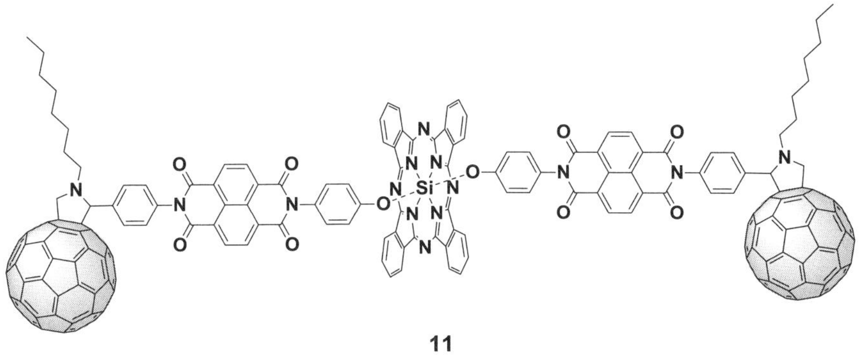

FIGURE 9.9 Photo- and electroactive pentad SiPc–(NDI)$_2$–(C$_{60}$)$_2$ (**11**).

showing that using a supramolecular approach does not sacrifice lifetime and results in a straightforward and more practical synthetic pathway.

D'Souza et al. [26] have developed a self-assembled tetrad with two C$_{60}$ and two Pcs where all components are assembled through supramolecular interactions. First, a cofacial Pc dimer is formed via potassium-induced dimerization and, then, facial coordination and ammonium complexation of a functionalized [60]fullerene complete the supramolecular assembly (Fig. 9.11, **13**). The binding constant of the Pc dimer is 5.05×10^{23} M^{-1}, showing a great stability but a moderate 2.5×10^3 M^{-1} for the final complex. Photophysical studies showed that, upon light irradiation, the charge separation occurs from the triplet excited state of the ZnPc moiety to the C$_{60}$ moiety, rather than the singlet excited state, to eventually form the charge-separated

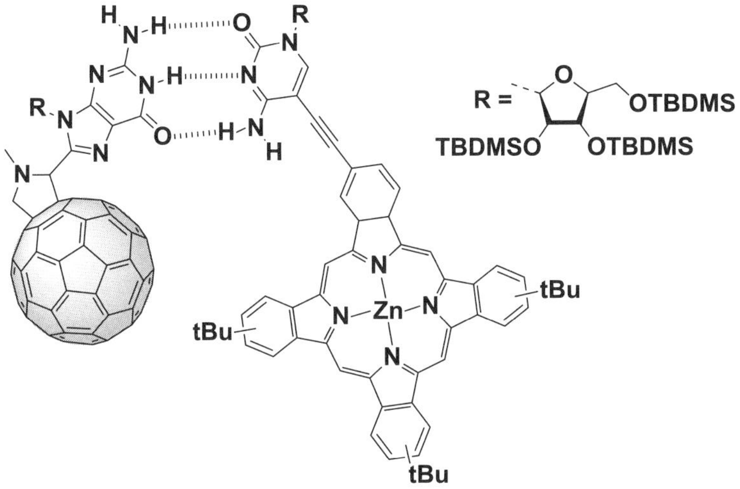

FIGURE 9.10 Supramolecular ZnPc–C$_{60}$ dyad (**12**).

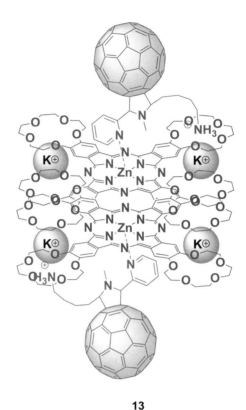

13

FIGURE 9.11 (ZnPc)$_2$–(C$_{60}$)$_2$ supramolecular tetrad (**13**). (*For a color version of this figure, see the color plate section.*)

state of the supramolecular complex. Pc$_2$–(C$_{60}$)$_2$ supramolecular complex showed a longer charge-separated state lifetime (6.7 μs) compared to that of the simpler Pc–C$_{60}$ complex, (4.8 μs), thus illustrating the crucial role of the Pc dimer to stabilize the charge-separated state.

9.2.4 Covalent Cyanine–Fullerene Systems

Cyanines have initially been used for silver halide emulsion photography. However, more recently, these fascinating compounds have shown their potential biological, medical, laser and optoelectronic device applications. Cyanines possess important properties such as absorption in the NIR with high extinction coefficients, high stability, solution processing ability, and suitable electrochemical properties which render these materials as very relevant for the study of PET processes and photovoltaics [27].

Xiao et al. [28] reported the first cyanine–fullerene dyad consisting in a dumbbell-type C$_{60}$ dimer linked through a cyanine spacer. Meng et al. [29] have prepared

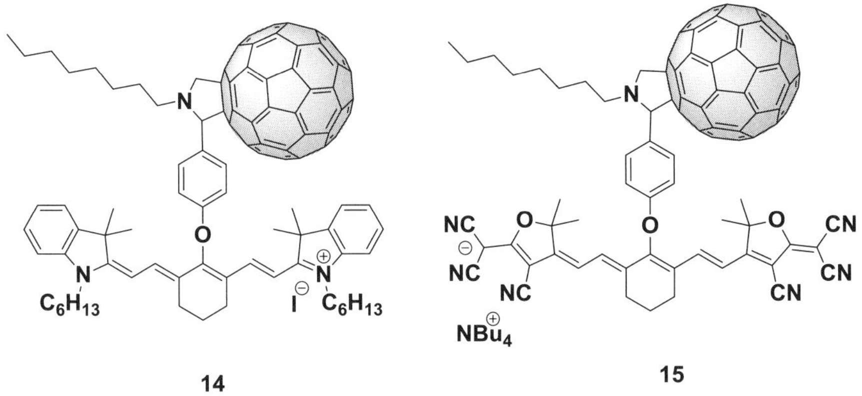

FIGURE 9.12 Cationic and anionic cyanine–C$_{60}$ derivatives **14** and **15**.

a cyanine–C$_{60}$ dyad which was used to construct a photovoltaic device reaching a modest 0.1% light energy conversion efficiency at low incident light intensity. Our research group [30] has recently described the preparation of heptamethine cyanine C$_{60}$ and C$_{70}$ conjugates (**14**, **15**) by reacting heptamethine cyanine aldehyde derivative with *N*-octylglycine, in the presence of the corresponding fullerene (Fig. 9.12). Time-resolved, transient absorption spectroscopy measurements revealed that photoexcitation is followed by a sequence of charge-transfer events from higher singlet excited states S_2 and the lowest singlet excited state S_1 to the fullerene. A similar trend was observed in a series of anionic heptamethine cyanines covalently connected to C$_{60}$ or C$_{70}$. [31]. These dyads are able to efficiently harvest sunlight in the NIR (900 nm) exhibiting quite remarkable extinction coefficients.

9.2.5 Covalent and Supramolecular BODIPY–Fullerene Systems

BODIPYs represent other kind of stable dyes showing interesting photophysical and electrochemical properties, which can be chemically modified by using organic synthesis [32]. Due to their high fluorescence quantum yields and relatively long singlet excited state lifetimes, they have been used as energy-absorbing and antenna fragments in artificial photosynthetic systems [33, 34].

Ziessel et al. [35], reported a BODIPY core covalently linked to a C$_{60}$ through a *N*-methylpyrrolidone diphenylacetylenic bridge (Fig. 9.13, **16**). The authors reported PET competing with intramolecular energy transfer from the BODIPY to the fullerene in polar solvents.

A more complex system based on ferrocene as primary electron donor, BODIPY as secondary donor, and C$_{60}$ as electron acceptor was later developed by Wijesinghe et al. [36] (Fig. 9.14, **17**). The lifetime of the charge-separated state was estimated to be 416 ps for $D_1^{•+}$–D_2–$A^{•-}$ showing efficient charge stabilization for the ferrocene-based system. Excitation of the central BODIPY of **17** results in abstraction of an

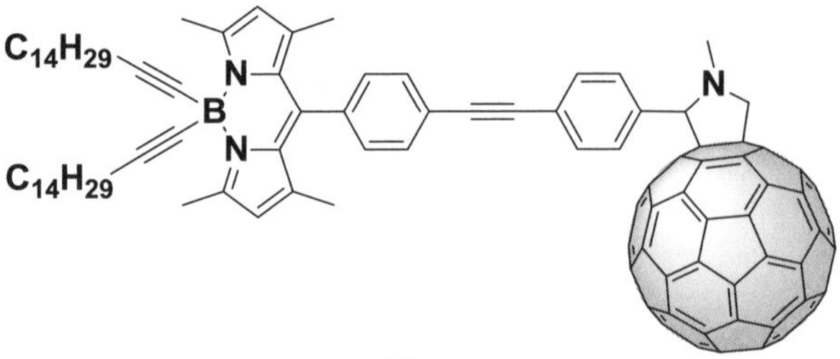

FIGURE 9.13 BODIPY–C_{60} dyad (**16**).

electron from the ferrocene moiety to produce boron dipyrrin anion radical which then undergoes a further electron shift.

Maligaspe et al. [37] have also described the preparation of supramolecular BODIPY–ZnP–C_{60} through the coordination of a fulleropyrrolidine appended with an imidazole ligand to the zinc porphyrin (Fig. 9.15, **18**). The binding constant decreases when the number of BODIPY units increases (from 3.0×10^4 to 3.9×10^3 M^{-1}) from one to four BODIPY units, respectively, probably due to steric hindrance. Photophysical studies showed an effective energy transfer from the BODIPY unit to the central porphyrin for the BDP–ZnP–C_{60} triad, although this effect was not so clear for BDP$_2$–ZnP–C_{60} and BDP$_4$–ZnP–C_{60} systems. Then, charge separation occurs with the [60]fullerene and the lifetime of the charge-separated state was estimated to be 4 ns in *o*-dichlorobenzene as solvent.

Rio et al. [38] have also developed a similar system using a Pc, instead of the porphyrin unit (Fig. 9.15, **19**). The lifetime of the charge-separated state was estimated to be 40 ns, which is longer than the same system without the BODIPY unit (<5 ns). Thus, the BODIPY unit seems to stabilize the charge-separated state. Indeed, the free-energy changes associated with the charge recombination are greater in the triad than in dyad. This shifts the driving force for charge recombination deeper into the Marcus-inverted region and serves to delay the rate of recombination of the ZnPc$^{\bullet+}$–$C_{60}^{\bullet-}$ species in the triad.

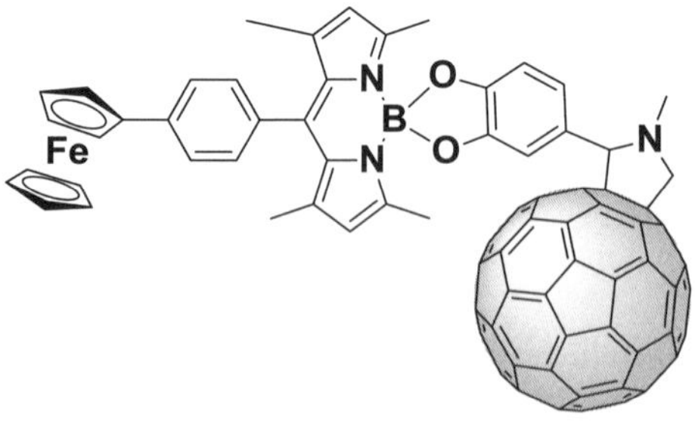

FIGURE 9.14 Fc–BDP–C_{60} triad (**17**).

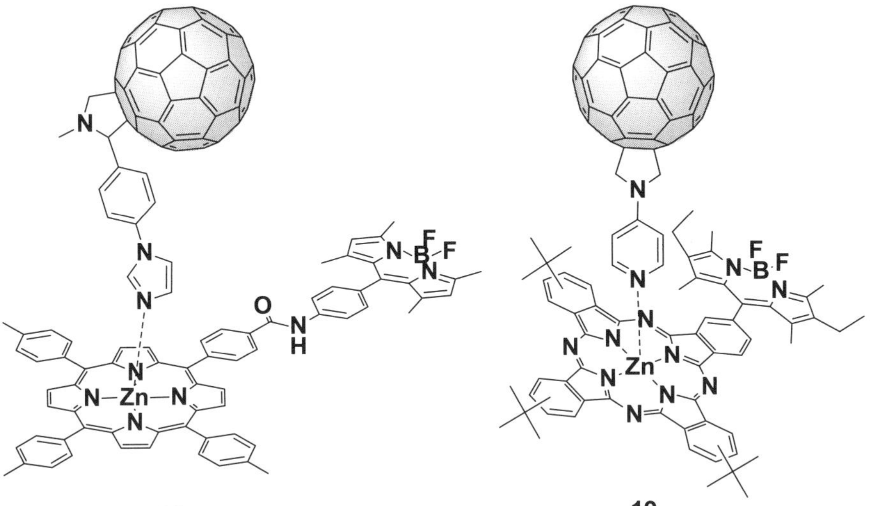

FIGURE 9.15 BDP–ZnP–C$_{60}$ supramolecular triad (**18**), and BDP–ZnPc–C$_{60}$ supramolecular triad (**19**).

The above examples reveal the great versatility of photo- and electroactive dyads, triads, and tetrads in which the donor and acceptor species can be changed and modified at will, thus controlling the photophysical features and lifetimes of the charge-separated states generated upon light irradiation. The great variety of chemical reactions available in the arsenal of the chemical synthesis allows the preparation of covalently and supramolecularly Donor/Acceptor systems able to carry out a function triggered by light irradiation. The aforementioned simple systems, when compared to those natural systems involved in the photosynthetic process, have allowed studying in a simpler manner the energy and electron processes involved upon light irradiation. Furthermore, they have also been used as a benchmark for the study of the electron transfer reactions occurring in the photovoltaic devices.

9.3 FULLERENES FOR ORGANIC PHOTOVOLTAICS

World dependence on fossil fuels is currently a key issue with important social consequences. A plausible alternative to overcome this problem could be the use of renewable energy sources, like solar energy, which could fulfill our energy requirements with environmentally clean procedures and low prices. Actually, the energy received from the Sun, calculated in 120,000 TW (5% ultraviolet; 43% visible, and 52% infrared), surpass in several thousands that consumed in the planet along a year [39]. Photovoltaic (PV) solar cells is currently a hot topic in science and since the former silicon-based device prepared by Chapin in 1954 exhibiting an efficiency around 6% [40], different semiconducting materials (inorganic, organic, molecular, polymeric, hybrids, quantum dots, etc.) have been used in order to transform sunlight into chemical energy. Among them, photo- and electroactive organic materials are promising due to key advantages such as the possibility of processing directly from solution, thus affording lighter, cheaper, and flexible all-organic PV devices.

Organic solar cells are constituted by semiconducting organic materials formed by contacting electron donor and acceptor compounds (p/n type). A further improvement in the construction of OPV devices consists on the realization of interpenetrated networks of the donor and acceptor materials. In such bulk heterojunction (BHJ) solar cells, the dramatic increase of the contact area between the D/A materials leads to a significant increase of the number of generated excitons (strongly bound electron–hole pairs formed after excitation with light) as well as their dissociation into free charge carriers and hence on the power conversion efficiency (PCE).

Transformation of solar energy into electricity occurs through a series of optical and electronic processes which basically involves (i) optical absorption of sunlight and formation of the exciton; (ii) exciton migration to the donor–acceptor interface; (iii) exciton dissociation into charges (electron and holes); and (iv) charge transport and collection at the electrodes. All these steps are not totally understood at present and a number of research groups are currently dedicated to unraveling essential aspects in the search for better energy transformation efficiencies. In contrast to inorganic semiconductors, which upon light excitation form free electrons and holes carriers, organic compounds form excitons whose dissociation into free carriers is not straightforward.

Due to their low dielectric constants, it needs to reach the donor–acceptor interface to dissociate into free charges; the driving force for this exciton dissociation being provided by the energy difference between the molecular orbitals of the donor and acceptor.

The transport of the generated free charges toward the electrodes represents another important issue. Organic materials typically show charge carrier mobility significantly lower (around 10^{-5} to $1\,cm^2\,V^{-1}\,s^{-1}$ ranging from amorphous to crystalline materials) than those of inorganic semiconductors (around 10^2–$10^3\,cm^2\,V^{-1}\,s^{-1}$). Since charge carrier mobility is strongly dependent on the molecular organization of the material, in order to achieve better efficiencies, a good control on the morphology of the donor–acceptor materials at a nanometer scale is necessary. Furthermore, an efficient charge carrier mobility is essential to prevent charge recombination (geminal or bimolecular) processes which result in lower energy conversion efficiencies.

In summary, an appropriate choice of the donor and acceptor materials is critical to ensure a good match between them in terms of optical, electronic, and morphological properties, which eventually determine the effective photocurrent and performance of the PV device.

In this regard, fullerenes and their derivatives possess a number of important properties such as small reorganization energy, high electron affinity, ability to transport charge, and stability that makes them one of the best candidates to act as electron acceptor components in photovoltaic devices [41]. Indeed, fullerenes have been demonstrated to be the ideal acceptor because of their singular electronic and geometrical properties and for the ability of their chemically functionalized derivatives

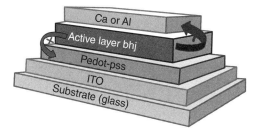

FIGURE 9.16 Typical sandwich-like architecture for a (OPV) solar cell. Polymers and fullerene derivatives are bended on the active layer. (*For a color version of this figure, see the color plate section.*)

to form a bicontinuous phase network with π-conjugated polymers acting as electron conducting (n type) material (Fig. 9.16).

In order to test the properties of these derivatives, many chemically modified fullerenes were initially synthesized and photovoltaic devices of these materials with semiconducting polymers were prepared. These fullerene derivatives were covalently linked to different chemical species such as electron acceptors, electron donors, and π-conjugated oligomers (Fig. 9.17). However, in general, the obtained blends resulted in PV devices exhibiting low energy conversion efficiencies [42].

The best known and most widely used fullerene derivative as acceptor for PV devices is [6,6]-phenyl-C61 butyric acid methyl ester (PCBM, **20**) [43]. Since its first reported application in solar cells [44], it has been by far the most widely used fullerene, being considered as a benchmark material for testing new devices. This initial report inspired the synthesis of many other PCBM analogs [45] in an attempt to increase the efficiencies of the cells by improving the stability or PV

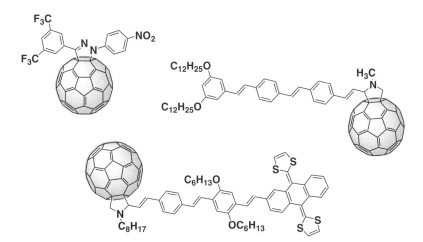

FIGURE 9.17 Some former examples of modified fullerenes bearing different organic addends used to prepare photovoltaic devices.

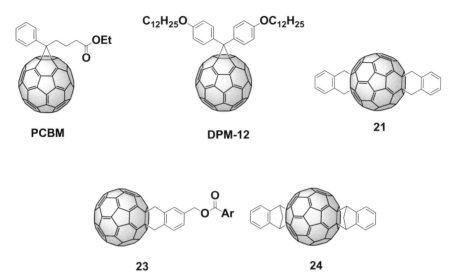

FIGURE 9.18 PCBM (**20**), DPM-12 (**22**), and other modified fullerenes used as successful acceptors for photovoltaic devices.

parameters such as the open-circuit voltage (Voc) by raising the LUMO energies of the fullerene acceptor. However, only small shifts (<100 meV) of the LUMO level have been obtained by attaching a single substituent on the fullerene sphere, even by using electron-donating groups. In contrast, significantly higher Voc values have been achieved through the polyaddition of organic addends to the fullerene cage (~100 mV raising the LUMO per saturated double bond). An externally verified PCE of 4.5% has been reported by Hummelen et al. employing a regioisomeric mixture of PCBM bisadducts as a result of an enhanced open-circuit voltage, while maintaining a high short-circuit current (Jsc) and fill factor (FF) values [46]. Recently, Meng et al. [47] described the preparation of a regioisomeric mixture of dihydronaphthyl-based [60]fullerene bisadducts (NC60BA, **21**, Fig. 9.18), displaying higher LUMO due to the saturation of two double bonds. NC60BA displayed higher Voc which was the key for [poly(3-hexylthiophene)] (P3HT)-based polymer solar cells to achieve better performance. A PCE of 5.37% was achieved from the P3HT:NC60BA solar cell with Voc of 0.82 V.

The cyclopropanation reaction of higher fullerenes to form PCBM analogs is more complex than for C_{60}. Indeed, the less symmetry and the presence of more than one reactive double bond are often responsible for the formation of regioisomeric mixtures. Nevertheless, the loss of symmetry of C_{70} induces a stronger absorption, even in the visible region. As a result, PC71BM [48] is considered a suitable candidate for more efficient polymer solar devices. Moreover, such devices performed the highest verified efficiency determined so far in a BHJ solar cell, with an internal quantum efficiency approaching 100% [49].

Although PCBMs are the acceptors that best performances guarantee to the moment, it does not mean that they are necessarily the optimum fullerene derivatives.

Therefore, a variety of other fullerene derivatives [50] have been synthesized in order to improve the device efficiency or to achieve a better understanding on the dependence of the cell parameters from the structure of the acceptor.

One of the most promising modified fullerenes prepared so far is diphenyl-methanofullerene (DPM12, **22**) prepared by Martín et al. [51] endowed with two alkyl chains to drastically improve the solubility of the acceptor in the blend and reaching efficiencies in the range of 3% (Fig. 9.18). Although the LUMO energy level for DPM12 is the same as that for PCBM, an increase for the Voc of 100 mV for the DPM12 or DPM6 over PCBM has been observed [52]. This is currently an important issue for improving the design of future fullerene-based acceptors [53].

Other fullerene derivatives such as the dihydronaphthylfullerene benzyl alcohol benzoic ester (**23**) described by Fréchet et al. [54] have also shown their ability to prepare efficient PV devices (PCE up to 4.5%).

He et al. [55] reported the preparation of a bisadduct fullerene derivative formed by two indene units covalently connected to the fullerene sphere of C_{60} (**24**). Interestingly, the presence of two aryl groups improves the visible absorption compared to the parent PCBM, as well as its solubility (>90 mg mL^{-1} in chloroform) and the LUMO energy level, which is 0.17 eV higher than PCBM. Photovoltaic devices formed with P3HT as semiconducting polymer revealed PCE values of 5.44% under illumination of AM1.5, 100 mW cm^{-2}, thus surpassing to PCBM which afforded an efficiency of 3.88% under the same experimental conditions.

Although some of the fullerene derivatives prepared to date exhibit outstanding performances in PV devices, the synthesis of new fullerene derivatives with stronger visible and NIR absorption and higher LUMO energy levels than PCBM is currently a challenge in OPVs.

In this regard Mikroyannidis et al. [56] described recently a simple and effective modification of PCBM by a two-step reaction. This modified PCBM (**25**, Fig. 9.19) showed broader and stronger absorption in the visible region of the solar spectrum than PCBM due to the presence of a cyanovinylene 4-nitrophenyl segment. PCE of **25** blended with P3HT reached 4.23%, under the illumination intensity of

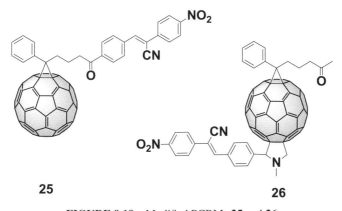

FIGURE 9.19 Modified PCBMs **25** and **26**.

100 mW cm^{-2}, which was significantly improved as compared to PCE of 2.93% of the device with PCBM as electron acceptor under the same experimental conditions. The same authors reported further improvement of this molecule through the addition of a second organic fragment to the fullerene cage [57]. PV devices fabricated with this compound (**26**, Fig. 9.19) displayed a better performance than those prepared with PCBM owing to a higher Voc and the stronger absorption of **26** in the visible region, which is missing for PCBM.

So far, we have seen the most relevant results obtained using different fullerene derivatives as accepting unit and semiconducting polymers as donor materials. Despite the outstanding PCE results described using semiconducting polymers, mainly P3HT. The use of polymers displays a number of troubles associated to their syntheses, purification, and inherent electronic properties [58]. Factors such as polydispersity, regioregularity, and molecular weight have also a strong influence on the final performance of the photovoltaic device. Concerning its electronic properties, P3HT can only absorb 46% of the available solar photons [59], owing to a band gap of 1.90 eV and a relatively narrow absorption band. A promising solution to overcome these limitations could be the use of soluble small conjugated molecules instead of polymers. This strategy would avoid the aforementioned structural problems inherent to polymers, and allow the use of several well-defined small molecules whose electronic properties can be tuned by chemical synthesis in order to obtain a better photovoltaic performance.

Following this approach, in the last years, a number of molecules such as oligomers, dendrimers, dyes, and porphyrins have been used as molecular donors blended with [60] and [70]fullerene or fullerene derivatives, in order to prepare new and efficient photovoltaic devices [60].

One of the most interesting approaches for the preparation of small molecular donors consists in the chemical modification of the Diketopyrrolopyrrole (DPP).core. Nguyen et al. [61] reported the preparation of an oligothiophene DPP (DPP(TBFu)$_2$ **27**, Fig. 9.20) which displays intense absorption bands covering a broad range of the

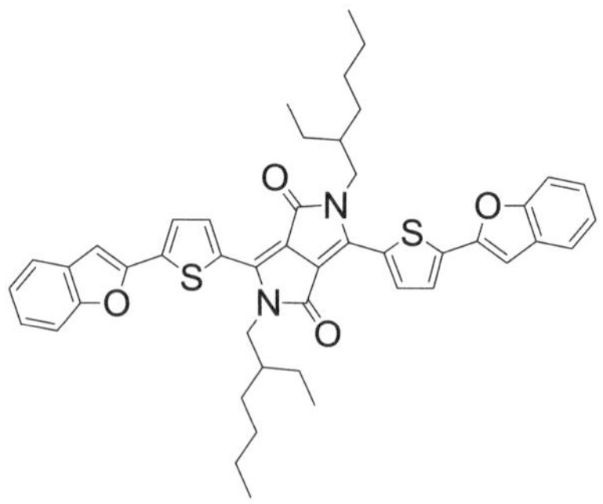

DPP(TBFu)$_2$
27

FIGURE 9.20 Chemical structure of DPP(TBFu)$_2$.

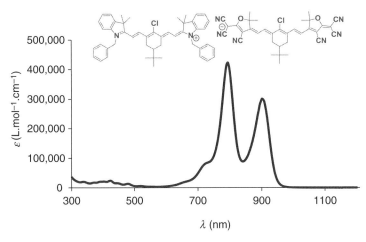

FIGURE 9.21 UV–vis spectra of the cyanine–cyanine salt.

visible spectrum. Blends of this material and C_{71}–PCBM were prepared in different ratios and the results indicated a PCE of 4.4%.

Following the use of well-defined molecules for the preparation of BHJ-PV devices, in the last years a new and interesting approach has emerged in the photovoltaic field. Organic dyes are single molecules which display high absorption extinction coefficients and electronic properties which can be easily tuned by chemical synthesis.

Our research group has recently described a new organic salt by association of two NIR-absorbing cyanine dyes, which exhibits a good solubility in organic solvents [62] (Fig. 9.21). As a result of the combination of the two cyanines, the complex is able to absorb light in a broad range of the solar spectrum (650–950 nm) displaying outstanding light-harvesting properties in the NIR region with giant extinction coefficients. Preliminary experiments showed a promising PCE obtained for the cyanine–cyanine complex:PCBM, indicating that this new dye is an appealing system for further studies as light harvester for photovoltaics.

Sun et al. [63] have recently described the preparation of solution-processed small molecule solar cells with 6.7% efficiency. The authors reported highly efficient solution-processed SM BHJ solar cells based on a new molecular donor, DTS(PTTh$_2$)$_2$ (**28**) which exhibits a broad optical absorption (Fig. 9.22). A record PCE of 6.7% under AM 1.5G irradiation (100 mW cm^{-2}) was achieved for small molecule BHJ devices from DTS(PTTh$_2$)$_2$VPC$_{70}$BM (donor to acceptor ratio of 7:3).

Other leading example in the field of molecular solar cells has recently been described by Chen et al. [64] achieving high PCEs by vacuum-deposition techniques. The authors prepared photovoltaic devices using C_{70} as acceptor and **29** (Fig. 9.22) as donor, obtaining impressively high PCE of 6.8% through appropriate balance between the photovoltage and the photocurrent. The optical properties of Ditolylaminothienyl-dicyanovinylen-benzothiadiazole (DTDCTB) (**29**), displaying

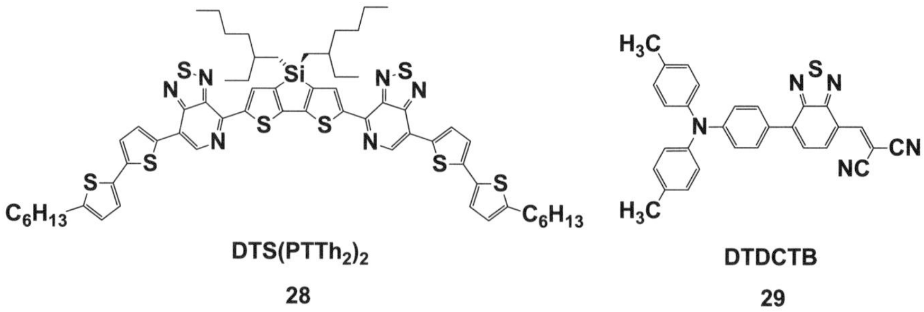

FIGURE 9.22 Small molecules used for the preparation of record PCE cells.

intense and broad absorption band in the long-wavelength region, played an important role in obtaining this remarkable PCE.

To conclude this section, it is important to remark that the field of molecular solar cells is under continuous development, and efficiencies over 10% have already been obtained [65]. The careful design of the materials and deposition techniques together with suitable control on the nanomorphology will allow the improvement of these cells in order to introduce them in the market as a reasonable alternative to complement silicon cells.

9.4 MOLECULAR WIRES

One of the main targets in the field of molecular electronics is the construction of molecules able to conduct electrical current between two electrodes [66]. In this sense, upon light irradiation, electron donor/bridge/electron acceptor architectures have been widely studied as suitable models for probing charge-transfer processes at the molecular level. In such cases, the molecular bridge is assumed to mediate charges between the donor and acceptor termini, behaving as a molecular wire. To evaluate the performance of a molecular wire in these donor/bridge/acceptor architectures it is necessary to measure the electron transfer processes (i.e., charge separation and charge recombination). The kinetics of both processes, namely, charge separation and/or charge recombination are then reflected by the electron transfer rate constant

$$K_{ET} = k_0 e^{-\beta R_{DA}},$$

where k_0 is the kinetic prefactor/pre-exponential factor, R_{DA} represents the donor–acceptor distance, and β is the so-called attenuation factor or dumping factor. The attenuation factor β is used to determine the charge-transfer capability of the molecular bridge and, therefore, becomes a bridge-specific parameter, which depends on the magnitude of the coupling between the donors and acceptors as well as the energy of the charge-transfer states localized on each site. Typical values for β may vary from 1.0 to 1.4 Å^{-1} for protein structures and from 0.001 to 0.004 Å^{-1} for highly π-conjugated bridge structures [67]. In vacuum, β values are relatively large in the range of 2.0–5.0 Å^{-1}. However, other parameters, such as the underlying driving forces ($-\Delta G^0$), the corresponding reorganization energies (λ), and the electronic couplings that exist between electron donor and electron acceptor (V), also exert an impact on the electron transfer rate.

Thus, the desirable features for obtaining an efficient wire-like behavior in donor/bridge/acceptor architectures would be as follows: (i) matching of energy levels between donor/acceptor and the bridge sites, (ii) strong electronic couplings between the electron donor and acceptor units via the bridge orbitals, and (iii) small attenuation factors.

To this end, two different strategies have been implemented. The first strategy is based on the covalent and noncovalent linkages of electron donors to electron acceptors by molecular wires. In this strategy, a certain degree of charge delocalization from the donor to the acceptor may occur in the ground state. Nevertheless, we will focus on electron donor/acceptor structures in solution, where irradiation with light induces the promotion of electrons from the donor to the acceptor, yielding a spatially separated radical ion pair. By studying the electron transfer process, the molecular wire behavior is evaluated through the attenuation factor values for these structures.

The second strategy, not considered in this chapter since no light harvesting is required, is the possibility to measure the conductance at unimolecular level. In this approach it is exclusively considered a conductive organic molecule that acts as a bridge connected to two metal electrodes. The molecular junction is a fairly immediate approach to the study of the behavior of the molecular components within an electronic circuit [68].

9.4.1 Covalent Donor–Bridge–Acceptor Systems

The covalent linkage of donor/acceptor electroactive units through the molecular wire approach has been widely developed in molecular electronics. This approach takes into account several important aspects such as (i) the covalent linkers that help to accelerate the electron transfer dynamics; (ii) the rigid molecular wire that allows a good donor–acceptor separation, orientation, and overlap, avoiding unrestrained and undesired rearrangements; (iii) the chemical nature of the bridge on the electronic nature of the donor/acceptor pair is supposed to be negligible, and finally; and (iv) the coupling of the bridge to the donor/acceptor sites which is proportional to the orbital overlaps.

It is well documented that a π-conjugated system represents a good candidate for a molecular wire due to its rigidity and directionality since it facilitates the charge transfer over larger distance and provides lower β values. A plethora of donor–bridge–acceptor (DBA) systems have been reported in the literature where the bridge is a π-conjugated system such as *p*-oligophenyleneethynylene (*p*-oPPE), *p*-oligophenylenevinylene (*p*-oPPV), or oligofluorene (oFL) with changeable lengths.

In these sense, DBA systems widely studied are based on porphyrins (ZnP) as donor and fullerene (C_{60}) as acceptor due to their good electroactive properties. These electroactive units were connected through a variety of different bridges such

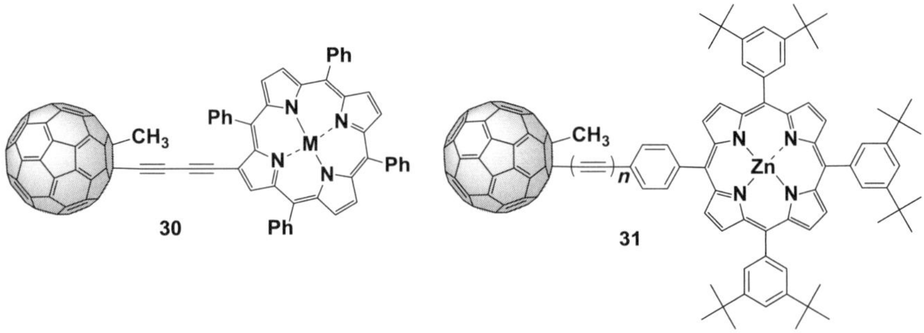

FIGURE 9.23 Chemical structures of some ZnP/C$_{60}$ conjugates (**30** and **31**).

as acetylene, polyacetylene, *p*-phenylenebutadiynilenes, and π-conjugated system (oPPV, oPPE, and oPFl) (Fig. 9.23). A representative example is that by Vail et al. [69] where a ZnP unit is covalently linked to the C$_{60}$ through a polyacetylene moiety (compounds **30** and **31**). Depending on the way of binding between the bridge and the donor unit, either through the β-position of ZnP or through a phenyl ring in the *meso* position of the porphyrin [70], a great variation in the electron transfer for charge separation (CS) and charge recombination (CR) were reported, with values of $k_{CS} = 1 \times 10^{11}$ s^{-1}, $k_{CR} = 2.5 \times 10^{10}$ s^{-1} (for compound **30**) and $7.5 \pm 2.4 \times 10^{9}$ s^{-1} and $1.6 \pm 0.2 \times 10^{6}$ s^{-1}, [1], (for compound **31**). Moreover, compound **31** presents a small attenuation factor (β) of 0.06 Å^{-1} in acetonitrile. This fact suggests that the presence of the phenyl ring in compound **31** retards the electron transfer processes, whereas the acetylene bridge in compound **30** boosts the orbital interactions between ZnP and C$_{60}$.

Figure 9.24 shows different DBA systems based on porphyrin (ZnP) as the electron donor and [60]fullerene as the electron acceptor covalently linked through oligomers such as *p*-oPPE, *p*-oPPV, and oFL of different length. For the β-substituted ZnP–oPPE–C$_{60}$ system (**32**), lifetime of the charge-separated state lasted longer with the increasing of the distance between the donor and acceptor moieties. A longer lifetime of pairs implies a through-bond mechanism where the bridge plays an important role whereas, the β value was in the order of 0.11 ± 0.05 Å^{-1} [71]. However, analogous systems in which the bridge has been substituted by an oPPV (**33**) lead to an improvement in the molecular wire behavior, reducing the values of β to (0.03 ± 0.005 Å^{-1}) [72]. However, a slight modification on the oPPV bridge, resulting in the incorporation of a [2,2′] paracyclophaneoligophenylenevinylene (pCp-oPPV), sparks off an increase of the damping factor approximately four times greater than that for fully conjugated oPPV bridges, due to the inhomogeneous and weaker coupling of the pCp units (pCP-oPPv) [73]. Furthermore, experimental and theoretical studies revealed that the pCP units facilitate the CT in one direction (from C$_{60}$ to ZnP via pCp) but disfavors in the opposite direction (from ZnP to C$_{60}$ via pCp).

On the other hand, Guldi et al. [74] reported a DBA system that contains π-conjugated oFLs of variable length between the two electroactive units (**34**, **35**). The studies confirm the dependence on the attenuation factor values to the connection pattern between the ZnP donor and the oFL. In compound **34**, the insertion of the

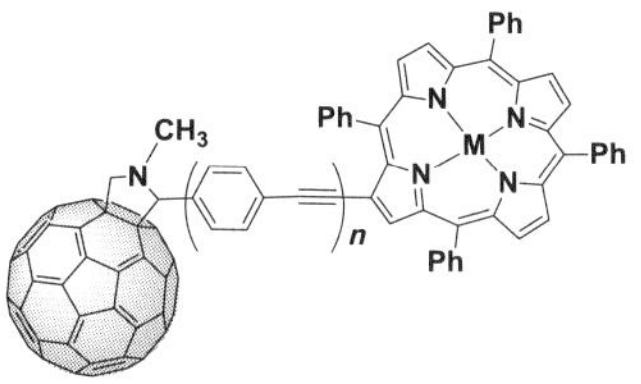

FIGURE 9.24 Chemical structures of some ZnP/π-conjugated/C$_{60}$ systems.

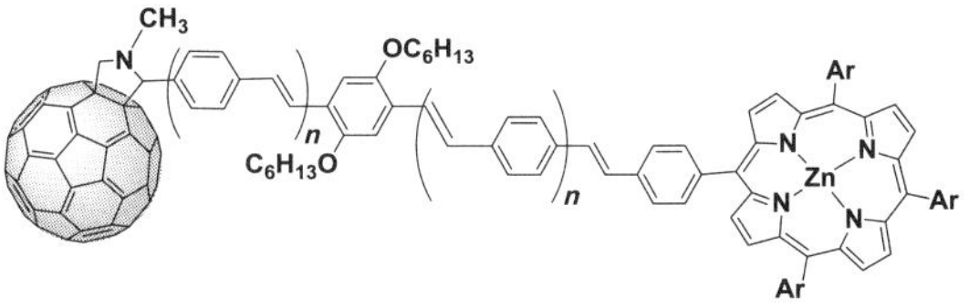

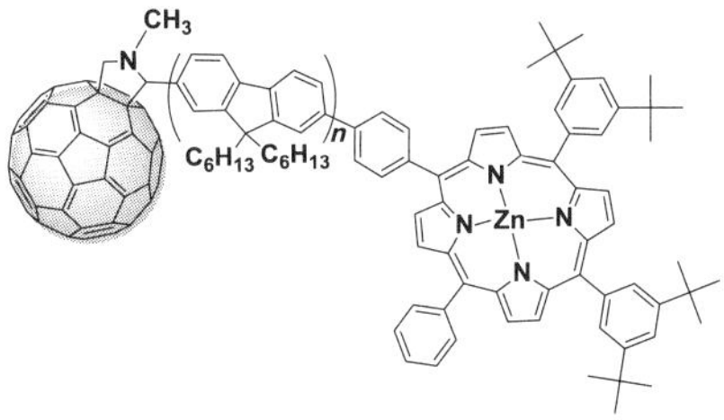

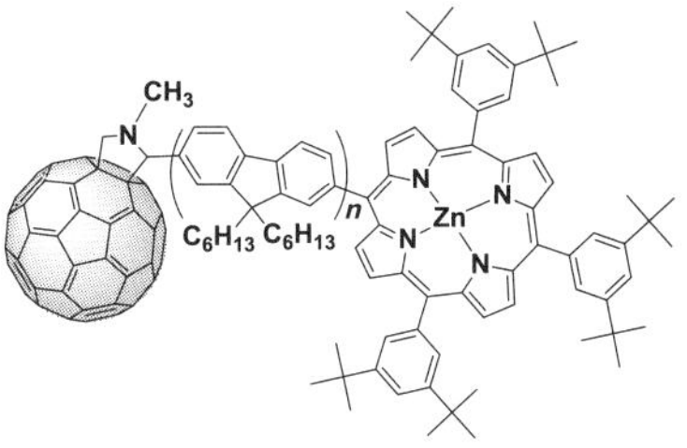

additional phenyl ring between the bridge and the donor perturbs the homogeneous oFL π-conjugation, creating a bottleneck for the electrons to pass. This fact corroborates that the charge separation process was slower and charge recombination process was faster, with β values of 0.11 Å^{-1}. However, compound **35** warrants a homogeneous distribution of π-electrons along the system, providing an excellent π-electron pathway for the charges to transfer between donor and acceptor. Hence the lower attenuation factor (β 0.07 Å^{-1}).

It is important to emphasize, however, how a subtle change on the bridge alters the electron transport process between the two electroactive units and the β values. Hence, another key aspect for achieving efficient molecular wire behavior in electron donor–acceptor system is the effective matching of orbital energies between the donor and the linker components and between the linker and acceptor components. It has been described in the literature on these kind of π-conjugated systems in DBA systems, where a π-exTTF has been used as electron donor and [60]fullerene as electron acceptor. In these cases, an extraordinary small attenuation factor of 0.01 ± 0.005 Å^{-1} was determined for compounds in which the donor exTTF and the acceptor C$_{60}$ are connected through a *p*-phenylenevinylene oligomer **38** (exTTF–oPPV–C$_{60}$, Fig. 9.25). This low β value has been explained attending to the π-conjugation of the bridge with the donor exTTF and the homogeneous distribution of the local electron affinity throughout the whole bridge. A remarkable value of ∼5.5 cm^{-1} of the coupling constant (V) was determined for these systems, unusually strong over a distance of 40 Å from the electron donor to the electron acceptor. This lower β value for compound **38**, in contrast to previously described using ZnP as electron donor, proved a better conjugation between the exTTF and the linker (oPPVs) [75].

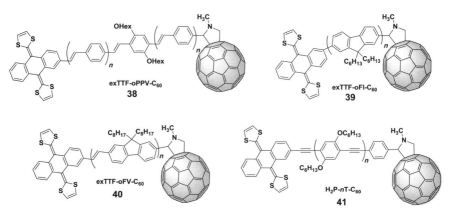

FIGURE 9.25 Chemical structures of some exTTF/π-conjugated wire/C_{60} systems.

The comparison of the charge-transfer characteristics of oFLs (**39**, Fig. 9.25) *versus* oligofluorenevinylenes (oFV, **40**, Fig. 9.25) shows that it is possible to tune the electronic properties by slight structural alterations of the compounds. Specifically, β values of 0.075 Å^{-1} are obtained for the oFV wires, lower than the value determined for oFL (0.09 Å^{-1}) [76]. The vinylene groups improve the π-conjugation of the wire in such a way that charge injection into the bridge is favored, facilitating CS and CR processes [77]. However, in exTTF-π-phenyleneethynylene–C_{60} (**41**, Fig. 9.25) the charge separation and charge recombination dynamics in THF provided an attenuation factor of 0.2 ± 0.05 Å^{-1} [78]. The higher values of the attenuation factor revealed that the HOMO for this system is completely localized on the exTTF and does not reach into the bridge. The presence of triple bonds in exTTF–oPPE–C_{60} disrupts the extended π-conjugation and influences in the charge transfer [79].

9.4.2 Supramolecular Donor–Bridge–Acceptor Systems

Recently, Hirsch et al. [80] have reported one of the first studies on supramolecular wires based on porphyrin/fullerene (Fig. 9.26). In this case, [60]fullerene is

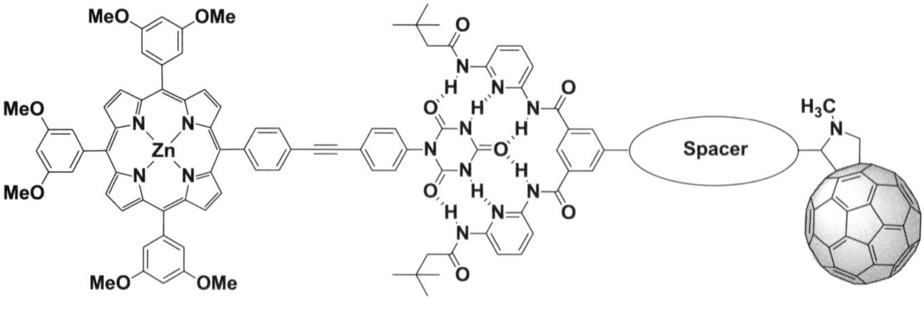

FIGURE 9.26 Supramolecular ZnP/C_{60} molecular wire (**42**).

connected to the Hamilton receptor through a series of π-conjugated spacers (i.e., *p*-phenylene-ethynylene, *p*-phenylene-vinylene, *p*-ethynylene, and *p*-fluorene), which provide wire-like behavior in terms of electron transfer/electron transport, and was used as complementary unit cyanuric acid-linked porphyrin.

In these studies, it was clearly demonstrated that the electronic communication between the donor and acceptor moieties in **42** is governed by the nature of the spacer. Thus, spacers with low *β* values facilitate the charge transfer along the supramolecular bridge. By analyzing the dependence of CS and CR on the D–A distance a hydrogen-bonding-mediated electron transfer afforded an attenuation factor of 0.11 Å^{-1}, a typical value found for covalent systems, was for the first time observed.

9.5 CONCLUSIONS

A variety of significant DBA organic compounds reported in the recent literature have been presented in a systematic way according to their use in different topics. In particular, we have focused on those functional molecules which undergo an electron transfer process upon light irradiation, namely in artificial photosynthetic systems, materials for photovoltaic applications in *all-organic* solar cells and in the study of molecules exhibiting a molecular wire behavior. Whereas artificial D–A molecules have allowed the study of some elemental steps of the more complex natural photosynthesis, BHJ plastic solar cells consisting of a semiconducting electron donor π-conjugated polymer and a fullerene as the acceptor moiety are currently promising devices for converting sunlight into electrical power, with energy conversion efficiencies reported so far reaching values as high as 7.4%.

It is important to note that organic solar cells are called to play an important role for satisfying the increasing energy demands of our society. Because of their features of low cost, flexibility, and lightness, these new PV cells—which should nicely complement the commercially available silicon cells—are currently closer to the market.

Finally, the study of molecular wires is critical for the design and better understanding of the so-called organic electronics. In particular, we have described only those wires in which the electron transfer process is triggered by light irradiation. The determination of the attenuation factor has allowed organizing the different π-conjugated systems according to their efficiency in charge transport from the donor to the acceptor unit and, therefore, their potential applications in the so-called organic electronics.

ACKNOWLEDGMENTS

This work has been supported by the MINECO of Spain (CTQ 2012-30668, CTQ 2011-27934, CTQ2011-24652, CTQ2009-08790, and Consolider-Ingenio CSD 2007-00010 on Molecular Nanoscience), Comunidad de Madrid (MADRISOLAR-2, S2009/PPQ-1533), and the EU project (FUNMOLS FP7-212942-1). J.L.D. and

C.A. thank the MINECO for a Ramón y Cajal Fellowship. D. J. thanks IMDEA-Nanociencia for a postdoctoral research grant.

REFERENCES

[1] (a) G. McDermott, S. M. Prince, A. A. Freer, A. M. Hawthornthwaite-Lawless, M. Z. Papiz, R. J. Cogdell, and N. W. Isaacs, "Crystal structure of an integral membrane light-harvesting complex from photosynthetic bacteria," *Nature*, **374**, 517–521 (1995). (b) A. M. van Oijen, M. Ketelaars, J. Köhler, T. J. Aartsma, and J. Schmidt, "Unraveling the electronic structure of individual photosynthetic pigment-protein complexes," *Science* **285**, 400–402 (1999).

[2] B. C. Popere, A. M. Della Pelle, A. Poe, and S. Thayumanavan, "Macromolecular architectures for organic photovoltaics," *Phys. Chem. Chem. Phys.* **14**, 4043–4057 (2012).

[3] (a) M. Häussler, Y. Phei Lok, M. Chen, J. Jasieniak, R. Adhikari, S. P. King, S. A. Haque, C. M. Forsyth, K. Winzenberg, S. E. Watkins, et al., "Benzothiadiazole-containing pendant polymers prepared by RAFT and their electro-optical properties," *Macromolecules* **43**, 7101–7110 (2010). (b) A. Harriman, L. J. Mallon, K. J. Elliot, A. Haefele, G. Ulrich, and R. Ziessel, "Length dependence for intramolecular energy transfer in three- and four-color donor–spacer–acceptor arrays," *J. Am. Chem. Soc.* **131**, 13375–13386 (2009).

[4] P. Furuta and J. M. J. Fréchet, "Controlling solubility and modulating peripheral function in dendrimer encapsulated dyes," *J. Am. Chem. Soc.* **125**, 13173–13181 (2003).

[5] R. Chen, J. Ling, and T. E. Hogen-Esch, "Synthesis and spectroscopic studies of macrocyclic polystyrene containing two fluorene units and single 9,10-anthracenylidene group," *Macromolecules* **42**, 6015–6022 (2009).

[6] (a) X. Zhang, Z. K. Chen, and K. P. Loh, "Coordination-assisted assembly of 1-D nanostructured light-harvesting antenna," *J. Am. Chem. Soc.* **131**, 7210–7211 (2009). (b) B. Branchi, P. Ceroni, V. Balzani, F. G. Klärner, and F. Vögtle, "A light-harvesting antenna resulting from the self-assembly of five luminescent components: a dendrimer, two clips, and two lanthanide ions," *Chem. Eur. J.* **16**, 6048–6055 (2010).

[7] D. M. Eisele, C. W. Cone, E. A. Bloemsma, S. M. Vlaming, C. G. F. van der Kwaak, R. J. Silbey, M. G. Bawendi, J. Knoester, J. P. Rabe, and D. A. Vanden, "Bout. Utilizing redox-chemistry to elucidate the nature of exciton transitions in supramolecular dye nanotubes," *Nat. Chem.* **4**, 655–662 (2012).

[8] (a) N. Martin, L. Sanchez, B. Illescas, and I. Perez, "C60-based electroactive organofullerenes,". *Chem. Rev.* **98**, 2527–2547 (1998). (b) H. Imahori and Y. Sakata, "Fullerenes as novel acceptors in photosynthetic electron transfer," *Eur. J. Org. Chem.* 2445–2457 (1999). (c) D. M. Guldi, B. M. Illescas, C. M. Atienza, M. Wielopolski, and N. Martin, "Fullerene for organic electronics," *Chem. Soc. Rev.* **38**, 1587–1597 (2009). (d) G. Bottari, G. de la Torre, D. M. Guldi, and T. Torres, "Covalent and Noncovalent Phthalocyanine-carbon nanostructure systems: synthesis, photoinduced electron transfer, and application to molecular photovoltaics," *Chem. Rev.* **110**, 6768–6816 (2010). (d) J. L. Delgado, N. Martin, P. de la Cruz, and F. Langa, "Pyrazolinofullerenes: a less known type of highly versatile fullerene derivatives," *Chem. Soc. Rev.* **40**, 5232–5241 (2011).

[9] H. Imahori and S, Fukuzumi, "Porphyrin- and fullerene-based molecular photovoltaic devices," *Adv. Funct. Mater.* **14**, 525–536 (2004).

[10] G. de Miguel, M. Wielopolski, D. I. Schuster, M. A. Fazio, O. P. Lee, C. K. Haley, A. L. Ortiz, L. Echegoyen, T. Clark, and D. M. Guldi, "Triazole bridges as versatile linkers in electron donor-acceptor conjugates," *J. Am. Chem. Soc.* **133**, 13036–13054 (2011).

[11] C. Villegas, J. L. Delgado, P.-A. Bouit, B. Grimm, W. Seitz, N. Martin, and D. M. Guldi, "Powering reductive charge shift reactions-linking fullerenes of different electron acceptor strength to secure an energy gradient," *Chem. Sci.* **2**, 1677–1681 (2011).

[12] H. Imahori, D. M. Guldi, K. Tamaki, Y. Yoshida, C. Luo, Y. Sakata, and S. Fukuzumi, "Charge separation in a novel artificial photosynthetic reaction center Lives 380 ms," *J. Am. Chem. Soc.* **123**, 6617–6628 (2001).

[13] J. D. Megiatto Jr., D. I. Schuster, S. Abwandner, G. de Miguel, and D. M. Guldi, "[2]Catenanes decorated with porphyrin and [60]fullerene groups: design, convergent synthesis, and photoinduced processes," *J. Am. Chem. Soc.* **132**, 3847–3861 (2010).

[14] (a) M. Bendikov, F. Wudl, and D. F. Perepichka, "Tetrathiafulvalenes, oligoacenenes, and their buckminsterfullerene derivatives: the brick and mortar of organic electronics," *Chem. Rev.* **104**, 4891–4945 (2004). (b) J. L. Segura and N. Martín, "New concepts in tetrathiafulvalene chemistry," *Angew. Chem. Int. Ed.* **40**, 1372–1409 (2001).

[15] F. Wudl, D. Wobschall, and E. J. Hufnagel, "Electrical conductivity by the bis(1,3-dithiole)-bis(1,3-dithiolium) system," *J. Am. Chem. Soc.* **94**, 670–672 (1972).

[16] J. Ferraris, D. O. Cowan, V. V. Walatka Jr., and J. H. Perlstein, "Electron transfer in a new highly conducting donor-acceptor complex," *J. Am. Chem. Soc.* **95**, 948 (1973)

[17] J.-I. Yamada and T. Sugimoto, *TTF Chemistry: Fundamentals and Applications of Tetrathiafulvalene* (Kodansha Springer, 2004).

[18] F. G. Brunetti, J. L. Lopez, C. Atienza, and N. Martin, "π-Extended TTF: a versatile molecule for organic electronics," *J. Mater. Chem.* **22**, 4188–4205 (2012).

[19] N. Martín, L. Sánchez, and D. M. Guldi, "Stabilisation of charge-separated states *via* gain of aromaticity and planarity of the donor moiety in C_{60}-based dyads," *Chem. Commun.* **2**, 113–114 (2000).

[20] L. Sanchez, I. Perez, N. Martin, and D. M. Guldi, "Controlling short- and long-range electron transfer processes in molecular dyads and triads," *Chem. Eur. J.* **9**, 2457–2468 (2003).

[21] S. S. Gayathri, M. Wielopolski, E. M. Perez, F. L. Sanchez, R. Viruela, E. Orti, D. M. Guldi, and N. Martin, "Discrete supramolecular donor–acceptor complexes," *Angew. Chem. Int. Ed.* **48**, 815–819 (2009).

[22] D. M. Guldi, A. Gouloumis, P. Vázquez, and T. Torres, "Charge-transfer states in strongly coupled phthalocyanine fullerene ensembles," *Chem. Commun.* **18**, 2056–2057 (2002).

[23] A. Kahnt, D. M. Guldi, A. de la Escosura, M. V. Martinez-Diaz, and T. Torres, "[2.2]Paracyclophane: a pseudoconjugated spacer for long-lived electron transfer in phthalocyanine–C_{60} dyads," *J. Mater. Chem.* **18**, 77–82 (2008).

[24] M. E. El-Khouly, J. Hoon Kim, K.-Y. Kay, C. Soo Choi, O. Ito, and S. Fukuzumi, "Synthesis and photoinduced intramolecular processes of light-harvesting silicon phthalocyanine–naphthalenediimide–fullerene connected systems," *Chem. Eur. J.* **15**, 5301 (2009).

[25] T. Torres, A. Gouloumis, D. Sanchez-Garcia, J. Jayawickramarajah, W. Seitz, D. M. Guldi, and J. L. Sessler, "Photophysical characterization of a cytidine–guanosine tethered phthalocyanine–fullerene dyad," *Chem. Commun.* **3**, 292–294 (2007).

[26] F. D'Souza, E. Maligaspe, K. Ohkubo, M. E. Zandler, N. K. Subbaiyan, and S. Fukuzumi, "Photosynthetic reaction center mimicry: low reorganization energy driven charge stabilization in self-assembled cofacial Zinc phthalocyanine dimer-fullerene conjugate," *J. Am. Chem. Soc.* **131**, 8787–8797 (2009).

[27] A. Mishra and P. Bäuerle, "Small molecule organic semiconductors on the move: promises for future solar energy technology," *Angew. Chem. Int. Ed.* **51**, 2020–2067 (2012).

[28] S. Xiao, J.-H. Xu, Y.-S. Li, C.-M. Du, Y.-L. Li, L. Jiang, and D. Zhu, "Preparation and characterization of a novel dumbbell-type [60]fullerene dimer containing a cyanine dye," *New J. Chem.* **25**, 1610–1612 (2001).

[29] F. Meng, J. Hua, K. Chen, H. Tian, L. Zuppiroli, and F. Nüesch, "Synthesis of novel cyanine–fullerene dyads for photovoltaic devices," *J. Mater. Chem.* **15**, 979–986 (2005).

[30] P.-A. Bouit, F. Spänig, G. Kuzmanich, E. Krokos, C. Oelsner, M. A. Garcia-Garibay, J. L. Delgado, N. Martin, and D. M. Guldi, "Efficient utilization of higher-lying excited states to trigger charge-transfer events," *Chem. Eur. J.* **16**, 9638–9645 (2010).

[31] C. Villegas, E. Krokos, P.-A. Bouit, J. L. Delgado, D. M. Guldi, and N. Martin, "Efficient light harvesting anionic heptamethine cyanine–[60] and [70]fullerene hybrids," *Energy Environ. Sci.* **4**, 679–684 (2011).

[32] G. Ulrich, R. Ziessel, and A. Harriman, "The chemistry of fluorescent bodipy dyes: versatility unsurpassed," *Angew. Chem. Int. Ed.* **47**, 1184–1201 (2008).

[33] A. Loudet and K. Burgess, "BODIPY dyes and their derivatives: syntheses and spectroscopic properties," *Chem. Rev.* **107**, 4891–4932 (2007).

[34] R. Ziessel and A. Harriman, "Artificial light-harvesting antennae: electronic energy transfer by way of molecular funnels," *Chem. Commun.* **47**, 611–631 (2011).

[35] R. Ziessel, B. D. Allen, D. B. Rewinska, and A. Harriman, "Selective triplet-state formation during charge recombination in a fullerene/bodipy molecular dyad (Bodipy = Borondipyrromethene)," *Chem. Eur. J.* **15**, 7382–7393 (2009).

[36] C. A. Wijesinghe, M. E. El-Khouly, J. D. Blakemore, M. E. Zandler, S. Fukuzumi, and F. D'Souza, "Charge stabilization in a closely spaced ferrocene–boron dipyrrin–fullerene triad," *Chem. Commun.* **46**, 3301–3303 (2010).

[37] E. Maligaspe, T. Kumpulainen, N. K. Subbaiyan, M. E. Zandler, H. Lemmetyinen, N. V. Tkachenko, and F. D'Souza, "Electronic energy harvesting multi BODIPY-zinc porphyrin dyads accommodating fullerene as photosynthetic composite of antenna-reaction center," *Phys. Chem. Chem. Phys.* **12**, 7434–7444 (2010).

[38] Y. Rio, W. Seitz, A. Gouloumis, P. Vazquez, J. L. Sessler, D. M. Guldi, and T. Torres, "A panchromatic supramolecular fullerene-based donor–acceptor assembly derived from a peripherally substituted bodipy–Zinc phthalocyanine dyad," *Chem. Eur. J.* **16**, 1929–1940 (2010).

[39] (a) N. Armaroli and V. Balzani, "The future of energy supply: Challenges and opportunities," *Angew. Chem. Int. Ed.*, **46**, 52–66 (2007). (b) For further information about solar energy, see the International Energy Agency Photovoltaic Power Systems Program at: www.iea-pvps.org.

[40] D. M. Chapin, C. S. Fuller, and G. L. Pearson, "A new silicon p-n junction photocell for converting solar radiation into electrical power," *J. Appl. Chem.*, **25**, 676–678 (1954).

[41] B. C. Thompson and J. M. J. Frechet, "Polymer–fullerene composite solar cells," *Angew. Chem. Int. Ed.* **47**, 58–77 (2008).

[42] M. T. Rispens, and J. C. Hummelen, "Fullerenes: from synthesis to optoelectronic properties," in *Dordrecht, The Netherlands*, edited by D. M. Guldi and N. Martín (Kluwer Acad. Publishers, Ch. 12 Photovoltaic applications, 2002), pp. 387–435.

[43] J. C. Hummelen, B. W. Knight, F. LePeq, F. Wudl, J. Yao, and C. L. Wilkins, "Preparation and characterization of fulleroid and methanofullerene derivatives," *J. Org. Chem.* **60**, 532–538 (1995).

[44] G. Yu, J. Gao, J. C. Hummelen, F. Wudl, and A. J. Heeger, "Polymer photovoltaic cells: enhanced efficiencies via a network of internal donor–acceptor heterojunctions," *Science*, **270**, 1789–1791 (1995).

[45] (a) Y. Zhang, H. L. Yip, O. Acton, S. K. Hau, F. Huang, and A. K. Y. Jen, "A simple and effective way of achieving highly efficient and thermally stable bulk-heterojunction polymer solar cells using amorphous fullerene derivatives as electron acceptor," *Chem. Mater.* **21** 2598–2600 (2009). (b) C. Yang, J. Y. Kim, S. Cho, J. K. Lee, A. J. Heeger, and F. Wudl, "Functionalized methanofullerenes used as n-type materials in bulk-heterojunction polymer solar cells and in field-effect transistors," *J. Am. Chem. Soc.* **130**, 6444–6450 (2008).

[46] L. Lenes G. J. A. H. Wetzelaer, F. B. Kooistra, S. C. Veenstra, J. C. Hummelen, and P. W. Blom, "Fullerene bisadducts for enhanced open-circuit voltages and efficiencies in polymer solar cells," *Adv. Mater.* **20**, 2116–2119 (2008).

[47] X. Meng, W. Zhang, Z. Tan, C. Du, C. Li, Z. Bo, Y. Li, X. Yang, M. Zhen, F. Jiang, et al., "Dihydronaphthyl-based [60]fullerene bisadducts for efficient and stable polymer solar cells," *Chem. Commun.* **48**, 425–427 (2012).

[48] M. M. Wienk, J. M. Kroon, W. J. H. Verhees, J. Knol, J. C. Hummelen, P. A. van Hal, and R. A. J. Janssen, "Efficient methano[70]fullerene/MDMO-PPV bulk heterojunction photovoltaic cells," *Angew. Chem. Int. Ed.* **42**, 3371–3375 (2003).

[49] (a) S. H. Park, A. Roy, S. Beaupré, S. Cho, N. Coates, J. S. Moon, D. Moses, M. Leclerc, K. Lee, and A. J. Heeger, "Bulk heterojunction solar cells with internal quantum efficiency approaching 100%," *Nat. Photonics* **3**, 297–302 (2009). (b) H. Y. Chen, J. Hou, S. Zhang, Y. Liang, G. Yang, L. Yang, L. Yu, Y. Wu, and G. Li, "Polymer solar cells with enhanced open-circuit voltage and efficiency," *Nat. Photonics* **3**, 649–653 (2009).

[50] C. Z. Li, H. L. Yip, and A. K. Y. Jen, "Functional fullerenes for organic photovoltaics," *J. Mater. Chem.*, **22**, 4161–4177 (2012).

[51] I. Riedel, E. von Hauff, J. Parisi, N. Martín, F. Giacalone, and V. Diakonov, "Diphenylmethanofullerenes: new and efficient acceptors in bulk-heterojunction solar cells," *Adv. Funct. Mater.* **15**, 1979–1987 (2005).

[52] I. Riedel, N. Martín, F. Giacalone, J. L. Segura, D. Chirvase, and J. Parisi, "V. Polymer solar cells with novel fullerene-based acceptor," *Thin Solid Films* **451**, 43–47 (2004).

[53] (a) G. Garcia-Belmonte, P. P. Boix, J. Bisquert, M. Lenes, H. J. Bolink, A. La Rosa, S. Filippone, and N. Martín, "Influence of the intermediate density-of-states occupancy on open-circuit voltage of bulk heterojunction solar cells with different fullerene acceptors," *J. Phys. Chem. Lett.* **1** 2566–2571 (2010). (b) H. J. Bolink, E. Coronado, A. Forment-Aliaga, M. Lenes, A. La Rosa, S. Filippone, and N. Martín, "Polymer solar cells based on diphenylmethanofullerenes with reduced sidechain length," *J. Mater. Chem.* **21**, 1382–1386 (2011). (c) A. Sánchez-Díaz, M. Izquierdo, S. Filippone, N. Martin, and E. Palomares, "The origin of the high voltage in DPM12/P3HT organic solar cells," *Adv. Funct. Mater.* **20**, 2695–2700 (2010). (d) M. Hallermann, E. Da Como, J. Feldmann, M. Izquierdo, S. Filippone, N. Martín, S. Jüchter, and E. von Hauff, "Correlation between

[54] S. Backer, K. Sivula, D. F. Kavulak, and J. M. J. Fréchet, "High efficiency organic photovoltaics incorporating a new family of soluble fullerene derivatives," *Chem. Mater.* **19**, 2927–2929 (2007).

[55] Y. He, H. Y. Chen, J. Hou, and Y. Li, "Indene−C60 bisadduct: a new acceptor for high-performance polymer solar cells," *J. Am. Chem. Soc.* **132**, 1377–1382 (2010).

[56] J. A. Mikroyannidis, A. N. Kabanakis, S. S. Sharma, and G. D. Sharma, "A simple and effective modification of PCBM for use as an electron acceptor in efficient bulk heterojunction solar cells," *Adv. Funct. Mater.* **21**, 746–755 (2011).

[57] J. A. Mikroyannidis, D. V. Tsagkournos, S. S. Sharma, and G. D. Sharma, "Synthesis of a broadly absorbing modified PCBM and application as electron acceptor with poly(3-hexylthiophene) as electron donor in efficient bulk heterojunction solar cells," *J. Phys. Chem. C* **115**, 7806–7816 (2011).

[58] J. Roncali, "Molecular bulk heterojunctions: an emerging approach to organic solar cells," *Acc. Chem. Res.* **42**, 1719–1730 (2009).

[59] C. Soci, I.-W. Hwang, D. Moses, Z. Zhu, D. Walter, R. Guadiana, C. J. Brabec and A. J. Heeger, "Photoconductivity of a low-bandgap conjugated polymer," *Adv. Funct. Mater.* **17**, 632–636 (2007).

[60] H. Burckstummer, E. V. Tulyakova, M. Deppisch, M. R. Lenze, N. M. Kronenberg, M. Gsanger, M. Stolte, K. Meerholz, and F. Wurthner, "Efficient solution-processed bulk heterojunction solar cells by antiparallel supramolecular arrangement of dipolar donor–acceptor dyes," *Angew. Chem. Int. Ed.* **50**, 11628–11632 (2011). (b) H. Bürckstümmer, N. M. Kronenberg, K. Meerholz, and F. Würthner, "Near-infrared absorbing merocyanine dyes for bulk heterojunction solar cells," *Org. Lett.* **12**, 3666–3669 (2010). (c) B. Walker, C. Kim, and T.-Q, "Nguyen, small molecule solution-processed bulk heterojunction solar cells," *Chem. Mater.* **23**, 470–482 (2011). (d) J. L. Delgado, P.-A. Bouit, M. A. Herranz, S. Filippone and N. Martin, "Organic photovoltaics: a chemical approach," *Chem. Commun.* **46**, 4853–4865 (2010).

[61] B. Walker, A. B. Tamayo, X.-D. Dang, P. Zalar, J. H. Seo, A. Garcia, M. Tantiwiwat, and T.-Q. Nguyen, "Nanoscale phase separation and high photovoltaic efficiency in solution-processed small-molecule bulk heterojunction solar cells," *Adv. Funct. Mater.* **19**, 3063–3069 (2009).

[62] P.-A. Bouit, D. Rauh, S. Neugebauer, J. L. Delgado, E. Di Piazza, S. Rigaut, O. Maury, C. Andraud, V. Dyakonov, and N. Martín, "Cyanine−cyanine salt exhibiting photovoltaic properties," *Org. Lett.* **11**, 4806–4809 (2009).

[63] Y. Sun, G. C. Welch, W. L. Leong, C. J. Takacs, G. C. Bazan, and A. J. Heeger, "Solution-processed small-molecule solar cells with 6.7% efficiency," *Nat. Mater.* **11**, 44–48 (2012).

[64] Y.-H. Chen, L.-Y. Lin, C.-W. Lu, F. Lin, Z.-Y. Huang, H.-W. Lin, P.-H. Wang, Y.-H. Liu, K.-T. Wong, J. Wen et al., "Vacuum-deposited small-molecule organic solar cells with high power conversion efficiencies by judicious molecular design and device optimization," *J. Am. Chem. Soc.* **134**, 13616–13623 (2012).

[65] (a) "Heliatek sets new world record efficiency for its organic tandem cell," 2012, www.heliatek.com. (b) A. Mishra and P. Bauerle, "Small molecule organic semiconductors on the move: promises for future solar energy technology," *Angew. Chem. Int. Ed.* **51**, 2020–2067 (2012).

[66] C. Joachim and M. A. Ratner, "Molecular electronics: some views on transport junctions and beyond," *Proc. Natl. Acad. Sci. USA* **102**, 8801–8808 (2005).
[67] E. A. Weiss, M. R. Wasielewski, and M. A. Ratner, "Molecules as wires: molecule-assisted movement of charge and energy," *Top. Curr. Chem.* **257**, 103–133 (2005).
[68] B. Xu and N. J. Tao, "Measurement of single-molecule resistance by repeated formation of molecular junctions," *Science* **301**, 1221–1223 (2003).
[69] S. A. Vail, D. I. Schuster, and D.M. Guldi, "Energy and electron transfer in beta-alkynyl-linked porphyrin-[60]fullerene dyads," *J. Phys. Chem. B* **110**, 14155–14166 (2006).
[70] S. A. Vail, P. J. Krawczuk, D. M. Guldi, A. Palkar, L. Echegoyen, J. P. C. Tom, M. A. Fazio, and D. I. Schuster, "Energy and electron transfer in polyacetylene-linked zinc–porphyrin–[60]fullerene molecular wires," *Chem. Eur. J.* **11**, 3375–3388 (2005).
[71] A. Lembo, P. Tagliatesta, D. M. Guldi, M. Wielopolski, and M. Nuccetelli, "Porphyrin-β-oligo-ethynylenephenylene-[60]fullerene triads: synthesis and electrochemical and photo-physical characterization of the new porphyrin-oligo-PPE-[60]fullerene systems," *J. Phys. Chem. A* **113**, 1779–1793 (2009).
[72] G. de la Torre, F. Giacalone, J. L. Segura, N. Martín, and D. M. Guldi, "Electronic communication through π-conjugated wires in covalently linked porphyrin/C_{60} ensembles," *Chem.-Eur. J.* **11**, 1267–1280 (2005).
[73] A. Molina-Ontoria, M. Wielopolski, J. Gebhardt, A. Gouloumis, T. Clark, D. M. Guldi, and N. Martín, "[2,2′]Paracyclophane-based π-conjugated molecular wires reveal molecular-junction behavior," *J. Am. Chem. Soc.* **133**, 2370–2373 (2011).
[74] M. Wielopolski, G. de Miguel Rojas, C. van der Pol, L. Brinkhaus, G. Katsukis, M. R. Bryce, T. Clark, and D. M. Guldi, "Control over Charge Transfer through Molecular Wires by Temperature and Chemical Structure Modifications," *ACS Nano* **4**, 6449–6462 (2010).
[75] F. Giacalone, J. L. Segura, N. Martín, and D.M. Guldi, "Exceptionally small attenuation factors in molecular wires," *J. Am. Chem. Soc.* **126**, 5340–5341 (2004).
[76] C. Atienza-Castellanos, M. Wielopolski, D.M. Guldi, C. Van der Pol, M. R. Bryce, S. Filippone, N. Martín, "Determination of the attenuation factor in fluorene-based molecular wires," *Chem. Commun.* **48**, 5164–5166 (2007).
[77] M. Wielopolski, J. Santos, B. M. Illescas, A. Ortiz, B. Insuasty, T. Bauer, T. Clark, D.M. Guldi, and N. Martín, "Vinyl spacers—tuning electron transfer through fluorene-based molecular wires," *Energy Environ. Sci.* **4**, 765–771 (2011).
[78] C. Atienza, N. Martín, M. Wielopolski, and D. M. Guldi, "Tuning electron transfer through p-phenyleneethynylene molecular wires," *Chem. Commun.* **30**, 3202–3204 (2006).
[79] M. Wielopolski, C. Atienza, T. Clark, D. M. Guldi, and N. Martín, "p-Phenyleneethynylene molecular wires: influence of structure on photoinduced electron-transfer properties," *Chem. –Eur. J.* **14**, 6379–6390 (2008).
[80] F. Wessendorf, B. Grimm, D. M. Guldi, and A. Hirsch, "Pairing fullerenes and porphyrins: supramolecular wires that exhibit charge transfer activity," *J. Am. Chem. Soc.* **132**, 10786–10795 (2010).

10

RECENT ADVANCES IN METAL OXIDE-BASED PHOTOELECTROCHEMICAL HYDROGEN PRODUCTION

BOB C. FITZMORRIS AND JIN Z. ZHANG

Department of Chemistry and Biochemistry, University of California, Santa Cruz, CA, USA

10.1 INTRODUCTION

10.1.1 Motivation

Energy is being consumed by human beings at an ever increasing rate with no end in sight. Several factors are contributing to increased consumption including population growth, rising consumerism in developing countries, and increased dependence on electronics. The main sources of energy utilized worldwide are fossil fuels (coal, petroleum, and natural gas), nuclear, hydroelectric, wind, geothermal, and solar. Due to the finite amount of fossil and nuclear fuel that can be harvested from the earth, it is important to develop an energy economy based off of renewable sources of energy, which will not be depleted over time [1]. Each form of renewable energy has advantages and drawbacks that require a diverse mix of the different energy sources [2].

One of the greatest challenges facing many renewable forms of energy is that they are not available on demand the way fossil fuels are. Times of high energy demand are not always times of high energy availability even in places where renewable energy sources, such as wind and sunlight, are plentiful. Because of these limitations, energy must be stored between times of high energy harvesting and times of high demand. Many methods for energy storage have been developed including flywheels, electrochemical storage (battery banks), and energy storage molecules [3, 4]. The most widely studied energy storage molecule is hydrogen gas. Molecular hydrogen has a high energy density and it is made from the most abundant element in the universe. Molecular hydrogen can be made using electricity through the electrolysis of water and then recombined with oxygen in an internal combustion engine or under more controlled conditions in a fuel cell to produce electricity [5]. Water splitting can be accomplished using any form of energy [6] but photoelectrochemical (PEC) water splitting is the preferred method because it directly splits water using sunlight as the only energy input. This method of energy conversion is especially attractive in places such as tropical islands where sunlight and saltwater are plentiful and fossil fuels must be imported.

10.1.2 Description of PEC Hydrogen Generation

The basic principles behind PEC water splitting are straightforward, a photoanode consisting of a semiconductor film on a conductive substrate is submerged in electrolyte solution and illuminated with sunlight. Photons are absorbed by the semiconductor producing electron–hole pairs, the holes migrate to the semiconductor/electrolyte interface and electrons move to the conductive substrate and travel through the circuit to the cathode. At the photoanode/electrolyte interface the holes react with water molecules to produce oxygen gas in what is called the oxygen evolution reaction (OER), which is usually the rate limiting step in water splitting. At the cathode, usually a platinum wire, electrons react with protons from the electrolyte solution to produce molecular hydrogen. In some cases the cathode and anode are submerged in different electrolyte solutions that are separated by a membrane, this also allows the hydrogen gas to be collected separate from the oxygen. Figure 10.1 shows a schematic of a typical PEC apparatus [7].

To compete with other sources of energy, PEC hydrogen production must be efficient. The largest loss of energy in PEC is due to solar photons not being absorbed by the semiconductor photoanode, for maximum efficiency the photoanode must absorb strongly in the visible region. The energy harvested from the light that is absorbed, or internal efficiency, is limited by several types of recombination summarized in Figure 10.2. After absorbing solar photons, the electrons and holes created must relax to the band edges of the semiconductor forming an exciton and they lose some energy as heat in the process. Once at the band edges the exciton may recombine, either radiatively or nonradiatively. Nonradiative recombination often occurs through trap states, which are introduced by defects in the crystal structure, dopant atoms, or dangling bonds on the crystal surface. Charge carrier recombination is a major source of energy loss in PEC. Improving the efficiency of hydrogen production requires the development of materials that maximize solar light absorption and minimize charge carrier recombination.

INTRODUCTION 345

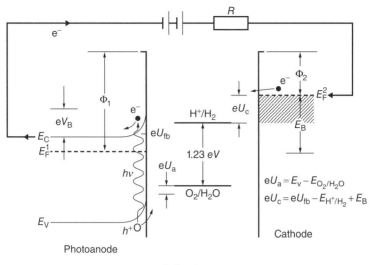

FIGURE 10.1 Diagram showing the movement of electrons through a PEC cell. E_F^1 is the Fermi level of the photoanode; E_F^2 is the Fermi level of the cathode; E_{CB} the energy of the CB edge; $\underline{E}_V$ the energy of the valence band edge; eU_a and eU_c are the overpotentials of the anode and cathode; R, resistance; e^-, electron; h^+, hole; Φ_1 and Φ_2 are the work functions of the anode and cathode materials respectively; U_{fb} flat band potential; E_B, potential energy related to the bias; V_B, surface energy related to band bending; EH^+/H_2, energy of the redox couple H^+/H_2; EO_2/H_2O, energy of the redox couple O_2/H_2O; h, Planck's constant; e, charge of an electron; and v, frequency. Electrons in the valence band of the (n-type) semiconductor are excited to the CB with photons of equal or greater energy than the bandgap of the semiconductor. The electron leaves a hole in its place, which migrates to the semiconductor/electrolyte interface and reacts with water to form oxygen. The excited electron migrates to the conductive substrate, into the circuit and to the cathode where it reacts with protons to form molecular hydrogen.

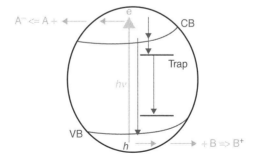

FIGURE 10.2 Diagram showing the journey of charge carriers in a semiconductor nanoparticle. Up arrow represents absorption of light with energy greater than the bandgap energy. Down arrows show intraband relaxation, carrier trapping, and carrier recombination. Horizontal arrows show separation of charge carriers. In a PEC photoanode the holes migrate to the electrolyte interface to react with water or a hole scavenger, the excited electrons go to the cathode to reduce protons to molecular hydrogen.

10.2 MATERIALS FOR PEC HYDROGEN PRODUCTION

In 1972, Fujishima and Honda [8] published their groundbreaking paper which first described PEC water splitting using a TiO_2 film as the photoanode. In the 40 years since its discovery there have been thousands of publications on this topic but industrial hydrogen production continues to be dominated by another more economical process: the steam reformation of methane (which releases CO_2 as a byproduct). In order for PEC to become economically feasible, a photoanode material must be developed which is inexpensive, abundant, and efficient at splitting water using solar radiation. The focus of recent PEC research has been on developing photoanode materials that meet these criteria.

10.2.1 Metal Oxides for PEC Water Splitting

Metal oxides are inexpensive, abundant materials that are stable in light, air, and water. The metal oxides which have shown promise for water splitting fall into one of the two groups: either they have a wide bandgap and can only absorb UV light but can split water without an applied potential; or, they have a narrower bandgap so they absorb visible light but require an external potential to split water. TiO_2 and ZnO have band positions that straddle the oxidation and reduction potentials for water splitting and therefore fall into the first group. The second group consists of $\alpha\text{-}Fe_2O_3$ and WO_3 which absorb visible light. These visible light absorbing MOs also suffer from poor stability in certain electrolytes. One advantage that all of these MOs share is low cost and high availability.

10.2.1.1 TiO₂ TiO_2 was the first material used for PEC water splitting and continues to be a focus of PEC research 40 years later [9]. TiO_2 is a white crystalline material that can be made in solution from reactive precursors such as titanium isopropoxide [10] or by electrochemical oxidation of titanium metal [11]. The two common crystal structures of TiO_2 are rutile and anatase which have slightly different bandgap energies and electrochemical properties. Different preparation methods will yield anatase, rutile, or a mixture of the two structures. The main advantages of TiO_2 as a PEC photoanode material are its ability to separate and transport charges, particularly electrons, efficiently. Another advantage is its stability in electrolyte solutions with a wide pH range, even under illumination. The large difference in electronegativity between titanium and oxygen causes the valence band, which has mostly O 2p character, to be of much lower energy than the unoccupied conduction band (CB) states which have mostly titanium d character [12, 13]. The wide bandgap of TiO_2 limits it to absorbing the highest energy solar photons which make up only about 3% of the overall spectrum. The highest reported PEC solar to hydrogen (STH) efficiency for a pure TiO_2 photoanode was 1.1% [14]. Time-resolved spectroscopy has been used to study the exciton dynamics of TiO_2 to better understand the sources of additional loss due to charge carrier recombination within the material [9, 15, 16]. Efforts to improve the PEC efficiency have focused on increasing visible light absorption through doping or combining TiO_2 with other narrow bandgap materials.

10.2.1.2 ZnO
ZnO, a direct wide bandgap semiconductor (3.37 eV), has been intensely investigated for use in PEC applications [17–19]. ZnO suffers from the same solar absorption issue as TiO_2 but has an advantage of higher electron mobility [20]. ZnO has responded well to sensitization with dyes and quantum dots for solar cells [21]. As ZnO is not stable in as wide a pH range as TiO_2, it must be studied in solutions with relatively neutral pH. ZnO nanoparticles are often fluorescent due to defect states in the crystal lattice. This fluorescence must be minimized in a PEC cell because it results in lost electron–hole pairs. This fluorescence permits the use of time-resolved fluorescence as an additional technique for studying exciton dynamics in ZnO photoanodes [16, 22–25].

10.2.1.3 Hematite (α-Fe_2O_3)
Hematite (α-Fe_2O_3) has recently garnered a great deal of attention in PEC research as some breakthroughs have been made in its performance [26, 27]. Hematite is extremely abundant; it is more commonly known as rust and forms spontaneously on iron metal in the presence of salt water. The red color of hematite is what makes it attractive for PEC. Hematite has a bandgap of 2.2 eV which allows it to absorb 16% of the solar spectrum. Compared to ZnO and TiO_2, the improved bandgap for solar light harvesting makes hematite attractive among the MOs but in practice the PEC efficiency lags behind those materials considerably. Although hematite is the most stable form of iron oxide at room temperature, it is not stable in acidic solutions so 1M NaOH is typically used as the electrolyte. Hematite has a CB energy below the reduction potential of hydrogen ion so applying an external potential is necessary to boost the energy of the excited electrons in the CB of Fe_2O_3 to allow the reaction to occur. Hematite also suffers from very low conductivity which limits the collection of excited electrons [27, 28]. The lifetime of excitons are very short in hematite [26, 29], if the electrons are not shuttled to the back contact quickly they will recombine with holes. In order to mitigate this high rate of exciton recombination, thin films of hematite can be deposited on a supporting structure with higher carrier mobility [30, 31]. Alternatively, dopant atoms can be incorporated into hematite to increase conductivity.

10.2.1.4 WO_3
Tungsten oxide, WO_3, is a material with properties intermediate between hematite and the wide bandgap MOs. WO_3 has a bandgap of 2.6 eV which allows it to absorb a small amount of visible light and makes it appear green in color. WO_3 also requires an external potential in order to reduce hydrogen ions at the counter electrode but it has higher electron mobility than hematite [30].

Photoanodes of hematite or WO_3 are studied with the required externally applied potential provided by a potentiostat, or electrochemical work station. For practical applications of these materials as PEC photoanodes the external potential would likely be applied by a photovoltaic PV cell. Different combinations of PV cells with PEC electrodes have been demonstrated that can be effective for water splitting [32]. Silicon PV cells or dye sensitized solar cells can work in tandem with a hematite or WO_3 photoanode, absorbing the red light that is of insufficient energy to be absorbed by the photoanode.

10.2.2 Composite Materials for PEC Water Splitting

The metal oxides introduced have advantages and drawbacks that have prompted solutions to overcome these problems through the incorporation of a second material or multiple materials together to exploit the advantages of each. It has proved too difficult to create a photoanode of a single material which can absorb visible light, efficiently separate charges, and transport charges. A more effective method is to use different materials for each of these purposes.

In designing a composite photoanode for PEC, it is important to consider the band positions of the semiconductors involved; this is often referred to as bandgap engineering. In bandgap engineering the goal is to deposit multiple materials with different electronic energy levels to promote transport of electrons from the electrolyte interface to the back contact and promote transport of holes to the electrolyte interface. A band diagram can often be helpful in visualizing this movement of charge carriers through the electronic levels of multiple materials. It is also important to consider how each of the materials will interact with the incident light in order to utilize the maximum number of solar photons.

The most common type of composite photoanode is one that combines a wide bandgap semiconductor, such as TiO_2 or ZnO, with a narrow bandgap semiconductor that can absorb visible light. A layered structure can be made by first depositing a film of the wide bandgap material followed by a layer of the narrow bandgap material. The narrow bandgap material must have a CB energy higher than that of the wide bandgap material in order to promote electron injection across the interface into the wide bandgap material. The photogenerated holes in the valence band of the small bandgap material are therefore of higher energy than the valence band of the wide bandgap material, preventing holes from moving from the small bandgap material into the large bandgap material. The interface of the two materials serves as an electron–hole pair separator, segregating the electrons in the wide bandgap material and holes in the small bandgap material. To improve the PEC efficiency of such a layered structure, the small bandgap material can be engineered to maximize absorption of solar light, the wide bandgap material engineered for improved charge transport, and the interface between the two materials engineered to maximize charge separation.

10.2.2.1 Mixed Metal Oxides

WO_3 has been combined with TiO_2 to produce mixed films [33] as well as to form core/shell nanorod arrays [34, 35]. In one study, nanorod arrays of quasi-core/shell TiO_2/WO_3 and WO_3/TiO_2 were compared [34], showing that WO_3 as the core material had superior performance. This was explained by considering the band positions of the two materials where WO_3 has lower valence and CB energies than TiO_2. Figure 10.3 shows a diagram of the band positions of WO_3 and TiO_2 with arrows indicating the transfer of electrons and holes across the interface. With TiO_2 on the surface, electrons move down in energy to travel into the core and holes generated in WO_3 travel up in energy moving into TiO_2. This is the preferred arrangement where the band positions should be higher for the material in contact with the electrolyte. In order to have efficient visible light absorption, it would be better to have a material with a narrow bandgap.

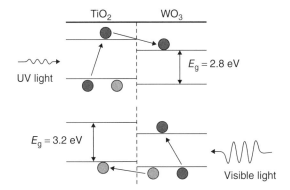

FIGURE 10.3 Band diagram showing the band edge positions of TiO_2 and WO_3 and the movement of electrons (dark grey) and holes (light grey) upon illumination with UV light (top) and visible light (bottom) [34]. The movement of charge carriers explains the superior water splitting performance of the WO_3/TiO_2 core/shell nanorod arrays. *(For a color version of this figure, see the color plate section.)*

Mixed metal oxides of WO_3 and Fe_2O_3 have also shown promise for water splitting. A comparison of nanoparticle films composed of WO_3, Fe_2O_3, and layered Fe_2O_3/WO_3 showed that the layered structure outperformed the single material photoanodes [31]. The increased performance was attributed to separation of charge carriers. Hematite absorbs visible light but has very poor conductivity which has prompted recent attempts to support iron oxide on different host materials. The Grätzel group has become a leader over the last decade in efforts to improve the PEC properties of hematite and one strategy they used was employing WO_3 as a supporting material for hematite nanoparticles [30]. The rough morphology of the WO_3 support allowed for significant absorption with a very thin layer of hematite. The WO_3 scaffold also improved charge separation, increasing the absorbed photon to current conversion efficiency (APCE) of Fe_2O_3 at 565 nm from 5.7% for Fe_2O_3 alone to 8% when supported on the WO_3 scaffold.

10.2.2.2 Quantum Dot Sensitization Quantum dots (QDs) are direct bandgap semiconductor nanocrystals which exhibit size-dependent bandgap energy. The ability to achieve desired bandgap energies by changing the size of the QDs makes these materials very attractive for applications in solar energy conversion. The high absorption coefficient of QDs allows them to absorb a great deal of visible light with even a very thin layer of QDs on the MO. This arrangement of QDs on a MO scaffold is often called QD sensitization. The most common quantum dot materials used for PEC applications are CdS, CdSe, and CdTe because they all absorb in the visible range. Supporting these QDs on MOs such as ZnO or TiO_2 promotes charge separation with the electrons moving from the QD into the MO and holes moving from the MO into the QD. Once the charges are separated in this manner there is an energy barrier for them to move back across the interface thereby decreasing the probability of recombination.

Quantum dot sensitization was inspired by the development of the dye sensitized solar cell by Dr. Michael Grätzel which uses molecular dyes to absorb visible light and inject electrons into a metal oxide support [36, 37]. Replacing molecular dyes with quantum dots in solar cells was a logical extension of this work because quantum dots have several advantages over dyes including tunable bandgaps and improved photostability; however, early attempts to use QDs for water splitting were challenging due to their poor stability in electrolyte solutions.

The poor stability of QDs in aqueous solutions is due to their valence band position being higher in energy than the potential for the OER. To prevent oxidative damage to the QDs a source of electrons must be provided in the electrolyte to react with the high energy holes in the QD valence band. A popular hole scavenging electrolyte is a mixture of S^{2-} and SO_3^{2-} which replaces the OER as the reaction at the photoanode while allowing the hydrogen reduction reaction to proceed at the cathode [38, 39]. Since complete water splitting is not accomplished when using a hole scavenging electrolyte, it is instead called PEC hydrogen generation. Due to the additional requirements of a hole scavenging electrolyte, PEC hydrogen generation using QDs must achieve an even higher efficiency to become economically competitive with current hydrogen production methods.

Quantum dot sensitization is very sensitive to the method of attaching the QDs to the MO. QDs are typically made in organic solvents and are coated in a monolayer of organic capping ligands that aid in both the synthesis and in maintaining solubility in nonpolar solvents. One of the simplest methods of adding QDs to the MO surface is by drop casting or spin coating the QD solution onto the MO film. Using this method, QD films of varying thicknesses can be prepared but there is the disadvantage of having insulating capping ligands surrounding the QDs which inhibit charge injection. Annealing the films to remove organic ligands can improve electrical contact between QDs and the MO film but this must be done in inert atmosphere to prevent oxidizing the QD surface and the temperature must be controlled to prevent melting the QDs together and losing the size-dependent properties.

Another popular method of QD attachment is to use a bifunctional linker molecule where one end of the molecule has an affinity for the MO surface (often a carboxylic acid group) and the other end has an affinity for the QD surface (usually a thiol group). This type of attachment usually results in a monolayer of the QDs which maximizes the charge injection from the excited QDs but will not absorb all available photons. Studies have also investigated the use of different linker molecules and have found that shorter linkers perform better [40, 41].

QD sensitized metal oxides have been studied extensively using time-resolved spectroscopy and these studies have aided in the development of more efficient PEC photoanodes. Many of these studies have investigated the dynamics of QDs with and without the presence of a metal oxide in order to determine the rate of electron injection from the QD into the MO. Dynamics studies of QD sensitized MOs began with coupled systems where the QD and MO were connected via a bifunctional linker molecule but still suspended in solution [42, 43]. By choosing a pump wavelength that would only excite the QD, these studies showed that the exciton lifetime of the QD was much shorter in the presence of a MO because of electron transfer from the

CB of the QD into the CB of the MO. By comparing the excited state lifetime of the isolated QDs (τ_{QD}) to the lifetime of the QD in the presence of a MO ($\tau_{QD,MO}$) the electron transfer rate can be estimated using Eq. (10.1):

$$k_{ET} = \frac{1}{\tau_{QD,MO}} - \frac{1}{\tau_{QD}}. \quad (10.1)$$

The drawback to studying the coupled colloidal system is that it is not equivalent to having the QD attached to MO electrode. One advantage of this colloidal coupling method is that the colloidal solutions often retain their fluorescence which allows the study of ns dynamics by time-resolved fluorescence. Studying the exciton dynamics of QDs attached to a MO film using a molecular linker allows the researcher to relate the dynamics to the performance of the film as a PEC electrode [44]. Determining the electron injection rate of QD sensitized films requires a control experiment of QDs where injection is not possible. Typically these control samples consist of a QD sensitized insulator such as ZrO_2 [45] or SiO_2 [46] which has similar dielectric constants but much higher CB energies which prevent electron injection from the QDs.

Efforts have been made to further improve the efficiency of charge injection from QDs into MOs by increasing the interfacial area between the two materials. Even with no capping ligands separating a QD from a film, the spherical shape of the QD dictates that it will contact the film at a single point. Chemical bath deposition is a method of growing QDs directly onto a film in order to maximize their electronic coupling. This method has been found to be very effective for sensitizing MOs such as TiO_2 with QDs of CdSe [47], CdS [45,48], and CdTe [49]. Atomic layer deposition is another method used to controllably coat thin layers of CdS onto TiO_2 nanotubes without using linker molecules [50,51]. Physical vapor deposition has also been used to create $CdSe/TiO_2$ homogenous nanorod arrays [52,53]. Each of these methods of growing QDs directly onto MOs has disadvantages because the QD growth is difficult to control and preparing equivalent control experiments without the MO can be difficult.

Time-resolved studies comparing TiO_2 films sensitized with QDs using a molecular linker or through direct attachment showed that the direct attachment method achieved higher photocurrent and exhibited a faster charge injection rate in the dynamic studies [40,54]. Chemical bath deposition has become a very popular choice for sensitizing metal oxides for PEC hydrogen production as well as for fundamental time-resolved studies [45].

The fast electron injection obtained from studies of chemical bath deposition of QDs on MOs has inspired new deposition techniques that maximize interfacial area. Physical vapor co-deposition was used to grow nanorod arrays that consisted of homogenous mixtures of CdSe and TiO_2 [52]. The co-deposited nanorod arrays exhibit higher efficiency for hydrogen generation than either pure CdSe or pure TiO_2 nanorod arrays made in the same way. Figure 10.4 shows plots of the recovery of the first exciton bleach feature for CdSe nanorods and $CdSe/TiO_2$ composite nanorods, as well as a diagram showing the relative band positions of CdSe and TiO_2. The exciton lifetime of the CdSe nanorod array was compared to that of the co-deposited

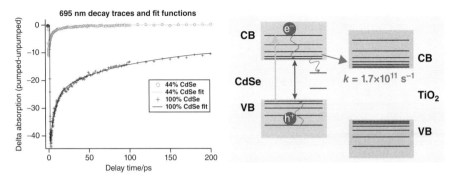

FIGURE 10.4 Time traces from Reference 52 showing the recovery of the CdSe transient bleach feature and a band diagram showing the relative band positions of CdSe and TiO_2. In the presence of TiO_2 the bleach feature recovers much faster. The band diagram to the right shows the excitation of CdSe by visible light, electrons and holes relaxing to the band edge, and electrons being injected from CdSe into TiO_2. The electron transfer rate, k_{ET}, is very fast due to the strong electronic coupling between the two materials. *(For a color version of this figure, see the color plate section.)*

sample and the resulting charge injection rate was 1.7×10^7 s^{-1}, one of the highest values reported in the literature. This very fast injection rate was attributed to the strong electronic coupling between CdSe and TiO_2 in the homogenous nanorods.

The optimal-sized CdSe QD to use as a sensitizer is an interesting question that has been investigated using time-resolved methods. Robel et al. [55] compared the injection rate as a function of QD size for CdSe QDs attached to TiO_2 porous films using mercaptopropionic acid as a linker. They found that with increasing size the QDs had a decreasing injection rate; they concluded that the larger CdSe QDs have lower CB energy, closer to the CB energy of TiO_2, and therefore a lower driving force for electron injection. The Kamat group continued their investigation of the importance of this driving force by determining the charge injection rate from CdSe QDs of various sizes into MOs with different CB energies [46]. They investigated three common metal oxides (ZnO, TiO_2, and SnO_2) and found that the injection was the slowest for ZnO and the fastest for SnO_2. This trend follows their previous results showing that a QD with a higher CB energy than the MO drives electron injection across the MO–QD interface.

To optimize charge injection into an MO a QD should have good electrical contact with the MO, which limits deposition to a monolayer of QDs. To achieve the optimal CB position for charge injection these QDs should also be quantum confined, making them very small. A monolayer of very small QDs will not be able to absorb light efficiently, so researchers have looked for ways to make large QDs that still have tunable bandgaps. Alloyed QDs have composition tunable bandgaps that allow for tunability throughout the visible spectrum and allows larger crystal sizes [56–58]. Alloyed CdSeS QDs and quantum rods (QRs) have been grown directly onto TiO_2 nanowires using CVD and were compared to similar structures composed of CdS and CdSe [59]. An SEM image of the CdSeS QRs on TiO_2 nanowires is shown

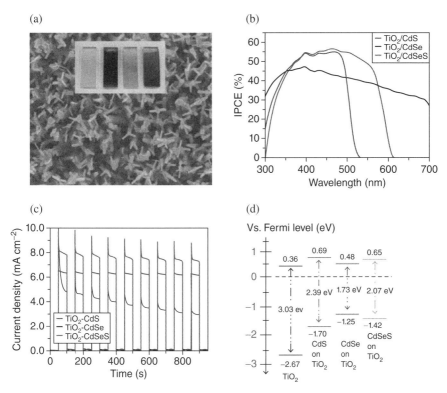

FIGURE 10.5 (a) SEM image of CdSeS rods grown on TiO_2 nanowires using CVD deposition, inset showing from left to right films of TiO_2, CdS nanorods on TiO_2, CdSe nanorods on TiO_2, and CdSeS nanorods on TiO_2. (b) IPCE plots for the sensitized samples showing that CdS has high efficiency but can only absorb wavelengths less than 520 nm, CdSe can absorb all of the visible spectrum, and CdSeS sensitized has high efficiency and intermediate visible light absorption extending to 600 nm. (c) White light on/off scans showing the photocurrent density from the TiO_2 films with different sensitizers. (d) Diagram showing the band positions of TiO_2, CdS, CdSe, and CdSeS alloy quantum dots. The CdSeS QDs were shown to have the optimal band positions for absorbing solar light and injecting charge into TiO_2. From Reference 59. *(For a color version of this figure, see the color plate section.)*

in Figure 10.5a. The CdS QRs had more efficient charge injection than the CdSe QRs because the CB of CdS is of higher energy. The narrow bandgap of CdSe allowed the CdSe QR sensitized TiO_2 to show photocurrent beyond 700 nm in IPCE (Fig. 10.5b) while the CdS sensitized TiO_2 showed no photocurrent beyond 520 nm. The CdSeS QRs showed intermediate behavior with photocurrent extending to 600 nm and intermediate IPCE efficiency. Under white light illumination the alloyed QRs showed the highest photocurrent density of the samples studied (Fig. 10.5c).

10.2.2.3 Metal Nanoparticle-Enhanced PEC One controversial strategy of enhancing the PEC performance of a MO photoanode is through the use of metal nanoparticles. In the field of photocatalytic water splitting where suspended nanoparticles or nanoparticle films are used to perform the complete water splitting reaction without using a circuit, decorating the surface of the nanoparticle with either Au or Pt nanoparticles has been shown to improve the efficiency [60–63]. Noble metals are typically chosen because they resist corrosion in aqueous solutions. The improvement in photocatalytic activity was attributed to stabilizing photoexcited electrons on the MO nanoparticle surface, reducing the probability of exciton recombination, and catalyzing the hydrogen evolution reaction. Based upon this mechanism it seems unlikely that metal nanoparticles would improve the efficiency of PEC water splitting where the excited electrons in the MO are transported to the back contact and the hydrogen evolution reaction occurs at a platinum cathode. There is still much debate about how the metal nanoparticles enhance the PEC performance of MO photoanodes.

Metal nanoparticles exhibit very strong extinction of visible light due to the collective oscillation of electrons on the particle surface. This surface plasmon resonance (SPR) is material, shape, and size dependent. Several studies have observed photocurrent at wavelengths corresponding to the SPR peak of AuNPs for TiO_2 films with gold nanoparticles (AuNPs) on the surface [63–68]. Some studies have suggested that this enhancement may be due to charge injection from the metal nanoparticles, analogous to the enhancement seen from adsorbing dyes or quantum dots [65]. This interpretation is problematic because the SPR is due to the collective excitation of many electrons that does not increase the energy of any single electron significantly. Some studies have attributed the increased performance to scattering from the AuNPs, which causes improved absorption of these wavelengths [69]. To maximize this scattering phenomenon it is advantageous to use AuNPs with diameters greater than 50 nm because at this size range scattering becomes dominant over absorption [70]. An alternative mechanism has been suggested that attributes the increased photocurrent to an enhanced electric field near the gold nanocrystal surface [71]. Localized electric field enhancement on the surface of metal nanoparticles has been thoroughly studied for application in surface enhanced Raman scattering [72]. In a PEC cell it is thought that the localized electric field on the nanoparticle surface can excite transitions in the MO, which would otherwise absorb very weakly. TiO_2 often exhibits a very small, nonzero photocurrent in the green region of the spectrum (due to defect related states) where gold exhibits an SPR peak for spherical particles and enhancing absorption of these green photons may result in the plasmonic enhancement seen for AuNPs on TiO_2.

Other metal nanoparticle/MO combinations have been investigated for PEC water splitting, inspired by the positive results found for AuNP/TiO_2 heterostructured photoanodes. Nitrogen doped TiO_2 photoanodes have been decorated with silver cubes with an average side length of 120 nm [73]. The silver cubes were chosen because they exhibit a strong SPR in the wavelength range (400–500 nm) where nitrogen doping causes visible light photocurrent in TiO_2. The visible light photocurrent was enhanced by a factor of 10 with the addition of the nanocubes. In a similar strategy, a hybrid photoanode was designed to take advantage of the spectral overlap between the SPR of AuNPs and the band edge transition in hematite [74]. Increased absorbance and PEC enhancement at the SPR wavelength were observed for AuNPs

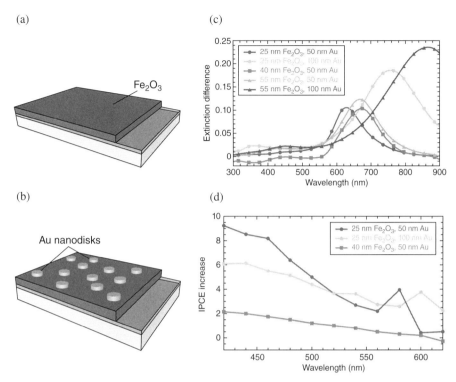

FIGURE 10.6 (a) Schematic drawing of hematite thin film deposited on conductive glass. Film thicknesses of 25, 40, and 55 nm were compared. (b) Schematic of lithographically deposited Au nanodisks on hematite thin films. Nanodisks were made with diameters of 50 and 100 nm. (c) Plots of the difference in extinction spectrum (with gold nanodisks–without gold nanodisks) for films of different thicknesses with different diameter gold nanodisks. (d) IPCE action difference spectra (IPCE with gold nanodisk–IPCE of hematite film) for three of the films showing enhancement of IPCE at the SPR wavelength of the gold nanodisks. From Reference 75. *(For a color version of this figure, see the color plate section.)*

deposited on the surface of hematite platelets but AuNPs embedded in hematite thin films were found to decrease the performance of the photoanode because they act as recombination centers for electrons and holes in the film. In another study, Au nanodisks of 50 and 100 nm diameter were lithographically deposited on hematite films of varying thicknesses. Schematic drawings of the gold nanodisks on hematite thin films are presented in Figure 10.6b and 10.6c. Gold nanodisks increased the absorption of the films at the SPR wavelength (Fig. 10.6c) and improved the IPCE at the SPR wavelength (Fig. 10.6d). Improved PEC performance under white light was observed and the thinnest (25 nm) film with 50 nm diameter Au nanodisks showed the greatest enhancement (factor of 5) [75]. These studies provide evidence that the plasmonic enhancement is due to a combination of electric field enhancement and scattering.

Time-resolved spectroscopy may also have the potential to provide a better understanding of the enhanced PEC behavior of noble metal decorated metal oxide films. Cooper et al. investigated the dynamics of gold particles decorated on ZnO nanowires to determine if there was any change in the gold excited state lifetime due to electron injection into ZnO [76]. By comparing the lifetime of AuNPs on ZnO to the lifetime of AuNPs in water, no significant difference was found in the dynamics. The excited state lifetime of ZnO also did not change with the addition of AuNPs. These AuNP decorated ZnO did not show any PEC performance enhancement over pristine ZnO so it is unsurprising that there was no evidence of charge injection.

Furube et al. [77] conducted a study of AuNPs on TiO_2 nanoparticle films. A 550 nm pump pulse was used to excite AuNPs without exciting the TiO_2 and a transient absorption signal was detected at 3500 nm. Control experiments of N3 dye on TiO_2 and AuNPs on ZrO_2 were used to better understand the nature of the signal. The N3 dye is known to inject electrons into the CB of TiO_2 from its extensive use as a sensitizer in solar cells and this electron injection can be monitored by transient absorption at 3500 nm [78]. Comparison of the AuNP decorated and N3 dye sensitized samples showed that the rise in the transient absorption signal is much faster for AuNP, it occurs within instrument response time of 240 fs. No transient absorption signal at 3500 nm was observed for AuNP decorated ZrO_2, proving that the signal observed in the AuNP decorated TiO_2 is not from the AuNPs. The authors concluded that hot electron injection from AuNPs into the CB of TiO_2 was the source of the 3500 nm transient absorption signal. Subsequent work, also from the Tachiya group, showed that the transient signal decays faster for smaller TiO_2 nanoparticles [79].

The work of the Tachiya group showed that illuminating AuNP decorated TiO_2 with 550 nm light results in electrons being promoted to the CB of TiO_2, which presumably leads to photocurrent in a PEC photoelectrode. Their findings did not conclusively show that electrons were injected from AuNPs into TiO_2; an alternative explanation is that the AuNP enhanced electric field causes increased absorption of 550 nm light by TiO_2. To determine if the CB electrons originate in AuNPs or in TiO_2, the excited state lifetime of the AuNP must be compared on TiO_2 and on an insulator. Charge injection from AuNPs into TiO_2 would result in faster recovery of the bleach feature at the SPR wavelength for AuNPs. Ideally, the transient bleach of the AuNP SPR near 550 nm and the 3500 nm TiO_2 CB electron transient absorption feature should both be monitored to determine whether charge transfer is occurring across the AuNP/TiO_2 interface.

10.2.3 Nanostructured Photoanodes

10.2.3.1 Nanoparticle Films One of the earliest advances in PEC water splitting was achieved by making photoanodes of nanocrystals instead of single crystal material. Researchers found that by making nanoporous or mesoporous films of nanoparticles they could increase the efficiency of a PEC photoanode [80]. The reason for the increase is that more active sites are available to do the OER and that the photoexcited hole has a shorter path length to the electrolyte. This allows a thicker film to be used in order to absorb more light.

10.2.3.2 1D Nanostructures

One disadvantage of nanoporous films is the complicated path for electrons to reach the back contact, having to hop from particle to particle. The solution to this issue of electron transport was solved with the advent of synthetic procedures for 1D nanostructures [19, 53, 81, 82]. 1D nanostructures (nanorods, nanowires, and nanotubes) can be grown vertically aligned on a substrate. Excitons generated in these oriented 1D structures are generated near the electrolyte interface and holes are able to migrate there with decreased likelihood of recombination. Excited electrons in a 1D structure have a clear path to the back contact; this vectorial transport of electrons improves their collection efficiency. A comparison of TiO_2 films with different nanostructures showed that nanoporous films were more efficient than planar films and a nanotube array outperformed both the planar and nanoporous films [83]. SEM images of the different film morphologies as well as plots of the photocurrent of each film are shown in Figure 10.7. In addition to the charge transport benefits of 1D structures, they also have internal reflection properties that improve the efficiency with which they absorb solar radiation. Additional

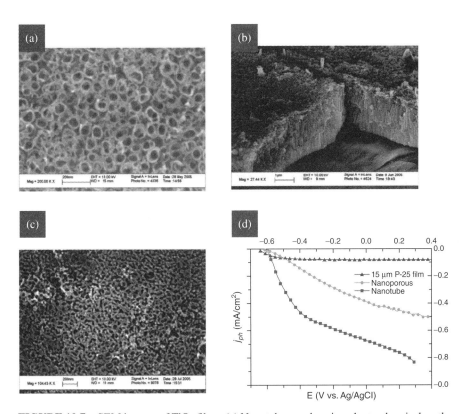

FIGURE 10.7 SEM images of TiO_2 films. (a) Nanotubes made using electrochemical methods. (b) Side view of nanotube film. (c) Nanoporous film. (d) Linear sweep voltammograms showing the improved photocurrent for nanoporous and nanotube films. From Reference 83.

shapes have been evaluated for PEC including dendritic or branched nanowires [84], cauliflower structures [85], and 2D nanodisks [86].

Some materials have anisotropic electron conductivities and this must be considered when designing photoanodes from these materials. An example of this is hematite which has much higher conductivity in the (001) direction than in crystal directions perpendicular to the (001) direction [27]. For optimal performance, 1D nanostructures should be grown in the direction with the highest conductivity.

1D nanostructured photoanodes also have advantages for fabricating composite materials. Previously we discussed how a layered structure, where a small bandgap material is deposited on a wide bandgap material (such as QD sensitized TiO_2 or ZnO), promotes charge separation at the interface of the two materials. An issue with this planar layered structure is that the narrow bandgap layer must be thin to maximize charge separation but a thin film will not absorb much visible light. In order to maximize absorption and charge separation, a 1D nanorod array of the wide bandgap material can be coated with a thin layer of the narrow bandgap material to produce a core/shell structure that allows for higher loading of the visible light absorbing material. 1D MO arrays are now commonly used in place of MO thin films for studies of QD sensitization for PEC [50, 87].

QD sensitization can also benefit from having the QDs themselves become 1D nanostructures. Quantum rods (QRs) have been synthesized with absorption spectra that can be tuned by changing the diameter of the QR but are independent of the QR length [88–90]. Because a QR contains more material than a QD of the same diameter, a monolayer of QRs would absorb more visible light when deposited on a MO surface. To take this a step further, MO nanowire arrays have been sensitized with QDs and QRs to compare the efficiency. Alloyed CdSeS QDs have been grown directly onto TiO_2 nanowires using CVD and were compared to similar structures using CdS and CdSe [59]. The QDs grew into QRs on the tips of the nanowires and these nanorods were shown to be superior in performance to the QD seeds they grew from.

10.2.4 Doped Metal Oxides

Nanoparticle films and 1D nanostructures have increased surface area, which can be advantageous for PEC applications. Nanostructured photoanodes have advantages when it comes to synthetic procedures as well. The high surface area to volume ratio facilitates post-synthesis treatments such as doping with chemical vapor deposition (CVD) and annealing treatments. In 1D nanostructures in particular, the superior electron transport provided by the 1D morphology can be further enhanced when the metal oxide is doped.

Doping is a versatile method for modifying the properties of a pure semiconductor material. Dopants are foreign atoms that are introduced into a crystal lattice, typically at low concentrations. Ionic compounds, such as MOs, can be doped with either anions or cations. These foreign ions can be incorporated into the crystal structure in place of one of the host ions as a substitutional dopant, or the dopant ion can occupy space between the host atoms in what is called interstitial doping. The purpose of adding dopants to the host material is to change the electronic structure of the material; this can lead to improved visible light absorption, conductivity, or electrochemical properties.

Doping can be carried out during the growth of a material or as a post-synthetic treatment. Post-synthetic doping has the advantage that materials made in the same way can be put through different doping treatments to determine the effect they have on the material; this provides ideal control samples. Incorporating a dopant into the synthetic procedure can sometimes interfere with the growth of the material, making it difficult to produce comparable control samples. One of the most popular methods of doping is using CVD to expose a premade film to a reactive source of the dopant atom in the gas phase at high temperature. Precursors include NH_3 for N-doping and CO for C-doping. CVD-based doping is especially effective for 1D nanostructures as the large surface area maximizes the interaction with the reactive dopant atoms.

10.2.4.1 Doping Hematite

Doping is effective at improving the PEC performance of Fe_2O_3 photoanodes. The strategy for doping Fe_2O_3 is different than the one employed for wide bandgap MOs. Instead of narrowing the bandgap, the goal is to increase the electrical conductivity and extend the excited state lifetime to promote the separation and collection of charge carriers. Both n- and p-type dopants have both been employed to increase the electrical conductivity of hematite [85, 91]. These efforts have shown promise and doping is now considered necessary to use hematite as a photoanode.

Time-resolved spectroscopy can be helpful in understanding the role of dopant states in promoting water splitting. Early efforts to dope γ-Fe_2O_3 with Zn, Co, Cr, and Cu in an attempt to extend the excited state lifetime had no significant effect on the excited state lifetime [29]. More recently, α-Fe_2O_3 (hematite) doped with Sn has been shown to have improved PEC photocurrent compared to pristine hematite [91–93]. Sn doped hematite formed by annealing hematite nanowires at 800°C on fluorine doped tin oxide (FTO) substrates was investigated using ultrafast transient absorption spectroscopy [91]. Both 1.1% Sn and 9.9% Sn doped hematite showed triple exponential decay with lifetimes of approximately 0.4, 2.4, and 70 ps but the sample with greater doping showed a faster overall decay with more recombination coming via the 0.4 and 2.4 ps decay pathway. This shows that the improved PEC performance in Sn doping does not originate from extended lifetimes, but rather is attributed to increased electrical conductivity and increased donor density found from electrochemical measurements.

10.2.4.2 Doping TiO_2 and ZnO

Many studies have investigated the effect of doping wide bandgap materials such as TiO_2 and ZnO in order to increase their absorption of visible light. The ideal modification to the electronic structure of these wide bandgap materials would add states that raise the valence band energy, thus decreasing the bandgap of the material. This can be accomplished by replacing a fraction of the oxygen atoms with less electronegative, but similar-sized dopant atoms such as: carbon, sulfur, or nitrogen. Annealing Ti metal in a hydrocarbon flame can produce carbon doped TiO_2 films with evidence of carbon incorporation provided by x-ray photoelectron spectroscopy [94]. TiO_2 nanotube arrays prepared through electrochemical anodization have also been carbon doped by annealing in an atmosphere of carbon monoxide [83,95]. The result of both methods was a reduction in the bandgap energy and visible light absorption. X-ray diffraction data showed that the crystal structure was still rutile TiO_2 after doping with carbon, which suggests that carbon is incorporated in place of oxygen in the crystal lattice. Density functional theory has shown that replacing oxygen atoms with carbon narrows the bandgap by increasing the density of states above the valence band [96]. Replacing an O^{2-} ion with a C^{4-} ion would result in the creation of an oxygen vacancy in order to retain charge balance. In a detailed review of the reports of carbon doped TiO_2, Dr. A. B. Murphy claims that most of these reports are flawed and that there is no evidence that carbon doping is the cause of the visible photocurrent. He instead attributes the visible absorption to the oxygen vacancy states introduced by annealing TiO_2 in a reducing environment [97].

Nitrogen doping has also been accomplished in a similar way by either incorporating nitrogen during the synthesis or by annealing in ammonia gas [98–101]. Examples of this are $N:TiO_2$ and $N:ZnO$ where N^{3-} ion is inserted in the place of O^{2-} ion in the crystal lattice and to maintain charge balance an oxygen vacancy is created. Theoretical calculations show that the oxygen vacancy creates new electronic states within the bandgap which may result in visible light absorption [101].

Further evidence of the importance of oxygen vacancies in doped ZnO or TiO_2 comes from studies of sulfur doping. Since sulfur, like carbon and nitrogen, is less electronegative than oxygen, replacing oxygen atoms with sulfur should narrow the bandgap of the material but because sulfide ions are isoelectronic with O, an oxygen vacancy is not produced by doping. Sulfur doped ZnO has strong visible fluorescence and faster charge carrier recombination compared to pristine ZnO [102]. Sulfur doped TiO_2 absorbs visible light but has been shown to have decreased PEC efficiency compared to pristine TiO_2, probably due to trap state mediated nonradiative recombination [103]. Sulfur doped TiO_2 has shown promise for photocatalysis [104,105]. The improved efficiency for photocatalysis and decreased PEC efficiency implies that sulfur doping decreases the electron mobility. It is possible that nitrogen and carbon doping would have a similar effect if they did not lead to the creation of oxygen vacancies.

10.2.4.3 Hydrogen Treatment To decouple the effect of anion doping from the creation of oxygen vacancies, MOs have been exposed to annealing treatments in either a vacuum or in a reducing atmosphere to generate oxygen vacancies without adding dopants. One versatile method for creating oxygen vacancies is hydrogen treatment where MO films are annealed in hydrogen gas. Hydrogen treatment has been shown to improve the PEC performance of TiO_2 [14], ZnO [76], and WO_3 [106]. The improved PEC performance is attributed to improved electron mobility and higher donor density due to V_O.

Hydrogen treatment causes TiO_2 and ZnO to change color from white, to brown, to black with increasing treatment time. The Mao group attributed the black color in hydrogenated TiO$_2$ to disorder in the crystal [107]. This type of disorder gives rise to band tails, a continuum of mid-gap defect states that can span from the valence to the conduction band. Evidence of these tails in hydrogen treated TiO$_2$ was found by XPS [107, 108]. These closely spaced energy levels permit transitions throughout the visible and into the near-IR range, which give rise to the color seen in these samples.

Hydrogen treated TiO$_2$ powders were used as photocatalysts for H$_2$ generation and it was found that the catalysts produce an impressive 10 mmol H$_2$ h^{-1} g^{-1} under white light but only 0.2 mmol H$_2$ h^{-1} g^{-1} under visible light (>400 nm) [107]. Similarly, hydrogen treated TiO$_2$ nanowire arrays showed dramatically improved photocurrent compared to pristine TiO$_2$, but IPCE measurements showed that the enhancement mostly occurs in the UV region [14]. The IPCE in the UV region was almost 100% for the hydrogen treated samples, and the overall STH efficiency was 1.1%, the highest achieved for a TiO$_2$ photoanode.

Recently, co-treatment of TiO$_2$ nanowire arrays with both hydrogen and ammonia have shown a synergistic effect [109]. The IPCE of the co-treated sample showed photocurrent out to 570 nm, compared to 550 nm for ammonia treated and 420 nm for hydrogen treated samples. The visible (>420 nm) photocurrent made up 41% of the overall white light photocurrent for the co-treated sample. The authors attribute the improved photocurrent to the interactions between substitutional nitrogen, and Ti^{3+} states introduced from hydrogen treatment.

10.2.4.4 QD Sensitized Doped Metal Oxides QD sensitization of TiO$_2$ has been shown to have a synergistic effect when combined with nitrogen doping of TiO$_2$ [99, 110]. This discovery highlights the importance of understanding the mechanism of enhancement between different methods of improving MO PEC. Nitrogen doping alone causes a slight increase in the PEC performance of TiO$_2$ due to increased visible light absorption and conductivity. CdSe QD sensitization dramatically improves the PEC performance, particularly in the visible region. When these two strategies were combined the enhancement was greater than the sum of the enhancement obtained by either treatment alone. The proposed model for this synergistic effect, shown schematically in Figure 10.8, is that the new electronic states created in TiO$_2$ above the valence band due to nitrogen doping allows the holes generated in the QD to be injected into TiO$_2$ rather than requiring them to be transferred to a sacrificial electrolyte. This study provided further evidence that PEC performance can be enhanced through thoughtful bandgap engineering.

A recent study compared the PEC efficiency of TiO$_2$ nanowire arrays subjected to different post-synthetic treatments designed to introduce oxygen vacancies into the MO: vacuum annealing, hydrogen treatment, and nitrogen doping [111]. All three methods were shown to have increased donor density and STH conversion efficiency, although vacuum annealing was less effective than the other two methods. The oxygen deficient TiO$_2$ films did not produce photocurrent under visible light so CdS QD sensitization was used to extend the photocurrent into the visible region for the hydrogen treated TiO$_2$. Photocurrent density of 7.2 mA cm^{-2} were produced by combining the superior electron transport of oxygen deficient TiO$_2$ nanowires

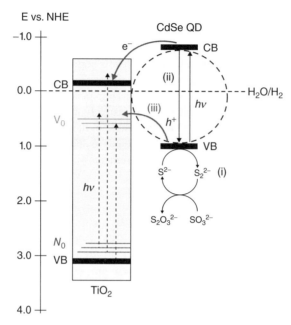

FIGURE 10.8 Band diagram of CdSe QD sensitized nitrogen doped TiO_2 where VB and CB are the valence and CB energies, the electronic states N_0 are those theoretically introduced by nitrogen doping and V_0 are the energies of states introduced by oxygen vacancies. In the proposed model, when CdSe and $N:TiO_2$ are excited by white light the visible light is absorbed by CdSe and the UV light (and some blue light) is absorbed by the $N:TiO_2$. Excited electrons are injected into the CB of TiO_2 and holes in CdSe can react with electrons in the oxygen vacancy states in TiO_2. This hole injection was used to explain the synergistic effect of combining CdSe QD sensitization and nitrogen doping on PEC performance.

with the efficient charge separation and improved light harvesting capabilities of QD sensitization.

10.3 CONCLUSION

We have presented a review of recent advancements in metal oxide-based PEC water splitting and hydrogen production. Significant progress has been made since the discovery of PEC water splitting 40 years ago and with the rising cost of fossil fuels water splitting is getting closer than ever to being a viable alternative to the steam reformation of methane. Metal oxide PEC performance has been improved through doping, combining different metal oxides, sensitizing with QDs, decorating with metal nanoparticles, and through combinations of these methods. Time-resolved spectroscopy has provided valuable insights into the sources of efficiency loss in PEC and the mechanisms by which efficiency is improved. In spite of all these advances, there is still a considerable amount of work to be done to explore new materials, geometries, combinations of materials and more work needs to be done to understand the fundamental science behind PEC.

REFERENCES

[1] S. Shafiee and E. Topal, "When will fossil fuel reserves be diminished?," *Energ. Policy* **37**(1), 181–189 (2009).

[2] M. Z. Jacobson and M. A. Delucchi, "Providing all global energy with wind, water, and solar power, part I: technologies, energy resources, quantities and areas of infrastructure, and materials," *Energ. Policy* **39**(3), 1154–1169 (2011).

[3] M. Murahara and K. Seki, "On-site Electrolysis Sodium Metal Production by Offshore Wind or Solar Energy for Hydrogen Storage and Hydrogen Fuel Cycle," Energy Conversion Congress and Exposition (ECCE), Atlanta, GA, September, 2010, (IEEE 2010), pp. 4264–4269.

[4] F. Díaz-González, A. Sumper, O. Gomis-Bellmunt, and R. Villafáfila-Robles, "A review of energy storage technologies for wind power applications," *Renew. Sust. Energ. Rev.* **16**(4), 2154–2171 (2012).

[5] H.-J. Neef, "International overview of hydrogen and fuel cell research," *Energy* **34**(3), 327–333 (2009).

[6] Ø. Ulleberg, T. Nakken, and A. Eté, "The wind/hydrogen demonstration system at Utsira in Norway: evaluation of system performance using operational data and updated hydrogen energy system modeling tools," *Int. J. Hydrogen Energy* **35**(5), 1841–1852 (2010).

[7] T. Bak, J. Nowotny, M. Rekas, and C.C. Sorrell, "Photo-electrochemical hydrogen generation from water using solar energy. Materials-related aspects," *Int. J. Hydrogen Energy* **27**(10), 991–1022 (2002).

[8] A. Fujishima and K. Honda, "Electrochemical photolysis of water at a semiconductor electrode," *Nature* **238**(5358), 37–38 (1972).

[9] P. Szymanski and M. A. El-Sayed, "Some recent developments in photoelectrochemical water splitting using nanostructured TiO_2: a short review," *Theor. Chem. Acc.* **131**(6), 1202 (2012).

[10] D. M. Antonelli and J. Y. Ying, "Synthesis of hexagonally packed mesoporous TiO_2 by a modified sol–gel method," *Angew. Chem. Int. Ed. Engl.* **34**(18), 2014–2017 (1995).

[11] V. Zwilling, E. Darque-Ceretti, A. Boutry-Forveille, D. David, M.-Y. Perrin, and M. Aucouturier, "Structure and physicochemistry of anodic oxide films on titanium and TA6V alloy," *Surf. Interface Anal.* **27**(7), 629–637 (1999).

[12] M. Sathish, B. Viswanathan, R. Viswanath, and C. S. Gopinath, "Synthesis, characterization, electronic structure, and photocatalytic activity of nitrogen-doped TiO_2 nanocatalyst," *Chem. Mater.* **17**(25), 6349–6353 (2005).

[13] Y. Gai, J. Li, S.-S. Li, J.-B. Xia, and S.-H. Wei, "Design of narrow-gap TiO_2: a passivated codoping approach for enhanced photoelectrochemical activity," *Phys. Rev. Lett.* **102**(3), 036402 (2009).

[14] G. Wang, H. Wang, Y. Ling, Y. Tang, X. Yang, R. C. Fitzmorris, C. Wang, J. Z. Zhang, and Y. Li, "Hydrogen-treated TiO_2 nanowire arrays for photoelectrochemical water splitting," *Nano Lett.* **11**(7), 3026–3033 (2011).

[15] P. D. Colombo, K. A. Roussel, J. Saeh, D. E. Skinner, J. J. Cavaleri, and R. M. Bowman, "Femtosecond study of the intensity dependence of electron-hole dynamics in TiO_2 nanoclusters," *Chem. Phys. Lett.* **232**(3), 207–214 (1995).

[16] R. L. Willis, C. Olson, B. O'Regan, T. Lutz, J. Nelson, and J. R. Durrant, "Electron dynamics in nanocrystalline ZnO and TiO_2 films probed by potential step chronoamperometry and transient absorption spectroscopy," *J. Phys. Chem. B* **106**(31), 7605–7613 (2002).

[17] S. Hotchandani and P. V. Kamat, "Photoelectrochemistry of semiconductor ZnO particulate films," *J. Electrochem. Soc.* **139**(6), 1630–1634 (1992).

[18] Y. Li and J. Z. Zhang, "Hydrogen generation from photoelectrochemical water splitting based on nanomaterials," *Lase. Phot. Rev.* **4**(4), 517–528 (2010).

[19] A. Wolcott, W. A. Smith, T. R. Kuykendall, Y. Zhao, and J. Z. Zhang, "Photoelectrochemical study of nanostructured ZnO thin films for hydrogen generation from water splitting," *Adv. Funct. Mater.* **19**(12), 1849–1856 (2009).

[20] E. M. C. Fortunato, P.M. C. Barquinha, A. C. Pimentel, A. M. F. Gonçalves, A. J. S. Marques, R. F. P. Martins, and L. M. N Pereira, "Wide-bandgap high-mobility ZnO thin-film transistors produced at room temperature," *Appl. Phys. Lett.* **85**(13), 2541–2543 (2004).

[21] J. Deng, Y. Zheng, Q. Hou, J.-F. Chen, W. Zhou, and X. Tao, "Solid-state dye-sensitized hierarchically structured ZnO solar cells," *Electrochim. Acta* **56**(11), 4176–4180 (2011).

[22] C. Bauer, G. Boschloo, E. Mukhtar, and A. Hagfeldt, "Ultrafast relaxation dynamics of charge carriers relaxation in ZnO nanocrystalline thin films," *Chem. Phys. Lett.* **387**(1), 176–181 (2004).

[23] J. K. Cooper, Y. Ling, C. Longo, Y. Li, and J. Z. Zhang, "Effects of hydrogen treatment and air annealing on ultrafast charge carrier dynamics in ZnO nanowires under in situ photoelectrochemical conditions," *J. Phys. Chem. C* **116**(33), 17360–17368 (2012).

[24] A. B. Djurišić and Y. H. Leung, "Optical properties of ZnO nanostructures," *Small* **2**(8–9), 944–961 (2006).

[25] R. P. Prasankumar, P. C. Upadhya, and A. J. Taylor, "Ultrafast carrier dynamics in semiconductor nanowires," *Phys. Status Solidi B* **246**(9), 1973–1995 (2009).

[26] D. A. Wheeler, G. Wang, Y. Ling, Y. Li, and J. Z. Zhang, "Nanostructured hematite: synthesis, characterization, charge carrier dynamics, and photoelectrochemical properties," *Energy Environ. Sci.* **5**(5), 6682–6702 (2012).

[27] K. Sivula, F. Le Formal, and M. Grätzel, "Solar water splitting: progress using hematite (α-Fe_2O_3) photoelectrodes," *Chem. Sus. Chem.* **4**(4), 432–449 (2011).

[28] K. L. Hardee and A. J. Bard, "Semiconductor electrodes V. The application of chemically vapor deposited iron oxide films to photosensitized electrolysis," *J. Electrochem. Soc.* **123**(7), 1024–1026 (1976).

[29] N. J. Cherepy, D. B. Liston, J. A. Lovejoy, H. Deng, and J. Z. Zhang, "Ultrafast studies of photoexcited electron dynamics in α- and γ-Fe_2O_3 semiconductor nanoparticles," *J. Phys. Chem. B* **102**(5), 770–776 (1998).

[30] K. Sivula, F. Le Formal, and M. Grätzel, "WO_3–Fe_2O_3 photoanodes for water splitting: a host scaffold, guest absorber approach," *Chem. Mater.* **21**(13), 2862–2867 (2009).

[31] W. Luo, T. Yu, Y. Wang, Z. Li, J. Ye, and Z. Zou, "Enhanced photocurrent–voltage characteristics of WO_3/Fe_2O_3 nano-electrodes," *J. Phys. D. Appl. Phys.* **40**(4), 1091 (2007).

[32] J. Brillet, M. Cornuz, F. Le Formal, J.-H. Yum, M. Grätzel, and K. Sivula, "Examining architectures of photoanode–photovoltaic tandem cells for solar water splitting," *J. Mater. Res.* **25**(01), 17–24 (2010).

[33] S. Higashimoto, M. Sakiyama, and M. Azuma, "Photoelectrochemical properties of hybrid WO_3/TiO_2 electrode. Effect of structures of WO_3 on charge separation behavior," *Thin Solid Films* **503**(1–2), 201–206 (2006).

[34] W. Smith, A. Wolcott, R. C. Fitzmorris, J. Z. Zhang, and Y. Zhao, "Quasi-core-shell TiO_2/WO_3 and WO_3/TiO_2 nanorod arrays fabricated by glancing angle deposition for solar water splitting," *J. Mater. Chem.* **21**(29), 10792–10800 (2011).

[35] J. H. Park, O. O. Park, and S. Kim, "Photoelectrochemical water splitting at titanium dioxide nanotubes coated with tungsten trioxide," *Appl. Phys. Lett.* **89**(16), 163106–163109 (2006).

[36] D. Cahen, M. Grätzel, J. F. Guillemoles, and G. Hodes, "Dye-sensitized solar cells: principles of operation," in *Electrochemistry of Nanomaterials*, edited by G. Hodes (Wiley-VCH Verlag GmbH, Weinheim, Germany, 2001), pp. 201–228.

[37] B. O'Regan and M. Grätzel, "A low-cost, high-efficiency solar cell based on dye-sensitized colloidal TiO_2 films," *Nature* **353**, 737–740 (1991).

[38] A. B. Ellis, S. W. Kaiser, and M. S. Wrighton, "Visible light to electrical energy conversion. Stable cadmium sulfide and cadmium selenide photoelectrodes in aqueous electrolytes," *J. Am. Chem. Soc.* **98**(6), 1635–1637 (1976).

[39] H. Minoura and M. Tsuiki, "Anodic reactions of several reducing agents on illuminated cadmium sulfide electrode," *Electrochim. Acta* **23**(12), 1377–1382 (1978).

[40] D. R. Pernik, K. Tvrdy, J. G. Radich, and P. V. Kamat, "Tracking the adsorption and electron injection rates of CdSe quantum dots on TiO_2: linked versus direct attachment," *J. Phys. Chem. C* **115**(27), 13511–13519 (2011).

[41] I. Mora-Seró, S. Giménez, T. Moehl, F. Fabregat-Santiago, T. Lana-Villareal, R. Gómez, and J. Bisquert, "Factors determining the photovoltaic performance of a CdSe quantum dot sensitized solar cell: the role of the linker molecule and of the counter electrode," *Nanotechnology* **19**(42), 424007 (2008).

[42] J. E. Evans, K. W. Springer, and J. Z. Zhang, "Femtosecond studies of interparticle electron transfer in a coupled CdS-TiO_2 colloidal system," *J. Chem. Phys.* **101**(7), 6222–6225 (1994).

[43] K. Gopidas, M. Bohorquez, and P. V. Kamat, "Photophysical and photochemical aspects of coupled semiconductors: charge-transfer processes in colloidal cadmium sulfide-titania and cadmium sulfide-silver (I) iodide systems," *J. Phys. Chem.* **94**(16), 6435–6440 (1990).

[44] I. Robel, V. Subramanian, M. Kuno, and P. V. Kamat, "Quantum dot solar cells. Harvesting light energy with CdSe nanocrystals molecularly linked to mesoscopic TiO_2 films," *J. Am. Chem. Soc.* **128**(7), 2385–2393 (2006).

[45] J. L. Blackburn, D. C. Selmarten, and A. J. Nozik, "Electron transfer dynamics in quantum dot/titanium dioxide composites formed by in situ chemical bath deposition," *J. Phys. Chem. B* **107**(51), 14154–14157 (2003).

[46] K. Tvrdy, P. A. Frantsuzov, and P. V. Kamat, "Photoinduced electron transfer from semiconductor quantum dots to metal oxide nanoparticles," *Proc. Natl. Acad. Sci. USA.* **108**(1), 29–34 (2011).

[47] Y.-L. Lee, B.-M. Huang, and H.-T. Chien, "Highly efficient CdSe-sensitized TiO_2 photoelectrode for quantum-dot-sensitized solar cell applications," *Chem. Mater.* **20**(22), 6903–6905 (2008).

[48] J. Piris, A. J. Ferguson, J. L. Blackburn, A. G. Norman, G. Rumbles, D. C. Selmarten, and N. Kopidakis, "Efficient photoinduced charge injection from chemical bath deposited CdS into mesoporous TiO2 probed with time-resolved microwave conductivity," *J. Phys. Chem. C* **112**(20), 7742–7749 (2008).

[49] X.-Y. Yu, B.-X. Lei, D.-B. Kuang, and C.-Y. Su, "Highly efficient CdTe/CdS quantum dot sensitized solar cells fabricated by a one-step linker assisted chemical bath deposition," *Chem. Sci.* **2**(7), 1396–1400 (2011).

[50] W.-T. Sun, Y. Yu, H.-Y. Pan, X.-F. Gao, Q. Chen, and L.-M. Peng, "CdS quantum dots sensitized TiO_2 nanotube-array photoelectrodes," *J. Am. Chem. Soc.* **130**(4), 1124–1125 (2008).

[51] W. Zhu, X. Liu, H. Liu, D. Tong, J. Yang, and J. Peng, "Coaxial heterogeneous structure of TiO_2 nanotube arrays with CdS as a superthin coating synthesized via modified electrochemical atomic layer deposition," *J. Am. Chem. Soc.* **132**(36), 12619–12626 (2010).

[52] R. C. Fitzmorris, G. Larsen, D. A. Wheeler, Y. Zhao, and J. Z. Zhang, "Ultrafast charge transfer dynamics in polycrystalline $CdSe/TiO_2$ nanorods prepared by oblique angle co-deposition," *J. Phys. Chem. C* **116**(8), 5033–5041 (2012).

[53] G. Larsen, R. C. Fitzmorris, J. Z. Zhang, and Y. Zhao, "Structural, optical, and photocatalytic properties of Cr: TiO_2 nanorod array fabricated by oblique angle co-deposition," *J. Phys. Chem. C* **115**(34), 16892–16903 (2011).

[54] N. Guijarro, Q. Shen, S. Giménez, I. Mora-Seró, J. Bisquert, T. Lana-Villarreal, T. Toyoda, and R. Gómez, "Direct correlation between ultrafast injection and photoanode performance in quantum dot sensitized solar cells," *J. Phys. Chem. C* **114**(50), 22352–22360 (2010).

[55] I. Robel, M. Kuno, and P. V. Kamat, "Size-dependent electron injection from excited CdSe quantum dots into TiO_2 nanoparticles," *J. Am. Chem. Soc.* **129**(14), 4136–4137 (2007).

[56] R. C. Fitzmorris, Y.-C. Pu, J. K. Cooper, Y.-F. Lin, Y.-J. Hsu, Y. Li, and J. Z. Zhang, "Optical properties and exciton dynamics of alloyed core/shell/shell $Cd_xZn_{1-x}Se/ZnSe/ZnS$ quantum dots," *ACS Appl. Mater. Interfaces* **5**(8), 2893–2900 (2013).

[57] J. Ouyang, C. I. Ratcliffe, D. Kingston, B. Wilkinson, J. Kuijper, X. Wu, J. A. Ripmeester, and K. Yu, "Gradiently alloyed $Zn_xCd_{1-x}S$ colloidal photoluminescent quantum dots synthesized via a noninjection one-pot approach," *J. Phys. Chem. C* **112**(13), 4908–4919 (2008).

[58] M. D. Regulacio and M. Y. Han, "Composition-tunable alloyed semiconductor nanocrystals," *Acc. Chem. Res.* **43**(5), 621–630 (2010).

[59] J. Luo, L. Ma, T. He, C. F. Ng, S. Wang, H. Sun, and H. J. Fan, "TiO_2/(CdS, CdSe, CdSeS) nanorod heterostructures and photoelectrochemical properties," *J. Phys. Chem. C* **116**(22), 11956–11963 (2012).

[60] G. Zhao, H. Kozuka, and T. Yoko, "Sol-gel preparation and photoelectrochemical properties of TiO_2 films containing Au and Ag metal particles," *Thin Solid Films* **277**(1), 147–154 (1996).

[61] M. Ni, M. K. H Leung, D. Y.C. Leung, and K. Sumathy, "A review and recent developments in photocatalytic water-splitting using TiO_2 for hydrogen production," *Renew. Sust. Energ. Rev.* **11**(3), 401–425 (2007).

[62] V. Subramanian, E. Wolf, and P. V. Kamat, "Semiconductor–metal composite nanostructures. To what extent do metal nanoparticles improve the photocatalytic activity of TiO_2 films?," *J. Phys. Chem. B* **105**(46), 11439–11446 (2001).

[63] Y. Tian and T. Tatsuma, "Mechanisms and applications of plasmon-induced charge separation at TiO_2 films loaded with gold nanoparticles," *J. Am. Chem. Soc.* **127**(20), 7632–7637 (2005).

[64] N. Chandrasekharan and P. V. Kamat, "Improving the photoelectrochemical performance of nanostructured TiO_2 films by adsorption of gold nanoparticles," *J. Phys. Chem. B* **104**(46), 10851–10857 (2000).

[65] Y. Nishijima, K. Ueno, Y. Yokota, K. Murakoshi, and H. Misawa, "Plasmon-assisted photocurrent generation from visible to near-infrared wavelength using a Au-nanorods/TiO_2 electrode," *J. Phys. Chem. L.* **1**(13), 2031–2036.

[66] Z. Zhang, L. Zhang, M. N. Hedhili, H. Zhang, and P. Wang, "Plasmonic gold nanocrystals coupled with photonic crystal seamlessly on TiO_2 nanotube photoelectrodes for efficient visible light photoelectrochemical water splitting," *Nano Lett.* **13**(1), 14–20 (2012).

[67] J. Lee, S. Mubeen, X. Ji, G. D. Stucky, and M. Moskovits, "Plasmonic photoanodes for solar water splitting with visible light," *Nano Lett.* **12**(9), 5014–5019 (2012).

[68] S. Linic, P. Christopher, and D. B. Ingram, "Plasmonic-metal nanostructures for efficient conversion of solar to chemical energy," *Nat. Mater.* **10**(12), 911–921 (2011).

[69] L. Liu, G. Wang, Y. Li, Y. Li, and J. Z. Zhang, "CdSe quantum dot-sensitized Au/TiO_2 hybrid mesoporous films and their enhanced photoelectrochemical performance," *Nano Research* **4**(3), 249–258 (2011).

[70] P. K. Jain, K. S. Lee, I. H. El-Sayed, and M. A. El-Sayed, "Calculated absorption and scattering properties of gold nanoparticles of different size, shape, and composition: applications in biological imaging and biomedicine," *J. Phys. Chem. B* **110**(14), 7238–7248 (2006).

[71] D. Schaadt, B. Feng, and E. Yu, "Enhanced semiconductor optical absorption via surface plasmon excitation in metal nanoparticles," *Appl. Phys. Lett.* **86**, 063106 (2005).

[72] A. M. Schwartzberg, C. D. Grant, A. Wolcott, C. E. Talley, T. R. Huser, R. Bogomolni, and J. Z. Zhang, "Unique gold nanoparticle aggregates as a highly active surface-enhanced Raman scattering substrate," *J. Phys. Chem. B* **108**(50), 19191–19197 (2004).

[73] D. B. Ingram and S. Linic, "Water splitting on composite plasmonic-metal/semiconductor photoelectrodes: evidence for selective plasmon-induced formation of charge carriers near the semiconductor surface," *J. Am. Chem. Soc.* **133**(14), 5202–5205 (2011).

[74] E. Thimsen, F. Le Formal, M. Grätzel, and S. C. Warren, "Influence of plasmonic Au nanoparticles on the photoactivity of Fe_2O_3 electrodes for water splitting," *Nano Lett.* **11**(1), 35–43 (2010).

[75] B. Iandolo and M. Zäch, "Enhanced water splitting on thin-film hematite photoanodes functionalized with lithographically fabricated Au nanoparticles," *Aust. J. Chem.* **65**(6), 633–637 (2012).

[76] J. K. Cooper, Y. Ling, Y. Li, and J. Z. Zhang, "Ultrafast charge carrier dynamics and photoelectrochemical properties of ZnO nanowires decorated with Au nanoparticles," in *Solar Hydrogen and Nanotechnology VI,* Proceedings of SPIE Volume 8109, San Diego, CA, September 2011, edited by Yasuhiro Tachibana (SPIE, 2011), pp. 81090M.

[77] A. Furube, L. Du, K. Hara, R. Katoh, and M. Tachiya, "Ultrafast plasmon-induced electron transfer from gold nanodots into TiO_2 nanoparticles," *J. Am. Chem. Soc.* **129**(48), 14852–14853 (2007).

[78] J. B. Asbury, Y.-Q. Wang, E. Hao, H. N. Ghosh, and T. Lian, "Evidences of hot excited state electron injection from sensitizer molecules to TiO_2 nanocrystalline thin films," *Res. Chem. Intermed.* **27**(4–5), 393–406 (2001).

[79] L. Du, A. Furube, K. Yamamoto, K. Hara, R. Katoh, and M. Tachiya, "Plasmon-induced charge separation and recombination dynamics in gold–TiO_2 nanoparticle systems: dependence on TiO_2 particle size," *J. Phys. Chem. C* **113**(16), 6454–6462 (2009).

[80] P. Hartmann, D.-K. Lee, B. M. Smarsly, and J. Janek, "Mesoporous TiO_2: comparison of classical sol–gel and nanoparticle based photoelectrodes for the water splitting reaction," *ACS Nano* **4**(6), 3147–3154 (2010).

[81] A. Wolcott, W. A. Smith, T. R. Kuykendall, Y. Zhao, and J. Z. Zhang, "Photoelectrochemical water splitting using dense and aligned TiO_2 nanorod arrays," *Small* **5**(1), 104–111 (2009).

[82] K. Sun, A. Kargar, N. Park, K. N. Madsen, P. W. Naughton, T. Bright, Y. Jing, and D. Wang, "Compound semiconductor nanowire solar cells," *IEEE J. Sel. Top. Quantum Electron.* **17**(4), 1033–1049 (2011).

[83] J. H. Park, S. Kim, and A. J. Bard, "Novel carbon-doped TiO_2 nanotube arrays with high aspect ratios for efficient solar water splitting," *Nano Lett.* **6**(1), 24–28 (2006).

[84] Y. Qiu, K. Yan, H. Deng, and S. Yang, "Secondary branching and nitrogen doping of ZnO nanotetrapods: building a highly active network for photoelectrochemical water splitting," *Nano Lett.* **12**(1), 407–413 (2011).

[85] Y.-S. Hu, A. Kleiman-Shwarstein, A. J. Forman, D. Hazen, J.-N. Park, and E. W. McFarland, "Pt-doped α-Fe_2O_3 thin films active for photoelectrochemical water splitting," *Chem. Mater.* **20**(12), 3803–3805 (2008).

[86] A. Wolcott, T. R. Kuykendall, W. Chen, S. Chen, and J. Z. Zhang, "Synthesis and characterization of ultrathin WO_3 nanodisks utilizing long-chain poly (ethylene glycol)," *J. Phys. Chem. B* **110**(50), 25288–25296 (2006).

[87] G. Wang, X. Yang, F. Qian, J. Z. Zhang, and Y. Li, "Double-sided CdS and CdSe quantum dot co-sensitized ZnO nanowire arrays for photoelectrochemical hydrogen generation," *Nano Lett.* **10**(3), 1088–1092 (2010).

[88] A. Wolcott, R. C. Fitzmorris, O. Muzaffery, and J. Z. Zhang, "CdSe quantum rod formation aided by in situ TOPO oxidation," *Chem. Mater.* **22**(9), 2814–2821 (2010).

[89] L.-s. Li J. Hu, W. Yang, and A. P. Alivisatos, "Band gap variation of size- and shape-controlled colloidal CdSe quantum rods," *Nano Lett.* **1**(7), 349–351 (2001).

[90] X. Peng, L. Manna, W. Yang, J. Wickham, E. Scher, A. Kadavanich, and A. P. Alivisatos, "Shape control of CdSe nanocrystals," *Nature* **404**(6773), 59–61 (2000).

[91] Y. Ling, G. Wang, D. A. Wheeler, J. Z. Zhang, and Y. Li, "Sn-doped hematite nanostructures for photoelectrochemical water splitting," *Nano Lett.* **11**(5), 2119–2125 (2011).

[92] N. T. Hahn and C. B. Mullins, "Photoelectrochemical performance of nanostructured Ti- and Sn-doped α-Fe_2O_3 photoanodes," *Chem. Mater.* **22**(23), 6474–6482 (2010).

[93] J. S. Jang, K. Y. Yoon, X. Xiao, F.-R. F. Fan, and A. J. Bard, "Development of a potential Fe_2O_3-based photocatalyst thin film for water oxidation by scanning electrochemical microscopy: effects of Ag−Fe_2O_3 nanocomposite and Sn doping," *Chem. Mater.* **21**(20), 4803–4810 (2009).

[94] S. U. M Khan, M. Al-Shahry, and W. B. Ingler Jr., "Efficient photochemical water splitting by a chemically modified n-TiO_2," *Science* **297**(5590), 2243–2245 (2002).

[95] C. Xu, Y. A. Shaban, W. B. Ingler Jr., and S. U. M Khan, "Nanotube enhanced photoresponse of carbon modified (CM)-n-TiO_2 for efficient water splitting," *Sol. Energy Mater. Sol. Cells* **91**(10), 938–943 (2007).

[96] H. Wang and J. P. Lewis, "Effects of dopant states on photoactivity in carbon-doped TiO_2," *J. Phys. Condens. Matter* **17**(21), L209–L213 (2005).

[97] A. Murphy, "Does carbon doping of TiO_2 allow water splitting in visible light? Comments on 'Nanotube enhanced photoresponse of carbon modified (CM)-*n*-TiO_2 for efficient water splitting'," *Sol. Energy Mater. Sol. Cells* **92**(3), 363–367 (2008).

[98] X. Yang, A. Wolcott, G. Wang, A. Sobo, R. C. Fitzmorris, F. Qian, J. Z. Zhang, and Y. Li, "Nitrogen-doped ZnO nanowire arrays for photoelectrochemical water splitting," *Nano Lett.* **9**(6), 2331–2336 (2009).

[99] T. López-Luke, A. Wolcott, L. Xu, S. Chen, Z. Wen, J. Li, E. De La Rosa, and J. Z. Zhang, "Nitrogen-doped and CdSe quantum-dot-sensitized nanocrystalline TiO_2 films for solar energy conversion applications," *J. Phys. Chem. C* **112**(4), 1282–1292 (2008).

[100] J. W. Tang, A. J. Cowan, J. R. Durrant, and D. R. Klug, "Mechanism of O_2 production from water splitting: nature of charge carriers in nitrogen doped nanocrystalline TiO_2 films and factors limiting O_2 production," *J. Phys. Chem. C* **115**(7), 3143–3150 (2011).

[101] J. Wang, D. N. Tafen, J. P. Lewis, Z. Hong, A. Manivannan, M. Zhi, M. Li, and N. Wu, "Origin of photocatalytic activity of nitrogen-doped TiO_2 nanobelts," *J. Am. Chem. Soc.* **131**(34), 12290–12297 (2009).

[102] J. V. Foreman, J. Li, H. Peng, S. Choi, H. O. Everitt, and J. Liu, "Time-resolved investigation of bright visible wavelength luminescence from sulfur-doped ZnO nanowires and micropowders," *Nano Lett.* **6**(6), 1126–1130 (2006).

[103] L. K. Randeniya, A. B. Murphy, and I. C. Plumb, "A study of S-doped TiO_2 for photoelectrochemical hydrogen generation from water," *J. Mater. Sci.* **43**(4), 1389–1399 (2008).

[104] J. C. Yu, W. Ho, J. Yu, H. Yip, P. K. Wong, and J. Zhao, "Efficient visible-light-induced photocatalytic disinfection on sulfur-doped nanocrystalline titania," *Environ. Sci. Tech.* **39**(4), 1175–1179 (2005).

[105] H. Wang and J.P. Lewis, "Second-generation photocatalytic materials: anion-doped TiO_2," *J. Phys. Condens. Matter* **18**(2), 421–434 (2006).

[106] G. Wang, Y. Ling, H. Wang, X. Yang, C. Wang, J. Z. Zhang, and Y. Li, "Hydrogen-treated WO_3 nanoflakes show enhanced photostability," *Energy Environ. Sci.* **5**(3), 6180–6187 (2012).

[107] X. Chen, L. Liu, P. Y. Yu, and S. S. Mao, "Increasing solar absorption for photocatalysis with black hydrogenated titanium dioxide nanocrystals," *Science* **331**(6018), 746–750 (2011).

[108] A. Naldoni, M. Allieta, S. Santangelo, M. Marelli, F. Fabbri, S. Cappelli, C. L. Bianchi, R. Psaro, and V. Dal Santo, "Effect of nature and location of defects on bandgap narrowing in black TiO_2 nanoparticles," *J. Am. Chem. Soc.* **134**(18), 7600–7603 (2012).

[109] S. Hoang, S. P. Berglund, N. T. Hahn, A. J. Bard, and C. B. Mullins, "Enhancing visible light photo-oxidation of water with TiO_2 nanowire arrays via cotreatment with H_2 and NH_3: synergistic effects between Ti^{3+} and N," *J. Am. Chem. Soc.* **134**(8), 3659–3662 (2012).

[110] J. Hensel, G. Wang, Y. Li, and J. Z. Zhang, "Synergistic effect of CdSe quantum dot sensitization and nitrogen doping of TiO_2 nanostructures for photoelectrochemical solar hydrogen generation," *Nano Lett.* **10**(2), 478–483 (2010).

[111] H. Wang, G. Wang, Y. Ling, M. Lepert, C. Wang, J. Z. Zhang, and Y. Li, "Photoelectrochemical study of oxygen deficient TiO_2 nanowire arrays with CdS quantum dot sensitization," *Nanoscale* **4**(5), 1463–1466 (2012).

11

OPTICAL CONTROL OF COLD ATOMS AND ARTIFICIAL ELECTROMAGNETISM

GEDIMINAS JUZELIŪNAS[1] AND PATRIK ÖHBERG[2]

[1] *Institute of Theoretical Physics and Astronomy, Vilnius University, Vilnius, Lithuania*
[2] *SUPA, Institute of Photonics and Quantum Sciences, Heriot-Watt University, Edinburgh, UK*

11.1 INTRODUCTION

Light is a versatile tool, which can be used to manipulate macroscopic objects and also individual atoms. With the advent of the laser in the 1950s, it became possible to accurately address the internal energy levels in atoms. This opened up the possibility to laser cool ensembles of atoms, and in combination with magnetic trapping, the first degenerate quantum gases were created. In particular, gases of bosonic and fermionic atoms are now routinely prepared in laboratories all over the world, where temperatures on the order of micro Kelvin and below are reached. The resulting Bose–Einstein Condensates (BECs) and degenerate Fermi gases have proven to be remarkably useful tools to study exotic phenomena known from almost any branch of physics. Perhaps, most importantly, the quantum gases have helped us understand many concepts from traditional condensed matter physics such as the superfluid to Mott phase transition and the Bardeen–Cooper–Schrieffer crossover in degenerate Fermi gases [1–5]. Also phenomena known from high energy physics can be mimicked in these ultracold settings. In particular, Higgs modes have been observed and studied [6, 7], confinement mechanisms for quarks [8–11], axion electrodynamics [12], and unconventional colour superconductivity [13] have been proposed. The

Photonics: Scientific Foundations, Technology and Applications, Volume II, First Edition.
Edited by David L. Andrews.
© 2015 John Wiley & Sons, Inc. Published 2015 by John Wiley & Sons, Inc.

possibility to study such diverse topics with ultracold gases is mainly due to the unprecedented level of control of the experimental parameters. This includes trap geometry, particle number and density of the cloud, and importantly also the collisional interaction strength between the atoms using Feshbach resonances [14].

In this chapter, we present an overview of ultracold atoms and optical forces acting on them, and subsequently discuss a specific aspect of controlling the dynamics of atoms optically, namely, how to optically induce synthetic gauge fields in ultracold gases, and how one can use this tool to study many new phenomena in quantum gases. Atomic quantum gases are charge neutral, and therefore, they are not affected by external electromagnetic fields the way electrons are. In particular, under normal circumstances, the charge-neutral atoms are not affected by the Lorentz force described by a vector potential with a non-zero curl. Therefore, the magnetic phenomena familiar for electrons in solids, such as the Quantum Hall effect [15], do not show up for ultracold atoms under usual circumstances. The atom–light coupling does however allow for the creation of versatile gauge potentials, which makes it possible to experimentally access many new phenomena at the quantum level. With this technology, atoms can be subject to static Abelian and non-Abelian [16] gauge fields, which in practice means that synthetic electric and magnetic fields can be experimentally tuned with lasers [17–19]. The non-Abelian gauge fields can not only reproduce Rashba-type spin-orbit couplings, but also mimic a variety of properties encountered in the context of high energy physics. The first experimental steps towards the realization of a two-dimensional spin-orbit-coupled atomic gas have been reported in References 20–27. Mimicking magnetic and spin-orbit effects in ultracold quantum gases also makes it possible to envisage quantum simulators of new kinds of exotic quantum matter [28–30] and realize Richard Feynman's vision [31] for constructing physical quantum emulators of systems or situations that are computationally or analytically intractable.

The techniques for creating synthetic gauge fields can also be extended to optical lattice potentials [28–30, 32, 33], where the link to quantum simulations of condensed matter phenomena is natural. Here, the artificial magnetic field results from a laser-induced tunneling between the lattice sites. Numerous theoretical proposals for simulating condensed matter models and realizing strongly correlated systems have been suggested. Recently, artificial gauge potentials corresponding to staggered [34–36] and uniform [37,38] magnetic fluxes have also been created in optical lattices.

The chapter is organized as follows. We will first briefly review the theoretical description of a Bose–Einstein condensate, followed by a discussion of the simplest possible scenario, based on a two-level atom subject to a laser field, for which scalar potentials, such as optical lattices, appear for the atoms. In the rest of the text, we will discuss the mechanisms for creating artificial gauge potential based on light–matter interaction.

11.2 ATOMIC BOSE–EINSTEIN CONDENSATES

In 1995, experimentalists were able to cool and trap gases of ^{87}Rb [39,40] and ^{7}Li [41] to such low temperatures that quantum effects started to play a major role. These gases consisted of bosonic atoms, which meant that they formed a Bose–Einstein condensate with exotic properties such as superfluidity and many-body effects at ultracold temperatures. It is fair to say that nobody anticipated the impact of the arrival of this new quantum fluid. Today, the Bose–Einstein condensate is used as a tool to study exotic properties of matter, where enormous stride has been made, in particular, experimentally, in creating macroscopic quantum states and using these for probing effects such as quantum phase transitions and nonequilibrium effects in quantum gases. One of the most appealing factors when using ultracold gases, be it bosonic or fermionic gases, is the fact that they amalgamate phenomena and techniques known from both quantum optics and condensed matter physics. In addition, these gases allow for an unprecedented control over the physical parameters in experiments, ranging from condensate density, trap geometry, and even collisional strength using Feshbach resonances [42]. In practice, this also means that theory and experiments are in a very fruitful symbiosis. The available theoretical tools often describe the dynamics of the gas extremely well. This has opened up the possibility to address the concept of quantum simulators in these systems, originally proposed by Richard Feynman, where the physical processes in an experiment are exactly described by some Hamiltonian, which may be too difficult to analyze on a classical computer due to the large Hilbert space one often encounters in quantum many-body phenomena. But the experiment, on the other hand, does not care about the Hilbert space. The experiment could allow the quantum fluid to relax to its ground states, for instance, and by doing so calculate the numerically inaccessible ground state for us.

In this chapter, we will review some standard theoretical tools for analyzing weakly interacting Bose–Einstein condensates. The emphasis is on the mean field treatment, and how we can obtain an effective equation of motion for the gas. This will be the celebrated Gross–Pitaevskii equation.

11.2.1 The Description of a Condensate

A Bose–Einstein condensate can be understood as a macroscopically occupied single quantum state. Such a non-interacting quantum gas can readily be analyzed using techniques from thermodynamics, but in most situations, collisions between the atoms are after all present, and will affect the properties of the gas significantly. We will consider a weakly interacting gas, which lends itself naturally to a mean field treatment of the condensate. The inclusion of collisions between the atoms will in fact not change too much the *onset* of BEC and the corresponding critical temperature, compared to the free gas [44]. At low temperatures, however, the collisions do affect the condensate, as we will show next.

Assuming zero temperature is a good approximation in many cases. Present cooling techniques makes it possible to reach temperatures far below the critical temperature which is typically in the micro Kelvin regime where any contribution from the remaining thermal component can be neglected, as one can see in Figure 11.1. For a dilute gas only two-body collisions take place. Because the gas is cold, we need to

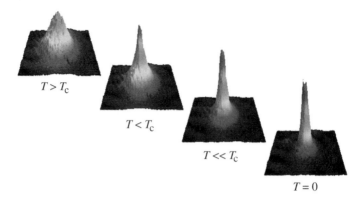

FIGURE 11.1 The onset of BEC is seen as a sharp peak in the density in the center of the trap. In the figures, the temperature is lowered from left to right. To the far right, an almost pure condensate with a negligible thermal component can be seen. Pictures courtesy of A. Arnold and are from the BEC experiment at University of Strathclyde, Glasgow, UK [43]. *(For a color version of this figure, see the color plate section.)*

consider only s-wave scattering as the mechanism for the interaction. The interaction potential is therefore of the form [44]

$$V_{\text{int}}(r - r') = \frac{4\pi\hbar^2 a}{m}\delta(r - r'), \tag{11.1}$$

where the interaction is described by the single parameter a called the s-wave scattering length. For a derivation of Eq. (11.1), we have to solve the two-body scattering problem in the limit of zero momentum. With these assumptions, we obtain a Hamiltonian of the form

$$\hat{H} = \int d\mathbf{r} \left\{ \hat{\Psi}^\dagger(\mathbf{r}) \left[-\frac{\hbar^2}{2m}\nabla^2 + V_{\text{ext}}(\mathbf{r}) \right] \hat{\Psi}(\mathbf{r}) + \frac{g}{2}\hat{\Psi}^\dagger(\mathbf{r})\hat{\Psi}^\dagger(\mathbf{r})\hat{\Psi}(\mathbf{r})\hat{\Psi}(\mathbf{r}) \right\}, \tag{11.2}$$

where $g = 4\pi\hbar^2 a/m$. The field operators $\hat{\Psi}(\mathbf{r},t)$ and $\hat{\Psi}^\dagger(\mathbf{r},t)$ destroys and creates respectively a particle at $\mathbf{r}$ at time t, and obeys the usual bosonic commutation rules

$$[\hat{\Psi}(\mathbf{r}), \hat{\Psi}^\dagger(\mathbf{r}')] = \delta(r - r')\delta(t - t') \tag{11.3}$$

and

$$[\hat{\Psi}(\mathbf{r},t), \hat{\Psi}(\mathbf{r}',t')] = [\hat{\Psi}^\dagger(\mathbf{r},t), \hat{\Psi}^\dagger(\mathbf{r}',t')] = 0. \tag{11.4}$$

With these commutation rules, we get the Heisenberg equation of motion for the field operator:

$$i\hbar\frac{\partial}{\partial t}\hat{\Psi} = [\hat{H}, \hat{\Psi}] = -\frac{\hbar^2}{2m}\nabla^2\hat{\Psi} + V_{\text{ext}}(\mathbf{r})\hat{\Psi} + g\hat{\Psi}^\dagger\hat{\Psi}^\dagger\hat{\Psi}. \tag{11.5}$$

The field operator $\hat{\Psi}$ is split into the operator for the lowest mode and a part representing the fluctuations and thermal excitations,

$$\hat{\Psi}(\mathbf{r}) = \hat{\Psi}_0(\mathbf{r}) + \delta\hat{\Psi}(\mathbf{r}). \tag{11.6}$$

At zero temperature, we can as a first approximation neglect the fluctuations $\delta\hat{\Psi}(\mathbf{r})$. In the presence of a condensate, the lowest mode is macroscopically populated. We therefore write

$$\hat{\Psi}(\mathbf{r}) = \Psi(\mathbf{r})\hat{a}_0 \approx \Psi(\mathbf{r})\sqrt{N}, \tag{11.7}$$

where we have replaced the annihilation operator $\hat{a}_0$ by $\sqrt{N}$, which is often called the Bogoliubov approximation [45]. This is a legitimate approximation provided that the number of atoms N in the condensate is sufficiently large. In other words, we have replaced the field operator by its average

$$\hat{\Psi}(\mathbf{r}) \approx \langle \hat{\Psi}(\mathbf{r}) \rangle = \Psi(\mathbf{r})\sqrt{N}. \tag{11.8}$$

The resulting equation of motion for the condensate wavefunction, $\Psi(\mathbf{r})$, then becomes

$$i\hbar \frac{\partial}{\partial t}\Psi(\mathbf{r},t) = \left[-\frac{\hbar^2}{2m}\nabla^2 + V_{\text{ext}}(\mathbf{r}) + g|\Psi(\mathbf{r},t)|^2 \right]\Psi(\mathbf{r},t). \tag{11.9}$$

This is the Gross–Pitaevskii equation [45,46], which is the true workhorse for describing the dynamics of a Bose–Einstein condensate. The Gross–Pitaevskii equation is based on mean field theory, in which each atom feels the presence of all the other atoms through the effective potential proportional to the density of the cloud. It describes the dynamics of a condensate remarkably well and has been used extensively in studies of Bose–Einstein condensates.

The time-independent version of Eq. (11.9),

$$\mu\varphi(\mathbf{r}) = \left[-\frac{\hbar^2}{2m}\nabla^2 + V_{\text{ext}}(\mathbf{r}) + g|\varphi(\mathbf{r})|^2 \right]\varphi(\mathbf{r}), \tag{11.10}$$

is obtained using the Ansatz $\Psi(\mathbf{r},t) = \varphi(\mathbf{r})e^{-i\mu t/\hbar}$, where μ is the chemical potential. The external potential $V_{\text{ext}}(\mathbf{r})$ can take many shapes. It can stem from magnetic trapping, and also optical traps as we will see in the next section. The flexibility in choosing the shape of the external trap makes it possible to study a broad spectrum of physical scenarios. Often the external potential is harmonic, but it can also be in the form of a lattice potential. Such optical lattices are created by standing wave laser beams, which allows for mimicking solid state effects where the ultracold atoms are playing the role of charge carriers in the lattice. The optical lattice has proven to be an important tool for studying quantum simulators in particular, where exotic Hamiltonians and their ground states are being modeled by well-controlled experiments.

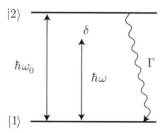

FIGURE 11.2 The two-level atom with its two states $|1\rangle$ and $|2\rangle$, and energy difference $\hbar\omega_0$. The frequency of the external laser which couples the two levels is ω, with detuning $\delta = \omega - \omega_0$. The decay rate of the excited state $|2\rangle$ is Γ.

11.3 OPTICAL FORCES ON ATOMS

From classical electromagnetism, it is known that light exerts a force on a classical dipole. The resulting force depends on the gradient of the amplitude and phase of the light [47],

$$m\ddot{\mathbf{R}} = \mathbf{d} \cdot (\nabla \mathbf{E}) = \mathbf{d} \cdot (\nabla \xi + \xi \nabla \theta) e^{i(\omega t + \theta)}, \tag{11.11}$$

where m is the mass of the dipole, $\mathbf{d}$ is the dipole moment, and $\mathbf{E}(\mathbf{R}, t) = \epsilon \xi(\mathbf{R}) e^{i(\omega t + \theta(\mathbf{R}))}$ is the electric field with the corresponding amplitude ξ, polarisation ϵ, phase θ, and frequency ω.

The two-level atom, shown in Figure 11.2, can also be seen as a dipole. Our approach is to describe the atom quantum mechanically, but allow the light field to be classical. The Hamiltonian for such a situation is

$$\hat{H} = \frac{\mathbf{p}^2}{2m} + \hat{H}_0 - \hat{\mathbf{d}} \cdot \mathbf{E}(\mathbf{R}, t), \tag{11.12}$$

where $\mathbf{p}^2/2m$ is the kinetic energy associated with the center of mass motion of the atom, $\hat{H}_0$ is the Hamiltonian for the unperturbed internal motion, and $\hat{\mathbf{d}} \cdot \mathbf{E}(\mathbf{R}, t)$ is the interaction between the atom and the light field, which is based on the electric dipole approximation. With the Hamiltonian from Eq. (11.12) and the Ehrenfest theorem, we obtain the expression for the force,

$$F = m\ddot{\mathbf{r}} = \langle \nabla(\hat{\mathbf{d}} \cdot \mathbf{E}) \rangle = \langle \hat{\mathbf{d}} \cdot \epsilon \rangle \nabla \xi(\mathbf{r}, t) \tag{11.13}$$

with $\mathbf{r} = \langle \mathbf{R} \rangle$ and $\xi(\mathbf{r}, t) = \xi(\mathbf{r}) e^{i(\omega t + \theta(\mathbf{r}))}$. On the right-hand side above, we have assumed that the force is uniform across the atomic wave packet. For a more detailed discussion, we refer the reader to References 48–50.

A two-level atom that is driven by a laser has been studied extensively, see for instance References 48, 49, 51, 52. In order to obtain an expression for the force acting on the atom, we need to calculate the response of the atom to the light, that is, the susceptibility or polarisation, $\langle \hat{\mathbf{d}} \cdot \boldsymbol{\epsilon} \rangle$. For this, we assume a monochromatic field of the form

$$\xi(\mathbf{r}, t) = \frac{1}{2} E(\mathbf{r}) e^{i(\theta(\mathbf{r}) + \omega t)}, \quad (11.14)$$

where $E(\mathbf{r})$ is the amplitude; ω, the laser frequency; and θ, a space-dependent phase factor. From the Schrödinger equation, we obtain the two coupled equations for the probability amplitudes C_1 and C_2 for the atom to be in state $|1\rangle$ and $|2\rangle$, respectively. By choosing a rotating frame with

$$C_1 = D_1 e^{i\frac{1}{2}(\delta t + \theta)} \quad (11.15)$$

$$C_2 = D_2 e^{-i\frac{1}{2}(\delta t + \theta)}, \quad (11.16)$$

we obtain the equations [48]

$$i\dot{D}_1 = \frac{1}{2}(\delta + \dot{\theta})D_1 - \frac{\Omega}{2}D_2 \quad (11.17)$$

$$i\dot{D}_2 = -\frac{1}{2}(\delta + \dot{\theta})D_2 - \frac{\Omega}{2}D_1. \quad (11.18)$$

In deriving the equations for D_1 and D_2, we have introduced the detuning $\delta = \omega - \omega_0$ and used the rotating wave approximation where rapidly oscillating terms are neglected [52]. The dipole moment d for the transition between state 1 and 2 is given by $d = \langle 1|\hat{\mathbf{d}} \cdot \boldsymbol{\epsilon}|1 \rangle$ and the Rabi frequency is defined by

$$\Omega = \frac{dE(t)}{\hbar}. \quad (11.19)$$

It is convenient to introduce the density matrix at this stage, which is defined as $\rho_{nm} = C_n C_m^*$ or $\sigma_{nm} = D_n D_m^*$ with

$$\rho_{11} = \sigma_{11} \quad (11.20)$$

$$\rho_{22} = \sigma_{22} \quad (11.21)$$

$$\rho_{12} = \sigma_{12} e^{i(\theta + \omega t)} \quad (11.22)$$

$$\rho_{21} = \sigma_{21} e^{-i(\theta + \omega t)}. \quad (11.23)$$

From the Eqs. (11.17) and (11.18), we see that the matrix elements of the density matrix obey

$$\dot{\sigma}_{11} = -\frac{i}{2}\Omega(\sigma_{12} - \sigma_{21}) + \Gamma\sigma_{22} \quad (11.24)$$

$$\dot{\sigma}_{22} = \frac{i}{2}\Omega(\sigma_{12} - \sigma_{21}) - \Gamma\sigma_{22} \quad (11.25)$$

$$\dot{\sigma}_{12} = -i(\delta + \dot{\theta})\sigma_{12} + \frac{i}{2}\Omega(\sigma_{22} - \sigma_{11}) - \frac{1}{2}\Gamma\sigma_{12}, \quad (11.26)$$

where we have introduced the spontaneous emission rate Γ to incorporate decay processes [52].

With the density matrix, we can calculate the expectation value for the dipole moment, which is given by

$$\langle \hat{\mathbf{d}} \cdot \boldsymbol{\epsilon} \rangle = d(\rho_{12} + \rho_{21}) = d(\sigma_{12} e^{i(\theta + \omega t)} + \sigma_{21} e^{-i(\theta + \omega t)}). \quad (11.27)$$

With this expression, and again utilizing the rotating wave approximation, we get from Eqs. (11.13) and (11.14) the force

$$\mathbf{F} = \frac{d}{2}(\sigma_{12} + \sigma_{21} - i(\sigma_{12} - \sigma_{21})) = \frac{\hbar}{2}(U \nabla \Omega + V \Omega \nabla \theta), \quad (11.28)$$

where we have introduced the notation $U = \sigma_{12} + \sigma_{21}$ and $V = i(\sigma_{12} - \sigma_{21})$, and used the fact that $\dot{\theta} = \nabla \theta(\mathbf{r}) \cdot \dot{\mathbf{r}}$. If the atomic motion is slow, such that the phase of the atomic state, θ, does not change much during the life time $1/\Gamma$ of the excited state, we can restrict ourselves to the steady state solution of the density matrix and put the time derivatives of the left-hand side equal to zero in Eqs. (11.24)–(11.26). The solutions for the corresponding U and V are then

$$U = \frac{\delta}{\Omega} \frac{s}{s+1} \quad (11.29)$$

$$V = \frac{\Gamma}{2\Omega} \frac{s}{s+1}, \quad (11.30)$$

where s is the saturation parameter

$$s = \frac{\Omega^2/2}{(\delta + \dot{\theta})^2 + \Gamma^2/4}. \quad (11.31)$$

The resulting force acting on the atom consists of two parts, which are referred to as the dipole force and the radiation force respectively,

$$\mathbf{F} = \mathbf{F}_{\text{dip}} + \mathbf{F}_{\text{pr}}, \quad (11.32)$$

where

$$\mathbf{F}_{\text{dip}} = -\frac{\hbar(\delta + \dot{\theta})}{2} \frac{\nabla s}{s+1} \quad (11.33)$$

$$\mathbf{F}_{\text{pr}} = -\frac{\hbar \Gamma}{2} \frac{s}{s+1} \nabla \theta. \quad (11.34)$$

For plane waves, the radiation force $\mathbf{F}_{\text{pr}}$, also called radiation pressure, is proportional to the wave vector $\mathbf{k} = \nabla \theta$. For trapping purposes, on the other hand, the dipole force $\mathbf{F}_{\text{dip}}$ is more important. The force $\mathbf{F}_{\text{dip}}$ is determined by the intensity of the laser field. If $s \ll 1$ and $|\delta| \ll \Omega$, we get the corresponding potential using $\mathbf{F}_{\text{dip}} = \nabla W_{\text{dip}}$, where

$$W_{\text{dip}} = \frac{\hbar \Omega^2}{4\delta} = \frac{d^2 E^2}{4\delta \hbar}. \tag{11.35}$$

From this expression, we see that if the intensity of the light is inhomogeneous, we obtain a non-zero force whose direction depends on the sign of the detuning. For a focused Gaussian beam, this means that the atoms are attracted to the high intensity if the laser is red detuned ($\delta < 0$), that is the atoms are high field seekers. On the other hand, if the laser is blue detuned ($\delta > 0$), the atoms are low field seekers and are repelled from the center of the beam.

Although Eq. (11.35) has been obtained using semiclassical arguments based on optical forces acting on atoms, the same form of the dipole potential W_{dip} is obtained for a quantum atomic center of mass motion, see, for example, Reference 53. This is important in the ultra cold regime where the atomic de Broglie wavelength is no longer small compared to the characteristic distance of variation of the dipole potential.

The dipole potential W_{dip} can be used as quite a versatile tool for controlling the atoms. In particular, the dipole potential created by two pairs of laser beams counter-propagating at a right angle, provides a square optical lattice [54], whereas hexagonal or triangular optical lattices are produced using three laser beams propagating at 120° [55]. More unusual geometries can be created by choosing suitable laser configurations and standing wave patterns. For example, an edge-centered honeycomb lattice, which features a five-band structure, can be created by choosing six lasers as

$$\mathbf{E}_1 = E(0, 1)e^{i\mathbf{k}\cdot\mathbf{a}_1},$$

$$\mathbf{E}_2 = E(\sqrt{3}/2, 1/2)e^{-i\mathbf{k}\cdot\mathbf{a}_2}, \tag{11.36}$$

$$\mathbf{E}_3 = E(\sqrt{3}/2, -/2)e^{i\mathbf{k}\cdot\mathbf{a}_3},$$

$$\mathbf{E}_4 = E(0, -1)e^{-i\mathbf{k}\cdot\mathbf{a}_1}, \tag{11.37}$$

$$\mathbf{E}_5 = E(-\sqrt{3}/2, -1/2)e^{i\mathbf{k}\cdot\mathbf{a}_2},$$

$$\mathbf{E}_6 = E(-\sqrt{3}/2, 1/2)e^{-i\mathbf{k}\cdot\mathbf{a}_3}, \tag{11.38}$$

where $\mathbf{a}_1 = (1, 0)$, $\mathbf{a}_2 = (-1/2, \sqrt{3}/2)$, $\mathbf{a}_3 = (-1/2, -\sqrt{3}/2)$ are the three nearest-neighbor vectors of the honeycomb structure and k is the wave vector of the lasers— see Figure 11.3 [56]. The intensity profile $I(x, y) = |\mathbf{E}_{\text{tot}}(x, y)|^2$ from the total electric field $\mathbf{E}_{\text{tot}} = \sum_i \mathbf{E}_i$ produces a potential landscape as shown in Figure 11.3. It is also possible to use Spatial Light Modulators for shaping the intensity of a light beam such that the desired minima creates the required lattice (see, for instance, Whyte and Courtial [57] and references therein).

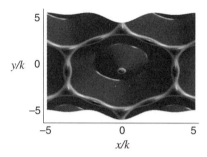

FIGURE 11.3 The intensity profile obtained from the six lasers in Eqs. (11.36)–(11.38) which creates an edge-centered honeycomb lattice for the cold atoms [56]. Red indicates high intensity and blue low intensity. (*For a color version of this figure, see the color plate section.*)

11.3.1 Artificial Magnetic Field and Gauge Potentials for Ultracold Atoms

In the previous section, we have shown that the intensity modulation of the light beams provides the dipole potential W_{dip} which can serve as an optical trap or an optical lattice for ultracold atoms. In this case, the phase of the light does not play a role. In this section, we shall turn to a situation where both intensity and the phase of the incident laser fields are important for the optical manipulation of atoms. We shall demonstrate how two or more laser fields can induce an effective vector potential for the atomic center of mass motion. This simulates an artificial magnetic field leading to a Lorentz-type force. The artificial magnetic field gives us a new tool for manipulating quantum gases composed of electrically neutral atoms, which are not affected by external electromagnetic fields like charged particle are. Furthermore, the atom–light coupling allows for the creation of not only the ordinary vector potential providing the usual Lorentz force, but also the non-Abelian gauge potentials familiar from high energy physics.

The traditional way to create an artificial magnetic field is to rotate an atomic cloud so that the vector potential appears in the rotating frame, see References 33, 53, 58. This reflects the equivalence between the Coriolis force in a rotating system and the Lorentz force acting on a charged particle in a uniform magnetic field. Yet stirring an ultracold cloud of atoms in a controlled manner is a rather demanding task. It is also difficult to create a magnetic field strong enough for observing phenomena such as the quantum Hall effect.

There have also been suggestions [53, 59–63] and recent experimental implementations [34, 37, 38, 64] to take advantage of a discrete periodic structure of an optical lattice to introduce assymetric atomic transitions between the lattice sites. Using this approach, one can induce a nonvanishing phase for the atoms moving along a closed path on the lattice, thus simulating a magnetic flux [60, 62]. Another strategy involves time-dependent optical lattices. An artificial magnetic flux can be generated by combining lattices and time-dependent quadrupolar potentials [65], by stirring [66] or shaking [32, 36, 67–70] optical lattices.

The light-induced geometric gauge potentials [53,63,71–78] do not rely on rotating the system or using the optical lattices. Using this technique, it is possible to generate not only the usual (Abelian) gauge potentials [71–78] but also non-Abelian Gauge potentials [79], whose Cartesian components do not commute (non-Abelian gauge potentials can also be created using the lattice schemes [80]). The light-induced Abelian scalar [73] and vector [17–19] gauge potential have already been implemented experimentally.

Geometric gauge potentials manifest themselves in different areas of physics [16, 81–90]. For ultracold atoms, they are formed when the atomic center of mass motion is coupled to its internal degrees of freedom [71–74, 77–79, 91–95]. In the following we shall provide a more detailed analysis of the light-induced gauge potentials for ultracold atoms.

11.3.1.1 Atom–Light Coupling

To understand the origin of the geometric potentials, let us consider a combined internal and center of mass motion dynamics of an atom described by a Hamiltonian

$$\hat{H} = \frac{p^2}{2m} + \hat{H}_{at}(\mathbf{r},t) + V(\mathbf{r}), \qquad (11.39)$$

where $\mathbf{p} \equiv -i\hbar \nabla$ is the momentum operator for the atom positioned at $\mathbf{r}$, m is the atomic mass, $\hat{H}_{at}(\mathbf{r},t)$ is the atomic Hamiltonian, $V(\mathbf{r})$ represents an external (state-independent) trapping potential, and the hats are placed on the operators acting on the atomic internal states. The Hamiltonian

$$\hat{H}_{at}(\mathbf{r},t) = \hat{H}_0 + \hat{H}_{int}(\mathbf{r},t) \qquad (11.40)$$

describes the electronic degrees of freedom of the atom, and parametrically depends on the atomic center of mass position $\mathbf{r}$. It is composed of the unperturbed atomic Hamiltonian $\hat{H}_0$ and the state-dependent atom–light interaction operator $\hat{H}_{int}(\mathbf{r},t)$. The former $\hat{H}_0$ describes the atomic internal dynamics and does not depend on the atomic center of mass coordinate $\mathbf{r}$. The latter $\hat{H}_{int}(\mathbf{r},t)$ is due to the position- and (possibly) time-dependent coupling of the atom to the external light fields. For resonant atom–light interaction, the coupling term reads in the electric dipole approximation $\hat{H}_{int}(\mathbf{r},t) = -\hat{\mathbf{d}} \cdot \mathbf{E}(\mathbf{r},t)$.

For non-resonant light beams, the atom–light coupling can be effectively represented in terms of the scalar and vector light shift operators acting on the atomic ground state manifold [53, 92, 96–99]. The scalar light shift is nothing but the state-independent dipole potential W_{dip} (presented in the previous Section, Eq. 11.35) which is attributed to the state-independent potential $V(\mathbf{r})$. Specifically, the scalar shift W_{dip} is proportional to the light intensity and is inversely proportional to the detuning: $W_{dip} \propto E^2/\delta$.

For alkali atoms, the vector light shift can be represented in terms of the interaction of the atomic magnetic moment $\boldsymbol{\mu}$ with a fictitious magnetic field $\mathbf{B}_{fict} \propto i\mathbf{E}^* \times \mathbf{E}\delta_{FS}/\delta^2$ [53, 92, 96–99], where δ_{FS} is the energy of the excited-state fine-structure splitting. The resulting operator of the vector light shift $\boldsymbol{\mu} \cdot \mathbf{B}_{\text{fict}}$ is state-dependent, so it is attributed to the interaction Hamiltonian $\hat{H}_{\text{int}}$ featured in the atomic Hamiltonian $\hat{H}_{\text{at}}$. Note that the scalar and vector light shifts can be independently controlled by changing the laser detuning δ and polarization.

It is instructive that the magnitude of the vector light shift operator drops off as $\delta_{\text{FS}}/\delta^2$ with the detuning δ, not like the scalar shift which goes as $1/\delta$. Since the amount of the off-resonant light scattering is also proportional to $1/\delta^2$, the ratio between the off-resonant scattering and the vector coupling is fixed and cannot be decreased by increasing detuning. This is a moderate problem for rubidium atoms characterized by 15 nm fine structure splitting. Yet the losses due to the off-resonant light scattering represent a more serious problem for alkali atoms with smaller fine structure splitting, such as potassium (≈ 4 nm) and lithium (≈ 0.02 nm).

11.3.1.2 Dressed-State Representation

For a fixed position $\mathbf{r}$, the atomic Hamiltonian $\hat{H}_{\text{at}}(\mathbf{r}, t)$ can be diagonalized to give a set of position- and time-dependent states $|\chi_n(\mathbf{r}, t)\rangle$ called dressed states, which are characterized by eigenvalues $\varepsilon_n(\mathbf{r}, t)$, with $n = 1, 2, \ldots, N$ (as illustrated in Figure 11.4), where N is a number of atomic internal states involved. Note that the position and time dependence of the dressed states comes from the position and time dependence of the atom–light coupling $\hat{H}_{\text{int}}(\mathbf{r}, t)$.

The full quantum state vector of the atom describing both internal and motional degrees of freedom can then be expanded in terms of the dressed states as

$$|\tilde{\Psi}\rangle = \sum_{n=1}^{N} \Psi_n(\mathbf{r}) |\chi_n(\mathbf{r}, t)\rangle, \qquad (11.41)$$

where $\Psi_n(\mathbf{r}) \equiv \Psi_n$ is a wavefunction for the center of mass motion of the atom in the internal state n. Substituting Eq. (11.41) into the Schrödinger equation $i\hbar \partial |\tilde{\Psi}\rangle / \partial t = \hat{H} |\tilde{\Psi}\rangle$, one arrives at a set of coupled equations for the multicomponent center of

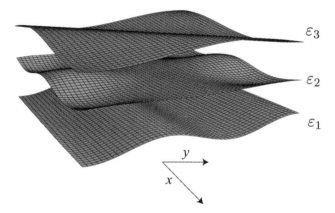

FIGURE 11.4 Schematic representation of the position dependence of the eigenenergies $\varepsilon_m \equiv \varepsilon_m(\mathbf{r}, t)$ for the atomic dressed states $|\chi_n(\mathbf{r}, t)\rangle$.

mass wavefunction Ψ_n. Introducing the N-dimensional column vector $\Psi = (\Psi_1, \Psi_2, \ldots, \Psi_N)^T$, it is convenient to represent these equations in a matrix form,

$$i\hbar \frac{\partial}{\partial t} \Psi = \left[\frac{1}{2m} (-i\hbar \nabla - \mathbf{A})^2 + \Phi + U \right] \Psi, \qquad (11.42)$$

where $\mathbf{A}$, U and Φ are the $N \times N$ matrices characterized by the elements

$$\mathbf{A}_{n,m} = i\hbar \langle \chi_n(\mathbf{r}) | \nabla \chi_m(\mathbf{r}) \rangle, \qquad (11.43)$$

$$U_{n,m} = [\varepsilon_n(\mathbf{r}) + V(\mathbf{r})] \delta_{n,m}, \qquad (11.44)$$

and

$$\Phi_{nm} = -i\hbar \langle \eta_n | \partial_t | \eta_m \rangle, \qquad (11.45)$$

the diagonal matrix U accommodating from both the internal atomic energies and also the external trapping potential. The matrix-vector $\mathbf{A}$ appears due to the position dependence of the atomic dressed states, whereas the matrix Φ stems from their time-dependence.

Up to now we have not made any approximation, so the set of Eqs. (11.42)–(11.45) are equivalent to the original Schrödinger equation for the combined atomic internal and center of mass motion within the manifold of N internal states. In the following we shall consider a reduced atomic dynamics where some atomic internal states can be eliminated.

11.3.1.3 Reduced Atomic Dynamics

If the off-diagonal elements of the matrices $\mathbf{A}$ and U are much smaller than the difference in the atomic energies $U_{nn} - U_{mm}$, the *adiabatic approximation* can be applied by neglecting the off-diagonal contributions. This leads to a separation of the dynamics in different dressed states. Atoms in any one of the dressed states evolve according to a separate Hamiltonian in which the gauge potentials $\mathbf{A}$ and V reduce to the 1×1 matrices, that is, the gauge potentials become Abelian. The simple adiabatic approximation fails if there are degenerate (or nearly degenerate) dressed states, so that the off-diagonal (non-adiabatic) couplings between the degenerate dressed states can no longer be ignored. In that case, the gauge potentials no longer reduce to the 1×1 matrices, and the gauge potentials are non-Abelian provided their Cartesian components do not commute.

Let us begin by considering such a more general situation where the first q atomic dressed states are degenerate (or nearly degenerate), and these levels are well separated from the remaining $N - q$ levels. Neglecting transitions to the remaining states, one can project the full Hamiltonian onto the selected subspace. As a result, one arrives at the closed Schrödinger equation for the reduced column vector $\tilde{\Psi} = (\Psi_1, \ldots, \Psi_q)^T$

$$i\hbar \frac{\partial}{\partial t} \tilde{\Psi} = \left[\frac{1}{2m} (-i\hbar \nabla - \mathbf{A})^2 + \Phi + U + W \right] \tilde{\Psi}, \qquad (11.46)$$

where $\mathbf{A}$, Φ, and U are the truncated $q \times q$ matrices. The projection of the term $\mathbf{A}^2$ to the q-dimensional subspace cannot entirely be expressed in terms of a truncated matrix $\mathbf{A}$. This gives rise to a geometric scalar potential W which is again a $q \times q$ matrix,

$$W_{n,m} = \frac{1}{2m} \sum_{l=q+1}^{N} \mathbf{A}_{n,l} \cdot \mathbf{A}_{l,m} \qquad (11.47)$$

with $n, m \in (1, \ldots, q)$. The potential W can be interpreted as a kinetic energy of the atomic micro-trembling due to off-resonance non-adiabatic transitions to the omitted dressed states with $l > q$ [100, 101].

It is worth emphasizing that the geometric vector and scalar potentials $\mathbf{A}$ and W emerge due to the spatial dependence of the atomic dressed states, whereas the potential Φ comes from their time dependence. For $q > 1$, the latter Φ describes the population transfer between the atomic levels due to the temporal dependence of the external fields [102, 103].

The reduced $q \times q$ matrix $\mathbf{A}$ represents the Mead–Berry connection [82, 88] also known as the effective vector potential. It is related to a curvature (an effective "magnetic" field) $\mathbf{B}$ as [53, 63, 89]:

$$B_i = \frac{1}{2}\epsilon_{ikl} F_{kl}, \quad F_{kl} = \partial_k A_l - \partial_l A_k - \frac{i}{\hbar}[A_k, A_l]. \qquad (11.48)$$

Alternatively, one can write in the index-free notations

$$\mathbf{B} = \nabla \times \mathbf{A} - \frac{i}{\hbar}\mathbf{A} \times \mathbf{A}. \qquad (11.49)$$

The term $\frac{1}{2}\epsilon_{ikl}[A_k, A_l] = (\mathbf{A} \times \mathbf{A})_i$ does not vanish in general, because the Cartesian components of $\mathbf{A}$ do not necessarily commute. This reflects the non-Abelian character of the vector potential.

Using Eq. (11.43) for $\mathbf{A}$ together with the completeness relation, the matrix elements of the curvature can be represented as a sum over the eliminated states:

$$\mathbf{B}_{nm} = -\frac{i}{\hbar} \sum_{l=q+1}^{N} \mathbf{A}_{nl} \times \mathbf{A}_{lm}, \quad n, m = 1, \ldots, q. \qquad (11.50)$$

Hence the Berry curvature $\mathbf{B}$ is non-zero only for the reduced atomic dynamics ($q < N$) when some of the atomic states are eliminated. The same applies to the geometric scalar potential W_{nm} given by Eq. (11.47).

Along with the effective magnetic field $\mathbf{B}$, the atom is affected by an effective electric field [53, 89]

$$\mathbf{E} = -\partial_t \mathbf{A} - \nabla(\Phi + U) + \frac{i}{\hbar}[\mathbf{A}, \Phi] \qquad (11.51)$$

containing a commutator $[\mathbf{A}, \Phi]$ in addition to the usual gradient $\nabla(\Phi + U)$ and induced $\partial_t \mathbf{A}$ contributions.

When the Cartesian components of $\mathbf{A}$ commute, the vector potential is called Abelian. This is always the case when $q = 1$. If additionally $[\mathbf{A}, \Phi] = 0$, the induced magnetic and electric fields acquire a standard form

$$\mathbf{B} = \nabla \times \mathbf{A}, \quad \text{and} \quad \mathbf{E} = -\nabla(\Phi + U) - \partial_t \mathbf{A}.$$

In particular, if the truncated space includes a single dressed state ($q = 1$) well separated from the others, the vector and scalar potentials reduce to the ordinary commuting vector and scalar potentials,

$$\mathbf{A} = i\hbar \langle \kappa_1 | \nabla | \kappa_1 \rangle \quad \text{and} \quad W = \frac{1}{2m} \sum_{l=2}^{N} \mathbf{A}_{n,l} \cdot \mathbf{A}_{l,m}. \quad (11.52)$$

An Abelian geometric vector potential $\mathbf{A}$ has been already engineered for the ultracold atoms in the case of $q = 1$ [17–19] and $q = 2$ [20]. For a single adiabatic state ($q = 1$) the spatial dependence of $\mathbf{A}$ yielded an artificial magnetic field [17, 18] and its temporal dependence generated an effective electric field [19].

In the next section, we shall consider a situation where two laser beams are coupled to the atoms in the so-called Λ configuration. In this scheme, there is a single non-degenerate electronic state (known as dark state). Thus the atomic center of mass undergoes the adiabatic motion influenced by the (Abelian) vector and trapping potentials. In Section 11.3.3, we shall analyze a tripod scheme of laser–atom interactions that provide two degenerate dark states. In that case, one has non-Abelian light-induced gauge potentials.

In what follows, the atomic Hamiltonian is taken to be time-independent, $\hat{H}_{at}(\mathbf{r}, t) \equiv \hat{H}_{at}(\mathbf{r})$. Consequently the dressed states are also time-independent, with the potential Φ given by Eq. (11.45) equal to zero.

11.3.2 Λ Setup for Creating Abelian Gauge Potentials

Let us begin our specific analysis of the light-induced gauge potentials with the atoms characterized by the Λ type level structure depicted in Figure 11.5. In this scheme, the laser beams couple two atomic internal states $|1\rangle$ and $|2\rangle$ with the third one $|0\rangle$, the coupling being characterized by the Rabi frequencies Ω_1 and Ω_2. The atom light

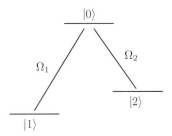

FIGURE 11.5 The Λ setup of the atom–light coupling. The laser beams induce atomic transitions $|1\rangle \to |0\rangle$ and $|2\rangle \to |0\rangle$ characterized by the Rabi frequencies Ω_1 and Ω_2.

coupling operator $\hat{H}_{at}(\mathbf{r})$ featured in the full atomic Hamiltonian (11.39) has then the following form:

$$\hat{H}_{at}(\mathbf{r}) = \hbar\delta(|2\rangle\langle 2| - |1\rangle\langle 1|) + \hbar\Delta|2\rangle\langle 2| + \hbar(\Omega_1|0\rangle\langle 1| + \Omega_2|0\rangle\langle 2| + \text{H.c.})/2, \quad (11.53)$$

where the frequencies δ and Δ characterize detuning between the atomic states. The Rabi frequencies $\Omega_1 \equiv \Omega_1(\mathbf{r})$ and $\Omega_2 \equiv \Omega_2(\mathbf{r})$ are generally complex and position-dependent.

In the context of the light-induced gauge potential for ultracold atoms, the Λ scheme was first considered in the late 1990s [71–73]. Subsequently, it was shown that the scheme can yield a non-zero artificial magnetic field using non-trivial arangements of laser fields [74–78, 93, 94, 101, 104, 105] or position-dependent detuning [18, 95]. In most of these treatments, the state $|0\rangle$ is assumed to be the atomic excited state coupled resonantly to the ground states $|1\rangle$ and $|2\rangle$ by laser beams, as shown in Figure 11.5. The involvement of the excited states is inevitably associated with a substantial dissipation due to the spontaneous emission. To avoid such losses, one can choose $|0\rangle$ to be an atomic ground state coupled to the other two ground states $|1\rangle$ and $|2\rangle$ via the Raman transitions [95, 105]. In that case Ω_1 and Ω_2 featured in the coupling operator (11.53) represent the Raman Rabi frequencies.

Assuming exact resonance between the atomic levels 1 and 3 ($\delta = 0$), the atomic operator $\hat{H}_{at}(\mathbf{r})$ given by Eq. (11.53) has a single dressed eigenstate $|D\rangle$ known as the dark (or uncoupled) state, containing no contribution by the excited state $|0\rangle$ and characterized by a zero eigenvalue $\varepsilon_D = 0$,

$$|D\rangle = \frac{|1\rangle - \zeta|2\rangle}{1 + \zeta^2}, \quad \zeta = \frac{\Omega_1}{\Omega_2} = |\zeta|e^{iS}, \quad (11.54)$$

where S is a relative phase between the two Rabi frequencies. Dark states are frequently encountered in quantum optics. They play an important role in applications such as electromagnetically induced transparency (EIT) [106–109] and stimulated Raman adiabatic passage (STIRAP) [110–112]. These phenomena rely on the fact that the excited level $|0\rangle$ is not populated for the dark state atoms and spontaneous decay is therefore suppressed.

In addition to the dark states, there are two other dressed states separated from the dark state by the Rabi frequency $\Omega = \sqrt{|\Omega_1|^2 + |\Omega_2|^2}$. If this separation exceeds the characteristic kinetic energy of the atomic motion, the dark state atoms undergo an adiabatic motion. As we have seen in the previous Section, the (Abelian) geometric vector potential emerges for such an adiabatic center of mass motion. For the dark state atoms, the vector potential $\mathbf{A} = i\hbar\langle D|\nabla|D\rangle$ and the associated artificial magnetic field $\mathbf{B} = \nabla \times \mathbf{A}$ are

$$\mathbf{A} = -\hbar\frac{|\zeta|^2}{1+|\zeta|^2}\nabla S, \quad (11.55)$$

$$\mathbf{B} = \hbar\frac{\nabla S \times \nabla|\zeta|^2}{(1+|\zeta|^2)^2}. \quad (11.56)$$

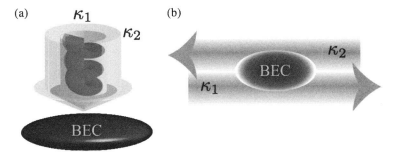

FIGURE 11.6 (a) The light beams carrying optical vortices [74–77, 104, 105] or (b) the counterpropagating beams with spatially shifted profiles [78, 93, 101, 113] provide a non-zero artificial magnetic field for the atoms in the laser-dressed states of the Λ scheme.

The geometric scalar potential for the dark state atoms is given by

$$W = \frac{\hbar^2}{2m} \frac{(\nabla|\zeta|)^2 + |\zeta|^2(\nabla S)^2}{(1+|\zeta|^2)^2}. \tag{11.57}$$

One easily recognizes that the gauge potential **A** gives a non-vanishing artificial magnetic **B** only if the gradients of the relative intensity and the relative phase are both non-zero and not parallel to each other. Therefore, the artificial magnetic field cannot be produced for the dark state atoms using the plane-waves for the Λ scheme [71,72]. However, plane waves can indeed be used in more complex tripod [79, 114–119] setups to generate the non-Abelian gauge potential for a pair of degenerate internal states, as we shall see in the next section.

Equation (11.56) has a very intuitive interpretation [78]. The vector $\nabla[|\zeta|^2/(1+|\zeta|^2)]$ connects the "center of mass" of the two light beams and ∇S is proportional to the vector of the relative momentum of the two light beams. Thus a nonvanishing **B** requires a *relative orbital angular momentum* of the two light beams. This is the case for light beams carrying optical vortices [74–77, 104] or if one uses two counterpropagating light beams of finite diameter with an axis offset [78, 101], as depicted in Figure 11.6. A more detailed analysis of these schemes is presented in the reviews [63, 120].

11.3.3 Tripod Scheme for Creating Non-Abelian Gauge Potentials

11.3.3.1 General Let us now turn to the non-Abelian geometric gauge potentials introduced in Section 11.3.1. They emerge when the center of mass motion of atoms takes place in a manifold of degenerate or quasi-degenerate internal dressed states. Note that the non-Abelian gauge potentials provide coupling between the center of mass motion and the internal (spin or quasi-spin) degrees of freedom, thus causing the spin–orbit coupling.

To illustrate the non-Abelian gauge potentials for ultracold atoms, let us consider a tripod setup for the atom–light coupling shown in Figure 11.7. The tripod scheme is the most commonly used theoretical model in which non-Abelian gauge fields

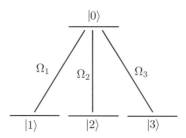

FIGURE 11.7 The tripod configuration of the atom–light coupling in which the atomic ground states $|j\rangle$ (with $j = 1, 2, 3$) are coupled to the excited state $|0\rangle$ via the Rabi frequencies Ω_j of the laser fields.

appear. It comprises three lower atomic levels coupled with an excited level via the laser fields with the Raman frequencies Ω_j. Compared to the Λ setup analyzed in the previous section, now there is an additional third laser driving the transitions between an extra ground state 3 and the excited state 0. The tripod scheme has two degenerate dark states representing the superpositions of the three ground states immune to the atom–light coupling.

The tripod scheme was initially considered by Olshanii and coauthors [121, 122] in the context of the velocity selective coherent population trapping. Subsequently, the group of Klaas Bergmann theoretically [102, 103, 123] and experimentally [124, 125] explored the adiabatic evolution of the dark-state atoms in the time-dependent laser fields (the tripod-STIRAP), governed by the matrix valued geometrix phase [16]. On the other hand, Ruseckas et al. considered the adiabatic motion of the dark-state atoms in the spatially varying laser fields [79] providing non-Abelian gauge potentials for the center of mass motion using the tripod scheme.

Assuming exact resonance for the atom–light coupling, the Hamiltonian of the tripod atom reads in the interaction representation:

$$\hat{H}_{at}(\mathbf{r}) = \frac{\hbar}{2}(\Omega_1 |0\rangle\langle 1| + \Omega_2 |0\rangle\langle 2| + \Omega_3 |0\rangle\langle 3|) + H.c. = \frac{\hbar}{2}\Omega |0\rangle\langle B| + H.c., \quad (11.58)$$

The Hamiltonian $\hat{H}_{at}(\mathbf{r})$ has two eigenstates $|D_j\rangle$ ($j = 1, 2$) characterized by zero eigen-energies $\hat{H}_{at}(\mathbf{r})|D_j\rangle = 0$. The eigenstates $|D_j\rangle$ are the dark states containing no excited-state contribution, as one can see in Eqs. (11.60)–(11.61).

Parameterizing the Rabi-frequencies Ω_μ with angle and phase variables according to

$$\begin{aligned}\Omega_1 &= \Omega \sin\theta \cos\phi\, e^{iS_1}, \\ \Omega_2 &= \Omega \sin\theta \sin\phi\, e^{iS_2}, \\ \Omega_3 &= \Omega \cos\theta\, e^{iS_3},\end{aligned} \quad (11.59)$$

where $\Omega = \sqrt{|\Omega_1|^2 + |\Omega_2|^2 + |\Omega_3|^2}$, the adiabatic dark states read

$$|D_1\rangle = \sin\phi e^{iS_{31}}|1\rangle - \cos\phi e^{iS_{32}}|2\rangle, \tag{11.60}$$

$$|D_2\rangle = \cos\theta\cos\phi e^{iS_{31}}|1\rangle + \cos\theta\sin\phi e^{iS_{32}}|2\rangle - \sin\theta|3\rangle, \tag{11.61}$$

with $S_{ij} = S_i - S_j$ being the relative phases. The dark states again depend of the atomic position through the spatial dependence of the Rabi frequencies Ω_j. This leads to the appearance of the gauge potentials **A** and W presented below.

We are interested in a situation where the atoms are kept in their dark states. This can be done neglecting transitions from the dark states to the bright state $|B\rangle \sim \Omega_1^*|1\rangle + \Omega_2^*|2\rangle + \Omega_3^*|3\rangle$. The latter is coupled to the excited state $|0\rangle$ with the Rabi frequency Ω, so the two states $|B\rangle$ and $|0\rangle$ split into a doublet separated from the dark states by the energies $\pm\Omega$. The adiabatic approximation is justified if Ω is sufficiently large compared to the kinetic energy of the atomic motion. In that case, the internal state of an atom does indeed evolve within the dark state manifold. The atomic state vector $|\Phi\rangle$ can then be expanded in terms of the dark states according to $|\Phi\rangle = \sum_{j=1}^{2} \Psi_j(\mathbf{r})|D_j(\mathbf{r})\rangle$, where $\Psi_j(\mathbf{r})$ is the wavefunction for the center of mass motion of the atom in the jth dark state. Adapting the general treatment from Section 11.3.1, the atomic center of mass motion is described by a two-component wavefunction

$$\Psi = \begin{pmatrix} \Psi_1 \\ \Psi_2 \end{pmatrix} \tag{11.62}$$

which obeys the Schrödinger equation

$$i\hbar\frac{\partial}{\partial t}\Psi = \left[\frac{1}{2m}(-i\hbar\nabla - \mathbf{A})^2 + V + W\right]\Psi, \tag{11.63}$$

where the geometric vector and scalar potentials **A** and W are now 2×2 matrices, and V is the state-independent trapping potential. For the tripod scheme, the gauge potentials **A** and Φ are found to be [79]

$$\mathbf{A}_{11} = \hbar(\cos^2\phi\nabla S_{23} + \sin^2\phi\nabla S_{13}) \tag{11.64}$$

$$\mathbf{A}_{12} = \hbar\cos\theta\left(\frac{1}{2}\sin(2\phi)\nabla S_{12} - i\nabla\phi\right) \tag{11.65}$$

$$\mathbf{A}_{22} = \hbar\cos^2\theta(\cos^2\phi\nabla S_{13} + \sin^2\phi\nabla S_{23}) \tag{11.66}$$

and

$$W_{11} = \frac{\hbar^2}{2m}\sin^2\theta\left(\frac{1}{4}\sin^2(2\phi)(\nabla S_{12})^2 + (\nabla\phi)^2\right) \tag{11.67}$$

$$W_{12} = \frac{\hbar^2}{2m} \sin\theta \left(\frac{1}{2} \sin(2\phi) \nabla S_{12} - i\nabla\phi \right)$$
$$\left(\frac{1}{2} \sin(2\theta)(\cos^2\phi \nabla S_{13} + \sin^2\phi \nabla S_{23}) - i\nabla\theta \right) \quad (11.68)$$

$$W_{22} = \frac{\hbar^2}{2m} \left(\frac{1}{4} \sin^2(2\theta)(\cos^2\phi \nabla S_{13} + \sin^2\phi \nabla S_{23})^2 + (\nabla\theta)^2 \right). \quad (11.69)$$

Next we shall turn to some specific configurations of the laser fields.

11.3.3.2 Specific Configurations of the Laser Fields for the Tripod Setup If all three beams copropagate with two of them carrying opposite vortices and the third beam without a vortex, then persistent currents can be generated in the Bose gas [126]. On the other hand, the tripod setup can also provide a vector potential which represents a magnetic monopole [79, 127]. For this, let us consider two copropagating and circularly polarized fields $\Omega_{1,2}$ with opposite orbital angular momenta $\pm\hbar$ along the propagation axis z, whereas the third field Ω_3 propagates in x direction and is linearly polarized along the y-axis:

$$\Omega_{1,2} = \Omega_0 \frac{\rho}{R} e^{i(kz\mp\varphi)}, \qquad \Omega_3 = \Omega_0 \frac{z}{R} e^{ik'x}. \quad (11.70)$$

Here ρ is the cylindrical radius and φ the azimuthal angle. It should be noted that these fields have a vanishing divergence and obey the Helmholtz equation. The total intensity of the laser fields (Eq. 11.70) vanishes at the origin which is a singular point.

The vector potential associated with these fields can be calculated using Eq. (11.65):

$$\mathbf{A} = -\hbar \frac{\cos\vartheta}{r\sin\vartheta} \hat{\mathbf{e}}_\varphi \begin{pmatrix} 0 & 1 \\ 1 & 0 \end{pmatrix} + \frac{\hbar}{2}(k\hat{\mathbf{e}}_z - k'\hat{\mathbf{e}}_x)$$
$$\times \left[(1 + \cos^2\vartheta) \begin{pmatrix} 1 & 0 \\ 0 & 1 \end{pmatrix} + (1 - \cos^2\vartheta) \begin{pmatrix} 1 & 0 \\ 0 & -1 \end{pmatrix} \right]. \quad (11.71)$$

The first term proportional to σ_x corresponds to a magnetic monopole of the unit strength at the origin. This is easily seen by calculating the magnetic field

$$\mathbf{B} = \frac{\hbar}{r^2} \hat{\mathbf{e}}_r \begin{pmatrix} 0 & 1 \\ 1 & 0 \end{pmatrix} + \cdots. \quad (11.72)$$

The dots indicate non-monopole field contributions proportional to the Pauli matrices σ_z and σ_y, and to the unity matrix.

Returning to the general case, by properly choosing the laser fields, the Cartesian components of the vector potential $\mathbf{A}$ can be made not to commute, leading to non-Abelian gauge potentials. It is instructive to note that this can happen even if the Rabi frequecies Ω_j of the tripod setup are (non-collinear) plane waves [115, 116, 119]

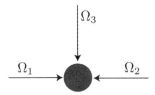

FIGURE 11.8 The three laser beams acting on atoms in the tripod configuration.

or properly chosen standing waves [114, 117, 118]. In that case, one can produce a uniform vector potential with the Cartesian components proportional to the Pauli matrices. This generates the spin–orbit coupling (SOC) of the Rashba–Dresselhaus type for the ultracold atoms.

Following References 115 and 116 let us consider in more detail the generation of Rashba–Dresselhaus spin–orbit coupling using the tripod scheme. For this, we assume that the first two lasers have the same intensities and counterpropagate in the x direction while the third one propagates in the negative y direction, as depicted in Figure 11.8. Specifically we have $\Omega_1 = \Omega \sin\theta e^{-i\kappa x}/\sqrt{2}$, $\Omega_2 = \Omega \sin\theta e^{i\kappa x}/\sqrt{2}$ and $\Omega_3 = \Omega \cos\theta e^{-i\kappa y}$, where $\Omega = \sqrt{|\Omega_1|^2 + |\Omega_2|^2 + |\Omega_3|^2}$ is the total Rabi frequency, and θ is the mixing angle defining the relative intensity of the third laser field. Furthermore, we take the external trapping potential $\hat{V}$ to be state-dependent and diagonal in the bare states with elements $V_1 = V_2$, and V_3.

Under these conditions, the geometric vector and scalar potentials, and the external trapping potential are in the dark state basis given by [115, 116]

$$\mathbf{A} = \hbar\kappa \begin{pmatrix} \mathbf{e}_y & -\mathbf{e}_x \cos\theta \\ -\mathbf{e}_x \cos\theta & \mathbf{e}_y \cos^2\theta \end{pmatrix}, \quad (11.73)$$

$$W = \begin{pmatrix} \hbar^2\kappa^2 \sin^2\theta/2m & 0 \\ 0 & \hbar^2\kappa^2 \sin^2(2\theta)/8m \end{pmatrix}, \quad (11.74)$$

$$V = \begin{pmatrix} V_1 & 0 \\ 0 & V_1 \cos^2\theta + V_3 \sin^2\theta \end{pmatrix}, \quad (11.75)$$

with $\mathbf{e}_x$ and $\mathbf{e}_y$ being unit Cartesian vectors.

By choosing $V_3 - V_1 = \hbar^2\kappa^2 \sin^2(\theta)/2m$, the overall trapping potential simplifies to $V + W = V_1 I$, where I is the unit matrix. In that case, both dark states are affected by the same trapping potential $V_1 \equiv V_1(\mathbf{r})$. This can be achieved by detuning the third laser from the two-photon resonance by the frequency $\Delta\omega_3 = \hbar\kappa^2 \sin^2\theta/2m$.

Additionally, by choosing the mixing angle $\theta = \theta_0$ to be such that $\sin^2\theta_0 = 2\cos\theta_0$, one has $\cos\theta_0 = \sqrt{2} - 1$. In that case, the vector potential can be represented in a symmetric way in terms of the Pauli matrices σ_x and σ_z

$$\mathbf{A} = \hbar\kappa'(-\mathbf{e}_x\sigma_x + \mathbf{e}_y\sigma_z) + \hbar\kappa_0\mathbf{e}_y I, \quad (11.76)$$

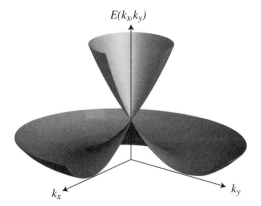

FIGURE 11.9 The dispersion of the spin-orbit coupled system with the vector potential **A** given by Eq. (11.76). Here **k** denotes momentum in the $\mathbf{e}_x - \mathbf{e}_y$ plane shifted by $\hbar \kappa_0 \mathbf{e}_y$. The lowest band is minimal on a ring at $|\mathbf{k}| = \kappa'$. The intersection between the upper and lower dispersion branches gives rise to a quasi-relativistic behavior (a Dirac cone), which is also familiar from graphene [128].

where $\kappa' = \kappa \cos \theta_0 \approx 0.414\kappa$ and $\kappa_0 = \kappa(1 - \cos \theta_0)$. Although the produced vector potential **A** is constant, it cannot be eliminated via a gauge transformation, because the Cartesian components A_x and A_y do not commute. Thus the potential **A** is non-Abelian.

The light-induced vector potential given by Eq. (11.76) describes the spin–orbit coupling of the Rashba–Dressehaus type, which has been widely explored in recent years. In particular, the lowest dispersion branch for such spin-orbit coupled atoms has a Mexican hat shape shown in Figure 11.9. This leads to an unconventional Bose–Einstein condensation of atoms at the energy minimum ring in the momentum space. On the other hand, the quasi-relativisitc (Dirac) dispersion (like the one emerging for electrons in graphene [128]) is formed at the intersection between the two dispersion branches. For more details on the spin-orbit coupled atomic gases, see recent reviews [53, 129, 130].

The Rashba–Dresselhaus spin–orbit coupling can also be generated using another slightly simpler "closed loop" scheme involving three or four sequentially coupled levels with properly chosen coupling phases [131]. Unlike the tripod setup, the degenerate dressed states of the "closed loop" scheme are the atomic ground states. This prevents losses due to transitions to other dressed states.

11.3.4 Concluding Remarks

Optical manipulation of single atoms and ensembles of atoms is likely to be a key ingredient in any future quantum technology. Large strides have been made in this respect since the advent of the laser. Today we routinely see exotic quantum gases being prepared in laboratories with the aid of light. In particular, the possibility to induce artificial gauge fields in these gases has opened up new research avenues.

Optical lattices, for instance, have proven to be particularly well suited for addressing the concept of quantum simulators. With the inclusion of gauge fields, the possibility to address many open questions in condensed matter physics is certainly an appealing endeavor.

It is interesting to speculate what the future will hold in its hands in this context. Orbital magnetism for charge-neutral atoms could indeed be important for quantum information processing, where concepts such as atomtronics, the atomic version of spintronics, might give some advantages over more standard techniques for manipulating quantum states. Recent advances in trapping and manipulating atoms in cavities may also provide the mechanisms to prepare large-scale strongly correlated quantum states. Also here artificial gauge fields could play an important role. Due to the back-action between the atoms and the cavity fields, a gauge field in this setup is likely to show at least some degree of dynamical aspects [132–134], which would open up the possibility to emulate much more sophisticated gauge theories in ultracold settings.

REFERENCES

[1] M. Greiner, C. A. Regal, and D. S. Jin, "Emergence of a molecular Bose–Einstein condensate from a Fermi gas," *Nature*, **426**, 537–540 (2003).

[2] T. Bourdel, L. Khaykovich, J. Cubizolles, J. Zhang, F. Chevy, M. Teichmann, L. Tarruell, S. Kokkelmans, and C. Salomon, "Experimental study of the BEC-BCS crossover region in Lithium 6," *Phys. Rev. Lett.* **93**(5), 050401 (2004).

[3] M. Bartenstein, A. Altmeyer, S. Riedl, S. Jochim, C. Chin, J. Denschlag, and R. Grimm, "Crossover from a molecular Bose-Einstein condensate to a degenerate Fermi gas," *Phys. Rev. Lett.* **92**(12), 120401 (2004).

[4] M. Zwierlein, C. Stan, C. Schunck, S. Raupach, A. Kerman, and W. Ketterle, "Condensation of pairs of fermionic atoms near a Feshbach resonance," *Phys. Rev. Lett.* **92**, 120403 (2004).

[5] J. Kinast, S. Hemmer, M. Gehm, A. Turlapov, and J. Thomas, "Evidence for superfluidity in a resonantly interacting Fermi gas," *Phys. Rev. Lett.* **92**(15), 150402 (2004).

[6] U. Bissbort, S. Götze, Y. Li, J. Heinze, J. S. Krauser, M. Weinberg, C. Becker, K. Sengstock, and W. Hofstetter, "Detecting the amplitude mode of strongly interacting lattice bosons by bragg scattering," *Phys. Rev. Lett.* **106**, 205303 (2011).

[7] M. Endres, T. Fukuhara, D. Pekker, M. Cheneau, P. Schau, C. Gross, E. Demler, S. Kuhr, and I. Bloch, "The 'Higgs' amplitude mode at the two-dimensional superfluid/Mott insulator transition," *Nature*, **487**(7408), 454–458 (2012).

[8] M. Merkl, A. Jacob, F. E. Zimmer, P. Öhberg, and L. Santos, "Chiral confinement in quasirelativistic Bose-Einstein condensates," *Phys. Rev. Lett.* **104**, 073603 (2010).

[9] E. Kapit and E. Mueller, "Optical-lattice Hamiltonians for relativistic quantum electrodynamics," *Phys. Rev. A* **83**, 033625 (2011).

[10] D. Banerjee, M. Bögli, M. Dalmonte, E. Rico, P. Stebler, U.-J. Wiese, and P. Zoller, "Atomic quantum simulation of $U(n)$ and SU(n) non-abelian lattice gauge theories," *Phys. Rev. Lett.* **110**, 125303 (2013).

[11] M. J. Edmonds, M. Valiente, G. Juzeliūnas, L. Santos, and P. Öhberg, "Simulating an interacting gauge theory with ultracold bose gases," *Phys. Rev. Lett.* **110**, 085301 (2013).

[12] A. Bermudez, L. Mazza, M. Rizzi, N. Goldman, M. Lewenstein, and M. Martin-Delgado, "Wilson fermions and axion electrodynamics in optical lattices," *Phys. Rev. Lett.* **105**(19), 190404 (2010).

[13] J. Williams, E. Hazlett, J. Huckans, R. Stites, Y. Zhang, and K. O'Hara, "Evidence for an excited-state Efimov trimer in a three-component Fermi gas," *Phys. Rev. Lett.* **103**(13), 130404 (2009).

[14] C. Chin, R. Grimm, P. Julienne, and E. Tiesinga, "Feshbach resonances in ultracold gasses," *Rev. Mod. Phys.* **82**, 1225 (2010).

[15] K. von Klitzing, "The quantized Hall effect, *Rev. Mod. Phys.* **58**, 519 (1986).

[16] F. Wilczek and A. Zee, "Appearance of gauge structure in simple dynamical systems," *Phys. Rev. Lett.* **52**(24), 2111–2114 (1984).

[17] Y.-J. Lin, R. L. Compton, A. R. Perry, W. D. Phillips, J. V. Porto, and I. B. Spielman, "Bose-Einstein condensate in a uniform light-induced vector potential," *Phys. Rev. Lett.* **102**, 130401 (2009).

[18] Y. J. Lin, R. L. Compton, K. Jimenez-Garcia, J. V. Porto, and I. B. Spielman, "Synthetic magnetic fields for ultracold neutral atoms," *Nature* **462**, 628–632 (2009).

[19] Y-J. Lin, R. L. Compton, K. Jimenez-Garcia, W. D. Phillips, J. V. Porto, and I. B. Spielman, "A synthetic electric force acting on neutral atoms," *Nature Phys.* **7**, 531–534 (2011).

[20] Y.-J. Lin, K. Jiménez-García, and I. B. Spielman, "Spin-orbit-coupled Bose-Einstein condensates," *Nature* **471**, 83–86 (2011).

[21] P. Wang, Z.-Q. Yu, Z. Fu, J. Miao, L. Huang, S. Chai, H. Zhai, and J. Zhang, "Spin-orbit coupled degenerate Fermi gases," *Phys. Phys. Lett.* **109**(9), 095301 (2012).

[22] L. W. Cheuk, A. T. Sommer, Z. Hadzibabic, T. Yefsah, W. S. Bakr, and M. W. Zwierlein, "Spin-injection spectroscopy of a spin-orbit coupled Fermi gas," *Phy. Rev. Lett.* **109**(9), 095302 (2012).

[23] J.-Y. Zhang, S.-C. Ji, Z. Chen, L. Zhang, Z.-D. Du, B. Yan, G.-S. Pan, B. Zhao, Y. Deng, H. Zhai, et al., "Collective dipole oscillations of a spin-orbit coupled Bose-Einstein condensate," *Phys. Lett. Lett.* **109**(11), 115301 (2012).

[24] Z. Fu, L. Huang, Z. Meng, P. Wang, L. Zhang, S. Zhang, H. Zhai, P. Zhang, and J. Zhang, "Production of Feshbach molecules induced by spin–orbit coupling in Fermi gases," *Nat. Phys.* **10**, 110–115 (2014).

[25] L. Zhang, J.-Y. Zhang, S.-C. Ji, Z.-D. Du, H. Zhai, Y. Deng, S. Chen, P. Zhang, and J.-W. Pan, "Stability of excited dressed states with spin-orbit coupling," *Phys. Rev. A* **87**(1), 011601(R) (2013).

[26] C. Qu, C. Hamner, M. Gong, C. Zhang, and P. Engels, "Observation of Zitterbewegung in a spin-orbit-coupled Bose-Einstein condensate," *Phys. Rev. A* **88**, 021604(R) (2013).

[27] L. J. LeBlanc, M. C. Beeler, K. Jimenez-Garcia, A. R. Perry, S. Sugawa, R. A. Williams, and I. B. Spielman, "Direct observation of Zitterbewegung in a Bose-Einstein condensate," *New. J. Phys.* **15**, 073011 (2013).

[28] M. Lewenstein, A. Sanpera, V. Ahufinger, B. Damski, A. Sen De, and U. Sen, "Ultracold atomic gases in optical lattices: mimicking condensed matter physics and beyond," *Adv. Phys.* **56**, 243–379 (2007).

[29] I. Bloch, J. Dalibard, and W. Zwerger, "Many-body physics with ultracold gases," *Rev. Mod. Phys.* **80**, 885 (2008).

[30] M. Lewenstein, S. Anna, and A. Verònica, *Ultracold Atoms in Optical Lattices: Simulating Quantum Many-Body Systems* (Oxford University Press, 2012).

[31] R. P. Feynman, "Simulating physics with computers," *Int. J. Theor. Phys.* **21**, 467 (1982).

[32] A. Eckardt, C. Weiss, and M. Holthaus, "Superfluid-insulator transition in a periodically driven optical lattice," *Phys. Rev. Lett.* **95**, 260404 (2005).

[33] N. R. Cooper, "Rapidly rotating atomic gases," *Adv. Phys.* **57**(6), 539–616 (2008).

[34] M. Aidelsburger, M. Atala, S. Nascimbène, S. Trotzky, Y.-A. Chen, and I. Bloch, "Experimental realization of strong effective magnetic fields in an optical lattice," *Phys. Rev. Lett.* **107**, 255301 (2011).

[35] K. Jiménez-García, L. LeBlanc, R. Williams, M. Beeler, A. Perry, and I. Spielman, "Peierls substitution in an engineered lattice potential," *Phys. Rev. Lett.* **108**, 225303 (2012).

[36] J. Struck, C. Ölschläger, M. Weinberg, P. Hauke, J. Simonet, A. Eckardt, M. Lewenstein, K. Sengstock, and P. Windpassinger, "Tunable gauge potential for neutral and spinless particles in driven optical lattices," *Phys. Rev. Lett.* **108**, 225304 (2012).

[37] M. Aidelsburger, M. Atala, M. Lohse, J. T. Barreiro, B. Paredes, and I. Bloch, "Realization of the Hofstadter Hamiltonian with ultracold atoms in optical lattices," *Phys. Rev. Lett.* **111**, 185302 (2013).

[38] H. Miyake, G. A. Siviloglou, C. J. Kennedy, W. Cody Burton, and W. Ketterle, "Realizing the Harper Hamiltonian with laser-assisted tunneling in optical lattices," *Phys. Rev. Lett.* **111**, 185301 (2013).

[39] M. H. Anderson, J. R. Ensher, M. R. Matthews, C. E. Wieman, and E. A. Cornell, "Observation of Bose-Einstein condensation in a dilute atomic vapor," *Science* **269**(5221), 198–201 (1995).

[40] K. B. Davis, M.-O. Mewes, M.l A. Joffe, M. R. Andrews, and W. Ketterle, "Evaporative cooling of sodium atoms," *Phys. Rev. Lett.* **74**(26), 5202 (1995).

[41] C. C. Bradley, C. A. Sackett, J. J. Tollett, and R. G. Hulet, "Evidence of Bose-Einstein condensation in an atomic gas with attractive interactions," *Phys. Rev. Lett.* **75**, 1687 (1995).

[42] J. Stenger, S. Inouye, M. R. Andrews, H.-J. Miesner, D. M. Stamper-Kurn, and W. Ketterle, "Strongly enhanced inelastic collisions in a Bose-Einstein condensate near feshbach resonances," *Phys. Rev. Lett.* **82**, 2422 (1999).

[43] A. S. Arnold, C. S. Garvie, and E. Riis, "Large magnetic storage ring for Bose-Einstein condensates," *Phys. Rev. A* **73**, 041606 (2006).

[44] K. Huang, *Statistical Mechanics* (John Wiley & Sons, 1987).

[45] L. P. Pitaevskii and S. Stringari, *Bose-Einstein Condensation* (Clarendon Press, 2003).

[46] C. J. Pethick and H. Smith, *Bose-Einstein Condensation in Dilute Gases* (Cambridge University Press, 2001).

[47] S. Stenholm, "The semiclassical theory of laser cooling," *Rev. Mod. Phys.* **58**, 699 (1986).

[48] R. J. Cook, "Atomic motion in resonant radiation: an application of ehrenfest's theorem," *Phys. Rev. A* **20**, 224 (1979).

[49] C. Cohen-Tannoudji, J. Dupont-Roc, and G. Grynberg, *Atom-Photon Interactions* (John Wiley & Sons, 1998).

[50] P. Meystre, *Atom Optics* (American Institute of Physics, 2001).

[51] S. M. Barnett and P. M. Radmore, *Methods in Theoretical Quantum Optics* (Clarendon Press, Oxford, 1997).

[52] R. Loudon, *The Quantum Theory of Light* (Oxford University Press, 1979).

[53] N. Goldman, G. Juzeliūnas, P. Öhberg, and I. B. Spielman, "Light-induced gauge fields for ultracold atoms," arXiv:1308.6533 (2013).

[54] I. Bloch, "Ultracold quantum gases in optical lattices," *Nature Phys.* **1**(1), 23 (2005).

[55] C. Becker, P. Soltan-Panahi, J. Kronjäger, S. Dörscher, K. Bongs, and K. Sengstock, "Ultracold quantum gases in triangular optical lattices," *New. J. Phys.* **12**, 065025 (2010).

[56] Z. Lan, N. Goldman, and P. Öhberg, "Coexistence of spin-$\frac{1}{2}$ and spin-1 dirac-weyl fermions in the edge-centered honeycomb lattice," *Phys. Rev. B* **85**:155451 (2012).

[57] G. Whyte and J. Courtial, "Experimental demonstration of holographic three-dimensional light shaping using a gerchberg-saxton algorithm," *NJP* **7**, 117 (2005).

[58] A. L. Fetter, "Rotating trapped Bose-Einstein condensates," *Rev. Modern Phys.* **81**(2), 647 (2009).

[59] J. Ruostekoski, G. V. Dunne, and J. Javanainen, "Particle number fractionilization of an atomic fermi-dirac gas in an optical lattice," *Phys. Rev. Lett.* **88**, 180401 (2002).

[60] D. Jaksch and P. Zoller, "Creation of effective magnetic fields in optical lattices: the hofstadter butterfly for cold neutral atoms," *New J. Phys.* **5**, 56 (2003).

[61] E. J. Mueller, "Artificial electromagnetism for neutral atoms: Escher staircase and laughlin liquids," *Phys. Rev. A* **70**, 041603 (2004).

[62] F. Gerbier and J. Dalibard, "Gauge fields for ultracold atoms in optical superlattices," *New J. Phys.* **12**, 3007 (2010).

[63] J. Dalibard, F. Gerbier, G. Juzeliūnas, and P. Öhberg, "Colloquium : artificial gauge potentials for neutral atoms," *Rev. Mod. Phys.* **83**, 1523–1543 (2011).

[64] M. Aidelsburger, M. Atala, S. Nascimbène, S. Trotzky, Y.-A. Chen, and I. Bloch, "Experimental realization of strong effective magnetic fields in optical superlattice potentials," *Appl. Phys. B* doi:10.1007/s00340-013-5418-1 (2013).

[65] A. S. Sørensen, E. Demler, and M. D. Lukin, "Fractional quantum Hall states of atoms in optical lattices," *Phys. Rev. Lett.* **94**(8), 086803 (2005).

[66] T. Kitagawa, E. Berg, M. Rudner, and E. Demler, "Topological characterization of periodically driven quantum systems," *Phys. Rev. B* **82**(23), 235114 (2010).

[67] A. Eckardt, P. Hauke, P. Soltan-Panahi, C. Becker, K. Sengstock, and M. Lewenstein, "Frustrated quantum antiferromagnetism with ultracold bosons in a triangular lattice," *Europhys. Lett.* **89**, 10010 (2010).

[68] A. Hemmerich, "Effective time-independent description of optical lattices with periodic driving," *Phys. Rev. A* **81**, 063626 (2010).

[69] A. R Kolovsky, "Creating artificial magnetic fields for cold atoms by photon-assisted tunneling," *EPL (Europhysics Letters)* **93**, 20003 (2011).

[70] E. Arimondo, D. Ciampinia, A. Eckardtd, M. Holthause, and O. Morsch, "Kilohertz-driven bose–einstein condensates in optical lattices," *Adv. At. Molec. Opt. Phys.* **61**, 515 (2012).

[71] R. Dum and M. Olshanii, "Gauge structures in atom-laser interaction: Bloch oscillations in a dark lattice," *Phys. Rev. Lett.* **76**, 1788–1791 (1996).

[72] P. M. Visser and G. Nienhuis, "Geometric potentials for subrecoil dynamics," *Phys. Rev. A* **57**, 4581–4591 (1998).

[73] S. K. Dutta, B. K. Teo, and G. Raithel, "Tunneling dynamics and gauge potentials in optical lattices," *Phys. Rev. Lett.* **83**, 1934–1937 (1999).

[74] G. Juzeliūnas and P. Öhberg, "Slow light in degenerate Fermi gases," *Phys. Rev. Lett.* **93**(3), 033602 (2004).

[75] G. Juzeliūnas, P. Öhberg, J. Ruseckas, and A. Klein, "Effective magnetic fields in degenerate atomic gases induced by light beams with orbital angular momenta," *Phys. Rev. A* **71**(5), 053614 (2005).

[76] G. Juzeliūnas, J. Ruseckas, and P. Öhberg, "Effective magnetic fields induced by eit in ultra-cold atomic gases," *J. Phys. B. At. Mol. Opt. Phys.* **38**, 4171 (2005).

[77] P. Zhang, Y. Li, and C. P. Sun, "Induced magnetic monopole from trapped λ-type atom," *Eur. Phys. J. D* **36**, 229 (2005).

[78] G. Juzeliūnas, J. Ruseckas, P. Öhberg, and M. Fleischhauer, "Light-induced effective magnetic fields for ultracold atoms in planar geometries," *Phys. Rev. A* **73**, 025602 (2006).

[79] J. Ruseckas, G. Juzeliūnas, P. Öhberg, and M. Fleischhauer, "Non-abelian gauge potentials for ultracold atoms with degenerate dark states," *Phys. Rev. Lett.* **95**, 010404 (2005).

[80] K. Osterloh, M. Baig, L. Santos, P. Zoller, and M. Lewenstein, "Cold atoms in non-abelian gauge potentials: from the hofstadter "moth" to lattice gauge theory," *Phys. Rev. Lett.* **95**(1), 4 (2005).

[81] C. A. Mead and D. G. Truhlar, "On the determination of Born–Oppenheimer nuclear motion wave functions including complications due to conical intersections and identical nuclei," *J. Chem. Phys.* **70**(5), 2284–2296 (1979).

[82] M. Berry, "Quantal phase factors accompanying adiabatic changes," *Proc. R. Soc. London. Series A* **392**(1802), 45–57 (1984).

[83] J. Moody, A. Shapere, and F. Wilczek, "Realizations of magnetic-monopole gauge fields: diatoms and spin precession," *Phys. Rev. Lett.* **56**, 893 (1986).

[84] B. Zygelman, "Appearance of gauge potentials in atomic collision physics," *Phys. Lett A* **125**, 476 (1987).

[85] A. Zee, "Non-abelian gauge structure in nuclear quadrupole resonance," *Phys. Rev. A* **38**, 1 (1988).

[86] R. Jackiw, "Berry's phase-topological ideas from atomic, molecular and optical physics," *Comments At. Mol. Phys.* **21**, 71 (1988).

[87] A. Shapere and F. Wilczek, eds, *Geometric Phases in Physics* (World Scientific, Singapore, 1989).

[88] C. Mead, "The geometric phase in molecular systems," *Rev. Modern Phys.* **64**(1), 51–85 (1992).

[89] A. Bohm, A. Mostafazadeh, H. Koizumi, Q. Niu, and J. Zwanziger, *Geometric Phases in Quantum Systems* Springer, Berlin, Heidelberg, New York, 2003).

[90] D. Xiao, M.-C. Chang, and Q. Niu, "Berry phase effects on electronic properties," *Rev. Mod. Phys.* **82**, 1959 (2010).

[91] G. Grynberg and C. Robilliard, "Cold atoms in dissipative optical lattices," *Phys. Rep.* **355**, 335 (2001).

[92] A. M. Dudarev, R. B. Diener, I. Carusotto, and Q. Niu, "Spin-orbit coupling and Berry phase with ultracold atoms in 2D optical lattices," *Phys. Rev. Lett.* **92**(15), 153005 (2004).

[93] S.-L. Zhu, H. Fu, C.-J. Wu, S.-C. Zhang, and L.-M. Duan, "Spin Hall effects for cold atoms in a light-induced gauge potential," *Phys. Rev. Lett.* **97**, 240401 (2006).

[94] K. J. Günter, M. Cheneau, T. Yefsah, S. P. Rath, and J. Dalibard, "Practical scheme for a light-induced gauge field in an atomic Bose gas," *Phys. Rev. A* **79**(1), 011604 (2009).

[95] I. B. Spielman, "Raman processes and effective gauge potentials," *Phys. Rev. A* **79**, 063613 (2009).

[96] W. Happer and B. S. Mathur, "Effective operator formalism in optical pumping," *Phys. Rev.* **163**(1), 12 (1967).

[97] I. H. Deutsch and P. S. Jessen, "Quantum-state control in optical lattices," *Phys. Rev. A* **57**(3), 1972–1986 (1998).

[98] J. Sebby-Strabley, M. Anderlini, P. S. Jessen, and J. V. Porto, "Lattice of double wells for manipulating pairs of cold atoms," *Phys. Rev. A* **73**, 033605 (2006).

[99] G. Juzeliūnas and I. B. Spielman, "Flux lattices reformulated," *New. J. Phys.* **14**, 123022 (2012).

[100] Y. Aharonov and A. Stern, "Origin of the geometric forces accompanying Berry's geometric potentials," *Phys. Rev. Lett.* **69**(25), 3593–3597 (1992).

[101] M. Cheneau, S. P. Rath, T. Yefsah, K. J. Gunter, G. Juzeliūnas, and J. Dalibard, "Geometric potentials in quantum optics: a semi-classical interpretation," *EPL (Europhysics Letters)* **83**(6), 60001 (2008).

[102] R. G. Unanyan, M. Fleischhauer, B. W. Shore, and K. Bergmann, "Robust creation and phase-sensitive probing of superposition states via stimulated raman adiabatic passage (STIRAP) with degenerate dark states," *Opt. Commun.* **155**, 144 (1998).

[103] R. G. Unanyan, B. W. Shore, and K. Bergmann, "Laser-driven population transfer in four-level atoms: consequences of non-abelian geometrical adiabatic phase factors," *Phys. Rev. A* **59**, 2910 (1999).

[104] J.-J. Song, B. A. Foreman, X.-J. Liu, and C. H. Oh, "Wave function engineering and circulating spin current in trapped Bose-Einstein condensates," *Europhys. Lett.* **84**, 20012 (2008).

[105] N. R. Cooper and Z. Hadzibabic, "Measuring the superfluid fraction of an ultracold atomic gas," *Phys. Rev. Lett.* **104**(3), 030401 (2010).

[106] E. Arimondo, "Coherent population trapping in laser spectroscopy," in *Progress in Optics*, edited by E. Wolf (Elsevier, 1996), Vol. 35, p. 259.

[107] S. E. Harris, "Electromagnetically induced transparency," *Phys. Today* **50**, 36 (1997).

[108] M. D. Lukin, "Colloquium: trapping and manipulating photon states in atomic ensembles," *Rev. Mod. Phys.* **75**(2), 457–472 (2003).

[109] M. Fleischhauer, A. Imamoglu, and J. P. Marangos, "Electromagnetically induced transparency: optics in coherent media," *Rev. Mod. Phys.* **77**, 633 (2005).

[110] K. Bergmann, H. Theuer, and B. W. Shore, "Coherent population transfer among quantum states of atoms and molecules," *Rev. Mod. Phys.* **70**(3), 1003–1025 (1998).

[111] N. V. Vitanov, M. Fleischhauer, B. W. Shore, and K. Bergmann, "Coherent manipulation of atoms and molecules by sequential pulses," *Adv. At. Mol. Opt. Phys.* **46**, 55 (2001).

[112] P. Král, I. Thanopulos, and M. Shapiro, "Colloquium: coherently controlled adiabatic passage," *Rev. Mod. Phys.* **79**, 53 (2007).

[113] Y. Li, C. Bruder, and C. P. Sun, "Generalized stern-gerlach effect for chiral molecules," *Phys. Rev. Lett.* **99**, 130403 (2007).

[114] T. D. Stanescu and V. Galitski, "Spin relaxation in a generic two-dimensional spin-orbit coupled system," *Phys. Rev. B* **75**(12), 125307 (2007).

[115] A. Jacob, P. Öhberg, G. Juzeliūnas, and L. Santos, "Cold atom dynamics in non-abelian gauge fields," *Appl. Phys. B* **89**, 439 (2007).

[116] G. Juzeliūnas, J. Ruseckas, M. Lindberg, L. Santos, and P. Öhberg, "Quasirelativistic behavior of cold atoms in light fields," *Phys. Rev. A* **77**, 011802(R) (2008).

[117] T. D. Stanescu, B. Anderson, and V. Galitski, "Spin-orbit coupled Bose-Einstein condensates," *Phys. Rev. A* **78**, 023616 (2008).

[118] J. Y. Vaishnav and C. W. Clark, "Observing Zitterbewegung with ultracold atoms," *Phys. Phys. Lett.* **100**, 153002 (2008).

[119] G. Juzeliūnas, J. Ruseckas, and J. Dalibard, "Generalized Rashba-Dresselhaus spin-orbit coupling for cold atoms," *Phys. Rev. A* **81**, 053403 (2010).

[120] D. L. Andrews, ed., "Optical Manipulation of Ultracold Atoms," *in Structured Light and Its Applications: An Introduction to Phase-Structured Beams and Nanoscale Optical Forces* (Elsevier, Amsterdam 2008), pp. 295–333.

[121] M. A. Ol'shanii and V. G. Minogin, "Three-dimensional velocity-selective coherent population trapping of (3+1)-level atoms," *Quantum Opt.* **3**, 317 (1991).

[122] S. Kulin, Y. Castin, M. Ol'shanii, E. Peik, B. Saubaméa, M. Leduc, and C. Cohen-Tannoudji, "Exotic quantum dark states," *Eur. Phys. J. D* **7**, 279 (1999).

[123] R. G. Unanyan and M. Fleischhauer, "Geometric phase gate without dynamical phases," *Phy. Rev. A* **69**, 050302(R) (2004).

[124] H. Theuer, R. G. Unanyan, C. Habscheid, K. Klein, and K. Bergmann, "Novel laser controlled variable matter wave beamsplitter," *Opt. Express* **4**, 77 (1999).

[125] F. Vewinger, M. Heinz, R. G. Fernandez, N. V. Vitanov, and K. Bergmann, "Creation and measurement of a coherent superposition of quantum states," *Phys. Lett. Lett.* **91**, 213001 (2003).

[126] J.-J. Song and B. A. Foreman, "Persistent currents in cold atoms," *Phys. Rev. A* **80**, 033602 (2009).

[127] V. Pietilä and M. Möttönen, "Non-abelian magnetic monopole in a Bose-Einstein condensate," *Phys. Lett. Lett.* **102**, 080403 (2009).

[128] A. H. Castro Neto, F. Guinea, N. M. R. Peres, K. S. Novoselov, and A. K. Geim, "The electronic properties of graphene," *Rev. Modern Phys.* **81**, 109 (2009).

[129] H. Zhai, "Spin-orbit coupled quantum gases," *Int. J. Modern Phys. B* **26**, 1230001 (2012).

[130] V. Galitski and I. B. Spielman, "Spin–orbit coupling in quantum gases," *Nature* **494**, 49–54 (2013).

[131] D. L. Campbell, G. Juzeliūnas, and I. B. Spielman, "Realistic Rashba and Dresselhaus spin-orbit coupling for neutral atoms," *Phys. Rev. A* **84**, 025602 (2011).

[132] D. Banerjee, M. Dalmonte, M. Müller, E. Rico, P. Stebler, U.-J. Wiese, and P. Zoller, "Atomic quantum simulation of dynamical gauge fields coupled to fermionic matter: from string breaking to evolution after a quench," *Phys. Rev. Lett.* **109**, 175302 (2012).

[133] L. Tagliacozzo, A. Celi, A. Zamora, and M. Lewenstein, "Optical abelian lattice gauge theories," *Ann. Phys.* **330**, 160–191 (2013).

[134] E. Zohar, J. I. Cirac, and B. Reznik, "Cold-atom quantum simulator for su(2) yang-mills lattice gauge theory," *Phys. Rev. Lett.* **110**, 125304 (2013).

INDEX

Note: Page numbers followed by f indicate figures.

Abelian, 384
Acrylic acid (AA), 153
Active functions, 3
Active tuning, 12
Add/drop filters, 157–158
Adhesive bonding, 13
Adiabatic approximation, 383
Adiabatic hot-electron nanoscopy, 118, 121–125, 123f, 125f
    hot electrons, adiabatic concentration of, 118–121, 120f–121f
    optical energy, adiabatic concentration of, 118–121, 120f–121f
AFM, *see* Atomic force microscope (AFM)
Alcohol, 281
ALD, *see* Atomic layer deposition (ALD)
Alq$_3$, 224–226, 225f
    chemical structure, 224
    isotope effect in MEL, 229–232, 231f–232f
Amplification in NIMs, 57–57f
Amplitude regeneration, 30

Analogies, photonic crystals (PC)
    quantum mechanics, 136–137
    semiconductors, 137
Annealing, 358
Arrayed waveguide grating (AWG), 5–6f
Artificial electromagnetism, optical control, 371–392
Artificial magnetic flux, 380
Artificial photosynthesis, 297
Atomic Bose–Einstein condensates, 372–374
Atomic force microscope (AFM), 122–123f
Atomic Hamiltonian, 381
Atomic layer deposition (ALD), 264–264f
Atomic quantum gases, 372
Atom–light coupling, 381–382
AWG, *see* Arrayed waveguide grating (AWG)

Background limited performance (BLIP), 189, 197
Backward Cerenkov radiation, 58

Photonics: Scientific Foundations, Technology and Applications, Volume II, First Edition.
Edited by David L. Andrews.
© 2015 John Wiley & Sons, Inc. Published 2015 by John Wiley & Sons, Inc.

INDEX

Bardeen–Cooper–Schrieffer crossover, 371
Batteries, 266
BCP, see Block copolymers (BCPs)
BDP–ZnP–C_{60} supramolecular triad, 323
β-factor (spontaneous emission), 44
BLIP, see Background limited performance (BLIP)
Bloch–Floquet theorem, 135
Bloch modes, 40
Block copolymers (BCPs), 149
BODIPY–C_{60} dyad, 322
BODIPY–fullerene systems, 321–323
Bogoliubov approximation, 375
Boron-dipyrromethenes (BODIPYs), 313
Bose–Einstein condensates, 371–373
Bosonic atoms, 373
Bosonic commutation rules, 374
Bowtie spaser (nanolaser), 90
Bragg scattering, 133
Bragg stacks, 23
Brillouin zone, 142f–143f
Bulk heterojunction (BHJ) solar cells, 324

CamIRa (infrared FPA evaluation system), 185
Cancer therapy, 301–304
Carbon nanotubes, 261
Carrier-based silicon modulators, 8
Cavity-based switches, 26–32, 27f
Cavity decays, 24, 30, 35–39
Cavity-enhanced LED, 42–42f
Cavity modes, 24–26
  and Green's tensor, 33
Cavity optomechanics, 21–22
Cavity photonics
  cavity-based switches, 26–32, 27f
  emitters
    Purcell effect (weak coupling), 32–34
    vacuum Rabi oscillations (strong coupling), 35–42, 35f, 40f
  fundamentals, 22–26
  light emitting diodes (LEDs), 42–46
  nanocavity lasers, 42–46
C_{60}-based films, spectroscopy of, 242–246, 244f–246f
Center for Quantum Devices (CQD), 176
Chemical vapor deposition (CVD), 358
CMT, see Coupled mode theory (CMT)
Colloidal crystals, 146–148, 147f

Colloidal gels, 148
Colloidal quantum dots (CQD), 215–216
Colloidal semiconductor nanocrystal, 215
Composite materials for PEC water splitting
  metal nanoparticle-enhanced PEC, 353–356
  mixed metal oxides, 348–349
  quantum dot sensitization, 349–353
Conduction band (CB), 346
Λ configuration, 385
Continuous wave (CW)
  equations, 98–101
  spasers in, 102–104, 103f
  spasing, 90
Conversion efficiency, see Quantum efficiency (QE)
Co-propagating Gaussian beam, 77–77f
Coupled colloidal system, 351
Coupled mode theory (CMT), 28, 31f
Coupled-resonator optical waveguide (CROW), 155
Covalent cyanine–fullerene systems, 320–321
CROW, see Coupled-resonator optical waveguide (CROW)
Cyanine–C_{60} derivatives, 321
Cyanine–cyanine salt, 329f
Cyanine–fullerene dyad, 320
Cyanines, 313, 320
Cyclopropanation reaction, 326

0D, see Zero-dimensional (0D)
1D, see One dimension (1D)
2D, see Two dimensions (2D)
3D, see Three dimensions (3D)
4,4′-Diaminodiphenyl ether (DDE), 265
1D nanostructures, 357–358
Dark current, 173–173f
Dark state, 385–386
DBGS, see Direct bandgap semiconductors (DBGSs)
Degree of electron localization, 273
Delocalization parameter of polaron wavefunction, 275f
Densities of state (DOS), 43–44, 44f, 137, 140
  quantum dots, 169–170, 170f
Detector metrics, 174–175
DFB spaser, 117

Diketopyrrolopyrrole (DPP), 328
Diphenylmethanofullerene (DPM), 327
Direct bandgap semiconductors (DBGSs), 101–101f
Directional couplers, 158
Direct laser writing (DLW), 150
Direct write assembly, photonic crystals (PC), 148
Distributed feedback (DFB)
  laser, 117
  modes, 26
Ditolylaminothienyl-dicyanovinylen-benzothiadiazole (DTDCTB), 329
Donor–bridge–acceptor (DBA) systems, 331
Doped metal oxides, 358–359
  doping hematite, 359
  doping TiO_2 and ZnO, 359–360
  hydrogen treatment, 360
  QD sensitized doped metal oxides, 361–362
Doping, 358–359
  hematite, 359
  nitrogen, 360
  TiO_2 and ZnO, 359–360
DOS, see Densities of state (DOS)
Dot-in-a-well (DWELL), 209
Double-negative materials, 54
Double-positive materials, 54
$DPP(TBFu)_2$, 328f
Dressed states, 382
Drude–Lorentz model, 62
Dye molecules, 278
Dynamic molecular assembling, 263

Echelle grating, 6f
Edge-centered honeycomb lattice, 379
Edge couplings, 7–7f
Effective mode volume, 25
Effective vector potential, 384
EFS, see Equifrequency surface (EFS)
EL, see Electroluminescence (EL) intensity
Electroluminescence (EL) intensity, 221–222
Electromagnetically induced transparency (EIT), 386
Electromagnetic (EM) concentrator, 72–73
Electromagnetism, 376

Electron injection efficiency, 281
Electron transfer, 332
π-electron wavefunctions, 269
Electro-optic conversion, 8–9
Electro-optic thin films, 300–301
Electrostatic potential, 317
Emitters in cavities
  Purcell effect (weak coupling), 32–34
  vacuum Rabi oscillations (strong coupling), 35–42
Energy cascade architecture, 312
Equifrequency surface (EFS), 143–144, 144f
$ExTTF-C_{60}$ dyad, 315f
$ExTTF_1-exTTF_2-C_{60}$ triad, 316f
$ExTTF/\pi$-conjugated wire/C_{60} systems, 333f

Fabry–Perot cavity, 23f, 25–26
Fano effects, 32
$Fc-BDP-C_{60}$ triad, 322
$Fc-ZnP-H_2P-C_{60}$ tetrad, 314f
FDTD method, 286
Fermi's Golden Rule, 32
Feshbach resonances, 373
Fiber couplings, 6–8, 7f
Fiber optics, 59–60
Fibers, 59
Figure of merit (FOM), 56, 67–68, 68f
Film-based integrated solar cell, 296
Fishnet
  MM, 66–67f
  NIM, 56–56f
  structure, 56
Flat photonics, 75
Focal plane array (FPA), 175, 208–212, 213f
  fabrication, 185
  imagers, 175
FOM, see Figure of merit (FOM)
Fourier transform infrared (FTIR) spectrometer, 184–185
Fowler's formula, probability, 124
Fullerenes, 334
  for organic photovoltaics, 323–330
Fundamental frequency (FF), NIM, 58–59

Gain (or active) medium, spaser, 87–88f, 89
Generation–recombination (GR) noise, 174

Germanium, 13
Glancing angle deposition (GLAD), 146
Δg mechanism, 222
Gold nanoparticles (AuNPs), 354
Graded-index
  metasurface, 76–77
  transition metamaterials, 62–70
Graphene, 261
Green's tensor, 33–34
Gross–Pitaevskii equation, 375
Group velocity effects, photonic crystals (PC), 143–145

2-Hydroxy-ethyl methacrylate (HEMA), 153
Half width at half maximum (HWHM), 226, 247
Halogen reactions, 265
Hamiltonian, 374, 376, 385–386
Hamilton receptor, 335
Helmholtz coil-pair, 233f
Helmholtz equation, 390
HEMA, *see* 2-Hydroxy-ethyl methacrylate (HEMA)
Hematite (α-Fe_2O_3), 347
  thin film, 355f
HFI, *see* Hyperfine interaction (HFI)
Higgs modes, 371
Highest occupied molecular orbital (HOMO), 292, 312
High operating temperature FPA, 212–215
High operating temperature quantum dot detector, 208–212
HWHM, *see* Half width at half maximum (HWHM)
Hydrogen treatment, 360
Hyperfine interaction (HFI), 222, 226, 228–230, 234
Hyperlens, 57f, 58

IL, *see* Interference lithography (IL)
InAs/InP QDIP, 199–203, 200f–202f
InGaAs/InGaP QDIP, 186–187, 186f–187f, 190–193, 190f–191f
InGaN nanospaser, 92–93f, 94
In-line coupled cavity, 28
Interconnect applications, 1–2, 2f
Interference lithography (IL), 150

Interstitial doping, 358
Intersubband (ISB) devices, 171
  dark current mechanisms, 173–173f
  quantum dots benefits for, 176–179
    high gain, 177
    low dark current, 177–178
    normal incidence absorption, 178–178f
    phonon bottleneck, 177
    versatility, 178–179
Inverse opal hydrogel PCs, 153

Jaynes–Cummings model, 37

Kerr coefficient, 28

Large-core waveguides, 4–4f
Lasers, 9–10, 13–14, 14f, 42f, 157
Lasing spaser, 86–87
LDOS, *see* Local density of optical states (LDOS)
Left-handed materials (LHM), 54
Light emitting diodes (LEDs), 42–46, 42f
Light-harvesting antennas, 298
Light-harvesting materials for organic electronics, 311–313
  fullerenes for organic photovoltaics, 323–330
  molecular wires, 330–331
    covalent donor–bridge–acceptor systems, 331–334
    supramolecular donor–bridge–acceptor systems, 334–335
  PET in artificial photosynthetic systems, 313
    covalent and supramolecular BODIPY–fullerene systems, 321–323
    covalent and supramolecular phthalocyanine–fullerene systems, 317–320
    covalent and supramolecular porphyrin–fullerene systems, 313–315
    covalent and supramolecular TTF–fullerene systems, 315–317
    covalent cyanine–fullerene systems, 320–321
Liquid-phase MLD (LP-MLD), 265

Local density of optical states (LDOS), 32–33
Long-wavelength infrared (LWIR) photodetectors, 175
Low dark current, 177–178
Lowest unoccupied molecular orbital (LUMO), 312
LP-MLD with molecule-switching sequences, 277f
Luminescence-assisted SOLNET (LA-SOLNET), 285

2-Methyl-4-nitroaniline (MNA) single crystal, 300
N-methylpyrrolidone diphenylacetylenic bridge, 321
Mach–Zehnder interferometers (MZIs), 5–6f
Magnetic field effects (MFE), 222, 227–228
    in Alq_3-based OLED devices, 229–231
    on excited state spectroscopies in organic semiconductor films, 236–246, 237f
        PP in MEH-PPV system, 238–242, 239f, 241f
        spectroscopy of C_{60}-based films, 242–246, 244f–246f
        TE in MEH-PPV system, 238–242, 239f, 241f
Magnetic metamaterials, 59–62, 61f
Magneto-conductance (MC), 222, 233f
    compass action in, 234–235
Magneto-electroluminescence (MEL), 222
    compass action in, 234–235
    defined, 223
    isotope effect in, 226–232
    measurements, experimental setup, 233f
    in organic light emitting diodes, 222–232
        π-conjugated polymer (polyfluorene), 224–226
        poly(dioctyloxy) phenyl vinylene (DOO-PPV), 226–229, 227f–229f
        small molecule (Alq_3), 224–226, 224f–225f, 231f–232f
        at small magnetic fields, 232–236, 234f–235f
Manley–Rowe relation, 59

Maxwell's equations, 62–63, 71
    and photonic band gap (PBG), 134
MBE, see Molecular beam epitaxy (MBE)
MC, see Magneto-conductance (MC)
MEH-PPV, 238
    backbone structure, 237f
MEL, see Magneto-electroluminescence (MEL)
Meta-atoms, 53
Metallic PCs, 151–152
Metal nanoparticle-enhanced PEC, 353–356
Metal oxide-based photoelectrochemical hydrogen production, 343–344
    composite materials for PEC water splitting
        metal nanoparticle-enhanced PEC, 353–356
        mixed metal oxides, 348–349
        quantum dot sensitization, 349–353
    doped metal oxides, 358–359
        doping hematite, 359
        doping TiO_2 and ZnO, 359–360
        hydrogen treatment, 360
        QD sensitized doped metal oxides, 361–362
    materials for, 346
        Hematite (α-Fe_2O_3), 347
        TiO_2, 346
        WO_3, 347
        ZnO, 347
    materials for PEC water splitting
        WO_3, 347
    nanostructured photoanodes, 356
        nanoparticle films, 356–358
    PEC hydrogen generation, 344–346
Metal-oxide semiconductor field effect transistor (MOSFET), 86
Metamaterials (MMs)
    and fiber optics, 59–60
    graded-index transition metamaterials, 62–70
    introduction, 53–54
    magnetic metamaterials, 59–62, 61f
    metasurfaces, 75–78, 76f–77f
    negative-index materials, 54–59, 55f–56f
    transformation optics, 70–74, 71f
Metasurfaces, 75–78, 76f–77f
    use, 77f, 78

Microelectromechanical systems (MEMSs), 266
Microfabrication techniques, photonic crystals, 145–146
Micropillar cavity, 22–23, 23f
Microring structures, 21
Mid-wavelength infrared (MWIR) photodetectors, 175
MIT Photonic Bands (MPB) package, 135
Mixed metal oxides, 348–349
MLD, see Molecular layer deposition (MLD)
MM, see Metamaterials (MMs)
Modified PCBM, 327f
Modulators, 11–12
Moecular orbital (MO) method, 269
Molecular assembling
  dynamic molecular assembling, 263
  static molecular assembling, 262–263
Molecular beam epitaxy (MBE), 264
Molecular bonding, 13
Molecular hydrogen, 344
Molecular layer deposition (MLD), 261, 264–266, 264f–266f
  carrier-gas type, 267f
Molecular MQDs, 265–283
  sensitization, 295f
Molecular wires, 330–331
  covalent donor–bridge–acceptor systems, 331–334
  supramolecular donor–bridge–acceptor systems, 334–335
Monostable spaser, 105–106
  dynamics, 107f
MOSFET, see Metal-oxide semiconductor field effect transistor (MOSFET)
Moss rule, semiconductors, 146
MQD, see Multiple quantum dot (MQD)
Multiple quantum dot (MQD), 261–262
Multi-step excitation of electrons, 294
Multistep heterostructure, 40
MZI, see Mach–Zehnder interferometers (MZIs)

Nanocavity lasers, 42–46
Nanocavity LED, 45–46f
Nanofocusing, 122
Nanolasers, see Surface plasmon polaritons (SPPs), spasers
Nanolens spaser, 90
Nanoparticle films, 356–358
Nanoplasmonics, 85
  applications, 85–86
Nanoscale amplifier, 86
Nanoscale pattern formation, 264f
Nanoshell spaser, 87–88f, 89
Nanospasers, 86
Nanostructured photoanodes, 356
  nanoparticle films, 356–358
Naphthalenediimide (NDI), 318
Near-field optical microscopy, 59
Near-field scanning optical microscope (NSOM)
  aperture, 118
  apertureless, 118
NEDT, see Noise equivalent difference temperature (NEDT)
Negative-index materials (NIMs), 54–59, 55f–57f
  amplification in, 57–57f
  transition layer, 63f
NIM, see Negative-index materials (NIMs)
Nitrogen doping, 360
Noise, 174
Noise equivalent difference temperature (NEDT), 175, 213–214, 214f
Non-Abelian gauge potentials, 387–392
Nonlinear cavity, 29–30
  bistability in, 29–30, 29f
Nonlinear concentrator, 72–73, 72f
Nonlinearity, spaser equations, 97
Nonperiodic defect region, 23f, 24
Normal incidence absorption, QDIPs, 178–178f
NSOM, see Near-field scanning optical microscope (NSOM)
N-type dye, 276

p-Oligophenyleneethynylene (p-oPPE), 331
p-Oligophenylenevinylene (p-oPPV), 331
OLED, see Organic light emitting diodes (OLEDs)
Oligofluorene (oFL), 331
On-chip spectrometer, 3–3f
One dimension (1D), 133
  photonic crystals, 137–140, 138f
Optical antenna array, 77f

Optical field, cavity, 24
Optical forces on atoms, 376–380
  Abelian gauge potentials, 385–387
    artificial magnetic field and gauge potentials for ultracold atoms, 380–381
      atom–light coupling, 381–382
      dressed-state representation, 382–383
      reduced atomic dynamics, 383–384
  non-Abelian gauge potentials, 387–392
Opto-electronic conversion, 8–9
Optoelectronic (OE) boards, 298
Organic electronics, light-harvesting materials for, 311–313
  fullerenes for organic photovoltaics, 323–330
  molecular wires, 330–331
    covalent donor–bridge–acceptor systems, 331–334
    supramolecular donor–bridge–acceptor systems, 334–335
  PET in artificial photosynthetic systems, 313
    covalent and supramolecular BODIPY–fullerene systems, 321–323
    covalent and supramolecular phthalocyanine–fullerene systems, 317–320
    covalent and supramolecular porphyrin–fullerene systems, 313–315
    covalent and supramolecular TTF–fullerene systems, 315–317
    covalent cyanine–fullerene systems, 320–321
Organic light emitting diodes (OLEDs), 221, 266
  magneto-electroluminescence (MEL) in, 222–232
    π-conjugated polymer (polyfluorene), 224–226, 224f–225f
    poly(dioctyloxy) phenyl vinylene (DOO-PPV), 226–229, 227f–229f
    small molecule (Alq_3), 224–226, 229–232
  structure, 222–223f

Organic MQD, 267
  molecular MQDs, 265–283
  polymer MQDs, 267–275
Organic MQD sensitization, 293–294
Organic thin films
  dimensionality in, 262f
  electron wavefunction shapes in, 263f
Oxygen evolution reaction (OER), 344

[6,6]-phenyl-C61 butyric acid methyl ester (PCBM), 325
p-Phenylenediamine (PPDA), 267
Passive functions, 3
Passive silicon waveguide circuits, 10–11, 10f
PBG, see Photonic band gap (PBG)
PC, see Photonic crystals (PC)
PC fibers, 158–159
π-Conjugated polymer (polyfluorene), 224–226
PC slab, 154
PC waveguide bend, 154f
PDMS, see Poly(dimethylsiloxane) (PDMS)
PEC hydrogen generation, 344–346
PET in artificial photosynthetic systems, 313
  covalent and supramolecular BODIPY–fullerene systems, 321–323
  covalent and supramolecular phthalocyanine–fullerene systems, 317–320
  covalent and supramolecular porphyrin–fullerene systems, 313–315
  covalent and supramolecular TTF–fullerene systems, 315–317
  covalent cyanine–fullerene systems, 320–321
Phonon bottleneck, 170–171, 177
Photo- and electroactive pentad $SiPc–(NDI)_2–(C_{60})_2$, 319f
Photo-assisted processes, 265
Photocurrent, 172–173, 172f
  generation in ZnO, 281f
Photodetectors, 13
Photoelectrochemical (PEC) water, 344
Photo-induced charge separation process, 312f

408 INDEX

Photoinduced electron transfer (PET), 312
Photoluminescence (PL) measurements, 194
Photonic band gap (PBG), 133, 138f, 140–143
  transverse electric (TE) modes, 143f
  transverse magnetic (TM) modes, 143f
Photonic crystal defect cavity, 23–24, 23f, 26
Photonic crystals (PC), 133
  applications, 154–159
    add/drop filters, 157–158
    directional couplers, 158
    fundamental effects, 155–157
    lasers, 157
    PC fibers, 158–159
    sensors, 157
  chemical techniques (for fabrication)
    colloidal crystals, 146–148
    direct write assembly, 148
    polymers, 149
  fundamentals, 134–145
    analogies, 136–137
    1D PCs, 137–140, 138f
    2D PCs, 140–143, 142f
    3D PCs, 140–143, 142f
    group velocity effects, 143–145
  lithography techniques (for fabrication)
    direct laser writing (DLW), 150
    interference lithography (IL), 150
    soft lithography, 149
  metallic PCs, 151–152
  microfabrication techniques, 145–146
  one dimension, 134f, 137–140, 138f
  physical techniques (for fabrication), 146
  structure, 24, 26, 134–134f
  three dimensions, 134f, 140–143
  tunable PCs, 152–154
  two dimensions, 134f, 140–143, 142f, 154f
Photonic crystal switches, 26–32, 27f
Photonic–electronic integration, 14–15
Photonic integrated circuits (PICs), 1
Photonic sensors, 2–3, 3f
Photonic wires, 4–4f
Photopolymer, waveguide construction in, 286f
Photosynthesis, 297–298
Photovoltaic (PV) solar cells, 323

Phthalocyanine–fullerene systems, 317–320
Phthalocyanines (Pcs), 313, 317
Physical vapor deposition (PVD), 146
PIC, *see* Photonic integrated circuits (PICs)
PIM, *see* Positive-index materials (PIMs)
Planar concave grating, 6f
Polarization, 382
Polaron pair (PP), 222
  in MEH-PPV system, 238–242
Poly-AM quantum wire, 271f
Poly-diacetylene (PDA) backbones, 301
Poly(dimethylsiloxane) (PDMS), 149
Poly(dioctyloxy) phenyl vinylene (DOO-PPV), 226–229, 227f–229f
Polymer MQDs, 267–275
  absorption spectra of poly-AM quantum wire and, 270f
  growth of, 268f
  MLD process for, 268f
  PL spectra of a poly-AM quantum wire and, 264f
  poly-AM quantum wire and, 269f
  quantum confinement effect in, 269–272
Polymers, 149
Poly-oxadiazole (OXD), 304
Poly(phenylene vinylene) (PPV), 238
Polyvinyl alcohol (PVA), 288
Porphyrins, 313–314
Porphyrin–fullerene systems, 313–315
Positive-index materials (PIMs), 54, 58–59
  transition layer, 63f
Power conversion efficiency (PCE), 324
Poynting vector, 54, 66
PP, *see* Polaron pair (PP)
PPV, *see* Poly(phenylene vinylene) (PPV)
Pristine MEH-PPV films, 238–239f
Probe pulse transmission, 31f
Propagating wave, 62–63, 63f
p-Type dye, 276
p-Type QWIPs, 176
Purcell effect (weak coupling), 32–34
Purcell-enhanced emission, 37, 44–45
Purcell factor, 32–34
PVD, *see* Physical vapor deposition (PVD)
Pyromellitic dianhydride (PMDA), 265

QD, *see* Quantum dots (QDs)
3QD, 272
  molecular structure and electron density for, 273f

QDIP, *see* Quantum dot infrared detectors
QD sensitized doped metal oxides, 361–362
QE, *see* Quantum efficiency (QE)
Quality *(Q)* factor, cavity, 24, 26, 39–40
Quantum boxes, *see* Quantum dots (QDs)
Quantum density matrix equations (optical Bloch equations), spasers, 95–98
Quantum dots (QDs), 34, 40–41, 41f, 43, 262
   absorption peak position, 270f
   colloidal quantum dots (CQD), 215–216
   densities of state (DOS), 169–170, 170f
   device fabrication, 184–185
   electron localization in, 272–275
   fabrication techniques, 170–171
   gallium arsenide–based detectors
     first QDIP FPA, 187–193, 188f–191f
     InGaAs/InGaP QDIP, 186–187, 186f–187f, 190–193, 190f–191f
     two temperature barrier growth for morphology, 193–198, 195f–196f
   growth, 179–184
     SK grown dots effect on QDIP, 180–184
     SK growth mode, 180
   indium phosphide-based detectors, 198–215
     focal plane array, 208–212, 213f
     high operating temperature FPA, 212–215
     high operating temperature quantum dot detector, 208–212
     InAs/InP QDIP, 199–203, 200f–202f
     wavelength tuning detection using quantum dot engineering, 203–208, 204f
   for infrared detection, 175–179
     intersubband detectors, 176–179
     potential of QDIPs, 179
   introduction, 169–175
   measurement procedures, 184–185
   with poly-diacetylene backbones, 301f
   properties
     density, 181–183
     shape, 181
     size, 181
     uniformity, 183–184
   sensitization, 349–353, 358

Quantum dot infrared detectors (QDIPs), 171–172, 176, 187–193, 188f–189f
   band structure, 172f
   dark current, 173–173f
   detector metrics, 174–175
   device structure, 171f
   noise, 174
   photocurrent, 172–173, 172f
   potential, 179
Quantum efficiency (QE), 174
Quantum electrodynamic (QED) effects, 155–156
Quantum Hall effect, 372, 380
Quantum mechanical models, 246–254
   spin-dependent recombination, 251–253, 252f
   spin quantum mechanical approach, 248–251, 250f–251f
   triplet–triplet annihilation (TTA), 253–254
Quantum plasmonics
   adiabatic hot-electron nanoscopy
     adiabatic hot-electron nanoscope, 121–125
     hot electrons, adiabatic concentration of, 118–121
     introduction, 118
     optical energy, adiabatic concentration of, 118–121
   spasers
     active development, 90–95
     compensation of loss by gain and spasing, 109–112
     conditions of loss compensation by gain and spasing, 112–118
     in CW regime, 102–104
     equations, 95–101
     fundamentals, 87–90
     introduction, 86–87
     as ultrafast quantum nanoamplifier, 104–109
Quantum rods (QRs), 352
Quantum well infrared photodetectors (QWIPs), 171–172, 171f, 175–176
   band structure, 172f
   dark current, 173–173f
   detector metrics, 174–175
   device structure, 171f

Quantum well infrared photodetectors (QWIPs) (*Continued*)
  noise, 174
  photocurrent, 172–173, 172f
Quantum wells (QWL), 169
Quantum wires (QWR), 169
Quasiclassical approximation, *see* Wentzel-Kramers-Brillouin (WKB)
Q values, cavity, 24, 35
QWIP, *see* Quantum well infrared photodetectors (QWIPs)
QWL, *see* Quantum wells (QWL)

Rabi frequency, 377
Radiation pressure, 378
Rashba–Dresselhaus spin–orbit coupling, 391
Readout integrated circuit (ROIC), 185
Refractive index, 8, 29–31, 53, 73
  negative, 55–56
  silicon, 8
Relative orbital angular momentum, 387
Resonator, 87
Reversed Doppler effect, 58
Richard Feynman's vision, 372
Right-handed materials (RHMs), 54
Ring-resonator-based biosensor, 2–3, 3f
Ring resonators, 5–6f
ROIC, *see* Readout integrated circuit (ROIC)
Rotating wave approximation (RWA), 96
Routing, optical functions, 3–5
R-SOLNET, 284–291
  with luminescent targets, 285, 287–288
  optical solder of, 299
  simulation for, with luminescent targets, 286f
  with wavelength filters, 284–285, 291
Rust, 347
RWA, *see* Rotating wave approximation (RWA)
Ryzhii's QDIP model, 183, 190

Sandwich-like architecture for (OPV) solar cell, 325
Scanning electron microscopy (SEM), 123f, 124
Schawlow–Towns theory (laser-line width), 97
Schottky (metal-semiconductor) junction, 122
Schrödinger equation, 382, 389
Schrödinger wave equation, 136
Second harmonic generation (SHG), 58–59
Second harmonic (SH) propagation, 58–59
Self-assembled monolayers (SAMs), 266
Self-organized lightwave network (SOLNET), 261–262, 283
  concept, 283–285, 283f–284f
  targeting reflective/luminescent sites, 288–292
  theoretical analysis, 285–288
SEM, *see* Scanning electron microscopy (SEM)
Semiconductor nanoparticle, 345f
Semiconductors, 137
Sensitized solar cells, 292–298
Sensors, 157
SERS, *see* Surface-enhanced Raman scattering (SERS)
SHG, *see* Second harmonic generation (SHG)
Silicon-on insulator (SOI), 10–11, 15
Silicon photonics
  applications
    interconnects, 1–2, 2f
    sensors, 2–3
    spectroscopy, 2–3
  optical functions
    coupling to fiber, 6–8, 7f
    electro-optic conversion, 8–9
    lasers, 9–10
    opto-electronic conversion, 8–9
    routing, 3–5
    waveguides, 3–5, 4f
    wavelength filtering, 5–6f
  technology
    active tuning, 12
    lasers, 13–14
    modulators, 11–12
    passive circuits, 10–11, 10f
    photodetectors, 13
    photonic–electronic integration, 14–15
Silicon waveguides, 3–5, 4f
Single mode fiber (SMF-28), 60

Single-negative materials, 54
Small-core photonic wires, 4–5
Small-polaron absorption, 272
Small-polaron absorption in WO_3, 274f
Snell's law, 75, 144
Soft lithography, 149
SOI, *see* Silicon-on insulator (SOI)
SOLNET, *see* Self-organized lightwave network (SOLNET)
SP, *see* Surface plasmons (SPs)
Spaser Hamiltonian, 96
Spasers
   active development, 90–95
      nanospaser with semiconductor gain media, 92–95
   compensation of loss by gain and spasing, 109–112
      introduction, 109–110
      permittivity of nanoplasmonic metamaterial, 110–111
      plasmonic eigenmodes, 111–112
      resonant permittivity of metamaterials, 111–112
   conditions of loss compensation by gain and spasing, 112–118
   in CW regime, 102–104, 103f
   equations
      for CW regime, 98–101
      quantum density matrix equations (optical Bloch equations), 95–98
   fundamentals, 87–90
   gain (or active) medium, 87–88f, 89
   geometry, 87f–88f
   introduction, 86–87
   ultrafast dynamics, 107f
   as ultrafast quantum nanoamplifier, 104–109
      bistable spaser with saturable absorber, 106–109, 107f
      monostable spaser as nanoamplifier in transient regime, 105–106, 107f
      setting problem (spaser as amplifier), 103f, 104–105
Spasing curve, 102
Spatial Light Modulators, 379
Specific detectivity D^*, 174–175, 185
Spectroscopy, 2–3, 3f
Spin-dependent recombination, 252f

Spin-mixing mechanism
   Δg mechanism, 222
   hyperfine interaction (HFI), 222
   triplet excitons (TE), 222
Spin–orbit coupling (SOC), 391
SPP, *see* Surface plasmon polaritons (SPPs)
SPP Schottky nanoscopy, 123f, 124
Static molecular assembling, 262–263
Steady state density, 238
Stimulated Raman adiabatic passage (STIRAP), 386
Stranski–Krastanow (SK) growth mode, 179–180, 180f
   properties and effect on QDIP, 180–184
      quantum dot density, 181–183
      quantum dot shape, 181
      quantum dot size, 181
      quantum dot uniformity, 183–184
   quantum dots formation in, 180
Strong coupling regime, 32, 35–42, 35f
Styrylpyridinium cyanine dye (SPCD) thin film, 300
Superlens, 57–58, 57f
Supramolecular ZnPc–C_{60} dyad, 319f
Supramolecular ZnP/C_{60} molecular wire, 334f
Surface-enhanced Raman scattering (SERS), 109
Surface plasmon polaritons (SPPs), 55
   spasers, 86, 90–91
Surface plasmon resonance (SPR), 354
Surface plasmons (SPs), 85
Surface potential of ZnO layers, 279f

Terephthalaldehyde (TPA), 267
Tetrahydrofuran (THF), 314
Tetrathiafulvalenes (TTF), 313
Thin-film molecular nanophotonics, 261–262
   applications
      cancer therapy, 301–304
      electro-optic thin films, 300–301
      sensitized solar cells, 292–298
      three-dimensional integrated optical interconnects, 298–299
   molecular assembling
      dynamic molecular assembling, 263
      static molecular assembling, 262–263

Thin-film molecular nanophotonics (*Continued*)
  molecular layer deposition, 264–266
  organic MQD, 267
    molecular MQDs, 265–283
    polymer MQDs, 267–275
    self-organized lightwave network, 283
      concept, 283–285
      targeting reflective/luminescent sites, 288–292
      theoretical analysis, 285–288
Thin films, 279
Thin-film transistors (TFTs), 266
Three-dimensional integrated optical interconnects, 298–299
Three dimensions (3D), 133
  photonic crystals (PC), 134f, 140–143, 147
Through-silicon vias (TSV), 15
Time-resolved spectroscopy, 356
TIR, *see* Total internal reflection (TIR)
Titanium dioxide (TiO_2), 346
TO, *see* Transformation optics (TO)
Top-down methods, 170
Total internal reflection (TIR), 154–155
TPA, *see* Two-photon absorption (TPA)
Transformation optics (TO), 70–74, 71f
Transformation of solar energy, 324
Transition layer, 63–63f
Transverse electric (TE), 140
  band gaps, 140–144, 142f–143f
  in MEH-PPV system, 238–242
Transverse magnetic (TM), 140
  band gaps, 140–143, 142ff, 143f
Triazole ring as spacer, 314f
Triplet excitons (TE), 222
Triplet–triplet annihilation (TTA), 222–223, 253–254
Tris(8-hydroxyquinolinato)aluminum (Alq3), 288
TSV, *see* Through-silicon vias (TSV)
TTA, *see* Triplet–triplet annihilation (TTA)
TTF–fullerene systems, 315–317
Tunable PCs, 152–154
  chemical stimuli, 153–154
  physical stimuli, 152–153, 153f
Tungsten oxide (WO_3), 347
Two dimensions
  photonic crystals (PC), 134f, 140–143, 142f, 154f
Two dimensions (2D), 133
Two-photon absorption (TPA), 28–30

USMPA, 240–242

Vacuum Rabi oscillations (strong coupling), 35–42, 35f, 40f
Varactor-loaded split-ring resonator, 59–60f
Versatility, QDIPs, 178–179
Vertical couplings, 7–7f
Vortex beam, 76–77f

Waveguides
  construction in photopolymer, 286f
  optical functions, 3–5, 4f
Waveguide-type sensitized solar cell, 296
Waveguide-type solar cells, 294–297
Wavelength-division multiplexing (WDM), 2–2f
Wavelength filtering, 5–6f
Wavelength tuning detection (using quantum dot engineering), 203–208
WDM, *see* Wavelength-division multiplexing (WDM)
Weak coupling regime, 32–34
Wentzel-Kramers-Brillouin (WKB), 119
WKB, *see* Wentzel-Kramers-Brillouin (WKB)

Zero-dimensional (0D), 169
  density of state, 169–170, 170f
Zinc oxide (ZnO), 347
Zinc phthalocyanine C_{60} dyads, 318
$(ZnPc)_2$–$(C_{60})_2$ supramolecular tetrad, 320f
ZnP–C_{60} dyad with a catenane spacer, 315f
ZnP/π-conjugated/C_{60} systems, 333f